AF477116

ELECTRICAL REQUIREMENTS FOR HAZARDOUS LOCATIONS

Peter J. Schram
Editor

National Fire Protection Association
Quincy, Massachusetts

Product Manager: Charles Durang
Composition: Modern Graphics
Cover Design: Cameron, Inc
Manufacturing Manager: Ellen Glisker
Printer: Courier/Westford

Copyright © 2005
National Fire Protection Association, Inc.
One Batterymarch Park
Quincy, Massachusetts 02169-7471

All rights reserved. No part of this material protected by this copyright notice may be reproduced or utilized in any form without acknowledgment of the copyright owner nor may it be used in any form for resale without written permission from the copyright owner.

® Registered trademarks of the National Fire Protection Association, Inc.:
National Electrical Code®
NEC®

NFPA No.: ERHL05
ISBN: 0-87765-622-3
Library of Congress Control No.: 2005924716

Printed in the United States of America

05 06 07 08 09 5 4 3 2 1

IMPORTANT NOTICES AND DISCLAIMERS OF LIABILITY

This volume is a compilation of extracts from a number of NFPA codes and standards. These extracts have been selected and arranged by the Editor of the book.

All NFPA codes and standards are documents that are developed through a consensus standards development process approved by the American National Standards Institute. This process brings together volunteers representing varied viewpoints and interests to achieve consensus on fire, electrical, and other safety issues. While the NFPA administers the process and establishes rules to promote fairness in the development of consensus, it does not independently test, evaluate, or verify the accuracy of any information or the soundness of any judgments contained in its codes and standards.

The NFPA and the Editor disclaim liability for any personal injury, property or other damages of any nature whatsoever, whether special, indirect, consequential or compensatory, directly or indirectly resulting from the publication, use of, or reliance on this volume or the regulations contained in it. The NFPA and the Editor also make no guaranty or warranty as to the accuracy or completeness of any information published herein.

This document is purely advisory as far as NFPA is concerned. It is made available for a wide variety of both public and private uses in the interest of life and property protection. These include both use in law and for regulatory purposes, and use in private self-regulation and standardization activities as insurance underwriting, building and facilities construction and management, and product testing and certification.

In issuing and making this book available, the NFPA and the Editor are not undertaking to render professional or other services for or on behalf of any person or entity. Nor are the NFPA or the Editor undertaking to perform any duty owed by any person or entity to someone else. Anyone using this document should rely on his or her own independent judgment or, as appropriate, seek the advice of a competent professional in determining the exercise of reasonable care in any given circumstances.

The NFPA has no power, nor does it undertake, to police or enforce compliance with the contents of this document. Nor does the NFPA list, certify, test or inspect products, designs, or installations for compliance with this document. Any certification or other statement of compliance with the requirements of this document shall not be attributable to the NFPA and is solely the responsibility of the certifier or maker of the statement.

For more important information and notices concerning the use of NFPA codes and standards, please consult the introductory materials accompanying all published versions of these documents.

NOTICE CONCERNING CODE INTERPRETATIONS

The introductory materials and the selection and arrangement of the extracts contained in this volume reflect the personal opinions and judgments of the Editor and do not necessarily represent the official position of the NFPA (which can only be obtained through Formal Interpretations processed in accordance with NFPA rules).

CONTENTS

Introduction

Part I: *NEC* Requirements

***NEC* Definitions** **1**

Section 100.1 1

Section 500.2 1

Section 504.2 2

Section 505.2 3

Section 506.2 4

***NEC* Text** **5**

Article 250, Grounding and Bonding 5

Article 500, Hazardous (Classified) Locations, Classes I, II, and III, Divisions 1 and 2 5

Article 501, Class I Locations 12

Article 505, Class I, Zone 0, 1, and 2 Locations 23

Article 502, Class II Locations 35

Article 503, Class III Locations 41

Article 506, Zone 20, 21, and 22 Locations for Combustible Dusts, Fibers, and Flyings 44

Article 504, Intrinsically Safe Locations 49

Article 510, Hazardous (Classified) Locations — Specific 51

Article 511, Commercial Garages, Repair and Storage 51

Article 513, Aircraft Hangars 54

Article 514, Motor Fuel Dispensing Facilities 56

Article 515, Bulk Storage Plants 61

Article 516, Spray Application, Dipping, and Coating Processes 65

Article 517, Health Care Facilities 71

Part II: Requirements from Other NFPA Documents

NFPA 30, *Flammable and Combustible Liquids Code* (2003) 73

NFPA 30A, *Code for Motor Fuel Dispensing Facilities and Repair Garages* (2003) 78

NFPA 30B, *Code for the Manufacture and Storage of Aerosol Products* (2002) 80

NFPA 32, *Standard for Drycleaning Plants* (2004) 84

NFPA 33, *Standard for Spray Application Using Flammable or Combustible Materials* (2003) 84

NFPA 34, *Standard for Dipping and Coating Processes Using Flammable or Combustible Liquids* (2003) 89

NFPA 35, *Standard for the Manufacture of Organic Coatings* (1999) 95

NFPA 36, *Standard for Solvent Extraction Plants* (2004) 96

NFPA 40, *Standard for the Storage and Handling of Cellulose Nitrate Film* (2001) 98

NFPA 45, *Standard on Fire Protection for Laboratories Using Chemicals* (2004) 98

NFPA 50A, *Standard for Gaseous Hydrogen Systems at Consumer Sites* (1999) 102

NFPA 50B, *Standard for Liquefied Hydrogen Systems at Consumer Sites* (1999) 103

NFPA 51, *Standard for the Design and Installation of Oxygen–Fuel Gas Systems for Welding, Cutting, and Allied Processes* (2002) 103

NFPA 51A, *Standard for Acetylene Cylinder Charging Plants* (2001) 104

NFPA 52, *Compressed Natural Gas (CNG) Vehicular Fuel Systems Code* (2002) 105

NFPA 53, *Recommended Practice on Materials, Equipment, and Systems Used in Oxygen-Enriched Atmospheres* (2004) 106

NFPA 54, *National Fuel Gas Code* (2002) 106

NFPA 55, *Standard for the Storage, Use, and Handling of Compressed Gases and Cryogenic Fluids in Portable and Stationary Containers, Cylinders, and Tanks* (2003) 107

NFPA 57, *Liquefied Natural Gas (LNG) Vehicular Fuel Systems Code* (2002) 108

NFPA 58, *Liquefied Petroleum Gas Code* (2004) 110

NFPA 59, *Utility LP-Gas Plant Code* (2004) 114
NFPA 59A, *Standard for the Production, Storage, and Handling of Liquefied Natural Gas (LNG)* (2001) 117
NFPA 61, *Standard for the Prevention of Fires and Dust Explosions in Agricultural and Food Processing Facilities* (2002) 120
NFPA 70B, *Recommended Practice for Electrical Equipment Maintenance* (2002) 121
NFPA 70E, *Standard for Electrical Safety in the Workplace* (2004) 123
NFPA 77, *Recommended Practice on Static Electricity* (2000) 127
NFPA 99B, *Standard for Hypobaric Facilities* (2005) 187
NFPA 120, *Standard for Fire Prevention and Control in Coal Mines* (2003) 188
NFPA 122, *Standard for Fire Prevention and Control in Metal/Nonmetal Mining and Metal Mineral Processing Facilities* (2004) 190
NFPA 303, *Fire Protection Standard for Marinas and Boatyards* (2000) 192
NFPA 318, *Standard for the Protection of Semiconductor Fabrication Facilities* (2002) 193
NFPA 407, *Standard for Aircraft Fuel Servicing* (2001) 193
NFPA 409, *Standard on Aircraft Hangars* (2004) 194
NFPA 423, *Standard for Construction and Protection of Aircraft Engine Test Facilities* (2004) 194
NFPA 484, *Standard for Combustible Metals* (2002) 195
NFPA 495, *Explosive Materials Code* (2001) 196
NFPA 496, *Standard for Purged and Pressurized Enclosures for Electrical Equipment* (2003) 197
NFPA 497, *Recommended Practice for the Classification of Flammable Liquids, Gases, or Vapors and of Hazardous (Classified) Locations for Electrical Installations in Chemical Process Areas* (2004) 216
NFPA 499, *Recommended Practice for the Classification of Combustible Dusts and of Hazardous (Classified) Locations for Electrical Installations in Chemical Process Areas* (2004) 275
NFPA 505, *Fire Safety Standard for Powered Industrial Trucks Including Type Designations, Areas of Use, Conversions, Maintenance, and Operation* (2002) 296
NFPA 560, *Standard for the Storage, Handling, and Use of Ethylene Oxide for Sterilization and Fumigation* (2002) 296
NFPA 654, *Standard for the Prevention of Fire and Dust Explosions from the Manufacturing, Processing, and Handling of Combustible Particulate Solids* (2002) 297
NFPA 655, *Standard for Prevention of Sulfur Fires and Explosions* (2001) 298
NFPA 664, *Standard for the Prevention of Fires and Explosions in Wood Processing and Woodworking Facilities* (2002) 298
NFPA 820, *Standard for Fire Protection in Wastewater Treatment and Collection Facilities* (2003) 299
NFPA 1124, *Code for the Manufacture, Transportation, Storage, and Retail Sales of Fireworks and Pyrotechnic Articles* (2003) 324
NFPA 1125, *Code for the Manufacture of Model Rocket and High Power Rocket Motors* (2001) 324
Index 327
About the Editor 337

INTRODUCTION

Electrical Requirements for Hazardous Locations brings together, in one volume, requirements for electrical work in hazardous locations drawn from numerous NFPA codes and standards. It is meant for use by both designers and installers of these specialized systems.

The book is divided into two parts: requirements found in the 2005 *National Electrical Code® (NEC®)* and requirements drawn from other NFPA codes and standards.

Part I begins with definitions from *NEC* Article 100 and other parts of the *NEC* that relate to hazardous locations. Following is the complete text of *NEC* articles covering installations in hazardous locations. (The order of topics has been slightly modified so that Article 505, containing requirements for "Zone Classified" installations where flammable gases and vapors may be present, follows Article 501, which covers general requirements for similar Class I locations. In a like manner, Article 506, covering "Zone Classified" installations where combustible dusts, fibers, and flyings may be present, follows Article 503, rules for Class III locations.) Finally, text applicable to hazardous locations in health care facilities is presented from *NEC* Article 517.

Part II consists of material from a number of other NFPA documents relating to hazardous locations, some of which are referenced in *NEC* Articles 500 through 516.

The requirements in Articles 500 through 516 of the *NEC* supplement or modify the requirements in *NEC* Chapters 1 through 4. It is therefore necessary to apply all of the applicable *Code* requirements to installations in hazardous locations, not just those in Articles 500 through 516. Similarly, material excerpted from each of the other codes should be considered in the context of the entire code. If there is a question as to the requirements in any of the various NFPA publications identified, it is recommended that the reader obtain a complete copy of the publication.

State and local agencies may—and often do—adopt regulations above and beyond the *National Electrical Code*. Local authorities may adopt the *NEC* exactly as written, or they may amend the *Code* or supplement it with more stringent regulations. The same is true for other NFPA documents. The reader should obtain a copy of all applicable electrical requirements for the city, county, and/or state in which the work is to be performed, as well as the electrical requirements of other organizations that may have jurisdiction, such as OSHA, the Department of Defense, and applicable insurance carriers.

PART I *NEC* REQUIREMENTS

NEC DEFINITIONS

Editor's Note: *Some of the definitions in Articles 500–506 are duplicated because the Article 500 and Article 504 definitions do not apply to Articles 505 and 506 unless repeated or referenced in these articles.*

ARTICLE 100

I. General

Volatile Flammable Liquid. A flammable liquid having a flash point below 38°C (100°F), or a flammable liquid whose temperature is above its flash point, or a Class II combustible liquid that has a vapor pressure not exceeding 276 kPa (40 psia) at 38°C (100°F) and whose temperature is above its flash point.

ARTICLE 500
Hazardous (Classified) Locations, Classes I, II, and III, Divisions 1 and 2

500.2 Definitions. For purposes of Articles 500 through 504 and Articles 510 through 516, the following definitions apply.

Associated Nonincendive Field Wiring Apparatus. Apparatus in which the circuits are not necessarily nonincendive themselves but that affect the energy in nonincendive field wiring circuits and are relied upon to maintain nonincendive energy levels. Associated nonincendive field wiring apparatus may be either of the following:

(1) Electrical apparatus that has an alternative type of protection for use in the appropriate hazardous (classified) location
(2) Electrical apparatus not so protected that shall not be used in a hazardous (classified) location

FPN: Associated nonincendive field wiring apparatus has designated associated nonincendive field wiring apparatus connections for nonincendive field wiring apparatus and may also have connections for other electrical apparatus.

Combustible Gas Detection System. A protection technique utilizing stationary gas detectors in industrial establishments.

Control Drawing. A drawing or other document provided by the manufacturer of the intrinsically safe or associated apparatus, or of the nonincendive field wiring apparatus or associated nonincendive field wiring apparatus, that details the allowed interconnections between the intrinsically safe and associated apparatus or between the nonincendive field wiring apparatus or associated nonincendive field wiring apparatus.

Dust-Ignitionproof. Equipment enclosed in a manner that excludes dusts and does not permit arcs, sparks, or heat otherwise generated or liberated inside of the enclosure to cause ignition of exterior accumulations or atmospheric suspensions of a specified dust on or in the vicinity of the enclosure.

FPN: For further information on dust-ignitionproof enclosures, see Type 9 enclosure in ANSI/NEMA 250-1991, *Enclosures for Electrical Equipment,* and ANSI/UL 1203-1994, *Explosionproof and Dust-Ignitionproof Electrical Equipment for Hazardous (Classified) Locations.*

Dusttight. Enclosures constructed so that dust will not enter under specified test conditions.

FPN: See ANSI/ISA 12.12.01-2000, *Nonincendive Electrical Equipment for Use in Class I and II, Division 2, and Class III, Divisions 1 and 2 Hazardous (Classified) Locations,* and UL 1604-1994, *Electrical Equipment for Use in Class I and II, Division 2 and Class III Hazardous (Classified) Locations.*

Electrical and Electronic Equipment. Materials, fittings, devices, appliances, and the like that are part of, or in connection with, an electrical installation.

FPN: Portable or transportable equipment having self-contained power supplies, such as battery-operated equipment, could potentially become an ignition source in hazardous (classified) locations. See ISA-RP12.12.03-2002, *Portable Electronic Products Suitable for Use in Class I and II, Division 2, Class I Zone 2 and Class III, Division 1 and 2 Hazardous (Classified) Locations.*

Explosionproof Apparatus. Apparatus enclosed in a case that is capable of withstanding an explosion of a specified gas or vapor that may occur within it and of preventing the ignition of a specified gas or vapor surrounding the enclosure by sparks, flashes, or explosion of the gas or vapor within, and that operates at such an external temperature that a surrounding flammable atmosphere will not be ignited thereby.

FPN: For further information, see ANSI/UL 1203-1994, *Explosion-Proof and Dust-Ignition-Proof Electrical Equipment for Use in Hazardous (Classified) Locations.*

Hermetically Sealed. Equipment sealed against the entrance of an external atmosphere where the seal is made by fusion, for example, soldering, brazing, welding, or the fusion of glass to metal.

FPN: For further information, see ANSI/ISA 12.12.01-2000, *Nonincendive Electrical Equipment for Use in Class I and II, Division 2, and Class III, Division 1 and 2 Hazardous (Classified) Locations.*

Nonincendive Circuit. A circuit, other than field wiring, in which any arc or thermal effect produced under intended operating conditions of the equipment is not capable, under specified test conditions, of igniting the flammable gas–air, vapor–air, or dust–air mixture.

FPN: Conditions are described in ANSI/ISA 12.12.01-2000, *Nonincendive Electrical Equipment for Use in Class I and II, Division 2, and Class III, Divisions 1 and 2 Hazardous (Classified) Locations.*

Nonincendive Component. A component having contacts for making or breaking an incendive circuit and the contacting mechanism is constructed so that the component is incapable of igniting the specified flammable gas–air or vapor–air mixture. The housing of a nonincendive component is not intended to exclude the flammable atmosphere or contain an explosion.

FPN: For further information, see UL 1604-1994, *Electrical Equipment for Use in Class I and II, Division 2, and Class III Hazardous (Classified) Locations.*

Nonincendive Equipment. Equipment having electrical/electronic circuitry that is incapable, under normal operating conditions, of causing ignition of a specified flammable gas–air, vapor–air, or dust–air mixture due to arcing or thermal means.

FPN: For further information, see ANSI/ISA 12.12.01-2000, *Nonincendive Electrical Equipment for Use in Class I and II, Division 2, and Class III, Divisions 1 and 2 Hazardous (Classified) Locations.*

Nonincendive Field Wiring. Wiring that enters or leaves an equipment enclosure and, under normal operating conditions of the equipment, is not capable, due to arcing or thermal effects, of igniting the flammable gas–air, vapor–air, or dust–air mixture. Normal operation includes opening, shorting, or grounding the field wiring.

Nonincendive Field Wiring Apparatus. Apparatus intended to be connected to nonincendive field wiring.

FPN: For further information see ANSI/ISA 12.12.01-2000, *Nonincendive Electrical Equipment for Use in Class I and II, Division 2, and Class III, Divisions 1 and 2 Hazardous (Classified) Locations.*

Oil Immersion. Electrical equipment immersed in a protective liquid in such a way that an explosive atmosphere that may be above the liquid or outside the enclosure cannot be ignited.

FPN: For further information, see ANSI/UL 698-1995, *Industrial Control Equipment for Use in Hazardous (Classified) Locations.*

Purged and Pressurized. The process of (1) purging, supplying an enclosure with a protective gas at a sufficient flow and positive pressure to reduce the concentration of any flammable gas or vapor initially present to an acceptable level; and (2) pressurization, supplying an enclosure with a protective gas with or without continuous flow at sufficient pressure to prevent the entrance of a flammable gas or vapor, a combustible dust, or an ignitible fiber.

FPN: For further information, see ANSI/ NFPA 496-2003, *Purged and Pressurized Enclosures for Electrical Equipment.*

Unclassified Locations. Locations determined to be neither Class I, Division 1; Class I, Division 2; Class I, Zone 0; Class I, Zone 1; Class I, Zone 2; Class II, Division 1; Class II, Division 2; Class III, Division 1; Class III, Division 2; or any combination thereof.

ARTICLE 504
Intrinsically Safe Systems

504.2 Definitions.

Associated Apparatus. Apparatus in which the circuits are not necessarily intrinsically safe themselves but that affect the energy in the intrinsically safe circuits and are relied on to maintain intrinsic safety. Associated apparatus may be either of the following:

(1) Electrical apparatus that has an alternative-type protection for use in the appropriate hazardous (classified) location
(2) Electrical apparatus not so protected that shall not be used within a hazardous (classified) location

FPN No. 1: Associated apparatus has identified intrinsically safe connections for intrinsically safe apparatus and also may have connections for nonintrinsically safe apparatus.

FPN No. 2: An example of associated apparatus is an intrinsic safety barrier, which is a network designed to limit the energy (voltage and current) available to the protected circuit in the hazardous (classified) location, under specified fault conditions.

Control Drawing. See definition in 500.2.

Different Intrinsically Safe Circuits. Intrinsically safe circuits in which the possible interconnections have not been evaluated and identified as intrinsically safe.

Intrinsically Safe Apparatus. Apparatus in which all the circuits are intrinsically safe.

Intrinsically Safe Circuit. A circuit in which any spark or thermal effect is incapable of causing ignition of a mixture of flammable or combustible material in air under prescribed test conditions.

FPN: Test conditions are described in ANSI/UL 913-1997, *Standard for Safety, Intrinsically Safe Apparatus and Associated Apparatus for Use in Class I, II, and III, Division 1, Hazardous (Classified) Locations.*

Intrinsically Safe System. An assembly of interconnected intrinsically safe apparatus, associated apparatus, and interconnecting cables in that those parts of the system that may be used in hazardous (classified) locations are intrinsically safe circuits.

FPN: An intrinsically safe system may include more than one intrinsically safe circuit.

Simple Apparatus. An electrical component or combination of components of simple construction with well-defined electrical parameters that does not generate more than 1.5 volts, 100 milliamps, and 25 milliwatts, or a passive component that does not dissipate more than 1.3 watts and is compatible with the intrinsic safety of the circuit in which it is used.

FPN: The following apparatus are examples of simple apparatus:

(a) Passive components, for example, switches, junction boxes, resistance temperature devices, and simple semiconductor devices such as LEDs
(b) Sources of generated energy, for example, thermocouples and photocells, which do not generate more than 1.5 V, 100 mA, and 25 mW

ARTICLE 505
Class I, Zone 0, 1, and 2 Locations

505.2 Definitions. For purposes of this article, the following definitions apply.

Combustible Gas Detection System. A protection technique utilizing stationary gas detectors in industrial establishments.

Electrical and Electronic Equipment. Materials, fittings, devices, appliances, and the like that are part of, or in connection with, an electrical installation.

FPN: Portable or transportable equipment having self-contained power supplies, such as battery-operated equipment, could potentially become an ignition source in hazardous (classified) locations.

Encapsulation "m." Type of protection where electrical parts that could ignite an explosive atmosphere by either sparking or heating are enclosed in a compound in such a way that this explosive atmosphere cannot be ignited.

FPN: See ANSI/ISA 12.23.01-2002, *Electrical Apparatus for Use in Class I, Zone 1 Hazardous (Classified) Locations, Type of Protection — Encapsulation "m"*; and ANSI/UL 60079–18, *Electrical apparatus for explosive gas atmospheres — Part 18: Encapsulation "m."*

Flameproof "d." Type of protection where the enclosure will withstand an internal explosion of a flammable mixture that has penetrated into the interior, without suffering damage and without causing ignition, through any joints or structural openings in the enclosure, of an external explosive gas atmosphere consisting of one or more of the gases or vapors for which it is designed.

FPN: See ANSI/ISA 12.22.01-2002, *Electrical Apparatus for Use in Class I, Zone 1 and 2 Hazardous (Classified) Locations, Type of Protection — Flameproof "d"*; and ANSI/UL 60079–1, *Electrical apparatus for explosive gas atmospheres — Part 1: Flameproof enclosures "d."*

Increased Safety "e." Type of protection applied to electrical equipment that does not produce arcs or sparks in normal service and under specified abnormal conditions, in which additional measures are applied so as to give increased security against the possibility of excessive temperatures and of the occurrence of arcs and sparks.

FPN: See ANSI/ISA — 12.16.01-2002, *Electrical Apparatus for Use in Class I, Zone 1 Hazardous (Classified) Locations, Type of Protection — Increased Safety "e"*; and ANSI/UL 60079-7, *Electrical apparatus for explosive gas atmospheres — Part 7: Increased Safety "e."*

Intrinsic Safety "i." Type of protection where any spark or thermal effect is incapable of causing ignition of a mixture of flammable or combustible material in air under prescribed test conditions.

FPN No. 1: See ANSI/UL 913-1997, *Intrinsically Safe Apparatus and Associated Apparatus for Use in Class I, II, and III, Hazardous Locations*; ISA 12.02.01-1999, *Electrical Apparatus for Use in Class I, Zones 0, 1 and 2 Hazardous (Classified) Locations — Intrinsic Safety "i"*; and ANSI/UL 60079-11, *Electrical apparatus for explosive gas atmospheres — Part II: Intrinsic safety "i."*

FPN No. 2: Intrinsic safety is designated type of protection "ia" for use in Zone 0 locations. Intrinsic safety is designated type of protection "ib" for use in Zone 1 locations.

FPN No. 3: Intrinsically safe associated apparatus, designated by [ia] or [ib], is connected to intrinsically safe apparatus ("ia" or "ib," respectively) but is located outside the hazardous (classified) location unless also protected by another type of protection (such as flameproof).

Oil Immersion "o." Type of protection where electrical equipment is immersed in a protective liquid in such a way that an explosive atmosphere that may be above the liquid or outside the enclosure cannot be ignited.

FPN: See ISA 12.26.01-1998, *Electrical Apparatus for Use in Class I, Zone 1 Hazardous (Classified) Locations, Type of Protection — Oil-Immersion "o"*; and ANSI/UL 60079-6, *Electrical apparatus for explosive gas atmospheres — Part 6: Oil-immersion "o."*

Powder Filling "q." Type of protection where electrical parts capable of igniting an explosive atmosphere are fixed in position and completely surrounded by filling material (glass or quartz powder) to prevent the ignition of an external explosive atmosphere.

FPN: See ANSI/ISA-12.25.01-2002, *Electrical Apparatus for Use in Class I, Zone 1 Hazardous (Classified) Locations Type of Protection — Powder Filling "q"*; and ANSI/UL 60079-5, *Electrical apparatus for explosive gas atmospheres — Part 5: Powder filling "q."*

Purged and Pressurized. Type of protection for electrical equipment that uses the technique of guarding against the ingress of the external atmosphere, which may be explosive, into an enclosure by maintaining a protective gas therein at a pressure above that of the external atmosphere.

FPN No. 1: See NFPA 496-2003, *Standard for Purged and Pressurized Enclosures for Electrical Equipment.*

FPN No. 2: See IEC 60079-2-2000, *Electrical Apparatus for Explosive Gas Atmospheres — Part 2: Electrical Apparatus, Type of Protection "p"*; and IEC 60079-13-1982, *Electrical Apparatus for Explosive Gas Atmospheres — Part 13: Construction and Use of Rooms or Buildings Protected by Pressurization.*

Type of Protection "n." Type of protection where electrical equipment, in normal operation, is not capable of igniting a surrounding explosive gas atmosphere and a fault capable of causing ignition is not likely to occur.

FPN: See ANSI/UL 60079-15-2002, *Electrical apparatus for explosive gas atmospheres — Part 15: Type of protection "n"*; and ANSI/ISA 12.12.02-2003, *Electrical apparatus for use in Class I, Zone 2 Hazardous (Classified) Locations: Type of protection "n."*

Unclassified Locations. Locations determined to be neither Class I, Division 1; Class I, Division 2; Class I, Zone 0; Class I, Zone 1; Class I, Zone 2; Class II, Division 1; Class II, Division 2; Class III, Division 1; Class III, Division 2; or any combination thereof.

ARTICLE 506
Zone 20, 21, and 22 Locations for Combustible Dusts, Fibers, and Flyings

506.2 Definitions. For purposes of this article, the following definitions apply.

Associated Nonincendive Field Wiring Apparatus. Apparatus in which the circuits are not necessarily nonincendive themselves but that affect the energy in nonincendive field wiring circuits and are relied upon to maintain nonincendive energy levels. Associated nonincendive field wiring apparatus may be either of the following:

(1) Electrical apparatus that has an alternative type of protection for use in the appropriate hazardous (classified) location
(2) Electrical apparatus not so protected that shall not be used in a hazardous (classified) location

FPN: Associated nonincendive field wiring apparatus has designated associated nonincendive field wiring apparatus connections for nonincendive field wiring apparatus and may also have connections for other electrical apparatus.

Dust-Ignitionproof. Equipment enclosed in a manner that excludes dusts and does not permit arcs, sparks, or heat otherwise generated or liberated inside of the enclosure to cause ignition of exterior accumulations or atmospheric suspensions of a specified dust on or in the vicinity of the enclosure.

FPN: For further information on dust-ignitionproof enclosures, see Type 9 enclosure in ANSI/NEMA 250-1991, *Enclosures for Electrical Equipment*, and ANSI/UL 1203-1994, *Explosionproof and Dust-Ignitionproof Electrical Equipment for Hazardous (Classified) Locations.*

Dusttight. Enclosures constructed so that dust will not enter under specified test conditions.

FPN: See ANSI/ISA 12.12.01-2000, *Nonincendive Electrical Equipment for Use in Class I and II, Division 2, and Class III, Divisions 1 and 2 Hazardous (Classified) Locations*, and UL 1604-1994, *Electrical Equipment for Use in Class I and II, Division 2 and Class III Hazardous (Classified) Locations.*

Nonincendive Circuit. A circuit, other than field wiring, in which any arc or thermal effect produced under intended operating conditions of the equipment is not capable, under specified test conditions, of igniting the flammable gas–air, vapor–air, or dust–air mixture.

FPN: Conditions are described in ANSI/ISA 12.12.01-2000, *Nonincendive Electrical Equipment for Use in Class I and II, Division 2, and Class III, Divisions 1 and 2 Hazardous (Classified) Locations.*

Nonincendive Equipment. Equipment having electrical/electronic circuitry that is incapable, under normal operating conditions, of causing ignition of a specified flammable gas–air, vapor–air, or dust–air mixture due to arcing or thermal means.

FPN: Conditions are described in ANSI/ISA 12.12.01-2000, *Nonincendive Electrical Equipment for Use in Class I and II, Division 2, and Class III, Divisions 1 and 2 Hazardous (Classified) Locations.*

Nonincendive Field Wiring. Wiring that enters or leaves an equipment enclosure and, under normal operating conditions of the equipment, is not capable, due to arcing or thermal effects, of igniting the flammable gas–air, vapor–air, or dust–air mixture. Normal operation includes opening, shorting, or grounding the field wiring.

Nonincendive Field Wiring Apparatus. Apparatus intended to be connected to nonincendive field wiring.

FPN: Conditions are described in ANSI/ISA 12.12.01-2000, *Nonincendive Electrical Equipment for Use in Class I and II, Division 2, and Class III, Divisions 1 and 2 Hazardous (Classified) Locations.*

Pressurized. The process of supplying an enclosure with a protective gas with or without continuous flow at sufficient pressure to prevent the entrance of combustible dust, or an ignitible fiber or flying.

FPN: For further information, see ANSI/ NFPA 496-2003, *Purged and Pressurized Enclosures for Electrical Equipment.*

Zone 20 Hazardous (Classified) Location. An area where combustible dust or ignitible fibers and flyings are present continuously or for long periods of time in quantities sufficient to be hazardous, as classified by 506.5(B)(1).

Zone 21 Hazardous (Classified) Location. An area where combustible dust or ignitible fibers and flyings are likely to exist occasionally under normal operation in quantities sufficient to be hazardous, as classified by 506.5(B)(2).

Zone 22 Hazardous (Classified) Location. An area where combustible dust or ignitible fibers and flyings are not likely to occur under normal operation in quantities sufficient to be hazardous, as classified by 506.5(B)(3).

NEC TEXT

ARTICLE 250
Grounding and Bonding

Editor's Note: *The text of NEC Section 250.100 is included because it relates specifically to wiring systems in other than hazardous locations but supplying power to hazardous locations.*

250.100 Bonding in Hazardous (Classified) Locations. Regardless of the voltage of the electrical system, the electrical continuity of non–current-carrying metal parts of equipment, raceways, and other enclosures in any hazardous (classified) location as defined in Article 500 shall be ensured by any of the methods specified in 250.92(B)(2) through (B)(4) that are approved for the wiring method used. One or more of these bonding methods shall be used whether or not supplementary equipment grounding conductors are installed.

ARTICLE 500
Hazardous (Classified) Locations, Classes I, II, and III, Divisions 1 and 2

Editor's Note: *The text of Section 500.2 is not included because it is included in Definitions. Article 500 does not apply if the zone classification system of Article 505 or 506 is used.*

FPN: Rules that are followed by a reference in brackets contain text that has been extracted from NFPA 497, *Recommended Practice for the Classification of Flammable Liquids, Gases, or Vapors and of Hazardous (Classified) Locations for Electrical Installations in Chemical Process Areas*, 2004 edition, and NFPA 499, *Recommended Practice for the Classification of Combustible Dusts and of Hazardous (Classified) Locations for Electrical Installation in Chemical Process Areas,* 2004 edition. Only editorial changes were made to the extracted text to make it consistent with this *Code.*

500.1 Scope — Articles 500 Through 504. Articles 500 through 504 cover the requirements for electrical and electronic equipment and wiring for all voltages in Class I, Divisions 1 and 2; Class II, Divisions 1 and 2; and Class III, Divisions 1 and 2 locations where fire or explosion hazards may exist due to flammable gases or vapors, flammable liquids, combustible dust, or ignitible fibers or flyings.

FPN No. 1: The unique hazards associated with explosives, pyrotechnics, and blasting agents are not addressed in this article.

FPN No. 2: For the requirements for electrical and electronic equipment and wiring for all voltages in Class I, Zone 0, Zone 1, and Zone 2 hazardous (classified) locations where fire or explosion hazards may exist due to flammable gases or vapors or flammable liquids, refer to Article 505.

FPN No. 3: For the requirements for electrical and electronic equipment and wiring for all voltages in Zone 20, Zone 21, and Zone 22 hazardous (classified) locations where fire or explosion hazards may exist due to combustible dusts or ignitible fibers or flyings, refer to Article 506.

500.3 Other Articles. Except as modified in Articles 500 through 504, all other applicable rules contained in this *Code* shall apply to electrical equipment and wiring installed in hazardous (classified) locations.

500.4 General.

(A) Documentation. All areas designated as hazardous (classified) locations shall be properly documented. This documentation shall be available to those authorized to design, install, inspect, maintain, or operate electrical equipment at the location.

(B) Reference Standards. Important information relating to topics covered in Chapter 5 may be found in other publications.

FPN No. 1: It is important that the authority having jurisdiction be familiar with recorded industrial experience as well as with the standards of the National Fire Protection Association (NFPA), the American Petroleum Institute (API), and the Instrumentation, Systems, and Automation Society (ISA) that may be of use in the classification of various locations, the determination of adequate ventilation, and the protection against static electricity and lightning hazards.

FPN No. 2: For further information on the classification of locations, see NFPA 30-2003, *Flammable and Combustible Liquids Code*; NFPA 32-2004, *Standard for Drycleaning Plants*; NFPA 33-2003, *Standard for Spray Application Using Flammable or Combustible Materials*; NFPA 34-2003, *Standard for Dipping and Coating Processes Using Flammable or Combustible Liquids*; NFPA 35-1999, *Standard for the Manufacture of Organic Coatings*; NFPA 36-2004, *Standard for Solvent Extraction Plants*; NFPA 45-2004, *Standard on Fire Protection for Laboratories Using Chemicals*; NFPA 50A-1999, *Standard for Gaseous Hydrogen Systems at Consumer Sites*; NFPA 50B-1999, *Standard for Liquefied Hydrogen Systems at Consumer Sites*; NFPA 58-2004, *Liquefied Petroleum Gas Code*; NFPA 59-2004, *Utility LP-Gas Plant Code*; NFPA 497-2004, *Recommended Practice for the Classification of Flammable Liquids, Gases, or Vapors and of Hazardous (Classified) Locations for Electrical Installations in Chemical Process Areas*; NFPA 499-2004, *Recommended Practice for the Classification of Combustible Dusts and of Hazardous (Classified) Locations for Electrical Installations in Chemical Process Areas*; NFPA 820-2003, *Standard for Fire Protection in Wastewater Treatment and Collection Facilities*; ANSI/API RP500-1997, *Recommended Practice for Classification of Locations of Electrical Installations at Petroleum Facilities Classified as Class I, Division 1 and Division 2*; ISA 12.10-1988, *Area Classification in Hazardous (Classified) Dust Locations*.

FPN No. 3: For further information on protection against static electricity and lightning hazards in hazardous (classified) locations, see NFPA 77-2000, *Recommended Practice on Static Electricity*; NFPA 780-2004, *Standard for the Installation of Lightning Protection Systems*; and API RP 2003-1998, *Protection Against Ignitions Arising Out of Static Lightning and Stray Currents*.

FPN No. 4: For further information on ventilation, see NFPA 30-2003, *Flammable and Combustible Liquids Code*; and API RP 500-1997, *Recommended Practice for Classification of Locations for Electrical Installations at Petroleum Facilities Classified as Class I, Division 1 and Division 2*.

FPN No. 5: For further information on electrical systems for hazardous (classified) locations on offshore oil- and gas-producing platforms, see ANSI/API RP 14F-1999, *Recommended Practice for Design and Installation of Electrical Systems for Fixed and Floating Offshore Petroleum Facilities for Unclassified and Class I, Division 1 and Division 2 Locations*.

500.5 Classifications of Locations.

(A) Classifications of Locations. Locations shall be classified depending on the properties of the flammable vapors, liquids, or gases, or combustible dusts or fibers that may be present, and the likelihood that a flammable or combustible concentration or quantity is present. Where pyrophoric materials are the only materials used or handled, these locations shall not be classified. Each room, section, or area shall be considered individually in determining its classification.

FPN: Through the exercise of ingenuity in the layout of electrical installations for hazardous (classified) locations, it is frequently possible to locate much of the equipment in a reduced level of classification or in an unclassified location and, thus, to reduce the amount of special equipment required.

Rooms and areas containing ammonia refrigeration systems that are equipped with adequate mechanical ventilation may be classified as "unclassified" locations.

FPN: For further information regarding classification and ventilation of areas involving ammonia, see ANSI/ASHRAE 15-1994, *Safety Code for Mechanical Refrigeration*, and ANSI/CGA G2.1-1989, *Safety Requirements for the Storage and Handling of Anhydrous Ammonia*.

(B) Class I Locations. Class I locations are those in which flammable gases or vapors are or may be present in the air in quantities sufficient to produce explosive or ignitible mixtures. Class I locations shall include those specified in 500.5(B)(1) and (B)(2).

(1) Class I, Division 1. A Class I, Division 1 location is a location

(1) In which ignitible concentrations of flammable gases or vapors can exist under normal operating conditions, or
(2) In which ignitible concentrations of such gases or vapors may exist frequently because of repair or maintenance operations or because of leakage, or
(3) In which breakdown or faulty operation of equipment or processes might release ignitible concentrations of flammable gases or vapors and might also cause simultaneous failure of electrical equipment in such a way as to directly cause the electrical equipment to become a source of ignition.

FPN No. 1: This classification usually includes the following locations:

(1) Where volatile flammable liquids or liquefied flammable gases are transferred from one container to another

(2) Interiors of spray booths and areas in the vicinity of spraying and painting operations where volatile flammable solvents are used
(3) Locations containing open tanks or vats of volatile flammable liquids
(4) Drying rooms or compartments for the evaporation of flammable solvents
(5) Locations containing fat- and oil-extraction equipment using volatile flammable solvents
(6) Portions of cleaning and dyeing plants where flammable liquids are used
(7) Gas generator rooms and other portions of gas manufacturing plants where flammable gas may escape
(8) Inadequately ventilated pump rooms for flammable gas or for volatile flammable liquids
(9) The interiors of refrigerators and freezers in which volatile flammable materials are stored in open, lightly stoppered, or easily ruptured containers
(10) All other locations where ignitible concentrations of flammable vapors or gases are likely to occur in the course of normal operations

FPN No. 2: In some Division 1 locations, ignitible concentrations of flammable gases or vapors may be present continuously or for long periods of time. Examples include the following:

(1) The inside of inadequately vented enclosures containing instruments normally venting flammable gases or vapors to the interior of the enclosure
(2) The inside of vented tanks containing volatile flammable liquids
(3) The area between the inner and outer roof sections of a floating roof tank containing volatile flammable fluids
(4) Inadequately ventilated areas within spraying or coating operations using volatile flammable fluids
(5) The interior of an exhaust duct that is used to vent ignitible concentrations of gases or vapors

Experience has demonstrated the prudence of avoiding the installation of instrumentation or other electric equipment in these particular areas altogether or where it cannot be avoided because it is essential to the process and other locations are not feasible [see 500.5(A), FPN] using electric equipment or instrumentation approved for the specific application or consisting of intrinsically safe systems as described in Article 504.

(2) Class I, Division 2. A Class I, Division 2 location is a location

(1) In which volatile flammable liquids or flammable gases are handled, processed, or used, but in which the liquids, vapors, or gases will normally be confined within closed containers or closed systems from which they can escape only in case of accidental rupture or breakdown of such containers or systems or in case of abnormal operation of equipment, or
(2) In which ignitible concentrations of gases or vapors are normally prevented by positive mechanical ventilation and which might become hazardous through failure or abnormal operation of the ventilating equipment, or
(3) That is adjacent to a Class I, Division 1 location, and to which ignitible concentrations of gases or vapors might occasionally be communicated unless such communication is prevented by adequate positive-pressure ventilation from a source of clean air and effective safeguards against ventilation failure are provided.

FPN No. 1: This classification usually includes locations where volatile flammable liquids or flammable gases or vapors are used but that, in the judgment of the authority having jurisdiction, would become hazardous only in case of an accident or of some unusual operating condition. The quantity of flammable material that might escape in case of accident, the adequacy of ventilating equipment, the total area involved, and the record of the industry or business with respect to explosions or fires are all factors that merit consideration in determining the classification and extent of each location.

FPN No. 2: Piping without valves, checks, meters, and similar devices would not ordinarily introduce a hazardous condition even though used for flammable liquids or gases. Depending on factors such as the quantity and size of the containers and ventilation, locations used for the storage of flammable liquids or liquefied or compressed gases in sealed containers may be considered either hazardous (classified) or unclassified locations. See NFPA 30-2003, *Flammable and Combustible Liquids Code*, and NFPA 58-2004, *Liquefied Petroleum Gas Code*.

(C) Class II Locations. Class II locations are those that are hazardous because of the presence of combustible dust. Class II locations shall include those specified in 500.5(C)(1) and (C)(2).

(1) Class II, Division 1. A Class II, Division 1 location is a location

(1) In which combustible dust is in the air under normal operating conditions in quantities sufficient to produce explosive or ignitible mixtures, or
(2) Where mechanical failure or abnormal operation of machinery or equipment might cause such explosive or ignitible mixtures to be produced, and might also provide a source of ignition through simultaneous failure of electric equipment, through operation of protection devices, or from other causes, or
(3) In which Group E combustible dusts may be present in quantities sufficient to be hazardous.

FPN: Dusts containing magnesium or aluminum are particularly hazardous, and the use of extreme precaution is necessary to avoid ignition and explosion.

(2) Class II, Division 2. A Class II, Division 2 location is a location

(1) In which combustible dust due to abnormal operations may be present in the air in quantities sufficient to produce explosive or ignitible mixtures; or

(2) Where combustible dust accumulations are present but are normally insufficient to interfere with the normal operation of electrical equipment or other apparatus, but could as a result of infrequent malfunctioning of handling or processing equipment become suspended in the air; or
(3) In which combustible dust accumulations on, in, or in the vicinity of the electrical equipment could be sufficient to interfere with the safe dissipation of heat from electrical equipment, or could be ignitible by abnormal operation or failure of electrical equipment.

FPN No. 1: The quantity of combustible dust that may be present and the adequacy of dust removal systems are factors that merit consideration in determining the classification and may result in an unclassified area.

FPN No. 2: Where products such as seed are handled in a manner that produces low quantities of dust, the amount of dust deposited may not warrant classification.

(D) Class III Locations. Class III locations are those that are hazardous because of the presence of easily ignitible fibers or flyings, but in which such fibers or flyings are not likely to be in suspension in the air in quantities sufficient to produce ignitible mixtures. Class III locations shall include those specified in 500.5(D)(1) and (D)(2).

(1) Class III, Division 1. A Class III, Division 1 location is a location in which easily ignitible fibers or materials producing combustible flyings are handled, manufactured, or used.

FPN No. 1: Such locations usually include some parts of rayon, cotton, and other textile mills; combustible fiber manufacturing and processing plants; cotton gins and cottonseed mills; flax-processing plants; clothing manufacturing plants; woodworking plants; and establishments and industries involving similar hazardous processes or conditions.

FPN No. 2: Easily ignitible fibers and flyings include rayon, cotton (including cotton linters and cotton waste), sisal or henequen, istle, jute, hemp, tow, cocoa fiber, oakum, baled waste kapok, Spanish moss, excelsior, and other materials of similar nature.

(2) Class III, Division 2. A Class III, Division 2 location is a location in which easily ignitible fibers are stored or handled other than in the process of manufacture.

500.6 Material Groups. For purposes of testing, approval, and area classification, various air mixtures (not oxygen-enriched) shall be grouped in accordance with 500.6(A) and 500.6(B).

Exception: Equipment identified for a specific gas, vapor, or dust.

FPN: This grouping is based on the characteristics of the materials. Facilities are available for testing and identifying equipment for use in the various atmospheric groups.

(A) Class I Group Classifications. Class I groups shall be according to 500.6(A)(1) through (A)(4).

FPN No. 1: FPN Nos. 2 and 3 apply to 500.6(A).

FPN No. 2: The explosion characteristics of air mixtures of gases or vapors vary with the specific material involved. For Class I locations, Groups A, B, C, and D, the classification involves determinations of maximum explosion pressure and maximum safe clearance between parts of a clamped joint in an enclosure. It is necessary, therefore, that equipment be identified not only for class but also for the specific group of the gas or vapor that will be present.

FPN No. 3: Certain chemical atmospheres may have characteristics that require safeguards beyond those required for any of the Class I groups. Carbon disulfide is one of these chemicals because of its low ignition temperature [100°C (212°F)] and the small joint clearance permitted to arrest its flame.

(1) Group A. Acetylene. [NFPA 497:3.3]

(2) Group B. Flammable gas, flammable liquid–produced vapor, or combustible liquid–produced vapor mixed with air that may burn or explode, having either a maximum experimental safe gap (MESG) value less than or equal to 0.45 mm or a minimum igniting current ratio (MIC ratio) less than or equal to 0.40. [NFPA 497:3.3]

FPN: A typical Class I, Group B material is hydrogen.

Exception No. 1: Group D equipment shall be permitted to be used for atmospheres containing butadiene, provided all conduit runs into explosionproof equipment are provided with explosionproof seals installed within 450 mm (18 in.) of the enclosure.

Exception No. 2: Group C equipment shall be permitted to be used for atmospheres containing allyl glycidyl ether, n-butyl glycidyl ether, ethylene oxide, propylene oxide, and acrolein, provided all conduit runs into explosionproof equipment are provided with explosionproof seals installed within 450 mm (18 in.) of the enclosure.

(3) Group C. Flammable gas, flammable liquid–produced vapor, or combustible liquid–produced vapor mixed with air that may burn or explode, having either a maximum experimental safe gap (MESG) value greater than 0.45 mm and less than or equal to 0.75 mm, or a minimum igniting current ratio (MIC ratio) greater than 0.40 and less than or equal to 0.80. [NFPA 497:3.3]

FPN: A typical Class I, Group C material is ethylene.

(4) Group D. Flammable gas, flammable liquid–produced vapor, or combustible liquid–produced vapor mixed with air that may burn or explode, having either a maximum experimental safe gap (MESG) value greater than 0.75 mm or a minimum igniting current ratio (MIC ratio) greater than 0.80. [NFPA 497:3.3]

FPN No. 1: A typical Class I, Group D material is propane.

FPN No. 2: For classification of areas involving ammonia atmospheres, see ANSI/ASHRAE 15-1994, *Safety Code for Mechanical Refrigeration*, and ANSI/CGA G2.1-1989, *Safety Requirements for the Storage and Handling of Anhydrous Ammonia*.

(B) Class II Group Classifications. Class II groups shall be in accordance with 500.6(B)(1) through (B)(3).

(1) Group E. Atmospheres containing combustible metal dusts, including aluminum, magnesium, and their commercial alloys, or other combustible dusts whose particle size, abrasiveness, and conductivity present similar hazards in the use of electrical equipment. [NFPA 499:3.3]

FPN: Certain metal dusts may have characteristics that require safeguards beyond those required for atmospheres containing the dusts of aluminum, magnesium, and their commercial alloys. For example, zirconium, thorium, and uranium dusts have extremely low ignition temperatures [as low as 20°C (68°F)] and minimum ignition energies lower than any material classified in any of the Class I or Class II groups.

(2) Group F. Atmospheres containing combustible carbonaceous dusts that have more than 8 percent total entrapped volatiles (see ASTM D 3175-89, *Standard Test Method for Volatile Material in the Analysis Sample for Coal and Coke*, for coal and coke dusts) or that have been sensitized by other materials so that they present an explosion hazard. Coal, carbon black, charcoal, and coke dusts are examples of carbonaceous dusts. [NFPA 499:3.3]

(3) Group G. Atmospheres containing combustible dusts not included in Group E or F, including flour, grain, wood, plastic, and chemicals.

FPN No. 1: For additional information on group classification of Class II materials, see NFPA 499-2004, *Recommended Practice for the Classification of Combustible Dusts and of Hazardous (Classified) Locations for Electrical Installations in Chemical Process Areas*.

FPN No. 2: The explosion characteristics of air mixtures of dust vary with the materials involved. For Class II locations, Groups E, F, and G, the classification involves the tightness of the joints of assembly and shaft openings to prevent the entrance of dust in the dust-ignitionproof enclosure, the blanketing effect of layers of dust on the equipment that may cause overheating, and the ignition temperature of the dust. It is necessary, therefore, that equipment be identified not only for the class, but also for the specific group of dust that will be present.

FPN No. 3: Certain dusts may require additional precautions due to chemical phenomena that can result in the generation of ignitible gases. See ANSI C2-2002, *National Electrical Safety Code*, Section 127A, Coal Handling Areas.

500.7 Protection Techniques. Section 500.7(A) through 500.7(L) shall be acceptable protection techniques for electrical and electronic equipment in hazardous (classified) locations.

(A) Explosionproof Apparatus. This protection technique shall be permitted for equipment in Class I, Division 1 or 2 locations.

(B) Dust Ignitionproof. This protection technique shall be permitted for equipment in Class II, Division 1 or 2 locations.

(C) Dusttight. This protection technique shall be permitted for equipment in Class II, Division 2 or Class III, Division 1 or 2 locations.

(D) Purged and Pressurized. This protection technique shall be permitted for equipment in any hazardous (classified) location for which it is identified.

(E) Intrinsic Safety. This protection technique shall be permitted for equipment in Class I, Division 1 or 2; or Class II, Division 1 or 2; or Class III, Division 1 or 2 locations. The provisions of Articles 501 through 503 and Articles 510 through 516 shall not be considered applicable to such installations, except as required by Article 504, and installation of intrinsically safe apparatus and wiring shall be in accordance with the requirements of Article 504.

(F) Nonincendive Circuit. This protection technique shall be permitted for equipment in Class I, Division 2; Class II, Division 2; or Class III, Division 1 or 2 locations.

(G) Nonincendive Equipment. This protection technique shall be permitted for equipment in Class I, Division 2; Class II, Division 2; or Class III, Division 1 or 2 locations.

(H) Nonincendive Component. This protection technique shall be permitted for equipment in Class I, Division 2; Class II, Division 2; or Class III, Division 1 or 2 locations.

(I) Oil Immersion. This protection technique shall be permitted for current-interrupting contacts in Class I, Division 2 locations as described in 501.115(B)(1)(2).

(J) Hermetically Sealed. This protection technique shall be permitted for equipment in Class I, Division 2; Class II, Division 2; or Class III, Division 1 or 2 locations.

(K) Combustible Gas Detection System. A combustible gas detection system shall be permitted as a means of protection in industrial establishments with restricted public access and where the conditions of maintenance and supervision ensure that only qualified persons service the installation. Gas detection equipment shall be listed for detection of the specific gas or vapor to be encountered. Where such a system is installed, equipment specified in 500.7(K)(1), (K)(2), or (K)(3) shall be permitted.

The type of detection equipment, its listing, installation

location(s), alarm and shutdown criteria, and calibration frequency shall be documented when combustible gas detectors are used as a protection technique.

FPN No. 1: For further information, see ANSI/ISA-12.13.01, *Performance Requirements, Combustible Gas Detectors.*

FPN No. 2: For further information, see ANSI/API RP 500, *Recommended Practice for Classification of Locations for Electrical Installations at Petroleum Facilities Classified as Class I, Division I or Division 2.*

FPN No. 3: For further information, see ISA-RP12.13.02, *Installation, Operation, and Maintenance of Combustible Gas Detection Instruments.*

(1) Inadequate Ventilation. In a Class I, Division 1 location that is so classified due to inadequate ventilation, electrical equipment suitable for Class I, Division 2 locations shall be permitted.

(2) Interior of a Building. In a building located in, or with an opening into, a Class I, Division 2 location where the interior does not contain a source of flammable gas or vapor, electrical equipment for unclassified locations shall be permitted.

(3) Interior of a Control Panel. In the interior of a control panel containing instrumentation utilizing or measuring flammable liquids, gases, or vapors, electrical equipment suitable for Class I, Division 2 locations shall be permitted.

(L) Other Protection Techniques. Other protection techniques used in equipment identified for use in hazardous (classified) locations.

500.8 Equipment. Articles 500 through 504 require equipment construction and installation that ensure safe performance under conditions of proper use and maintenance.

FPN No. 1: It is important that inspection authorities and users exercise more than ordinary care with regard to installation and maintenance.

FPN No. 2: Since there is no consistent relationship between explosion properties and ignition temperature, the two are independent requirements.

FPN No. 3: Low ambient conditions require special consideration. Explosionproof or dust-ignitionproof equipment may not be suitable for use at temperatures lower than −25°C (−13°F) unless they are identified for low-temperature service. However, at low ambient temperatures, flammable concentrations of vapors may not exist in a location classified as Class I, Division 1 at normal ambient temperature.

(A) Approval for Class and Properties.

(1) Equipment shall be identified not only for the class of location but also for the explosive, combustible, or ignitible properties of the specific gas, vapor, dust, fiber, or flyings that will be present. In addition, Class I equipment shall not have any exposed surface that operates at a temperature in excess of the ignition temperature of the specific gas or vapor. Class II equipment shall not have an external temperature higher than that specified in 500.8(C)(2). Class III equipment shall not exceed the maximum surface temperatures specified in 503.5.

FPN: Luminaires (lighting fixtures) and other heat-producing apparatus, switches, circuit breakers, and plugs and receptacles are potential sources of ignition and are investigated for suitability in classified locations. Such types of equipment, as well as cable terminations for entry into explosionproof enclosures, are available as listed for Class I, Division 2 locations. Fixed wiring, however, may utilize wiring methods that are not evaluated with respect to classified locations. Wiring products such as cable, raceways, boxes, and fittings, therefore, are not marked as being suitable for Class I, Division 2 locations. Also see 500.8(B)(6)(a).

Suitability of identified equipment shall be determined by any of the following:

(1) Equipment listing or labeling
(2) Evidence of equipment evaluation from a qualified testing laboratory or inspection agency concerned with product evaluation
(3) Evidence acceptable to the authority having jurisdiction such as a manufacturer's self-evaluation or an owner's engineering judgment

(2) Equipment that has been identified for a Division 1 location shall be permitted in a Division 2 location of the same class, group, and temperature class and shall comply with (a) or (b) as applicable.

(a) Intrinsically safe apparatus having a control drawing requiring the installation of associated apparatus for a Division 1 installation shall be permitted to be installed in a Division 2 location if the same associated apparatus is used for the Division 2 installation.

(b) Equipment that is required to be explosionproof shall incorporate seals per 501.15(A) or 501.15(D) when the wiring methods of 501.10(B) are employed.

(3) Where specifically permitted in Articles 501 through 503, general-purpose equipment or equipment in general-purpose enclosures shall be permitted to be installed in Division 2 locations if the equipment does not constitute a source of ignition under normal operating conditions.

(4) Equipment that depends on a single compression seal, diaphragm, or tube to prevent flammable or combustible fluids from entering the equipment shall be identified for a Class I, Division 2 location even if installed in an unclassified location. Equipment installed in a Class I, Division 1 location shall be identified for the Class I, Division 1 location.

FPN: Equipment used for flow measurement is an example of equipment having a single compression seal, diaphragm, or tube.

(5) Unless otherwise specified, normal operating conditions for motors shall be assumed to be rated full-load steady conditions.

(6) Where flammable gases or combustible dusts are or may be present at the same time, the simultaneous presence of both shall be considered when determining the safe operating temperature of the electrical equipment.

FPN: The characteristics of various atmospheric mixtures of gases, vapors, and dusts depend on the specific material involved.

(B) Marking. Equipment shall be marked to show the environment for which it has been evaluated. Unless otherwise specified or allowed in (B)(6), the marking shall include the information specified in (B)(1) through (B)(5).

(1) Class. The marking shall specify the class(es) for which the equipment is suitable.

(2) Division. The marking shall specify the division if the equipment is suitable for Division 2 only. Equipment suitable for Division 1 shall be permitted to omit the division marking.

FPN: Equipment not marked to indicate a division, or marked "Division 1" or "Div. 1," is suitable for both Division 1 and 2 locations; see 500.8(A)(2). Equipment marked "Division 2" or "Div. 2" is suitable for Division 2 locations only.

(3) Material Classification Group. The marking shall specify the applicable material classification group(s) in accordance with 500.6.

Exception: Fixed luminaires (lighting fixtures) marked for use only in Class I, Division 2 or Class II, Division 2 locations shall not be required to indicate the group.

(4) Equipment Temperature. The marking shall specify the temperature class or operating temperature at a 40°C ambient temperature, or at the higher ambient temperature if the equipment is rated and marked for an ambient temperature of greater than 40°C. The temperature class, if provided, shall be indicated using the temperature class (T Codes) shown in Table 500.8(B). Equipment for Class I and Class II shall be marked with the maximum safe operating temperature, as determined by simultaneous exposure to the combinations of Class I and Class II conditions.

Exception: Equipment of the non–heat-producing type, such as junction boxes, conduit, and fittings, and equipment of the heat-producing type having a maximum temperature not more than 100°C shall not be required to have a marked operating temperature or temperature class.

Table 500.8(B) Classification of Maximum Surface Temperature

Maximum Temperature		Temperature Class
C°	F°	(T Code)
450	842	T1
300	572	T2
280	536	T2A
260	500	T2B
230	446	T2C
215	419	T2D
200	392	T3
180	356	T3A
165	329	T3B
160	320	T3C
135	275	T4
120	248	T4A
100	212	T5
85	185	T6

FPN: More than one marked temperature class or operating temperature, for gases and vapors, dusts, and different ambient temperatures, may appear.

(5) Ambient Temperature Range. For equipment rated for a temperature range other than −25°C to +40°C, the marking shall specify the special range of ambient temperatures. The marking shall include either the symbol "Ta" or "Tamb."

FPN: As an example, such a marking might be "−30°C ≤ Ta ≤ +40°C."

(6) Special Allowances.

(a) General Purpose Equipment. Fixed general-purpose equipment in Class I locations, other than fixed luminaires (lighting fixtures), that is acceptable for use in Class I, Division 2 locations shall not be required to be marked with the class, division, group, temperature class, or ambient temperature range.
(b) Dusttight Equipment. Fixed dusttight equipment, other than fixed luminaires (lighting fixtures), that is acceptable for use in Class II, Division 2 and Class III locations shall not be required to be marked with the class, division, group, temperature class, or ambient temperature range.
(c) Associated Apparatus. Associated intrinsically safe apparatus and associated nonincendive field wiring apparatus that are not protected by an alternative type of

protection shall not be marked with the class, division, group, or temperature class. Associated intrinsically safe apparatus and associated nonincendive field wiring apparatus shall be marked with the class, division, and group of the apparatus to which it is to be connected.

(d) Simple Apparatus. "Simple apparatus" as defined in Article 504, shall not be required to be marked with class, division, group, temperature class, or ambient temperature range.

(C) Temperature.

(1) Class I Temperature. The temperature marking specified in 500.8(B) shall not exceed the ignition temperature of the specific gas or vapor to be encountered.

FPN: For information regarding ignition temperatures of gases and vapors, see NFPA 497-2004, *Recommended Practice for the Classification of Flammable Liquids, Gases, or Vapors, and of Hazardous (Classified) Locations for Electrical Installations in Chemical Process Areas.*

(2) Class II Temperature. The temperature marking specified in 500.8(B) shall be less than the ignition temperature of the specific dust to be encountered. For organic dusts that may dehydrate or carbonize, the temperature marking shall not exceed the lower of either the ignition temperature or 165°C (329°F).

FPN: See NFPA 499-2004, *Recommended Practice for the Classification of Combustible Dusts and of Hazardous (Classified) Locations for Electrical Installations in Chemical Process Areas,* for minimum ignition temperatures of specific dusts.

The ignition temperature for which equipment was approved prior to this requirement shall be assumed to be as shown in Table 500.8(C)(2).

Table 500.8(C)(2) Class II Temperatures

Class II Group	Equipment Not Subject to Overloading		Equipment (Such as Motors or Power Transformers) That May Be Overloaded: Normal Operation		Equipment (Such as Motors or Power Transformers) That May Be Overloaded: Abnormal Operation	
	°C	°F	°C	°F	°C	°F
E	200	392	200	392	200	392
F	200	392	150	302	200	392
G	165	329	120	248	165	329

(D) Threading. All NPT threaded conduit and fittings referred to herein shall be threaded with a National (American) Standard Pipe Taper (NPT) thread that provides a taper of 1 in 16 (¾-in. taper per foot). Conduit and fittings shall be made wrenchtight to prevent sparking when fault current flows through the conduit system, and to ensure the explosionproof integrity of the conduit system where applicable. Equipment provided with threaded entries for field wiring connections shall be installed in accordance with 500.8(D)(1) or (D)(2). Threaded entries into explosionproof equipment shall be made up with at least five threads fully engaged.

Exception: For listed explosionproof equipment, factory threaded NPT entries shall be made up with at least 4½ threads fully engaged.

(1) Equipment Provided with Threaded Entries for NPT Threaded Conduit or Fittings. For equipment provided with threaded entries for NPT threaded conduit or fittings, listed conduit, conduit fittings, or cable fittings shall be used.

FPN: Thread form specifications for NPT threads are located in ANSI/ASME B1.20.1-1983, *Pipe Threads, General Purpose (Inch).*

(2) Equipment Provided with Threaded Entries for Metric Threaded Conduit or Fittings. For equipment with metric threaded entries, such entries shall be identified as being metric, or listed adapters to permit connection to conduit or NPT-threaded fittings shall be provided with the equipment. Adapters shall be used for connection to conduit or NPT-threaded fittings. Listed cable fittings that have metric threads shall be permitted to be used.

FPN: Threading specifications for metric threaded entries are located in ISO 965/1-1980, *Metric Screw Threads,* and ISO 965/3-1980, *Metric Screw Threads.*

(E) Fiber Optic Cable Assembly. Where a fiber optic cable assembly contains conductors that are capable of carrying current, the fiber optic cable assembly shall be installed in accordance with the requirements of Articles 500, 501, 502, or 503, as applicable.

500.9 Specific Occupancies. Articles 510 through 517 cover garages, aircraft hangars, motor fuel dispensing facilities, bulk storage plants, spray application, dipping and coating processes, and health care facilities.

ARTICLE 501
Class I Locations

Editor's Note: *Article 501 does not apply if the zone classification system of Article 505 is used.*

I. General

501.1 Scope. Article 501 covers the requirements for electrical and electronic equipment and wiring for all voltages in Class I, Division 1 and 2 locations where fire or explosion hazards may exist due to flammable gases or vapors or flammable liquids.

FPN: For the requirements for electrical and electronic equipment and wiring for all voltages in Class I, Zone 0, Zone 1, or Zone 2 hazardous (classified) locations where fire or explosion hazards may exist due to flammable gases or vapors or flammable liquids, refer to Article 505.

501.5 General. The general rules of this *Code* shall apply to the electric wiring and equipment in locations classified as Class I in 500.5.

Exception: As modified by this article.

Equipment listed and marked in accordance with 505.9(C)(2) for use in Class I, Zone 0, 1, or 2 locations shall be permitted in Class I, Division 2 locations for the same gas and with a suitable temperature class. Equipment listed and marked in accordance with 505.9(C)(2) for use in Class I, Zone 0 locations shall be permitted in Class I, Division 1 or Division 2 locations for the same gas and with a suitable temperature class.

II. Wiring

501.10 Wiring Methods. Wiring methods shall comply with 501.10(A) or 501.10(B).

(A) Class I, Division 1.

(1) General. In Class I, Division 1 locations, the wiring methods in (a) through (d) shall be permitted.

(a) Threaded rigid metal conduit or threaded steel intermediate metal conduit.

Exception: Rigid nonmetallic conduit complying with Article 352 shall be permitted where encased in a concrete envelope a minimum of 50 mm (2 in.) thick and provided with not less than 600 mm (24 in.) of cover measured from the top of the conduit to grade. The concrete encasement shall be permitted to be omitted where subject to the provisions of 514.8, Exception No. 2; and 515.8(A). Threaded rigid metal conduit or threaded steel intermediate metal conduit shall be used for the last 600 mm (24 in.) of the underground run to emergence or to the point of connection to the aboveground raceway. An equipment grounding conductor shall be included to provide for electrical continuity of the raceway system and for grounding of non–current-carrying metal parts.

(b) Type MI cable with termination fittings listed for the location. Type MI cable shall be installed and supported in a manner to avoid tensile stress at the termination fittings.

(c) In industrial establishments with restricted public access, where the conditions of maintenance and supervision ensure that only qualified persons service the installation, Type MC-HL cable, listed for use in Class I, Division 1 locations, with a gas/vaportight continuous corrugated metallic sheath, an overall jacket of suitable polymeric material, separate grounding conductors in accordance with 250.122, and provided with termination fittings listed for the application.

FPN: See 330.12 for restrictions on use of Type MC cable.

(d) In industrial establishments with restricted public access, where the conditions of maintenance and supervision ensure that only qualified persons service the installation, Type ITC-HL cable, listed for use in Class I, Division 1 locations, with a gas/vaportight continuous corrugated metallic sheath, an overall jacket of suitable polymeric material and provided with termination fittings listed for the application.

(2) Flexible Connections. Where necessary to employ flexible connections, as at motor terminals, flexible fittings listed for Class I, Division 1 locations or flexible cord in accordance with the provisions of 501.140 shall be permitted.

(3) Boxes and Fittings. All boxes and fittings shall be approved for Class I, Division 1.

(B) Class I, Division 2.

(1) General. In Class I, Division 2 locations, the following wiring methods shall be permitted:

(1) All wiring methods permitted in Article 501.10(A).
(2) Threaded rigid metal conduit, threaded steel intermediate metal conduit.
(3) Enclosed gasketed busways, enclosed gasketed wireways.
(4) Type PLTC cable in accordance with the provisions of Article 725, or in cable tray systems. PLTC shall be installed in a manner to avoid tensile stress at the termination fittings.
(5) Type ITC cable as permitted in 727.4.
(6) Type MI, MC, MV, or TC cable with termination fittings, or in cable tray systems and installed in a manner to avoid tensile stress at the termination fittings. Single conductor Type MV cables shall be shielded or metallic armored.

(2) Flexible Connections. Where provision must be made for limited flexibility, one or more of the following shall also be permitted:

(1) Flexible metal fittings
(2) Flexible metal conduit with listed fittings
(3) Liquidtight flexible metal conduit with listed fittings
(4) Liquidtight flexible nonmetallic conduit with listed fittings
(5) Flexible cord listed for extra-hard usage and provided with listed bushed fittings. An additional conductor for grounding shall be included in the flexible cord.

FPN: See 501.30(B) for grounding requirements where flexible conduit is used.

(3) Nonincendive Field Wiring. Nonincendive field wiring shall be permitted using any of the wiring methods permitted for unclassified locations. Nonincendive field wiring systems shall be installed in accordance with the control drawing(s). Simple apparatus, not shown on the control drawing, shall be permitted in a nonincendive field wiring circuit, provided the simple apparatus does not interconnect the nonincendive field wiring circuit to any other circuit.

FPN: Simple apparatus is defined in 504.2.

Separate nonincendive field wiring circuits shall be installed in accordance with one of the following:

(1) In separate cables
(2) In multiconductor cables where the conductors of each circuit are within a grounded metal shield
(3) In multiconductor cables, where the conductors of each circuit have insulation with a minimum thickness of 0.25 mm (0.01 in.)

(4) Boxes and Fittings. Boxes and fittings shall not be required to be explosionproof except as required by 501.105(B)(1), 501.115(B)(1), and 501.150(B)(1).

501.15 Sealing and Drainage. Seals in conduit and cable systems shall comply with 501.15(A) through 501.15(F). Sealing compound shall be used in Type MI cable termination fittings to exclude moisture and other fluids from the cable insulation.

FPN No. 1: Seals are provided in conduit and cable systems to minimize the passage of gases and vapors and prevent the passage of flames from one portion of the electrical installation to another through the conduit. Such communication through Type MI cable is inherently prevented by construction of the cable. Unless specifically designed and tested for the purpose, conduit and cable seals are not intended to prevent the passage of liquids, gases, or vapors at a continuous pressure differential across the seal. Even at differences in pressure across the seal equivalent to a few inches of water, there may be a slow passage of gas or vapor through a seal and through conductors passing through the seal. See 501.15(E)(2). Temperature extremes and highly corrosive liquids and vapors can affect the ability of seals to perform their intended function. See 501.15(C)(2).

FPN No. 2: Gas or vapor leakage and propagation of flames may occur through the interstices between the strands of standard stranded conductors larger than 2 AWG. Special conductor constructions, for example, compacted strands or sealing of the individual strands, are means of reducing leakage and preventing the propagation of flames.

(A) Conduit Seals, Class I, Division 1. In Class I, Division 1 locations, conduit seals shall be located in accordance with 501.15(A)(1) through (A)(4).

(1) Entering Enclosures. In each conduit entry into an explosionproof enclosure where either of the following apply:

(1) The enclosure contains apparatus, such as switches, circuit breakers, fuses, relays, or resistors, that may produce arcs, sparks, or high temperatures that are considered to be an ignition source in normal operation.
(2) The entry is metric designator 53 (trade size 2) or larger and the enclosure contains terminals, splices, or taps.

For the purposes of this section, high temperatures shall be considered to be any temperatures exceeding 80 percent of the autoignition temperature in degrees Celsius of the gas or vapor involved.

Exception to 501.15(A)(1)(1): Seals shall not be required for conduit entering an enclosure where such switches, circuit breakers, fuses, relays, or resistors comply with one of the following:

(1) Are enclosed within a chamber hermetically sealed against the entrance of gases or vapors
(2) Are immersed in oil in accordance with 501.115(B)(1)(2)
(3) Are enclosed within a factory-sealed explosionproof chamber located within the enclosure, identified for the location, and marked "factory sealed" or equivalent, unless the enclosure entry is metric designator 53 (trade size 2) or larger
(4) Are in nonincendive circuits

Factory-sealed enclosures shall not be considered to serve as a seal for another adjacent explosionproof enclosure that is required to have a conduit seal.

Conduit seals shall be installed within 450 mm (18 in.) from the enclosure. Only explosionproof unions, couplings, reducers, elbows, capped elbows, and conduit bodies similar to L, T, and Cross types that are not larger than the trade size of the conduit shall be permitted between the sealing fitting and the explosionproof enclosure.

(2) Pressurized Enclosures. In each conduit entry into a pressurized enclosure where the conduit is not pressurized as part of the protection system. Conduit seals shall be installed within 450 mm (18 in.) from the pressurized enclosure.

FPN No. 1: Installing the seal as close as possible to the enclosure will reduce problems with purging the dead airspace in the pressurized conduit.

FPN No. 2: For further information, see NFPA 496-2003, *Standard for Purged and Pressurized Enclosures for Electrical Equipment.*

(3) Two or More Explosionproof Enclosures. Where two or more explosionproof enclosures for which conduit seals are required under 501.15(A)(1) are connected by nipples or by runs of conduit not more than 900 mm (36 in.) long, a single conduit seal in each such nipple connection or run of conduit shall be considered sufficient if located not more than 450 mm (18 in.) from either enclosure.

(4) Class I, Division 1 Boundary. In each conduit run leaving a Class I, Division 1 location. The sealing fitting shall be permitted on either side of the boundary of such location within 3.05 m (10 ft) of the boundary and shall be designed and installed so as to minimize the amount of gas or vapor within the Division 1 portion of the conduit from being communicated to the conduit beyond the seal. Except for listed explosionproof reducers at the conduit seal, there shall be no union, coupling, box, or fitting between the conduit seal and the point at which the conduit leaves the Division 1 location.

Exception No. 1: Metal conduit that contains no unions, couplings, boxes, or fittings, and passes completely through a Class I, Division 1 location with no fittings less than 300 mm (12 in.) beyond each boundary, shall not require a conduit seal if the termination points of the unbroken conduit are in unclassified locations.

Exception No. 2: For underground conduit installed in accordance with 300.5 where the boundary is beneath the ground, the sealing fitting shall be permitted to be installed after the conduit leaves the ground, but there shall be no union, coupling, box, or fitting, other than listed explosionproof reducers at the sealing fitting, in the conduit between the sealing fitting and the point at which the conduit leaves the ground.

(B) Conduit Seals, Class I, Division 2. In Class I, Division 2 locations, conduit seals shall be located in accordance with 501.15(B)(1) and (B)(2).

(1) Entering Enclosures. For connections to enclosures that are required to be explosionproof, a conduit seal shall be provided in accordance with 501.15(A)(1) and (A)(3). All portions of the conduit run or nipple between the seal and such enclosure shall comply with 501.10(A).

(2) Class I, Division 2 Boundary. In each conduit run passing from a Class I, Division 2 location into an unclassified location. The sealing fitting shall be permitted on either side of the boundary of such location within 3.05 m (10 ft) of the boundary. Rigid metal conduit or threaded steel intermediate metal conduit shall be used between the sealing fitting and the point at which the conduit leaves the Division 2 location, and a threaded connection shall be used at the sealing fitting. Except for listed reducers at the conduit seal, there shall be no union, coupling, box, or fitting between the conduit seal and the point at which the conduit leaves the Division 2 location. Conduits shall be sealed to minimize the amount of gas or vapor within the Division 2 portion of the conduit from being communicated to the conduit beyond the seal. Such seals shall not be required to be explosionproof but shall be identified for the purpose of minimizing passage of gases under normal operating conditions and shall be accessible.

Exception No. 1: Metal conduit that contains no unions, couplings, boxes, or fittings, and passes completely through a Class I, Division 2 location with no fittings less than 300 mm (12 in.) beyond each boundary, shall not be required to be sealed if the termination points of the unbroken conduit are in unclassified locations.

Exception No. 2: Conduit systems terminating at an unclassified location where a wiring method transition is made to cable tray, cablebus, ventilated busway, Type MI cable, or cable not installed in any cable tray or raceway system, shall not be required to be sealed where passing from the Class I, Division 2 location into the unclassified location. The unclassified location shall be outdoors or, if the conduit system is all in one room, it shall be permitted to be indoors. The conduits shall not terminate at an enclosure containing an ignition source in normal operation.

Exception No. 3: Conduit systems passing from an enclosure or room that is unclassified as a result of pressurization into a Class I, Division 2 location shall not require a seal at the boundary.

FPN: For further information, refer to NFPA 496-2003, *Standard for Purged and Pressurized Enclosures for Electrical Equipment.*

Exception No. 4: Segments of aboveground conduit systems shall not be required to be sealed where passing from a Class I, Division 2 location into an unclassified location if all of the following conditions are met:

(1) No part of the conduit system segment passes through a Class I, Division 1 location where the conduit contains unions, couplings, boxes, or fittings within 300 mm (12 in.) of the Class I, Division 1 location.
(2) The conduit system segment is located entirely in outdoor locations.
(3) The conduit system segment is not directly connected to canned pumps, process or service connections for flow, pressure, or analysis measurement, and so forth, that depend on a single compression seal, diaphragm,

or tube to prevent flammable or combustible fluids from entering the conduit system.

(4) The conduit system segment contains only threaded metal conduit, unions, couplings, conduit bodies, and fittings in the unclassified location.

(5) The conduit system segment is sealed at its entry to each enclosure or fitting housing terminals, splices, or taps in Class I, Division 2 locations.

(C) Class I, Divisions 1 and 2. Seals installed in Class I, Division 1 and Division 2 locations shall comply with 501.15(C)(1) through (C)(6).

Exception: Seals not required to be explosionproof by 501.15(B)(2) or 504.70.

(1) Fittings. Enclosures for connections or equipment shall be provided with an integral means for sealing, or sealing fittings listed for the location shall be used. Sealing fittings shall be listed for use with one or more specific compounds and shall be accessible.

(2) Compound. The compound shall provide a seal against passage of gas or vapors through the seal fitting, shall not be affected by the surrounding atmosphere or liquids, and shall not have a melting point of less than 93°C (200°F).

(3) Thickness of Compounds. Except for listed cable sealing fittings, the thickness of the sealing compound in a completed seal shall not be less than the metric designator (trade size) of the sealing fitting expressed in the units of measurement employed, and in no case less than 16 mm (⅝ in.).

(4) Splices and Taps. Splices and taps shall not be made in fittings intended only for sealing with compound, nor shall other fittings in which splices or taps are made be filled with compound.

(5) Assemblies. In an assembly where equipment that may produce arcs, sparks, or high temperatures is located in a compartment separate from the compartment containing splices or taps, and an integral seal is provided where conductors pass from one compartment to the other, the entire assembly shall be identified for the location. Seals in conduit connections to the compartment containing splices or taps shall be provided in Class I, Division 1 locations where required by 501.15(A)(1)(2).

(6) Conductor Fill. The cross-sectional area of the conductors permitted in a seal shall not exceed 25 percent of the cross-sectional area of a rigid metal conduit of the same trade size unless it is specifically identified for a higher percentage of fill.

(D) Cable Seals, Class I, Division 1. In Class I, Division 1 locations, cable seals shall be located according to 501.15(D)(1) through (D)(3).

(1) At Terminations. Cable shall be sealed at all terminations. The sealing fitting shall comply with 501.15(C). Multiconductor Type MC-HL cables with a gas/vaportight continuous corrugated metallic sheath and an overall jacket of suitable polymeric material shall be sealed with a listed fitting after removing the jacket and any other covering so that the sealing compound surrounds each individual insulated conductor in such a manner as to minimize the passage of gases and vapors.

Exception: Shielded cables and twisted pair cables shall not require the removal of the shielding material or separation of the twisted pairs, provided the termination is by an approved means to minimize the entrance of gases or vapors and prevent propagation of flame into the cable core.

(2) Cables Capable of Transmitting Gases or Vapors. Cables in conduit with a gas/vaportight continuous sheath capable of transmitting gases or vapors through the cable core shall be sealed in the Division 1 location after removing the jacket and any other coverings so that the sealing compound will surround each individual insulated conductor and the outer jacket.

Exception: Multiconductor cables with a gas/vaportight continuous sheath capable of transmitting gases or vapors through the cable core shall be permitted to be considered as a single conductor by sealing the cable in the conduit within 450 mm (18 in.) of the enclosure and the cable end within the enclosure by an approved means to minimize the entrance of gases or vapors and prevent the propagation of flame into the cable core, or by other approved methods. For shielded cables and twisted pair cables, it shall not be required to remove the shielding material or separate the twisted pair.

(3) Cables Incapable of Transmitting Gases or Vapors. Each multiconductor cable in conduit shall be considered as a single conductor if the cable is incapable of transmitting gases or vapors through the cable core. These cables shall be sealed in accordance with 501.15(A).

(E) Cable Seals, Class I, Division 2. In Class I, Division 2 locations, cable seals shall be located in accordance with 501.15(E)(1) through (E)(4).

(1) Terminations. Cables entering enclosures that are required to be explosionproof shall be sealed at the point of entrance. The sealing fitting shall comply with 501.15(B)(1). Multiconductor cables with a gas/vaportight continuous sheath capable of transmitting gases or vapors through the cable core shall be sealed in a listed fitting in the Division

2 location after removing the jacket and any other coverings so that the sealing compound surrounds each individual insulated conductor in such a manner as to minimize the passage of gases and vapors. Multiconductor cables in conduit shall be sealed as described in 501.15(D).

Exception No. 1: Cables passing from an enclosure or room that is unclassified as a result of Type Z pressurization into a Class I, Division 2 location shall not require a seal at the boundary.

Exception No. 2: Shielded cables and twisted pair cables shall not require the removal of the shielding material or separation of the twisted pairs, provided the termination is by an approved means to minimize the entrance of gases or vapors and prevent propagation of flame into the cable core.

(2) Cables That Do Not Transmit Gases or Vapors. Cables that have a gas/vaportight continuous sheath and do not transmit gases or vapors through the cable core in excess of the quantity permitted for seal fittings shall not be required to be sealed except as required in 501.15(E)(1). The minimum length of such cable run shall not be less than that length that limits gas or vapor flow through the cable core to the rate permitted for seal fittings [200 cm;s3/hr (0.007 ft;s3/hr) of air at a pressure of 1500 pascals (6 in. of water)].

FPN No. 1: See ANSI/UL 886-1994, *Outlet Boxes and Fittings for Use in Hazardous (Classified) Locations.*

FPN No. 2: The cable core does not include the interstices of the conductor strands.

(3) Cables Capable of Transmitting Gases or Vapors. Cables with a gas/vaportight continuous sheath capable of transmitting gases or vapors through the cable core shall not be required to be sealed except as required in 501.15(E)(1), unless the cable is attached to process equipment or devices that may cause a pressure in excess of 1500 pascals (6 in. of water) to be exerted at a cable end, in which case a seal, barrier, or other means shall be provided to prevent migration of flammables into an unclassified location.

Exception: Cables with an unbroken gas/vaportight continuous sheath shall be permitted to pass through a Class I, Division 2 location without seals.

(4) Cables Without Gas/Vaportight Sheath. Cables that do not have gas/vaportight continuous sheath shall be sealed at the boundary of the Division 2 and unclassified location in such a manner as to minimize the passage of gases or vapors into an unclassified location.

(F) Drainage.

(1) Control Equipment. Where there is a probability that liquid or other condensed vapor may be trapped within enclosures for control equipment or at any point in the raceway system, approved means shall be provided to prevent accumulation or to permit periodic draining of such liquid or condensed vapor.

(2) Motors and Generators. Where the authority having jurisdiction judges that there is a probability that liquid or condensed vapor may accumulate within motors or generators, joints and conduit systems shall be arranged to minimize the entrance of liquid. If means to prevent accumulation or to permit periodic draining are judged necessary, such means shall be provided at the time of manufacture and shall be considered an integral part of the machine.

(3) Canned Pumps, Process, or Service Connections, etc. For canned pumps, process, or service connections for flow, pressure, or analysis measurement, and so forth, that depend on a single compression seal, diaphragm, or tube to prevent flammable or combustible fluids from entering the electrical raceway or cable system capable of transmitting fluids, an additional approved seal, barrier, or other means shall be provided to prevent the flammable or combustible fluid from entering the raceway or cable system capable of transmitting fluids beyond the additional devices or means, if the primary seal fails. The additional approved seal or barrier and the interconnecting enclosure shall meet the temperature and pressure conditions to which they will be subjected upon failure of the primary seal, unless other approved means are provided to accomplish this purpose. Drains, vents, or other devices shall be provided so that primary seal leakage will be obvious.

FPN: See also the fine print notes to 501.15.

Process-connected equipment that is listed and marked "Dual Seal" shall not require additional process sealing when used within the manufacturer's ratings.

FPN: For construction and testing requirements for dual seal process connected equipment, refer to ISA 12.27.01, *Requirements for Process Sealing Between Electrical Systems and Potentially Flammable or Combustible Process Fluids.*

501.20 Conductor Insulation, Class I, Divisions 1 and 2. Where condensed vapors or liquids may collect on, or come in contact with, the insulation on conductors, such insulation shall be of a type identified for use under such conditions; or the insulation shall be protected by a sheath of lead or by other approved means.

501.25 Uninsulated Exposed Parts, Class I, Divisions 1 and 2. There shall be no uninsulated exposed parts, such as electric conductors, buses, terminals, or components, that operate at more than 30 volts (15 volts in wet locations). These parts shall additionally be protected by a protection technique according to 500.7(E), 500.7(F), or 500.7(G) that is suitable for the location.

501.30 Grounding and Bonding, Class I, Divisions 1 and 2. Wiring and equipment in Class I, Division 1 and 2 locations shall be grounded as specified in Article 250 and with the requirements in 501.30(A) and 501.30(B).

(A) Bonding. The locknut-bushing and double-locknut types of contacts shall not be depended on for bonding purposes, but bonding jumpers with proper fittings or other approved means of bonding shall be used. Such means of bonding shall apply to all intervening raceways, fittings, boxes, enclosures, and so forth between Class I locations and the point of grounding for service equipment or point of grounding of a separately derived system.

Exception: The specific bonding means shall be required only to the nearest point where the grounded circuit conductor and the grounding electrode are connected together on the line side of the building or structure disconnecting means as specified in 250.32(A), (B), and (C), provided the branch-circuit overcurrent protection is located on the load side of the disconnecting means.

FPN: See 250.100 for additional bonding requirements in hazardous (classified) locations.

(B) Types of Equipment Grounding Conductors. Where flexible metal conduit or liquidtight flexible metal conduit is used as permitted in 501.10(B) and is to be relied on to complete a sole equipment grounding path, it shall be installed with internal or external bonding jumpers in parallel with each conduit and complying with 250.102.

Exception: In Class I, Division 2 locations, the bonding jumper shall be permitted to be deleted where all of the following conditions are met:

(1) Listed liquidtight flexible metal conduit 1.8 m (6 ft) or less in length, with fittings listed for grounding, is used.
(2) Overcurrent protection in the circuit is limited to 10 amperes or less.
(3) The load is not a power utilization load.

501.35 Surge Protection.

(A) Class I, Division 1. Surge arresters, transient voltage surge suppressors (TVSS), and capacitors shall be installed in enclosures identified for Class I, Division 1 locations. Surge-protective capacitors shall be of a type designed for specific duty.

(B) Class I, Division 2. Surge arresters and TVSS shall be nonarcing, such as metal-oxide varistor (MOV) sealed type, and surge-protective capacitors shall be of a type designed for specific duty. Enclosures shall be permitted to be of the general-purpose type. Surge protection of types other than described in this paragraph shall be installed in enclosures identified for Class I, Division 1 locations.

501.40 Multiwire Branch Circuits. In a Class I, Division 1 location, a multiwire branch circuit shall not be permitted.

Exception: Where the disconnect device(s) for the circuit opens all ungrounded conductors of the multiwire circuit simultaneously.

III. Equipment

501.100 Transformers and Capacitors.

(A) Class I, Division 1. In Class I, Division 1 locations, transformers and capacitors shall comply with 501.100(A)(1) and (A)(2).

(1) Containing Liquid That Will Burn. Transformers and capacitors containing a liquid that will burn shall be installed only in vaults that comply with 450.41 through 450.48 and with (1) through (4) as follows:

(1) There shall be no door or other communicating opening between the vault and the Division 1 location.
(2) Ample ventilation shall be provided for the continuous removal of flammable gases or vapors.
(3) Vent openings or ducts shall lead to a safe location outside of buildings.
(4) Vent ducts and openings shall be of sufficient area to relieve explosion pressures within the vault, and all portions of vent ducts within the buildings shall be of reinforced concrete construction.

(2) Not Containing Liquid That Will Burn. Transformers and capacitors that do not contain a liquid that will burn shall be installed in vaults complying with 501.100(A)(1) or be approved for Class I locations.

(B) Class I, Division 2. In Class I, Division 2 locations, transformers and capacitors shall comply with 450.21 through 450.27.

501.105 Meters, Instruments, and Relays.

(A) Class I, Division 1. In Class I, Division 1 locations, meters, instruments, and relays, including kilowatt-hour meters, instrument transformers, resistors, rectifiers, and thermionic tubes, shall be provided with enclosures identified for Class I, Division 1 locations. Enclosures for Class I, Division 1 locations include explosionproof enclosures and purged and pressurized enclosures.

FPN: See NFPA 496-2003, *Standard for Purged and Pressurized Enclosures for Electrical Equipment.*

(B) Class I, Division 2. In Class I, Division 2 locations, meters, instruments, and relays shall comply with 501.105(B)(1) through (B)(6).

(1) Contacts. Switches, circuit breakers, and make-and-break contacts of pushbuttons, relays, alarm bells, and horns shall have enclosures identified for Class I, Division 1 locations in accordance with 501.105(A).

Exception: General-purpose enclosures shall be permitted if current-interrupting contacts comply with one of the following:

(1) Are immersed in oil
(2) Are enclosed within a chamber that is hermetically sealed against the entrance of gases or vapors
(3) Are in nonincendive circuits
(4) Are listed for Division 2

(2) Resistors and Similar Equipment. Resistors, resistance devices, thermionic tubes, rectifiers, and similar equipment that are used in or in connection with meters, instruments, and relays shall comply with 501.105(A).

Exception: General-purpose-type enclosures shall be permitted if such equipment is without make-and-break or sliding contacts [other than as provided in 501.105(B)(1)] and if the maximum operating temperature of any exposed surface will not exceed 80 percent of the ignition temperature in degrees Celsius of the gas or vapor involved or has been tested and found incapable of igniting the gas or vapor. This exception shall not apply to thermionic tubes.

(3) Without Make-or-Break Contacts. Transformer windings, impedance coils, solenoids, and other windings that do not incorporate sliding or make-or-break contacts shall be provided with enclosures. General-purpose-type enclosures shall be permitted.

(4) General-Purpose Assemblies. Where an assembly is made up of components for which general-purpose enclosures are acceptable as provided in 501.105(B)(1), (B)(2), and (B)(3), a single general-purpose enclosure shall be acceptable for the assembly. Where such an assembly includes any of the equipment described in 501.105(B)(2), the maximum obtainable surface temperature of any component of the assembly shall be clearly and permanently indicated on the outside of the enclosure. Alternatively, equipment shall be permitted to be marked to indicate the temperature class for which it is suitable, using the temperature class (T Code) of Table 500.8(B).

(5) Fuses. Where general-purpose enclosures are permitted in 501.105(B)(1) through (B)(4), fuses for overcurrent protection of instrument circuits not subject to overloading in normal use shall be permitted to be mounted in general-purpose enclosures if each such fuse is preceded by a switch complying with 501.105(B)(1).

(6) Connections. To facilitate replacements, process control instruments shall be permitted to be connected through flexible cord, attachment plug, and receptacle, provided all of the following conditions apply:

(1) A switch complying with 501.105(B)(1) is provided so that the attachment plug is not depended on to interrupt current.
(2) The current does not exceed 3 amperes at 120 volts, nominal.
(3) The power-supply cord does not exceed 900 mm (3 ft), is of a type listed for extra-hard usage or for hard usage if protected by location, and is supplied through an attachment plug and receptacle of the locking and grounding type.
(4) Only necessary receptacles are provided.
(5) The receptacle carries a label warning against unplugging under load.

501.115 Switches, Circuit Breakers, Motor Controllers, and Fuses.

(A) Class I, Division 1. In Class I, Division 1 locations, switches, circuit breakers, motor controllers, and fuses, including pushbuttons, relays, and similar devices, shall be provided with enclosures, and the enclosure in each case, together with the enclosed apparatus, shall be identified as a complete assembly for use in Class I locations.

(B) Class I, Division 2. Switches, circuit breakers, motor controllers, and fuses in Class I, Division 2 locations shall comply with 501.115(B)(1) through (B)(4).

(1) Type Required. Circuit breakers, motor controllers, and switches intended to interrupt current in the normal performance of the function for which they are installed shall be provided with enclosures identified for Class I, Division 1 locations in accordance with 501.105(A), unless general-purpose enclosures are provided and any of the following apply:

(1) The interruption of current occurs within a chamber hermetically sealed against the entrance of gases and vapors.
(2) The current make-and-break contacts are oil-immersed and of the general-purpose type having a 50-mm (2-in.) minimum immersion for power contacts and a 25-mm (1-in.) minimum immersion for control contacts.
(3) The interruption of current occurs within a factory-sealed explosionproof chamber.
(4) The device is a solid state, switching control without contacts, where the surface temperature does not exceed 80 percent of the ignition temperature in degrees Celsius of the gas or vapor involved.

(2) Isolating Switches. Fused or unfused disconnect and isolating switches for transformers or capacitor banks that are not intended to interrupt current in the normal performance of the function for which they are installed shall be permitted to be installed in general-purpose enclosures.

(3) Fuses. For the protection of motors, appliances, and lamps, other than as provided in 501.115(B)(4), standard plug or cartridge fuses shall be permitted, provided they are placed within enclosures identified for the location; or fuses shall be permitted if they are within general-purpose enclosures, and if they are of a type in which the operating element is immersed in oil or other approved liquid, or the operating element is enclosed within a chamber hermetically sealed against the entrance of gases and vapors, or the fuse is a nonindicating, filled, current-limiting type.

(4) Fuses Internal to Luminaires (Lighting Fixtures). Listed cartridge fuses shall be permitted as supplementary protection within luminaires (lighting fixtures).

501.120 Control Transformers and Resistors. Transformers, impedance coils, and resistors used as, or in conjunction with, control equipment for motors, generators, and appliances shall comply with 501.120(A) and 501.120(B).

(A) Class I, Division 1. In Class I, Division 1 locations, transformers, impedance coils, and resistors, together with any switching mechanism associated with them, shall be provided with enclosures identified for Class I, Division 1 locations in accordance with 501.105(A).

(B) Class I, Division 2. In Class I, Division 2 locations, control transformers and resistors shall comply with 501.120(B)(1) through (B)(3).

(1) Switching Mechanisms. Switching mechanisms used in conjunction with transformers, impedance coils, and resistors shall comply with 501.115(B).

(2) Coils and Windings. Enclosures for windings of transformers, solenoids, or impedance coils shall be permitted to be of the general-purpose type.

(3) Resistors. Resistors shall be provided with enclosures; and the assembly shall be identified for Class I locations, unless resistance is nonvariable and maximum operating temperature, in degrees Celsius, will not exceed 80 percent of the ignition temperature of the gas or vapor involved or has been tested and found incapable of igniting the gas or vapor.

501.125 Motors and Generators.

(A) Class I, Division 1. In Class I, Division 1 locations, motors, generators, and other rotating electric machinery shall be one of the following:

(1) Identified for Class I, Division 1 locations
(2) Of the totally enclosed type supplied with positive-pressure ventilation from a source of clean air with discharge to a safe area, so arranged to prevent energizing of the machine until ventilation has been established and the enclosure has been purged with at least 10 volumes of air, and also arranged to automatically de-energize the equipment when the air supply fails
(3) Of the totally enclosed inert gas-filled type supplied with a suitable reliable source of inert gas for pressurizing the enclosure, with devices provided to ensure a positive pressure in the enclosure and arranged to automatically de-energize the equipment when the gas supply fails
(4) Of a type designed to be submerged in a liquid that is flammable only when vaporized and mixed with air, or in a gas or vapor at a pressure greater than atmospheric and that is flammable only when mixed with air; and the machine is arranged so to prevent energizing it until it has been purged with the liquid or gas to exclude air, and also arranged to automatically de-energize the equipment when the supply of liquid or gas or vapor fails or the pressure is reduced to atmospheric

Totally enclosed motors of the types specified in 501.125(A)(2) or 501.125(A)(3) shall have no external surface with an operating temperature in degrees Celsius in excess of 80 percent of the ignition temperature of the gas or vapor involved. Appropriate devices shall be provided to detect and automatically de-energize the motor or provide an adequate alarm if there is any increase in temperature of the motor beyond designed limits. Auxiliary equipment shall be of a type identified for the location in which it is installed.

FPN: See D 2155-69, *ASTM Test Procedure.*

(B) Class I, Division 2. In Class I, Division 2 locations, motors, generators, and other rotating electric machinery in which are employed sliding contacts, centrifugal or other types of switching mechanism (including motor overcurrent, overloading, and overtemperature devices), or integral resistance devices, either while starting or while running, shall be identified for Class I, Division 1 locations, unless such sliding contacts, switching mechanisms, and resistance devices are provided with enclosures identified for Class I, Division 2 locations in accordance with 501.105(B). The exposed surface of space heaters used to prevent condensation of moisture during shutdown periods shall not exceed 80 percent of the ignition temperature in degrees Celsius of the gas or vapor involved when operated at rated voltage, and the maximum surface temperature [based on a 40°C (104°F) ambient] shall be permanently marked on a visible nameplate mounted on the motor. Otherwise, space heaters shall be identified for Class I, Division 2 locations. In Class I, Division 2 locations, the installation of open or nonexplosionproof enclosed motors, such as squirrel-cage induction motors without brushes, switching mechanisms, or similar arc-producing devices that are not identified for use in a Class I, Division 2 location, shall be permitted.

FPN No. 1: It is important to consider the temperature of internal and external surfaces that may be exposed to the flammable atmosphere.

FPN No. 2: It is important to consider the risk of ignition due to currents arcing across discontinuities and overheating of parts in multisection enclosures of large motors and generators. Such motors and generators may need equipotential bonding jumpers across joints in the enclosure and from enclosure to ground. Where the presence of ignitible gases or vapors is suspected, clean-air purging may be needed immediately prior to and during start-up periods.

FPN No. 3: For further information on the application of electric motors in Class I, Division 2 hazardous (classified) locations, see IEEE Std. 1349-2001, *IEEE Guide for the Application of Electric Motors in Class I, Division 2 Hazardous (Classified) Locations.*

501.130 Luminaires (Lighting Fixtures). Luminaires (lighting fixtures) shall comply with 501.130(A) or (B).

(A) Class I, Division 1. In Class I, Division 1 locations, luminaires (lighting fixtures) shall comply with 501.130(A)(1) through (A)(4).

(1) Luminaires (Lighting Fixtures). Each luminaire (lighting fixture) shall be identified as a complete assembly for the Class I, Division 1 location and shall be clearly marked to indicate the maximum wattage of lamps for which it is identified. Luminaires (lighting fixtures) intended for portable use shall be specifically listed as a complete assembly for that use.

(2) Physical Damage. Each luminaire (lighting fixture) shall be protected against physical damage by a suitable guard or by location.

(3) Pendant Luminaires (Lighting Fixtures). Pendant luminaires (lighting fixtures) shall be suspended by and supplied through threaded rigid metal conduit stems or threaded steel intermediate conduit stems, and threaded joints shall be provided with set-screws or other effective means to prevent loosening. For stems longer than 300 mm (12 in.), permanent and effective bracing against lateral displacement shall be provided at a level not more than 300 mm (12 in.) above the lower end of the stem, or flexibility in the form of a fitting or flexible connector identified for the Class I, Division 1 location shall be provided not more than 300 mm (12 in.) from the point of attachment to the supporting box or fitting.

(4) Supports. Boxes, box assemblies, or fittings used for the support of luminaires (lighting fixtures) shall be identified for Class I locations.

(B) Class I, Division 2. In Class I, Division 2 locations, luminaires (lighting fixtures) shall comply with 501.130(B)(1) through 501.130(B)(6).

(1) Luminaires (Lighting Fixtures). Where lamps are of a size or type that may, under normal operating conditions, reach surface temperatures exceeding 80 percent of the ignition temperature in degrees Celsius of the gas or vapor involved, fixtures shall comply with 501.130(A)(1) or shall be of a type that has been tested in order to determine the marked operating temperature or temperature class (T Code).

(2) Physical Damage. Luminaires (lighting fixtures) shall be protected from physical damage by suitable guards or by location. Where there is danger that falling sparks or hot metal from lamps or fixtures might ignite localized concentrations of flammable vapors or gases, suitable enclosures or other effective protective means shall be provided.

(3) Pendant Luminaires (Fixtures). Pendant luminaires (lighting fixtures) shall be suspended by threaded rigid metal conduit stems, threaded steel intermediate metal conduit stems, or other approved means. For rigid stems longer than 300 mm (12 in.), permanent and effective bracing against lateral displacement shall be provided at a level not more than 300 mm (12 in.) above the lower end of the stem, or flexibility in the form of an identified fitting or flexible connector shall be provided not more than 300 mm (12 in.) from the point of attachment to the supporting box or fitting.

(4) Portable Lighting Equipment. Portable lighting equipment shall comply with 501.130(A)(1).

Exception: Where portable lighting equipment is mounted on movable stands and is connected by flexible cords, as covered in 501.140, it shall be permitted, where mounted in any position, if it conforms to 501.130(B)(2).

(5) Switches. Switches that are a part of an assembled fixture or of an individual lampholder shall comply with 501.115(B)(1).

(6) Starting Equipment. Starting and control equipment for electric-discharge lamps shall comply with 501.120(B).

Exception: A thermal protector potted into a thermally protected fluorescent lamp ballast if the luminaire (lighting fixture) is identified for the location.

501.135 Utilization Equipment.

(A) Class I, Division 1. In Class I, Division 1 locations, all utilization equipment shall be identified for Class I, Division 1 locations.

(B) Class I, Division 2. In Class I, Division 2 locations, all utilization equipment shall comply with 501.135(B)(1) through (B)(3).

(1) Heaters. Electrically heated utilization equipment shall conform with either item (1) or (2):

(1) The heater shall not exceed 80 percent of the ignition temperature in degrees Celsius of the gas or vapor involved on any surface that is exposed to the gas or vapor when continuously energized at the maximum rated ambient temperature. If a temperature controller is not provided, these conditions shall apply when the heater is operated at 120 percent of rated voltage.

Exception No. 1: For motor-mounted anticondensation space heaters, see 501.125.

Exception No. 2: Where a current-limiting device is applied to the circuit serving the heater to limit the current in the heater to a value less than that required to raise the heater surface temperature to 80 percent of the ignition temperature.

(2) The heater shall be identified for Class I, Division 1 locations.

Exception: Electrical resistance heat tracing identified for Class I, Division 2 locations.

(2) Motors. Motors of motor-driven utilization equipment shall comply with 501.125(B).

(3) Switches, Circuit Breakers, and Fuses. Switches, circuit breakers, and fuses shall comply with 501.115(B).

501.140 Flexible Cords, Class I, Divisions 1 and 2.

(A) Permitted Uses. Flexible cord shall be permitted:

(1) For connection between portable lighting equipment or other portable utilization equipment and the fixed portion of their supply circuit.
(2) For that portion of the circuit where the fixed wiring methods of 501.10(A) cannot provide the necessary degree of movement for fixed and mobile electrical utilization equipment, and the flexible cord is protected by location or by a suitable guard from damage and only in an industrial establishment where conditions of maintenance and engineering supervision ensure that only qualified persons install and service the installation.
(3) For electric submersible pumps with means for removal without entering the wet-pit. The extension of the flexible cord within a suitable raceway between the wet-pit and the power source shall be permitted.
(4) For electric mixers intended for travel into and out of open-type mixing tanks or vats.

(B) Installation. Where flexible cords are used, the cords shall comply with all of the following:

(1) Be of a type listed for extra-hard usage
(2) Contain, in addition to the conductors of the circuit, a grounding conductor complying with 400.23
(3) Be connected to terminals or to supply conductors in an approved manner
(4) Be supported by clamps or by other suitable means in such a manner that there is no tension on the terminal connections
(5) Be provided with suitable seals where the flexible cord enters boxes, fittings, or enclosures of the explosionproof type

Exception to (5): Seals shall not be required as provided in 501.10(B) and 501.105(B)(6).

(6) Be of continuous length

FPN: See 501.20 for flexible cords exposed to liquids having a deleterious effect on the conductor insulation.

501.145 Receptacles and Attachment Plugs, Class I, Divisions 1 and 2. Receptacles and attachment plugs shall be of the type providing for connection to the grounding conductor of a flexible cord and shall be identified for the location.

Exception: As provided in 501.105(B)(6).

501.150 Signaling, Alarm, Remote-Control, and Communications Systems.

(A) Class I, Division 1. In Class I, Division 1 locations, all apparatus and equipment of signaling, alarm, remote-control, and communications systems, regardless of voltage, shall be identified for Class I, Division 1 locations, and all wiring shall comply with 501.10(A), 501.15(A), and 501.15(C).

(B) Class I, Division 2. In Class I, Division 2 locations, signaling, alarm, remote-control, and communications systems shall comply with 501.150(B)(1) through (B)(4).

(1) Contacts. Switches, circuit breakers, and make-and-break contacts of pushbuttons, relays, alarm bells, and horns shall have enclosures identified for Class I, Division 1 locations in accordance with 501.105(A).

Exception: General-purpose enclosures shall be permitted if current-interrupting contacts are one of the following:

(1) Immersed in oil
(2) Enclosed within a chamber hermetically sealed against the entrance of gases or vapors
(3) In nonincendive circuits
(4) Part of a listed nonincendive component

(2) Resistors and Similar Equipment. Resistors, resistance devices, thermionic tubes, rectifiers, and similar equipment shall comply with 501.105(B)(2).

(3) Protectors. Enclosures shall be provided for lightning protective devices and for fuses. Such enclosures shall be permitted to be of the general-purpose type.

(4) Wiring and Sealing. All wiring shall comply with 501.10(B), 501.15(B), and 501.15(C).

ARTICLE 505
Class I, Zone 0, 1, and 2 Locations

Editor's Note: *The text of Section 505.2 is not included because it is included in Definitions.*

FPN: Rules that are followed by a reference in brackets contain text that has been extracted from NFPA 497-2004, *Recommended Practice for the Classification of Flammable Liquids, Gases, or Vapors and of Hazardous (Classified) Locations for Electrical Installations in Chemical Process Areas.* Only editorial changes were made to the extracted text to make it consistent with this *Code.*

505.1 Scope. This article covers the requirements for the zone classification system as an alternative to the division classification system covered in Article 500 for electrical and electronic equipment and wiring for all voltages in Class I, Zone 0, Zone 1, and Zone 2 hazardous (classified) locations where fire or explosion hazards may exist due to flammable gases, vapors, or liquids.

FPN: For the requirements for electrical and electronic equipment and wiring for all voltages in Class I, Division 1 or Division 2; Class II, Division 1 or Division 2; and Class III, Division 1 or Division 2 hazardous (classified) locations where fire or explosion hazards may exist due to flammable gases or vapors, flammable liquids, or combustible dusts or fibers, refer to Articles 500 through 504.

505.3 Other Articles. All other applicable rules contained in this *Code* shall apply to electrical equipment and wiring installed in hazardous (classified) locations.

Exception: As modified by Article 504 and this article.

505.4 General.

(A) Documentation for Industrial Occupancies. All areas in industrial occupancies designated as hazardous (classified) locations shall be properly documented. This documentation shall be available to those authorized to design, install, inspect, maintain, or operate electrical equipment at the location.

FPN: For examples of area classification drawings, see ANSI/API RP 505-1997, *Recommended Practice for Classification of Locations for Electrical Installations at Petroleum Facilities Classified as Class I, Zone 0, Zone 1, or Zone 2*; ISA RP12.24.01-1998, *Recommended Practice for Classification of Locations for Electrical Installations Classified as Class I, Zone 0, Zone 1, or Zone 2*; IEC 60079-10-1995, *Electrical Apparatus for Explosive Gas Atmospheres, Classification of Hazardous Areas*; and *Model Code of Safe Practice in the Petroleum Industry, Part 15: Area Classification Code for Petroleum Installations*, IP 15, The Institute of Petroleum, London.

(B) Reference Standards. Important information relating to topics covered in Chapter 5 may be found in other publications.

FPN No. 1: It is important that the authority having jurisdiction be familiar with recorded industrial experience as well as with standards of the National Fire Protection Association (NFPA), the American Petroleum Institute (API), the Instrumentation, Systems, and Automation Society (ISA), and the International Electrotechnical Commission (IEC) that may be of use in the classification of various locations, the determination of adequate ventilation, and the protection against static electricity and lightning hazards.

FPN No. 2: For further information on the classification of locations, see ANSI/API RP 505-1997, *Recommended Practice for Classification of Locations for Electrical Installations at Petroleum Facilities Classified as Class I, Zone 0, Zone 1, or Zone 2*; ISA RP 12.24.01-1998, *Recommended Practice for Classification of Locations for Electrical Installations Classified as Class I, Zone 0, Zone 1, or Zone 2*; IEC 60079-10-1995, *Electrical Apparatus for Explosive Gas Atmospheres, Classification of Hazardous Areas*; and *Model Code of Safe Practice in the Petroleum Industry, Part 15: Area Classification Code for Petroleum Installations*, IP 15, The Institute of Petroleum, London.

FPN No. 3: For further information on protection against static electricity and lightning hazards in hazardous (classified) locations, see NFPA 77-2000, *Recommended Practice on Static Electricity*; NFPA 780-2004, *Standard for the Installation of Lightning Protection Systems*; and API RP 2003-1998, *Protection Against Ignitions Arising Out of Static Lightning and Stray Currents.*

FPN No. 4: For further information on ventilation, see NFPA 30-2003, *Flammable and Combustible Liquids Code*, and ANSI/API RP 505-1997, *Recommended Practice for Classification of Locations for Electrical Installations at Petroleum Facilities Classified as Class I, Zone 0, Zone 1, or Zone 2.*

FPN No. 5: For further information on electrical systems for hazardous (classified) locations on offshore oil and gas producing platforms, see ANSI/API RP 14FZ-2000, *Recommended Practice for Design and Installation of Electrical Systems for Fixed and Floating Offshore Petroleum Facilities for Unclassified and Class I, Zone 0, Zone 1, and Zone 2 Locations.*

FPN No. 6: For further information on the installation of electrical equipment in hazardous (classified) locations in general, see IEC 60079-14-1996, *Electrical Apparatus for Explosive Gas Atmospheres — Part 14: Electrical Installations in Explosive Gas Atmospheres (Other Than Mines)*, and IEC 60079-16-1990, *Electrical Apparatus for Explosive Gas Atmospheres — Part 16: Artificial Ventilation for the Protection of Analyzer(s) Houses.*

FPN No. 7: For further information on application of electrical equipment in hazardous (classified) locations in general, see ANSI/ISA 12.00.01-2002, *Electrical Apparatus for Use in Class I, Zones 0 and 1, Hazardous (Classified) Locations: General Requirements*; ISA 12.01.01-1999, *Definitions and Information Pertaining to Electrical Apparatus in Hazardous (Classified) Locations*; and ANSI/UL 60079-0, *Electrical apparatus for explosive gas atmospheres — Part 0: General requirements.*

505.5 Classifications of Locations.

(A) Classification of Locations. Locations shall be classified depending on the properties of the flammable vapors, liquids, or gases that may be present and the likelihood that a flammable or combustible concentration or quantity is present. Where pyrophoric materials are the only materials used or handled, these locations shall not be classified. Each room, section, or area shall be considered individually in determining its classification.

FPN No. 1: See 505.7 for restrictions on area classification.

FPN No. 2: Through the exercise of ingenuity in the layout of electrical installations for hazardous (classified) locations, it is frequently possible to locate much of the equipment in reduced level of classification or in an unclassified location and, thus, to reduce the amount of special equipment required.

Rooms and areas containing ammonia refrigeration systems that are equipped with adequate mechanical ventilation may be classified as "unclassified" locations.

FPN: For further information regarding classification and ventilation of areas involving ammonia, see ANSI/ASHRAE 15-1994, *Safety Code for Mechanical Refrigeration*; and ANSI/CGA G2.1-1989 (14-39), *Safety Requirements for the Storage and Handling of Anhydrous Ammonia.*

(B) Class I, Zone 0, 1, and 2 Locations. Class I, Zone 0, 1, and 2 locations are those in which flammable gases or vapors are or may be present in the air in quantities sufficient to produce explosive or ignitible mixtures. Class I, Zone 0, 1, and 2 locations shall include those specified in 505(B)(1), (B)(2), and (B)(3).

(1) Class I, Zone 0. A Class I, Zone 0 location is a location in which

(1) Ignitible concentrations of flammable gases or vapors are present continuously, or
(2) Ignitible concentrations of flammable gases or vapors are present for long periods of time.

FPN No. 1: As a guide in determining when flammable gases or vapors are present continuously or for long periods of time, refer to ANSI/API RP 505-1997, *Recommended Practice for Classification of Locations for Electrical Installations of Petroleum Facilities Classified as Class I, Zone 0, Zone 1 or Zone 2*; ISA 12.24.01-1998, *Recommended Practice for Classification of Locations for Electrical Installations Classified as Class I, Zone 0, Zone 1, or Zone 2*; IEC 60079-10-1995, *Electrical Apparatus for Explosive Gas Atmospheres, Classifications of Hazardous Areas*; and *Area Classification Code for Petroleum Installations, Model Code, Part 15*, Institute of Petroleum.

FPN No. 2: This classification includes locations inside vented tanks or vessels that contain volatile flammable liquids; inside inadequately vented spraying or coating enclosures, where volatile flammable solvents are used; between the inner and outer roof sections of a floating roof tank containing volatile flammable liquids; inside open vessels, tanks and pits containing volatile flammable liquids; the interior of an exhaust duct that is used to vent ignitible concentrations of gases or vapors; and inside inadequately ventilated enclosures that contain normally venting instruments utilizing or analyzing flammable fluids and venting to the inside of the enclosures.

FPN No. 3: It is not good practice to install electrical equipment in Zone 0 locations except when the equipment is essential to the process or when other locations are not feasible. [See 505.5(A) FPN No. 2.] If it is necessary to install electrical systems in a Zone 0 location, it is good practice to install intrinsically safe systems as described by Article 504.

(2) Class I, Zone 1. A Class I, Zone 1 location is a location

(1) In which ignitible concentrations of flammable gases or vapors are likely to exist under normal operating conditions; or
(2) In which ignitible concentrations of flammable gases or vapors may exist frequently because of repair or maintenance operations or because of leakage; or
(3) In which equipment is operated or processes are carried on, of such a nature that equipment breakdown or faulty operations could result in the release of ignitible concentrations of flammable gases or vapors and also cause simultaneous failure of electrical equipment in a mode to cause the electrical equipment to become a source of ignition; or
(4) That is adjacent to a Class I, Zone 0 location from which ignitible concentrations of vapors could be communicated, unless communication is prevented by adequate positive pressure ventilation from a source of clean air and effective safeguards against ventilation failure are provided.

FPN No. 1: Normal operation is considered the situation when plant equipment is operating within its design parameters. Minor releases of flammable material may be part of normal operations. Minor releases include the releases from mechanical packings on pumps. Failures that involve repair or shutdown (such as the breakdown of pump seals and flange gaskets, and spillage caused by accidents) are not considered normal operation.

FPN No. 2: This classification usually includes locations where volatile flammable liquids or liquefied flammable gases are transferred from one container to another. In areas in the vicinity of spraying and painting operations where flammable solvents are used; adequately ventilated drying rooms or compartments for evaporation of flammable solvents; adequately ventilated locations containing fat and oil extraction equipment using volatile flammable solvents; portions of cleaning and dyeing plants where volatile flammable liquids are used; adequately ventilated gas generator rooms and other portions of gas manufacturing plants where flammable gas may escape; inadequately ventilated pump rooms for flammable gas or for volatile flammable liquids; the interiors of refrigerators and freezers in which volatile flammable materials are stored in the open, lightly stoppered, or in easily ruptured containers; and other locations where ignitible concentrations of flammable vapors or gases are likely to occur in the course of normal operation but not classified Zone 0.

(3) Class I, Zone 2. A Class I, Zone 2 location is a location

(1) In which ignitible concentrations of flammable gases or vapors are not likely to occur in normal operation and, if they do occur, will exist only for a short period; or
(2) In which volatile flammable liquids, flammable gases, or flammable vapors are handled, processed, or used but in which the liquids, gases, or vapors normally are confined within closed containers of closed systems from which they can escape, only as a result of accidental rupture or breakdown of the containers or system, or as a result of the abnormal operation of the equipment with which the liquids or gases are handled, processed, or used; or
(3) In which ignitible concentrations of flammable gases or vapors normally are prevented by positive mechanical ventilation but which may become hazardous as a result of failure or abnormal operation of the ventilation equipment; or
(4) That is adjacent to a Class I, Zone 1 location, from which ignitible concentrations of flammable gases or vapors could be communicated, unless such communication is prevented by adequate positive-pressure ventilation from a source of clean air and effective safeguards against ventilation failure are provided.

FPN: The Zone 2 classification usually includes locations where volatile flammable liquids or flammable gases or vapors are used but which would become hazardous only in case of an accident or of some unusual operating condition.

505.6 Material Groups. For purposes of testing, approval, and area classification, various air mixtures (not oxygen enriched) shall be grouped as required in 505.6(A), (B), and (C).

FPN: Group I is intended for use in describing atmospheres that contain firedamp (a mixture of gases, composed mostly of methane, found underground, usually in mines). This *Code* does not apply to installations underground in mines. See 90.2(B).

Group II shall be subdivided into IIC, IIB, and IIA, as noted in 505.6(A), (B), and (C), according to the nature of the gas or vapor, for protection techniques "d," "ia," "ib," "[ia]," and "[ib]," and, where applicable, "n" and "o."

FPN No. 1: The gas and vapor subdivision as described above is based on the maximum experimental safe gap (MESG), minimum igniting current (MIC), or both. Test equipment for determining the MESG is described in IEC 60079-1A-1975, Amendment No. 1 (1993), *Construction and Verification Tests of Flameproof Enclosures of Electrical Apparatus*; and *UL Technical Report No. 58* (1993). The test equipment for determining MIC is described in IEC 60079-11-1999, *Electrical Apparatus for Explosive Gas Atmospheres — Part 11: Intrinsic Safety "i."* The classification of gases or vapors according to their maximum experimental safe gaps and minimum igniting currents is described in IEC 60079-12-1978, *Classification of Mixtures of Gases or Vapours with Air According to Their Maximum Experimental Safe Gaps and Minimum Igniting Currents.*

FPN No. 2: Verification of electrical equipment utilizing protection techniques "e," "m," "p," and "q," due to design technique, does not require tests involving MESG or MIC. Therefore, Group II is not required to be subdivided for these protection techniques.

FPN No. 3: It is necessary that the meanings of the different equipment markings and Group II classifications be carefully observed to avoid confusion with Class I, Divisions 1 and 2, Groups A, B, C, and D.

Class I, Zone 0, 1, and 2, groups shall be as follows:

(A) Group IIC. Atmospheres containing acetylene, hydrogen, or flammable gas, flammable liquid-produced vapor, or combustible liquid-produced vapor mixed with air that may burn or explode, having either a maximum experimental safe gap (MESG) value less than or equal to 0.50 mm or minimum igniting current ratio (MIC ratio) less than or equal to 0.45. [NFPA 497:3.3]

FPN: Group IIC is equivalent to a combination of Class I, Group A, and Class I, Group B, as described in 500.6(A)(1) and 500.6(A)(2).

(B) Group IIB. Atmospheres containing acetaldehyde, ethylene, or flammable gas, flammable liquid-produced vapor, or combustible liquid-produced vapor mixed with air that may burn or explode, having either maximum experimental safe gap (MESG) values greater than 0.50 mm and less than or equal to 0.90 mm or minimum igniting current ratio (MIC ratio) greater than 0.45 and less than or equal to 0.80. [NFPA 497:3.3]

FPN: Group IIB is equivalent to Class I, Group C, as described in 500.6(A)(3).

(C) Group IIA. Atmospheres containing acetone, ammonia, ethyl alcohol, gasoline, methane, propane, or flammable gas, flammable liquid-produced vapor, or combustible liquid-produced vapor mixed with air that may burn or explode, having either a maximum experiment safe gap (MESG) value greater than 0.90 mm or minimum igniting current ratio (MIC ratio) greater than 0.80. [NFPA 497:3.3]

FPN: Group IIA is equivalent to Class I, Group D as described in 500.6(A)(4).

505.7 Special Precaution. Article 505 requires equipment construction and installation that ensures safe performance under conditions of proper use and maintenance.

FPN No. 1: It is important that inspection authorities and users exercise more than ordinary care with regard to the installation and maintenance of electrical equipment in hazardous (classified) locations.

FPN No. 2: Low ambient conditions require special consideration. Electrical equipment depending on the protection techniques described by 505.8(A) may not be suitable for use at temperatures lower than −20°C (−4°F) unless they are identified for use at lower temperatures. However, at low ambient temperatures, flammable concentrations of vapors may not exist in a location classified Class I, Zones 0, 1, or 2 at normal ambient temperature.

(A) Supervision of Work. Classification of areas and selection of equipment and wiring methods shall be under the supervision of a qualified Registered Professional Engineer.

(B) Dual Classification. In instances of areas within the same facility classified separately, Class I, Zone 2 locations shall be permitted to abut, but not overlap, Class I, Division 2 locations. Class I, Zone 0 or Zone 1 locations shall not abut Class I, Division 1 or Division 2 locations.

(C) Reclassification Permitted. A Class I, Division 1 or Division 2 location shall be permitted to be reclassified as a Class I, Zone 0, Zone 1, or Zone 2 location, provided all of the space that is classified because of a single flammable gas or vapor source is reclassified under the requirements of this article.

(D) Solid Obstacles. Flameproof equipment with flanged joints shall not be installed such that the flange openings are closer than the distances shown in Table 505.7(D) to any solid obstacle that is not a part of the equipment (such as steelworks, walls, weather guards, mounting brackets, pipes, or other electrical equipment) unless the equipment is listed for a smaller distance of separation.

Table 505.7(D) Minimum Distance of Obstructions from Flameproof "d" Flange Openings

Gas Group	Minimum Distance mm	Minimum Distance in.
IIC	40	1 37/64
IIB	30	1 3/16
IIA	10	25/64

505.8 Protection Techniques. Acceptable protection techniques for electrical and electronic equipment in hazardous (classified) locations shall be as described in 505.8(A) through 505.8(I).

FPN: For additional information, see ANSI/ISA 12.00.01-2002, *Electrical Apparatus for Use in Class I, Zones 0 and 1 Hazardous (Classified) Locations, General Requirements*; ANSI/ISA 12.01.01-2002, *Definitions and Information Pertaining to Electrical Apparatus in Hazardous (Classified) Locations*; and ANSI/UL 60079–0, *Electrical apparatus for explosive gas atmospheres — Part 0: General requirements.*

(A) Flameproof "d". This protection technique shall be permitted for equipment in Class I, Zone 1 or Zone 2 locations.

(B) Purged and Pressurized. This protection technique shall be permitted for equipment in those Class I, Zone 1 or Zone 2 locations for which it is identified.

(C) Intrinsic Safety. This protection technique shall be permitted for apparatus and associated apparatus in Class I, Zone 0, Zone 1, or Zone 2 locations for which it is listed.

(D) Type of Protection "n". This protection technique shall be permitted for equipment in Class I, Zone 2 locations. Type of protection "n" is further subdivided into nA, nC, and nR.

FPN: See Table 505.9(C)(2)(4) for the descriptions of subdivisions for type of protection "n".

(E) Oil Immersion "o". This protection technique shall be permitted for equipment in Class I, Zone 1 or Zone 2 locations.

(F) Increased Safety "e". This protection technique shall be permitted for equipment in Class I, Zone 1 or Zone 2 locations.

(G) Encapsulation "m". This protection technique shall be permitted for equipment in Class I, Zone 1 or Zone 2 locations.

(H) Powder Filling "q". This protection technique shall be permitted for equipment in Class I, Zone 1 or Zone 2 locations.

(I) Combustible Gas Detection System. A combustible gas detection system shall be permitted as a means of protection in industrial establishments with restricted public access and where the conditions of maintenance and supervision ensure that only qualified persons service the installation. Gas detection equipment shall be listed for detection of the specific gas or vapor to be encountered. Where such a system is installed, equipment specified in 505.8(I)(1), I(2), or I(3) shall be permitted. The type of detection equipment, its listing, installation location(s), alarm and shutdown criteria, and calibration frequency shall be documented when combustible gas detectors are used as a protection technique.

FPN No. 1: For further information, see ANSI/ISA-12.13.01, *Performance Requirements, Combustible Gas Detectors.*

FPN No. 2: For further information, see ANSI/API RP 505, *Recommended Practice for Classification of Locations for Electrical Installations at Petroleum Facilities Classified as Class I, Zone 0, Zone 1, and Zone 2.*

FPN No. 3: For further information, see ISA-RP12.13.02, *Installation, Operation, and Maintenance of Combustible Gas Detection Instruments.*

(1) Inadequate Ventilation. In a Class I, Zone 1 location that is so classified due to inadequate ventilation, electrical equipment suitable for Class I, Zone 2 locations shall be permitted.

(2) Interior of a Building. In a building located in, or with an opening into, a Class I, Zone 2 location where the interior does not contain a source of flammable gas or vapor, electrical equipment for unclassified locations shall be permitted.

(3) Interior of a Control Panel. In the interior of a control panel containing instrumentation utilizing or measuring flammable liquids, gases, or vapors, electrical equipment suitable for Class I, Zone 2 locations shall be permitted.

505.9 Equipment.

(A) Suitability. Suitability of identified equipment shall be determined by one of the following:

(1) Equipment listing or labeling
(2) Evidence of equipment evaluation from a qualified testing laboratory or inspection agency concerned with product evaluation
(3) Evidence acceptable to the authority having jurisdiction such as a manufacturer's self-evaluation or an owner's engineering judgment

(B) Listing.

(1) Equipment that is listed for a Zone 0 location shall be permitted in a Zone 1 or Zone 2 location of the same gas or vapor, provided that it is installed in accordance with the requirements for the marked type of protection. Equipment that is listed for a Zone 1 location shall be permitted in a Zone 2 location of the same gas or vapor, provided that it is installed in accordance with the requirements for the marked type of protection.
(2) Equipment shall be permitted to be listed for a specific gas or vapor, specific mixtures of gases or vapors, or any specific combination of gases or vapors.

FPN: One common example is equipment marked for "IIB. + H2."

(C) Marking. Equipment shall be marked in accordance with 505.9(C)(1) or (C)(2).

(1) Division Equipment. Equipment identified for Class I, Division 1 or Class I, Division 2 shall, in addition to being marked in accordance with 500.8(B), be permitted to be marked with all of the following:

(1) Class I, Zone 1 or Class I, Zone 2 (as applicable)
(2) Applicable gas classification group(s) in accordance with Table 505.9(C)
(3) Temperature classification in accordance with 505.9(D)(1)

(2) Zone Equipment. Equipment meeting one or more of the protection techniques described in 505.8 shall be marked with all of the following in the order shown:

(1) Class
(2) Zone
(3) Symbol "AEx"
(4) Protection technique(s) in accordance with Table 505.9(C)(2)(4)

Table 505.9(C) Gas Classification Groups

Gas Group	Comment
IIC	See 505.6(A)(1)
IIB	See 505.6(A)(2)
IIA	See 505.6(A)(3)

Table 505.9(C)(2)(4) Types of Protection Designation

Designation	Technique	Zone*
d	Flameproof enclosure	1
e	Increased safety	1
ia	Intrinsic safety	0
ib	Intrinsic safety	1
[ia]	Associated apparatus	Unclassified
[ib]	Associated apparatus	Unclassified
m	Encapsulation	1
nA	Nonsparking equipment	2
nC	Sparking equipment in which the contacts are suitably protected other than by restricted breathing enclosure	2
nR	Restricted breathing enclosure	2
o	Oil immersion	1
p	Purged and pressurized	1 or 2
q	Powder filled	1

*Does not address use where a combination of techniques is used.

(5) Applicable gas classification group(s) in accordance with Table 505.9(C)
(6) Temperature classification in accordance with 505.9(D)(1)

Exception No. 1: Associated apparatus NOT suitable for installation in a hazardous (classified) locations shall be required to be marked only with (3), (4), and (5), but BOTH the symbol AEx (3) and the symbol for the type of protection (4) shall be enclosed within the same square brackets, for example, [AEx ia] IIC.

Exception No. 2: Simple apparatus as defined in 504.2 shall not be required to have a marked operating temperature or temperature class.

Electrical equipment of types of protection "e," "m," "p," or "q" shall be marked Group II. Electrical equipment of types of protection "d," "ia," "ib," "[ia]," or "[ib]" shall be marked Group IIA, IIB, or IIC, or for a specific gas or vapor. Electrical equipment of types of protection "n" shall be marked Group II unless it contains enclosed-break devices, nonincendive components, or energy-limited equipment or circuits, in which case it shall be marked Group IIA, IIB, or IIC, or a specific gas or vapor. Electrical equipment of other types of protection shall be marked Group II unless the type of protection utilized by the equipment requires that it be marked Group IIA, IIB, or IIC, or a specific gas or vapor.

FPN No. 1: An example of the required marking for intrinsically safe apparatus for installation in Class I, Zone 0 is "Class I, Zone 0, AEx ia IIC T6." An explanation of the marking that is required is shown in FPN Figure 505.9(C)(2).

FPN No. 2: An example of the required marking for intrinsically safe associated apparatus mounted in a flameproof enclosure for installation in Class I, Zone 1 is "Class I, Zone 1 AEx d[ia] IIC T4."

FPN No. 3: An example of the required marking for intrinsically safe associated apparatus NOT for installation in a hazardous (classified) location is "[AEx ia] IIC."

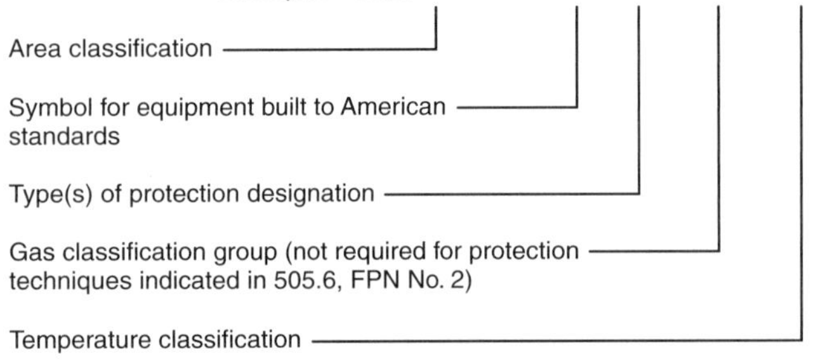

FPN Figure 505.9(C)(2) Zone Equipment Marking.

(D) Class I Temperature. The temperature marking specified below shall not exceed the ignition temperature of the specific gas or vapor to be encountered.

FPN: For information regarding ignition temperatures of gases and vapors, see NFPA 497-2004, *Recommended Practice for the Classification of Flammable Liquids, Gases, or Vapors and of Hazardous (Classified) Locations for Electrical Installations in Chemical Process Areas*; and IEC 60079-20-1996, *Electrical Apparatus for Explosive Gas Atmospheres, Data for Flammable Gases and Vapours, Relating to the Use of Electrical Apparatus.*

(1) Temperature Classifications. Equipment shall be marked to show the operating temperature or temperature class referenced to a 40°C (104°F) ambient. The temperature class, if provided, shall be indicated using the temperature class (T Code) shown in Table 505.9(D)(1).

Electrical equipment designed for use in the ambient temperature range between −20°C and +40°C shall require no additional ambient temperature marking.

Electrical equipment that is designed for use in a range of ambient temperatures other than −20°C and +40°C is considered to be special; and the ambient temperature range shall then be marked on the equipment, including either the symbol "Ta" or "Tamb" together with the special range of ambient temperatures. As an example, such a marking might be "−30°C ≤ Ta ≤ +40°C."

Electrical equipment suitable for ambient temperatures exceeding 40°C (104°F) shall be marked with both the maximum ambient temperature and the operating temperature or temperature class at that ambient temperature.

Exception No. 1: Equipment of the non–heat-producing type, such as conduit fittings, and equipment of the heat-producing type having a maximum temperature of not more than 100°C (212°F) shall not be required to have a marked operating temperature or temperature class.

Exception No. 2: Equipment identified for Class I, Division 1 or Division 2 locations as permitted by 505.20(B) and 505.20(D) shall be permitted to be marked in accordance with 500.8(B) and Table 500.8(B).

Table 505.9(D)(1) Classification of Maximum Surface Temperature for Group II Electrical Equipment

Temperature Class (T Code)	Maximum Surface Temperature (°C)
T1	≤450
T2	≤300
T3	≤200
T4	≤135
T5	≤100
T6	≤85

(E) Threading. All NPT threaded conduit and fittings referred to herein shall be threaded with a National (American) Standard Pipe Taper (NPT) thread that provides a taper of 1 in 16 (¾-in. taper per foot). Conduit and fittings shall be made wrenchtight to prevent sparking when fault current flows through the conduit system, and to ensure the explosionproof or flameproof integrity of the conduit system where applicable. Equipment provided with threaded entries for field wiring connections shall be installed in accordance with 505.9(E)(1) or 505.9(E)(2). Threaded entries into explosionproof or flameproof equipment shall be made up with at least five threads fully engaged.

Exception: For listed explosionproof or flameproof equipment, factory threaded NPT entries shall be made up with at least 4½ threads fully engaged.

(1) Equipment Provided with Threaded Entries for NPT Threaded Conduit or Fittings. For equipment provided with threaded entries for NPT threaded conduit or fittings, listed conduit fittings or cable fittings shall be used.

FPN: Thread form specifications for NPT threads are located in ANSI/ASME B1.20.1-1983, *Pipe Threads, General Purpose (Inch)*.

(2) Equipment Provided with Threaded Entries for Metric Threaded Conduit or Fittings. For equipment with metric threaded entries, such entries shall be identified as being metric, or listed adapters to permit connection to conduit or NPT-threaded fittings shall be provided with the equipment. Adapters shall be used for connection to conduit or NPT-threaded fittings. Listed cable fittings that have metric threads shall be permitted to be used.

FPN: Threading specifications for metric threaded entries are located in ISO 965/1-1980, *Metric Screw Threads*; and ISO 965/3-1980, *Metric Screw Threads*.

505.15 Wiring Methods. Wiring methods shall maintain the integrity of protection techniques and shall comply with 505.15(A) through 505.15(C).

(A) Class I, Zone 0. In Class I, Zone 0 locations, only intrinsically safe wiring methods in accordance with Article 504 shall be permitted.

FPN: Article 504 only includes protection technique "ia."

(B) Class I, Zone 1.

(1) General. In Class I, Zone 1 locations, the wiring methods in (B)(1)(a) through (B)(1)(f) shall be permitted.

(a) All wiring methods permitted by 505.15(A).

(b) In industrial establishments with restricted public access, where the conditions of maintenance and supervision ensure that only qualified persons service the installation, and where the cable is not subject to physical damage, Type MC-HL cable listed for use in Class I, Zone 1 or Division 1 locations, with a gas/vaportight continuous corrugated metallic sheath, an overall jacket of suitable polymeric material, separate grounding conductors in accordance with 250.122, and provided with termination fittings listed for the application.

FPN: See 330.12 for restrictions on use of Type MC cable.

(c) In industrial establishments with restricted public access, where the conditions of maintenance and supervision ensure that only qualified persons service the installation, and where the cable is not subject to physical damage, Type ITC-HL cable, listed for use in Class I, Zone 1 or Division 1 locations, with a gas/vaportight continuous corrugated metallic sheath, an overall jacket of suitable polymeric material and provided with termination fittings listed for the application.

(d) Type MI cable with termination fittings listed for Class I, Zone 1 or Division 1 locations. Type MI cable shall be installed and supported in a manner to avoid tensile stress at the termination fittings.

(e) Threaded rigid metal conduit, or threaded steel intermediate metal conduit.

(f) Rigid nonmetallic conduit complying with Article 352 shall be permitted where encased in a concrete envelope a minimum of 50 mm (2 in.) thick and provided with not less than 600 mm (24 in.) of cover measured from the top of the conduit to grade. Threaded rigid metal conduit or threaded steel intermediate metal conduit shall be used for the last 600 mm (24 in.) of the underground run to emergence or to the point of connection to the aboveground raceway. An equipment grounding conductor shall be included to provide for electrical continuity of the raceway system and for grounding of non–current-carrying metal parts.

(2) Flexible Connections. Where necessary to employ flexible connections, flexible fittings listed for Class I, Zone 1 or Division 1 locations or flexible cord in accordance with the provisions of 505.17 shall be permitted.

(C) Class I, Zone 2.

(1) General. In Class I, Zone 2 locations, the wiring methods in (C)(1)(a) through (C)(1)(g) shall be permitted.

(a) All wiring methods permitted by 505.15(B).

(b) Types MI, MC, MV, or TC cable with termination fittings, or in cable tray systems and installed in a manner to avoid tensile stress at the termination fittings. Single conductor Type MV cables shall be shielded or metallic-armored.

(c) Type ITC cable as permitted in 727.4.

(d) Type PLTC cable in accordance with the provisions of Article 725, or in cable tray systems. PLTC shall be installed in a manner to avoid tensile stress at the termination fittings.

(e) Enclosed gasketed busways, enclosed gasketed wireways.

(f) Threaded rigid metal conduit, threaded steel intermediate metal conduit.

(g) Nonincendive field wiring shall be permitted using any of the wiring methods permitted for unclassified locations. Nonincendive field wiring systems shall be installed in accordance with the control drawing(s). Simple apparatus, not shown on the control drawing, shall be permitted in a nonincendive field wiring circuit, provided the simple apparatus does not interconnect the nonincendive field wiring circuit to any other circuit.

FPN: Simple apparatus is defined in 504.2.

Separate nonincendive field wiring circuits shall be installed in accordance with one of the following:

(1) In separate cables
(2) In multiconductor cables where the conductors of each circuit are within a grounded metal shield
(3) In multiconductor cables where the conductors of each circuit have insulation with a minimum thickness of 0.25 mm (0.01 in.)

(2) Flexible Connections. Where provision must be made for limited flexibility, flexible metal fittings, flexible metal conduit with listed fittings, liquidtight flexible metal conduit with listed fittings, liquidtight flexible nonmetallic conduit with listed fittings, or flexible cord in accordance with the provisions of 505.17 shall be permitted.

FPN: See 505.25(B) for grounding requirements where flexible conduit is used.

505.16 Sealing and Drainage. Seals in conduit and cable systems shall comply with 505.16(A) through 505.16(E). Sealing compound shall be used in Type MI cable termination fittings to exclude moisture and other fluids from the cable insulation.

FPN No. 1: Seals are provided in conduit and cable systems to minimize the passage of gases and vapors and prevent the passage of flames from one portion of the electrical installation to another through the conduit. Such communication through Type MI cable is inherently prevented by construction of the cable. Unless specifically designed and tested for the purpose, conduit and cable seals are not intended to prevent the passage of liquids, gases, or vapors at a continuous pressure differential across the seal. Even at differences in pressure across the seal equivalent to a few inches of water, there may be a slow passage of gas or vapor through a seal and through conductors passing through the seal. See 505.16(C)(2)(b). Temperature extremes and highly corrosive liquids and vapors can affect the ability of seals to perform their intended function. See 505.16(D)(2).

FPN No. 2: Gas or vapor leakage and propagation of flames may occur through the interstices between the strands of standard stranded conductors larger than 2 AWG. Special conductor constructions, for example, compacted strands or sealing of the individual strands, are means of reducing leakage and preventing the propagation of flames.

(A) Zone 0. In Class I, Zone 0 locations, seals shall be located according to 505.16(A)(1), (A)(2), and (A)(3).

(1) Conduit Seals. Seals shall be provided within 3.05 m (10 ft) of where a conduit leaves a Zone 0 location. There shall be no unions, couplings, boxes, or fittings, except listed reducers at the seal, in the conduit run between the seal and the point at which the conduit leaves the location.

Exception: A rigid unbroken conduit that passes completely through the Zone 0 location with no fittings less than 300 mm (12 in.) beyond each boundary shall not be required to be sealed if the termination points of the unbroken conduit are in unclassified locations.

(2) Cable Seals. Seals shall be provided on cables at the first point of termination after entry into the Zone 0 location.

(3) Not Required to Be Explosionproof or Flameproof. Seals shall not be required to be explosionproof or flameproof.

(B) Zone 1. In Class I, Zone 1 locations, seals shall be located in accordance with 505.16(B)(1) through (B)(8).

(1) Type of Protection "d" or "e" Enclosures. Conduit seals shall be provided within 50 mm (2 in.) for each conduit entering enclosures having type of protection "d" or "e."

Exception No. 1: Where the enclosure having type of protection "d" is marked to indicate that a seal is not required.

Exception No. 2: For type of protection "e," conduit and fittings employing only NPT to NPT raceway joints or fittings listed for type of protection "e" shall be permitted between the enclosure and the seal, and the seal shall not be required to be within 50 mm (2 in.) of the entry.

FPN: Examples of fittings employing other than NPT threads include conduit couplings, capped elbows, unions, and breather drains.

Exception No. 3: For conduit installed between type of protection "e" enclosures employing only NPT to NPT raceway joints or conduit fittings listed for type of protection "e," a seal shall not be required.

(2) Explosionproof Equipment. Conduit seals shall be provided for each conduit entering explosionproof equipment according to (B)(2)(a), (B)(2)(b), and (B)(2)(c).

(a) In each conduit entry into an explosionproof enclosure where either (1) the enclosure contains apparatus, such as switches, circuit breakers, fuses, relays, or resistors, that may produce arcs, sparks, or high temperatures that are considered to be an ignition source in normal operation, or (2) the entry is metric designator 53 (trade size 2) or larger and the enclosure contains terminals, splices, or taps. For the purposes of this section, high temperatures shall be considered to be any temperatures exceeding 80 percent of the autoignition temperature in degrees Celsius of the gas or vapor involved.

Exception: Conduit entering an enclosure where such switches, circuit breakers, fuses, relays, or resistors comply with one of the following:

(1) Are enclosed within a chamber hermetically sealed against the entrance of gases or vapors.
(2) Are immersed in oil.
(3) Are enclosed within a factory-sealed explosionproof chamber located within the enclosure, identified for the location, and marked "factory sealed" or equivalent, unless the entry is metric designator 53 (trade size 2) or larger. Factory-sealed enclosures shall not be considered to serve as a seal for another adjacent explosionproof enclosure that is required to have a conduit seal.

(b) Conduit seals shall be installed within 450 mm (18 in.) from the enclosure. Only explosionproof unions, couplings, reducers, elbows, capped elbows, and conduit bodies similar to L, T, and cross types that are not larger than the trade size of the conduit shall be permitted between the sealing fitting and the explosionproof enclosure.

(c) Where two or more explosionproof enclosures for which conduit seals are required under 505.16(B)(2) are connected by nipples or by runs of conduit not more than 900 mm (36 in.) long, a single conduit seal in each such nipple connection or run of conduit shall be considered sufficient if located not more than 450 mm (18 in.) from either enclosure.

(3) Pressurized Enclosures. Conduit seals shall be provided in each conduit entry into a pressurized enclosure where the conduit is not pressurized as part of the protection system. Conduit seals shall be installed within 450 mm (18 in.) from the pressurized enclosure.

FPN No. 1: Installing the seal as close as possible to the enclosure reduces problems with purging the dead airspace in the pressurized conduit.

FPN No. 2: For further information, see NFPA 496-2003, *Standard for Purged and Pressurized Enclosures for Electrical Equipment.*

(4) Class I, Zone 1 Boundary. Conduit seals shall be provided in each conduit run leaving a Class I, Zone 1 location. The sealing fitting shall be permitted on either side of the boundary of such location within 3.05 m (10 ft) of the boundary and shall be designed and installed so as to minimize the amount of gas or vapor within the Zone 1 portion of the conduit from being communicated to the conduit beyond the seal. Except for listed explosionproof reducers at the conduit seal, there shall be no union, coupling, box, or fitting between the conduit seal and the point at which the conduit leaves the Zone 1 location.

Exception: Metal conduit containing no unions, couplings, boxes, or fittings and passing completely through a Class I, Zone 1 location with no fittings less than 300 mm (12 in.) beyond each boundary shall not require a conduit seal if the termination points of the unbroken conduit are in unclassified locations.

(5) Cables Capable of Transmitting Gases or Vapors. Conduits containing cables with a gas/vaportight continuous sheath capable of transmitting gases or vapors through the cable core shall be sealed in the Zone 1 location after removing the jacket and any other coverings so that the sealing compound surrounds each individual insulated conductor and the outer jacket.

Exception: Multiconductor cables with a gas/vaportight continuous sheath capable of transmitting gases or vapors through the cable core shall be permitted to be considered as a single conductor by sealing the cable in the conduit within 450 mm (18 in.) of the enclosure and the cable end within the enclosure by an approved means to minimize the entrance of gases or vapors and prevent the propagation of flame into the cable core, or by other approved methods. For shielded cables and twisted pair cables, it shall not be required to remove the shielding material or separate the twisted pair.

(6) Cables Incapable of Transmitting Gases or Vapors. Each multiconductor cable in conduit shall be considered as a single conductor if the cable is incapable of transmitting gases or vapors through the cable core. These cables shall be sealed in accordance with 505.16(D).

(7) Cables Entering Enclosures. Cable seals shall be provided for each cable entering flameproof or explosionproof enclosures. The seal shall comply with 505.16(D).

(8) Class I, Zone 1 Boundary. Cables shall be sealed at the point at which they leave the Zone 1 location.

Exception: Where cable is sealed at the termination point.

(C) Zone 2. In Class I, Zone 2 locations, seals shall be located in accordance with 505.16(C)(1) and (C)(2).

(1) Conduit Seals. Conduit seals shall be located in accordance with (C)(1)(a) and (C)(1)(b).

(a) For connections to enclosures that are required to be flameproof or explosionproof, a conduit seal shall be provided in accordance with 505.16(B)(1) and 505.16(B)(2). All portions of the conduit run or nipple between the seal and such enclosure shall comply with 505.16(B).

(b) In each conduit run passing from a Class I, Zone 2 location into an unclassified location. The sealing fitting shall be permitted on either side of the boundary of such location within 3.05 m (10 ft) of the boundary and shall be designed and installed so as to minimize the amount of gas or vapor within the Zone 2 portion of the conduit from being communicated to the conduit beyond the seal. Rigid metal conduit or threaded steel intermediate metal conduit shall be used between the sealing fitting and the point at which the conduit leaves the Zone 2 location, and a threaded connection shall be used at the sealing fitting. Except for listed explosionproof reducers at the conduit seal, there shall be no union, coupling, box, or fitting between the conduit seal and the point at which the conduit leaves the Zone 2 location.

Exception No. 1: Metal conduit containing no unions, couplings, boxes, or fittings and passing completely through a Class I, Zone 2 location with no fittings less than 300 mm (12 in.) beyond each boundary shall not be required to be sealed if the termination points of the unbroken conduit are in unclassified locations.

Exception No. 2: Conduit systems terminating at an unclassified location where a wiring method transition is made to cable tray, cablebus, ventilated busway, Type MI cable, or cable that is not installed in a raceway or cable tray system shall not be required to be sealed where passing from the Class I, Zone 2 location into the unclassified location. The unclassified location shall be outdoors or, if the conduit system is all in one room, it shall be permitted to be indoors. The conduits shall not terminate at an enclosure containing an ignition source in normal operation.

Exception No. 3: Conduit systems passing from an enclosure or room that is unclassified as a result of pressurization into a Class I, Zone 2 location shall not require a seal at the boundary.

FPN: For further information, refer to NFPA 496-2003, *Standard for Purged and Pressurized Enclosures for Electrical Equipment.*

Exception No. 4: Segments of aboveground conduit systems shall not be required to be sealed where passing from a Class I, Zone 2 location into an unclassified location if all the following conditions are met:

(1) No part of the conduit system segment passes through a Class I, Zone 0 or Class I, Zone 1 location where the conduit contains unions, couplings, boxes, or fittings within 300 mm (12 in.) of the Class I, Zone 0 or Class I, Zone 1 location.
(2) The conduit system segment is located entirely in outdoor locations.
(3) The conduit system segment is not directly connected to canned pumps, process or service connections for flow, pressure, or analysis measurement, and so forth, that depend on a single compression seal, diaphragm, or tube to prevent flammable or combustible fluids from entering the conduit system.
(4) The conduit system segment contains only threaded metal conduit, unions, couplings, conduit bodies, and fittings in the unclassified location.
(5) The conduit system segment is sealed at its entry to each enclosure or fitting housing terminals, splices, or taps in Class I, Zone 2 locations.

(2) Cable Seals. Cable seals shall be located in accordance with (C)(2)(a), (C)(2)(b), and (C)(2)(c).

(a) Explosionproof and Flameproof Enclosures. Cables entering enclosures required to be flameproof or explosionproof shall be sealed at the point of entrance. The seal shall comply with 505.16(D). Multiconductor cables with a gas/vaportight continuous sheath capable of transmitting gases or vapors through the cable core shall be sealed in the Zone 2 location after removing the jacket and any other coverings so that the sealing compound surrounds each individual insulated conductor in such a manner as to minimize the passage of gases and vapors. Multiconductor cables in conduit shall be sealed as described in 505.16(B)(4).

Exception No. 1: Cables passing from an enclosure or room that is unclassified as a result of Type Z pressurization into a Class I, Zone 2 location shall not require a seal at the boundary.

Exception No. 2: Shielded cables and twisted pair cables shall not require the removal of the shielding material or separation of the twisted pairs, provided the termination is by an approved means to minimize the entrance of gases or vapors and prevent propagation of flame into the cable core.

(b) Cables That Will Not Transmit Gases or Vapors. Cables with a gas/vaportight continuous sheath and that will not transmit gases or vapors through the cable core in excess of the quantity permitted for seal fittings shall not be required to be sealed except as required in 505.16(C)(2)(a). The minimum length of such cable run shall not be less than the length that limits gas or vapor flow through the cable core to the rate permitted for seal fittings [200 cm^3/hr (0.007 ft^3/hr) of air at a pressure of 1500 pascals (6 in. of water)].

FPN No. 1: See ANSI/UL 886-1994, *Outlet Boxes and Fittings for Use in Hazardous (Classified) Locations.*

FPN No. 2: The cable core does not include the interstices of the conductor strands.

(c) Cables Capable of Transmitting Gases or Vapors. Cables with a gas/vaportight continuous sheath capable of transmitting gases or vapors through the cable core shall not be required to be sealed except as required in 505.16(C)(2)(a), unless the cable is attached to process equipment or devices that may cause a pressure in excess of 1500 pascals (6 in. of water) to be exerted at a cable end, in which case a seal, barrier, or other means shall be provided to prevent migration of flammables into an unclassified area.

Exception: Cables with an unbroken gas/vaportight continuous sheath shall be permitted to pass through a Class I, Zone 2 location without seals.

(d) Cables Without Gas/Vaportight Continuous Sheath. Cables that do not have gas/vaportight continuous sheath shall be sealed at the boundary of the Zone 2 and unclassified location in such a manner as to minimize the passage of gases or vapors into an unclassified location.

FPN: The cable sheath may be either metal or a nonmetallic material.

(D) Class I, Zones 0, 1, and 2. Where required, seals in Class I, Zones 0, 1, and 2 locations shall comply with 505.16(D)(1) through (D)(5).

(1) Fittings. Enclosures for connections or equipment shall be provided with an integral means for sealing, or sealing fittings listed for the location shall be used. Sealing fittings shall be listed for use with one or more specific compounds and shall be accessible.

(2) Compound. The compound shall provide a seal against passage of gas or vapors through the seal fitting, shall not be affected by the surrounding atmosphere or liquids, and shall not have a melting point less than 93°C (200°F).

(3) Thickness of Compounds. In a completed seal, the minimum thickness of the sealing compound shall not be less than the trade size of the sealing fitting and, in no case, less than 16 mm (⅝ in.).

Exception: Listed cable sealing fittings shall not be required to have a minimum thickness equal to the trade size of the fitting.

(4) Splices and Taps. Splices and taps shall not be made in fittings intended only for sealing with compound, nor shall other fittings in which splices or taps are made be filled with compound.

(5) Conductor Fill. The cross-sectional area of the conductors permitted in a seal shall not exceed 25 percent of the cross-sectional area of a rigid metal conduit of the same trade size unless it is specifically listed for a higher percentage of fill.

(E) Drainage.

(1) Control Equipment. Where there is a probability that liquid or other condensed vapor may be trapped within enclosures for control equipment or at any point in the raceway system, approved means shall be provided to prevent accumulation or to permit periodic draining of such liquid or condensed vapor.

(2) Motors and Generators. Where the authority having jurisdiction judges that there is a probability that liquid or condensed vapor may accumulate within motors or generators, joints and conduit systems shall be arranged to minimize entrance of liquid. If means to prevent accumulation or to permit periodic draining are judged necessary, such means shall be provided at the time of manufacture and shall be considered an integral part of the machine.

(3) Canned Pumps, Process, or Service Connections, and So Forth. For canned pumps, process, or service connections for flow, pressure, or analysis measurement, and so forth, that depend on a single compression seal, diaphragm, or tube to prevent flammable or combustible fluids from entering the electrical conduit system, an additional approved seal, barrier, or other means shall be provided to prevent the flammable or combustible fluid from entering the conduit system beyond the additional devices or means if the primary seal fails.

The additional approved seal or barrier and the interconnecting enclosure shall meet the temperature and pressure conditions to which they will be subjected upon failure of the primary seal, unless other approved means are provided to accomplish the purpose in the preceding paragraph.

Drains, vents, or other devices shall be provided so that primary seal leakage is obvious.

FPN: See also the fine print notes to 505.16.

Process-connected equipment that is listed and marked "Dual Seal" shall not require additional process sealing when used within the manufacturer's ratings.

FPN: For construction and testing requirements for dual seal process, connected equipment, refer to ISA 12.27.01, *Requirements for Process Sealing Between Electrical Systems and Potentially Flammable or Combustible Process Fluids.*

505.17 Flexible Cords, Class I, Zones 1 and 2. A flexible cord shall be permitted for connection between portable lighting equipment or other portable utilization equipment and the fixed portion of their supply circuit. Flexible cord shall also be permitted for that portion of the circuit where the fixed wiring methods of 505.15(B) cannot provide the necessary degree of movement for fixed and mobile electrical utilization equipment, in an industrial establishment where conditions of maintenance and engineering supervision ensure that only qualified persons install and service the installation, and the flexible cord is protected by location or by a suitable guard from damage. The length of the flexible cord shall be continuous. Where flexible cords are used, the cords shall comply with all of the following:

(1) Be of a type listed for extra-hard usage
(2) Contain, in addition to the conductors of the circuit, a grounding conductor complying with 400.23
(3) Be connected to terminals or to supply conductors in an approved manner
(4) Be supported by clamps or by other suitable means in such a manner that there will be no tension on the terminal connections
(5) Be provided with listed seals where the flexible cord enters boxes, fittings, or enclosures that are required to be explosionproof or flameproof

Exception: As provided in 505.16.

Electric submersible pumps with means for removal without entering the wet-pit shall be considered portable utilization equipment. The extension of the flexible cord within a suitable raceway between the wet-pit and the power source shall be permitted.

Electric mixers intended for travel into and out of open-type mixing tanks or vats shall be considered portable utilization equipment.

FPN: See 505.18 for flexible cords exposed to liquids having a deleterious effect on the conductor insulation.

505.18 Conductors and Conductor Insulation.

(A) Conductors. For type of protection "e," field wiring conductors shall be copper. Every conductor (including spares) that enters Type "e" equipment shall be terminated at a Type "e" terminal.

(B) Conductor Insulation. Where condensed vapors or liquids may collect on, or come in contact with, the insulation on conductors, such insulation shall be of a type identified for use under such conditions, or the insulation shall be protected by a sheath of lead or by other approved means.

505.19 Uninsulated Exposed Parts. There shall be no uninsulated exposed parts, such as electric conductors, buses, terminals, or components that operate at more than 30 volts (15 volts in wet locations). These parts shall additionally be protected by type of protection ia, ib, or nA that is suitable for the location.

505.20 Equipment Requirements.

(A) Zone 0. In Class I, Zone 0 locations, only equipment specifically listed and marked as suitable for the location shall be permitted.

Exception: Intrinsically safe apparatus listed for use in Class I, Division 1 locations for the same gas, or as permitted by 505.9(B)(2), and with a suitable temperature class shall be permitted.

(B) Zone 1. In Class I, Zone 1 locations, only equipment specifically listed and marked as suitable for the location shall be permitted.

Exception No. 1: Equipment identified for use in Class I, Division 1 or listed for use in Class I, Zone 0 locations for the same gas, or as permitted by 505.9(B)(2), and with a suitable temperature class shall be permitted.

Exception No. 2: Equipment identified for Class I, Zone 1, or Zone 2 type of protection "p" shall be permitted.

(C) Zone 2. In Class I, Zone 2 locations, only equipment specifically listed and marked as suitable for the location shall be permitted.

Exception No. 1: Equipment listed for use in Class I, Zone 0 or Zone 1 locations for the same gas, or as permitted by 505.9(B)(2), and with a suitable temperature class, shall be permitted.

Exception No. 2: Equipment identified for Class I, Zone 1 or Zone 2 type of protection "p" shall be permitted.

Exception No. 3: Equipment identified for use in Class I, Division 1 or Division 2 locations for the same gas, or as permitted by 505.9(B)(2), and with a suitable temperature class shall be permitted.

Exception No. 4: In Class I, Zone 2 locations, the installation of open or nonexplosionproof or nonflameproof enclosed motors, such as squirrel-cage induction motors without brushes, switching mechanisms, or similar arc-producing devices that are not identified for use in a Class I, Zone 2 location shall be permitted.

FPN No. 1: It is important to consider the temperature of internal and external surfaces that may be exposed to the flammable atmosphere.

FPN No. 2: It is important to consider the risk of ignition due to currents arcing across discontinuities and overheating of parts in multisection enclosures of large motors and generators. Such motors and generators may need equipotential bonding jumpers across joints in the enclosure and from enclosure to ground. Where the presence of ignitible gases or vapors is suspected, clean air purging may be needed immediately prior to and during start-up periods.

(D) Manufacturer's Instructions. Electrical equipment installed in hazardous (classified) locations shall be installed in accordance with the instructions (if any) provided by the manufacturer.

505.21 Multiwire Branch Circuits. In a Class I, Zone 1 location, a multiwire branch circuit shall not be permitted.

Exception: Where the disconnect device(s) for the circuit opens all ungrounded conductors of the multiwire circuit simultaneously.

505.22 Increased Safety "e" Motors and Generators. In Class I, Zone 1 locations, Increased Safety "e" motors and generators of all voltage ratings shall be listed for Class I, Zone 1 locations, and shall comply with all of the following:

(1) Motors shall be marked with the current ratio, I_A/I_N, and time, t_E.
(2) Motors shall have controllers marked with the model or identification number, output rating (horsepower or kilowatt), full-load amperes, starting current ratio (I_A/I_N), and time (t_E) of the motors that they are intended to protect; the controller marking shall also include the specific overload protection type (and setting, if applicable) that is listed with the motor or generator.
(3) Connections shall be made with the specific terminals listed with the motor or generator.
(4) Terminal housings shall be permitted to be of substantial, nonmetallic, nonburning material, provided an internal grounding means between the motor frame and the equipment grounding connection is incorporated within the housing.
(5) The provisions of Part III of Article 430 shall apply regardless of the voltage rating of the motor.
(6) The motors shall be protected against overload by a separate overload device that is responsive to motor current. This device shall be selected to trip or shall be rated in accordance with the listing of the motor and its overload protection.
(7) Sections 430.32(C) and 430.44 shall not apply to such motors.
(8) The motor overload protection shall not be shunted or cut out during the starting period.

505.25 Grounding and Bonding. Grounding and bonding shall comply with Article 250 and the requirements in 505.25(A) and 505.25(B).

(A) Bonding. The locknut-bushing and double-locknut types of contacts shall not be depended on for bonding purposes, but bonding jumpers with proper fittings or other approved means of bonding shall be used. Such means of bonding shall apply to all intervening raceways, fittings, boxes, enclosures, and so forth, between Class I locations and the point of grounding for service equipment or point of grounding of a separately derived system.

Exception: The specific bonding means shall be required only to the nearest point where the grounded circuit conductor and the grounding electrode are connected together on the line side of the building or structure disconnecting means as specified in 250.32(A), (B), and (C), provided the branch-circuit overcurrent protection is located on the load side of the disconnecting means.

FPN: See 250.100 for additional bonding requirements in hazardous (classified) locations.

(B) Types of Equipment Grounding Conductors. Where flexible metal conduit or liquidtight flexible metal conduit is used as permitted in 505.15(C) and is to be relied on to complete a sole equipment grounding path, it shall be installed with internal or external bonding jumpers in parallel with each conduit and complying with 250.102.

Exception: In Class I, Zone 2 locations, the bonding jumper shall be permitted to be deleted where all of the following conditions are met:

(a) Listed liquidtight flexible metal conduit 1.8 m (6 ft) or less in length, with fittings listed for grounding, is used.
(b) Overcurrent protection in the circuit is limited to 10 amperes or less.
(c) The load is not a power utilization load.

ARTICLE 502
Class II Locations

Editor's Note: *Article 502 does not apply if the zone classification system of Article 506 is used.*

I. General

502.1 Scope. Article 502 covers the requirements for electrical and electronic equipment and wiring for all voltages in Class II, Division 1 and 2 locations where fire or explosion hazards may exist due to combustible dust.

502.5 General. The general rules of this *Code* shall apply to the electric wiring and equipment in locations classified as Class II locations in 500.5(C).

Exception: As modified by this article.

Equipment installed in Class II locations shall be able to function at full rating without developing surface temperatures high enough to cause excessive dehydration or gradual carbonization of any organic dust deposits that may occur.

FPN: Dust that is carbonized or excessively dry is highly susceptible to spontaneous ignition.

Explosionproof equipment and wiring shall not be required and shall not be acceptable in Class II locations unless identified for such locations.

II. Wiring

502.10 Wiring Methods. Wiring methods shall comply with 502.10(A) or 502.10(B).

(A) Class II, Division 1.

(1) General. In Class II, Division 1 locations, the wiring methods in (1) through (4) shall be permitted.

(1) Threaded rigid metal conduit, or threaded steel intermediate metal conduit.
(2) Type MI cable with termination fittings listed for the location. Type MI cable shall be installed and supported in a manner to avoid tensile stress at the termination fittings.
(3) In industrial establishments with limited public access, where the conditions of maintenance and supervision ensure that only qualified persons service the installation, Type MC cable, listed for use in Class II, Division 1 locations, with a gas/vaportight continuous corrugated metallic sheath, an overall jacket of suitable polymeric material, separate grounding conductors in accordance with 250.122, and provided with termination fittings listed for the application, shall be permitted.
(4) Fittings and boxes shall be provided with threaded bosses for connection to conduit or cable terminations and shall be dusttight. Fittings and boxes in which taps, joints, or terminal connections are made, or that are used in Group E locations, shall be identified for Class II locations.

(2) Flexible Connections. Where necessary to employ flexible connections, one or more of the following shall also be permitted:

(1) Dusttight flexible connectors
(2) Liquidtight flexible metal conduit with listed fittings
(3) Liquidtight flexible nonmetallic conduit with listed fittings
(4) Interlocked armor Type MC cable having an overall jacket of suitable polymeric material and provided with termination fittings listed for Class II, Division 1 locations.
(5) Flexible cord listed for extra-hard usage and provided with bushed fittings. Where flexible cords are used, they shall comply with 502.140.

FPN: See 502.30(B) for grounding requirements where flexible conduit is used.

(B) Class II, Division 2.

(1) General. In Class II, Division 2 locations, the following wiring methods shall be permitted:

(1) All wiring methods permitted in 502.10(A).
(2) Rigid metal conduit, intermediate metal conduit, electrical metallic tubing, dusttight wireways.
(3) Type MC or MI cable with listed termination fittings.
(4) Type PLTC in cable trays.
(5) Type ITC in cable trays.
(6) Type MC, MI, or TC cable installed in ladder, ventilated trough, or ventilated channel cable trays in a single layer, with a space not less than the larger cable diameter between the two adjacent cables, shall be the wiring method employed.

Exception to (6): Type MC cable listed for use in Class II, Division 1 locations shall be permitted to be installed without the spacings required by (6).

(2) Flexible Connections. Where provision must be made for flexibility, 502.10(A)(2) shall apply.

(3) Nonincendive Field Wiring. Nonincendive field wiring shall be permitted using any of the wiring methods permitted for unclassified locations. Nonincendive field wiring systems shall be installed in accordance with the control drawing(s). Simple apparatus, not shown on the control drawing, shall be permitted in a nonincendive field wiring circuit, provided the simple apparatus does not interconnect the nonincendive field wiring circuit to any other circuit.

FPN: Simple apparatus is defined in 504.2.

Separate nonincendive field wiring circuits shall be installed in accordance with one of the following:

(1) In separate cables
(2) In multiconductor cables where the conductors of each circuit are within a grounded metal shield
(3) In multiconductor cables where the conductors of each circuit have insulation with a minimum thickness of 0.25 mm (0.01 in.)

(4) Boxes and Fittings. All boxes and fittings shall be dusttight.

502.15 Sealing, Class II, Divisions 1 and 2. Where a raceway provides communication between an enclosure that is required to be dust-ignitionproof and one that is not, suitable means shall be provided to prevent the entrance of dust into the dust-ignitionproof enclosure through the raceway. One of the following means shall be permitted:

(1) A permanent and effective seal
(2) A horizontal raceway not less than 3.05 m (10 ft) long
(3) A vertical raceway not less than 1.5 m (5 ft) long and extending downward from the dust-ignitionproof enclosure
(4) A raceway installed in a manner equivalent to (2) or (3) that extends only horizontally and downward from the dust-ignition proof enclosures.

Where a raceway provides communication between an enclosure that is required to be dust-ignitionproof and an enclosure in an unclassified location, seals shall not be required.

Sealing fittings shall be accessible.

Seals shall not be required to be explosionproof.

FPN: Electrical sealing putty is a method of sealing.

502.25 Uninsulated Exposed Parts, Class II, Divisions 1 and 2. There shall be no uninsulated exposed parts, such as electric conductors, buses, terminals, or components, that operate at more than 30 volts (15 volts in wet locations). These parts shall additionally be protected by a protection technique according to 500.7(E), 500.7(F), or 500.7(G) that is suitable for the location.

502.30 Grounding and Bonding, Class II, Divisions 1 and 2. Wiring and equipment in Class II, Division 1 and 2 locations shall be grounded as specified in Article 250 and with the requirements in 502.30(A) and 502.30(B).

(A) Bonding. The locknut-bushing and double-locknut types of contact shall not be depended on for bonding purposes, but bonding jumpers with proper fittings or other approved means of bonding shall be used. Such means of bonding shall apply to all intervening raceways, fittings, boxes, enclosures, and so forth, between Class II locations and the point of grounding for service equipment or point of grounding of a separately derived system.

Exception: The specific bonding means shall only be required to the nearest point where the grounded circuit conductor and the grounding electrode conductor are connected together on the line side of the building or structure disconnecting means as specified in 250.32(A), (B), and (C), if the branch-circuit overcurrent protection is located on the load side of the disconnecting means.

FPN: See 250.100 for additional bonding requirements in hazardous (classified) locations.

(B) Types of Equipment Grounding Conductors. Where flexible conduit is used as permitted in 502.10, it shall be installed with internal or external bonding jumpers in parallel with each conduit and complying with 250.102.

Exception: In Class II, Division 2 locations, the bonding jumper shall be permitted to be deleted where all of the following conditions are met:

(1) Listed liquidtight flexible metal conduit 1.8 m (6 ft) or less in length, with fittings listed for grounding, is used.
(2) Overcurrent protection in the circuit is limited to 10 amperes or less.
(3) The load is not a power utilization load.

502.35 Surge Protection — Class II, Divisions 1 and 2. Surge arresters and transient voltage surge suppressors (TVSS) installed in a Class II, Division 1 location shall be in suitable enclosures. Surge-protective capacitors shall be of a type designed for specific duty.

502.40 Multiwire Branch Circuits. In a Class II, Division 1 location, a multiwire branch circuit shall not be permitted.

Exception: Where the disconnect device(s) for the circuit opens all ungrounded conductors of the multiwire circuit simultaneously.

III. Equipment

502.100 Transformers and Capacitors.

(A) Class II, Division 1. In Class II, Division 1 locations, transformers and capacitors shall comply with 502.100(A)(1) through (A)(3).

(1) Containing Liquid That Will Burn. Transformers and capacitors containing a liquid that will burn shall be installed only in vaults complying with 450.41 through 450.48, and, in addition, (1), (2), and (3) shall apply.

(1) Doors or other openings communicating with the Division 1 location shall have self-closing fire doors on both sides of the wall, and the doors shall be carefully fitted and provided with suitable seals (such as weather stripping) to minimize the entrance of dust into the vault.
(2) Vent openings and ducts shall communicate only with the outside air.
(3) Suitable pressure-relief openings communicating with the outside air shall be provided.

(2) Not Containing Liquid That Will Burn. Transformers and capacitors that do not contain a liquid that will burn shall be installed in vaults complying with 450.41 through 450.48 or be identified as a complete assembly, including terminal connections for Class II locations.

(3) Metal Dusts. No transformer or capacitor shall be installed in a location where dust from magnesium, aluminum, aluminum bronze powders, or other metals of similarly hazardous characteristics may be present.

(B) Class II, Division 2. In Class II, Division 2 locations, transformers and capacitors shall comply with 502.100(B)(1) through (B)(3).

(1) Containing Liquid That Will Burn. Transformers and capacitors containing a liquid that will burn shall be installed in vaults that comply with 450.41 through 450.48.

(2) Containing Askarel. Transformers containing askarel and rated in excess of 25 kVA shall be as follows:

(1) Provided with pressure-relief vents
(2) Provided with a means for absorbing any gases generated by arcing inside the case, or the pressure-relief vents shall be connected to a chimney or flue that will carry such gases outside the building
(3) Have an airspace of not less than 150 mm (6 in.) between the transformer cases and any adjacent combustible material

(3) Dry-Type Transformers. Dry-type transformers shall be installed in vaults or shall have their windings and terminal connections enclosed in tight metal housings without ventilating or other openings and shall operate at not over 600 volts, nominal.

502.115 Switches, Circuit Breakers, Motor Controllers, and Fuses.

(A) Class II, Division 1. In Class II, Division 1 locations, switches, circuit breakers, motor controllers, and fuses shall comply with 502.115(A)(1) through (A)(3).

(1) Type Required. Switches, circuit breakers, motor controllers, and fuses, including pushbuttons, relays, and similar devices that are intended to interrupt current during normal operation or that are installed where combustible dusts of an electrically conductive nature may be present, shall be provided with identified dust-ignitionproof enclosures.

(2) Isolating Switches. Disconnecting and isolating switches containing no fuses and not intended to interrupt current and not installed where dusts may be of an electrically conductive nature shall be provided with tight metal enclosures that shall be designed to minimize the entrance of dust and that shall (1) be equipped with telescoping or close-fitting covers or with other effective means to prevent the escape of sparks or burning material and (2) have no openings (such as holes for attachment screws) through which, after installation, sparks or burning material might escape or through which exterior accumulations of dust or adjacent combustible material might be ignited.

(3) Metal Dusts. In locations where dust from magnesium, aluminum, aluminum bronze powders, or other metals of similarly hazardous characteristics may be present, fuses, switches, motor controllers, and circuit breakers shall have enclosures identified for such locations.

(B) Class II, Division 2. In Class II, Division 2 locations, enclosures for fuses, switches, circuit breakers, and motor controllers, including pushbuttons, relays, and similar devices, shall be dusttight.

502.120 Control Transformers and Resistors.

(A) Class II, Division 1. In Class II, Division 1 locations, control transformers, solenoids, impedance coils, resistors, and any overcurrent devices or switching mechanisms associated with them shall have dust-ignitionproof enclosures identified for Class II locations. No control transformer, impedance coil, or resistor shall be installed in a location where dust from magnesium, aluminum, aluminum bronze powders, or other metals of similarly hazardous characteristics may be present unless provided with an enclosure identified for the specific location.

(B) Class II, Division 2. In Class II, Division 2 locations, transformers and resistors shall comply with 502.120(B)(1) through (B)(3).

(1) Switching Mechanisms. Switching mechanisms (including overcurrent devices) associated with control transformers, solenoids, impedance coils, and resistors shall be provided with dusttight enclosures.

(2) Coils and Windings. Where not located in the same enclosure with switching mechanisms, control transformers, solenoids, and impedance coils shall be provided with tight metal housings without ventilating openings.

(3) Resistors. Resistors and resistance devices shall have dust-ignitionproof enclosures identified for Class II locations.

Exception: Where the maximum normal operating temperature of the resistor will not exceed 120°C (248°F), nonadjustable resistors or resistors that are part of an automatically timed starting sequence shall be permitted to have enclosures complying with 502.120(B)(2).

502.125 Motors and Generators.

(A) Class II, Division 1. In Class II, Division 1 locations, motors, generators, and other rotating electrical machinery shall be in conformance with either of the following:

(1) Identified for Class II, Division 1 locations
(2) Totally enclosed pipe-ventilated, meeting temperature limitations in 502.5

(B) Class II, Division 2. In Class II, Division 2 locations, motors, generators, and other rotating electrical equipment shall be totally enclosed nonventilated, totally enclosed pipe-ventilated, totally enclosed water-air-cooled, totally enclosed fan-cooled or dust-ignitionproof for which maximum full-load external temperature shall be in accordance with 500.8(C)(2) for normal operation when operating in free air (not dust blanketed) and shall have no external openings.

Exception: If the authority having jurisdiction believes accumulations of nonconductive, nonabrasive dust will be moderate and if machines can be easily reached for routine cleaning and maintenance, the following shall be permitted to be installed:

(1) Standard open-type machines without sliding contacts, centrifugal or other types of switching mechanism (including motor overcurrent, overloading, and overtemperature devices), or integral resistance devices
(2) Standard open-type machines with such contacts, switching mechanisms, or resistance devices enclosed within dusttight housings without ventilating or other openings
(3) Self-cleaning textile motors of the squirrel-cage type

502.128 Ventilating Piping. Ventilating pipes for motors, generators, or other rotating electric machinery, or for enclosures for electric equipment, shall be of metal not less than 0.53 mm (0.021 in.) in thickness or of equally substantial noncombustible material and shall comply with all of the following:

(1) Lead directly to a source of clean air outside of buildings
(2) Be screened at the outer ends to prevent the entrance of small animals or birds
(3) Be protected against physical damage and against rusting or other corrosive influences

Ventilating pipes shall also comply with 502.128(A) and 502.128(B).

(A) Class II, Division 1. In Class II, Division 1 locations, ventilating pipes, including their connections to motors or to the dust-ignitionproof enclosures for other equipment, shall be dusttight throughout their length. For metal pipes, seams and joints shall comply with one of the following:

(1) Be riveted and soldered
(2) Be bolted and soldered
(3) Be welded
(4) Be rendered dusttight by some other equally effective means

(B) Class II, Division 2. In Class II, Division 2 locations, ventilating pipes and their connections shall be sufficiently tight to prevent the entrance of appreciable quantities of dust into the ventilated equipment or enclosure and to prevent the escape of sparks, flame, or burning material that might ignite dust accumulations or combustible material in the vicinity. For metal pipes, lock seams and riveted or welded joints shall be permitted; and tight-fitting slip joints shall be permitted where some flexibility is necessary, as at connections to motors.

502.130 Luminaires (Lighting Fixtures). Luminaires (lighting fixtures) shall comply with 502.130(A) and 502.130(B).

(A) Class II, Division 1. In Class II, Division 1 locations, luminaires (lighting fixtures) for fixed and portable lighting shall comply with 502.130(A)(1) through (A)(4).

(1) Fixtures. Each luminaire (fixture) shall be identified for Class II locations and shall be clearly marked to indicate the maximum wattage of the lamp for which it is designed. In locations where dust from magnesium, aluminum, aluminum bronze powders, or other metals of similarly hazardous characteristics may be present, luminaires (fixtures) for fixed or portable lighting and all auxiliary equipment shall be identified for the specific location.

(2) Physical Damage. Each luminaire (fixture) shall be protected against physical damage by a suitable guard or by location.

(3) Pendant Luminaires (Fixtures). Pendant luminaires (fixtures) shall be suspended by threaded rigid metal conduit stems, threaded steel intermediate metal conduit stems, by chains with approved fittings, or by other approved means. For rigid stems longer than 300 mm (12 in.), permanent and effective bracing against lateral displacement shall be provided at a level not more than 300 mm (12 in.) above the lower end of the stem, or flexibility in the form of a fitting or a flexible connector listed for the location shall be provided not more than 300 mm (12 in.) from the point of attachment to the supporting box or fitting. Threaded joints shall be provided with set-screws or other effective means to prevent loosening. Where wiring between an outlet box or fitting and a pendant luminaire (fixture) is not enclosed in conduit, flexible cord listed for hard usage shall be used, and suitable seals shall be provided where the cord enters the luminaire (fixture) and the outlet box or fitting. Flexible cord shall not serve as the supporting means for a fixture.

(4) Supports. Boxes, box assemblies, or fittings used for the support of luminaires (lighting fixtures) shall be identified for Class II locations.

(B) Class II, Division 2. In Class II, Division 2 locations, luminaires (lighting fixtures) shall comply with 502.130(B)(1) through (B)(5).

(1) Portable Lighting Equipment. Portable lighting equipment shall be identified for Class II locations. They shall be clearly marked to indicate the maximum wattage of lamps for which they are designed.

(2) Fixed Lighting. Luminaires (lighting fixtures) for fixed lighting, where not of a type identified for Class II locations, shall provide enclosures for lamps and lampholders that shall be designed to minimize the deposit of dust on lamps and to prevent the escape of sparks, burning material, or hot metal. Each fixture shall be clearly marked to indicate the maximum wattage of the lamp that shall be permitted without exceeding an exposed surface temperature in accordance with 500.8(C)(2) under normal conditions of use.

(3) Physical Damage. Luminaires (lighting fixtures) for fixed lighting shall be protected from physical damage by suitable guards or by location.

(4) Pendant Luminaires (Fixtures). Pendant luminaires (fixtures) shall be suspended by threaded rigid metal conduit stems, threaded steel intermediate metal conduit stems, by chains with approved fittings, or by other approved means. For rigid stems longer than 300 mm (12 in.), permanent and effective bracing against lateral displacement shall be provided at a level not more than 300 mm (12 in.) above the lower end of the stem, or flexibility in the form of an identified fitting or a flexible connector shall be provided not more than 300 mm (12 in.) from the point of attachment to the supporting box or fitting. Where wiring between an outlet box or fitting and a pendant luminaire (fixture) is not enclosed in conduit, flexible cord listed for hard usage shall be used. Flexible cord shall not serve as the supporting means for a fixture.

(5) Electric-Discharge Lamps. Starting and control equipment for electric-discharge lamps shall comply with the requirements of 502.120(B).

502.135 Utilization Equipment.

(A) Class II, Division 1. In Class II, Division 1 locations, all utilization equipment shall be identified for Class II locations. Where dust from magnesium, aluminum, aluminum bronze powders, or other metals of similarly hazardous characteristics may be present, such equipment shall be identified for the specific location.

(B) Class II, Division 2. In Class II, Division 2 locations, all utilization equipment shall comply with 502.135(B)(1) through (B)(4).

(1) Heaters. Electrically heated utilization equipment shall be identified for Class II locations.

Exception: Metal-enclosed radiant heating panel equipment shall be dusttight and marked in accordance with 500.8(B).

(2) Motors. Motors of motor-driven utilization equipment shall comply with 502.125(B).

(3) Switches, Circuit Breakers, and Fuses. Enclosures for switches, circuit breakers, and fuses shall be dusttight.

(4) Transformers, Solenoids, Impedance Coils, and Resistors. Transformers, solenoids, impedance coils, and resistors shall comply with 502.120(B).

502.140 Flexible Cords — Class II, Divisions 1 and 2. Flexible cords used in Class II locations shall comply with all of the following:

(1) Be of a type listed for extra-hard usage

Exception: Flexible cord listed for hard usage as permitted by 502.130(A)(3) and (B)(4).

(2) Contain, in addition to the conductors of the circuit, a grounding conductor complying with 400.23
(3) Be connected to terminals or to supply conductors in an approved manner
(4) Be supported by clamps or by other suitable means in such a manner that there will be no tension on the terminal connections
(5) Be provided with suitable seals to prevent the entrance of dust where the flexible cord enters boxes or fittings that are required to be dust-ignitionproof

502.145 Receptacles and Attachment Plugs.

(A) Class II, Division 1. In Class II, Division 1 locations, receptacles and attachment plugs shall be of the type providing for connection to the grounding conductor of the flexible cord and shall be identified for Class II locations.

(B) Class II, Division 2. In Class II, Division 2 locations, receptacles and attachment plugs shall be of the type that provides for connection to the grounding conductor of the flexible cord and shall be designed so that connection to the supply circuit cannot be made or broken while live parts are exposed.

502.150 Signaling, Alarm, Remote-Control, and Communications Systems; and Meters, Instruments, and Relays.

FPN: See Article 800 for rules governing the installation of communications circuits.

(A) Class II, Division 1. In Class II, Division 1 locations, signaling, alarm, remote-control, and communications systems; and meters, instruments, and relays shall comply with 502.150(A)(1) through (A)(6).

(1) Wiring Methods. The wiring method shall comply with 502.100(A).

(2) Contacts. Switches, circuit breakers, relays, contactors, fuses and current-breaking contacts for bells, horns, howlers, sirens, and other devices in which sparks or arcs may be produced shall be provided with enclosures identified for a Class II location.

Exception: Where current-breaking contacts are immersed in oil or where the interruption of current occurs within a chamber sealed against the entrance of dust, enclosures shall be permitted to be of the general-purpose type.

(3) Resistors and Similar Equipment. Resistors, transformers, choke coils, rectifiers, thermionic tubes, and other heat-generating equipment shall be provided with enclosures identified for Class II locations.

Exception: Where resistors or similar equipment are immersed in oil or enclosed in a chamber sealed against the entrance of dust, enclosures shall be permitted to be of the general-purpose type.

(4) Rotating Machinery. Motors, generators, and other rotating electric machinery shall comply with 502.125(A).

(5) Combustible, Electrically Conductive Dusts. Where dusts are of a combustible, electrically conductive nature, all wiring and equipment shall be identified for Class II locations.

(6) Metal Dusts. Where dust from magnesium, aluminum, aluminum bronze powders, or other metals of similarly hazardous characteristics may be present, all apparatus and equipment shall be identified for the specific conditions.

(B) Class II, Division 2. In Class II, Division 2 locations, signaling, alarm, remote-control, and communications systems; and meters, instruments, and relays shall comply with 502.150(B)(1) through (B)(5).

(1) Contacts. Enclosures shall comply with 502.150(A)(2), or contacts shall have tight metal enclosures designed to minimize the entrance of dust and shall have telescoping or tight-fitting covers and no openings through which, after installation, sparks or burning material might escape.

Exception: In nonincendive circuits, enclosures shall be permitted to be of the general-purpose type.

(2) Transformers and Similar Equipment. The windings and terminal connections of transformers, choke coils, and similar equipment shall be provided with tight metal enclosures without ventilating openings.

(3) Resistors and Similar Equipment. Resistors, resistance devices, thermionic tubes, rectifiers, and similar equipment shall comply with 502.130(A)(3).

Exception: Enclosures for thermionic tubes, nonadjustable resistors, or rectifiers for which maximum operating temperature will not exceed 120°C (248°F) shall be permitted to be of the general-purpose type.

(4) Rotating Machinery. Motors, generators, and other rotating electric machinery shall comply with 502.125(B).

(5) Wiring Methods. The wiring method shall comply with 502.10(B).

ARTICLE 503
Class III Locations

Editor's Note: *Article 503 does not apply if the zone classification system of Article 506 is used.*

I. General

503.1 Scope. Article 503 covers the requirements for electrical and electronic equipment and wiring for all voltages in Class III, Division 1 and 2 locations where fire or explosion hazards may exist due to ignitible fibers or flyings.

503.5 General. The general rules of this *Code* shall apply to electric wiring and equipment in locations classified as Class III locations in 500.5(D).

Exception: As modified by this article.

Equipment installed in Class III locations shall be able to function at full rating without developing surface temperatures high enough to cause excessive dehydration or gradual carbonization of accumulated fibers or flyings. Organic material that is carbonized or excessively dry is highly susceptible to spontaneous ignition. The maximum surface temperatures under operating conditions shall not exceed 165°C (329°F) for equipment that is not subject to overloading, and 120°C (248°F) for equipment (such as motors or power transformers) that may be overloaded.

FPN: For electric trucks, see NFPA 505-2002, *Fire Safety Standard for Powered Industrial Trucks Including Type Designations, Areas of Use, Conversions, Maintenance, and Operation.*

II. Wiring

503.10 Wiring Methods. Wiring methods shall comply with 503.10(A) or 503.10(B).

(A) Class III, Division 1. In Class III, Division 1 locations, the wiring method shall be rigid metal conduit, rigid nonmetallic conduit, intermediate metal conduit, electrical metallic tubing, dusttight wireways, or Type MC or MI cable with listed termination fittings.

(1) Boxes and Fittings. All boxes and fittings shall be dusttight.

(2) Flexible Connections. Where necessary to employ flexible connections, dusttight flexible connectors, liquidtight flexible metal conduit with listed fittings, liquidtight flexible nonmetallic conduit with listed fittings, or flexible cord in conformance with 503.140 shall be used.

FPN: See 503.30(B) for grounding requirements where flexible conduit is used.

(3) Nonincendive Field Wiring. Nonincendive field wiring shall be permitted using any of the wiring methods permitted for unclassified locations. Nonincendive field wiring systems shall be installed in accordance with the control drawing(s). Simple apparatus, not shown on the control drawing, shall be permitted in a nonincendive field wiring circuit, provided the simple apparatus does not interconnect the nonincendive field wiring circuit to any other circuit.

FPN: Simple apparatus is defined in 504.2.

Separate nonincendive field wiring circuits shall be installed in accordance with one of the following:

(1) In separate cables
(2) In multiconductor cables where the conductors of each circuit are within a grounded metal shield
(3) In multiconductor cables where the conductors of each circuit have insulation with a minimum thickness of 0.25 mm (0.01 in.)

(B) Class III, Division 2. In Class III, Division 2 locations, the wiring method shall comply with 503.10(A).

Exception: In sections, compartments, or areas used solely for storage and containing no machinery, open wiring on insulators shall be permitted where installed in accordance with Article 398, but only on condition that protection as required by 398.15(C) be provided where conductors are not run in roof spaces and are well out of reach of sources of physical damage.

503.25 Uninsulated Exposed Parts, Class III, Divisions 1 and 2. There shall be no uninsulated exposed parts, such as electric conductors, buses, terminals, or components, that operate at more than 30 volts (15 volts in wet locations). These parts shall additionally be protected by a protection technique according to 500.7(E), 500.7(F), or 500.7(G) that is suitable for the location.

Exception: As provided in 503.155.

503.30 Grounding and Bonding — Class III, Divisions 1 and 2. Wiring and equipment in Class III, Division 1 and 2 locations shall be grounded as specified in Article 250 and with the following additional requirements in 503.30(A) and 503.30(B).

(A) Bonding. The locknut-bushing and double-locknut types of contacts shall not be depended on for bonding purposes, but bonding jumpers with proper fittings or other approved means of bonding shall be used. Such means of bonding shall apply to all intervening raceways, fittings, boxes, enclosures, and so forth, between Class III locations and the point of grounding for service equipment or point of grounding of a separately derived system.

Exception: The specific bonding means shall only be required to the nearest point where the grounded circuit conductor and the grounding electrode conductor are connected together on the line side of the building or structure disconnecting means as specified in 250.32(A), (B), and (C), if the branch-circuit overcurrent protection is located on the load side of the disconnecting means.

FPN: See 250.100 for additional bonding requirements in hazardous (classified) locations.

(B) Types of Equipment Grounding Conductors. Where flexible conduit is used as permitted in 503.10, it shall be installed with internal or external bonding jumpers in parallel with each conduit and complying with 250.102.

Exception: In Class III, Division 1 and 2 locations, the bonding jumper shall be permitted to be deleted where all of the following conditions are met:

(1) Listed liquidtight flexible metal 1.8 m (6 ft) or less in length, with fittings listed for grounding, is used.
(2) Overcurrent protection in the circuit is limited to 10 amperes or less.
(3) The load is not a power utilization load.

III. Equipment

503.100 Transformers and Capacitors — Class III, Divisions 1 and 2. Transformers and capacitors shall comply with 502.100(B).

503.115 Switches, Circuit Breakers, Motor Controllers, and Fuses — Class III, Divisions 1 and 2. Switches, circuit breakers, motor controllers, and fuses, including pushbuttons, relays, and similar devices, shall be provided with dusttight enclosures.

503.120 Control Transformers and Resistors — Class III, Divisions 1 and 2. Transformers, impedance coils, and resistors used as or in conjunction with control equipment for motors, generators, and appliances shall be provided with dusttight enclosures complying with the temperature limitations in 503.5.

503.125 Motors and Generators — Class III, Divisions 1 and 2. In Class III, Divisions 1 and 2 locations, motors, generators, and other rotating machinery shall be totally enclosed nonventilated, totally enclosed pipe ventilated, or totally enclosed fan cooled.

Exception: In locations where, in the judgment of the authority having jurisdiction, only moderate accumulations of lint or flyings are likely to collect on, in, or in the vicinity of a rotating electric machine and where such machine is readily accessible for routine cleaning and maintenance, one of the following shall be permitted:

(1) Self-cleaning textile motors of the squirrel-cage type
(2) Standard open-type machines without sliding contacts, centrifugal or other types of switching mechanisms, including motor overload devices
(3) Standard open-type machines having such contacts, switching mechanisms, or resistance devices enclosed within tight housings without ventilating or other openings

503.128 Ventilating Piping — Class III, Divisions 1 and 2. Ventilating pipes for motors, generators, or other rotating electric machinery, or for enclosures for electric equipment, shall be of metal not less than 0.53 mm (0.021 in.) in thickness, or of equally substantial noncombustible material, and shall comply with the following:

(1) Lead directly to a source of clean air outside of buildings
(2) Be screened at the outer ends to prevent the entrance of small animals or birds
(3) Be protected against physical damage and against rusting or other corrosive influences

Ventilating pipes shall be sufficiently tight, including their connections, to prevent the entrance of appreciable quantities of fibers or flyings into the ventilated equipment or enclosure and to prevent the escape of sparks, flame, or burning material that might ignite accumulations of fibers or flyings or combustible material in the vicinity. For metal pipes, lock seams and riveted or welded joints shall be permitted; and tight-fitting slip joints shall be permitted where some flexibility is necessary, as at connections to motors.

503.130 Luminaires (Lighting Fixtures) — Class III, Divisions 1 and 2.

(A) Fixed Lighting. Luminaires (lighting fixtures) for fixed lighting shall provide enclosures for lamps and lampholders that are designed to minimize entrance of fibers and flyings and to prevent the escape of sparks, burning material, or hot metal. Each luminaire (fixture) shall be clearly marked to show the maximum wattage of the lamps that shall be permitted without exceeding an exposed surface temperature of 165°C (329°F) under normal conditions of use.

(B) Physical Damage. A luminaire (fixture) that may be exposed to physical damage shall be protected by a suitable guard.

(C) Pendant Luminaires (Fixtures). Pendant luminaires (fixtures) shall be suspended by stems of threaded rigid metal conduit, threaded intermediate metal conduit, threaded metal tubing of equivalent thickness, or by chains with approved fittings. For stems longer than 300 mm (12 in.), permanent and effective bracing against lateral displacement shall be provided at a level not more than 300 mm (12 in.) above the lower end of the stem, or flexibility in the form of an identified fitting or a flexible connector shall be provided not more than 300 mm (12 in.) from the point of attachment to the supporting box or fitting.

(D) Portable Lighting Equipment. Portable lighting equipment shall be equipped with handles and protected with substantial guards. Lampholders shall be of the unswitched type with no provision for receiving attachment plugs. There shall be no exposed current-carrying metal parts, and all exposed non–current-carrying metal parts shall be grounded. In all other respects, portable lighting equipment shall comply with 503.130(A).

503.135 Utilization Equipment — Class III, Divisions 1 and 2.

(A) Heaters. Electrically heated utilization equipment shall be identified for Class III locations.

(B) Motors. Motors of motor-driven utilization equipment shall comply with 503.125.

(C) Switches, Circuit Breakers, Motor Controllers, and Fuses. Switches, circuit breakers, motor controllers, and fuses shall comply with 503.115.

503.140 Flexible Cords — Class III, Divisions 1 and 2. Flexible cords shall comply with the following:

(1) Be of a type listed for extra-hard usage
(2) Contain, in addition to the conductors of the circuit, a grounding conductor complying with 400.23
(3) Be connected to terminals or to supply conductors in an approved manner
(4) Be supported by clamps or other suitable means in such a manner that there will be no tension on the terminal connections
(5) Be provided with suitable means to prevent the entrance of fibers or flyings where the cord enters boxes or fittings

503.145 Receptacles and Attachment Plugs — Class III, Divisions 1 and 2. Receptacles and attachment plugs shall be of the grounding type and shall be designed so as to minimize the accumulation or the entry of fibers or flyings, and shall prevent the escape of sparks or molten particles.

Exception: In locations where, in the judgment of the authority having jurisdiction, only moderate accumulations of lint or flyings will be likely to collect in the vicinity of a receptacle, and where such receptacle is readily accessible for routine cleaning, general-purpose grounding-type receptacles mounted so as to minimize the entry of fibers or flyings shall be permitted.

503.150 Signaling, Alarm, Remote-Control, and Local Loudspeaker Intercommunications Systems — Class III, Divisions 1 and 2. Signaling, alarm, remote-control, and local loudspeaker intercommunications systems shall comply with the requirements of Article 503 regarding wiring methods, switches, transformers, resistors, motors, luminaires (lighting fixtures), and related components.

503.155 Electric Cranes, Hoists, and Similar Equipment — Class III, Divisions 1 and 2. Where installed for operation over combustible fibers or accumulations of flyings, traveling cranes and hoists for material handling, traveling cleaners for textile machinery, and similar equipment shall comply with 503.155(A) through (D).

(A) Power Supply. Power supply to contact conductors shall be electrically isolated from all other systems, ungrounded, and shall be equipped with an acceptable ground detector that gives an alarm and automatically de-energizes the contact conductors in case of a fault to ground or gives a visual and audible alarm as long as power is supplied to the contact conductors and the ground fault remains.

(B) Contact Conductors. Contact conductors shall be located or guarded so as to be inaccessible to other than authorized persons and shall be protected against accidental contact with foreign objects.

(C) Current Collectors. Current collectors shall be arranged or guarded so as to confine normal sparking and prevent escape of sparks or hot particles. To reduce sparking, two or more separate surfaces of contact shall be provided for each contact conductor. Reliable means shall be provided to keep contact conductors and current collectors free of accumulations of lint or flyings.

(D) Control Equipment. Control equipment shall comply with 503.115 and 503.120.

503.160 Storage Battery Charging Equipment — Class III, Divisions 1 and 2. Storage battery charging equipment shall be located in separate rooms built or lined with substantial noncombustible materials. The rooms shall be constructed to prevent the entrance of ignitible amounts of flyings or lint and shall be well ventilated.

ARTICLE 506
Zone 20, 21, and 22 Locations for Combustible Dusts, Fibers, and Flyings

Editor's Note: *The text of Section 506.2 is not included because it is included in Definitions.*

506.1 Scope This article covers the requirements for the zone classification system as an alternative to the division classification system covered in Article 500, Article 502, and Article 503 for electrical and electronic equipment and wiring for all voltages in Zone 20, Zone 21, and Zone 22 hazardous (classified) locations where fire and explosion hazards may exist due to combustible dusts, or ignitible fibers or flyings. Combustible metallic dusts are not covered by the requirements of this article.

FPN No. 1: For the requirements for electrical and electronic equipment and wiring for all voltages in Class I, Division 1 or Division 2; Class II, Division 1 or Division 2; Class III, Division 1 or Division 2; and Class I, Zone 0 or Zone 1 or Zone 2 hazardous (classified) locations where fire or explosion hazards may exist due to flammable gases or vapors, flammable liquids, or combustible dusts or fibers, refer to Articles 500 through 505.

FPN No. 2: Zone 20, Zone 21, and Zone 22 area classifications are based on the modified IEC area classification system as defined in ISA 12.10.05, *Electrical Apparatus for Use in Zone 20, Zone 21, and Zone 22 Hazardous (Classified) Locations — Classification of Zone 20, Zone 21, and Zone 22 Hazardous (Classified) Locations (IEC61241-10 Mod).*

FPN No. 3: The unique hazards associated with explosives, pyrotechnics, and blasting agents are not addressed in this article.

506.4 General

(A) Documentation for Industrial Occupancies Areas designated as hazardous (classified) locations shall be properly documented. This documentation shall be available to those authorized to design, install, inspect, maintain or operate electrical equipment.

(B) Reference Standards Important information relating to topics covered in Chapter 5 are found in other publications.

FPN: It is important that the authority having jurisdiction be familiar with the recorded industrial experience as well as with standards of the National Fire Protection Association (NFPA), the ISA, International Society for Measurement and Control, and the International Electrotechnical Commission (IEC) that may be of use in the classification of various locations, the determination of adequate ventilation, and the protection against static electricity and lightning hazards.

506.5 Classification of Locations

(A) Classifications of Locations Locations shall be classified on the basis of the properties of the combustible dust, ignitible fibers or flyings that may be present, and the likelihood that a combustible or combustible concentration or quantity is present. Each room, section, or area shall be considered individually in determining its classification. Where pyrophoric materials are the only materials used or handled, these locations are outside of the scope of this article.

(B) Zone 20, Zone 21, and Zone 22 Locations Zone 20, Zone 21, and Zone 22 locations are those in which combustible dust, ignitible fibers, or flyings are or may be present in the air or in layers, in quantities sufficient to produce explosive or ignitible mixtures. Zone 20, Zone 21, and Zone 22 locations shall include those specified in 506.5(B)(1), (B)(2), and (B)(3).

FPN: Through the exercise of ingenuity in the layout of electrical installations for hazardous (classified) locations, it is frequently possible to locate much of the equipment in a reduced level of classification, and, thus, to reduce the amount of special equipment required.

(1) Zone 20 A Zone 20 location is a location in which

(a) Ignitible concentrations of combustible dust or ignitible fibers or flyings are present continuously.

(b) Ignitible concentrations of combustible dust or ignitible fibers or flyings are present for long periods of time.

FPN No. 1: As a guide to classification of Zone 20 locations, refer to ISA 12.10.05, *Electrical Apparatus for Use in Zone 20, Zone 21, and Zone 22 Hazardous (Classified) Locations — Classification of Zone 20, Zone 21, and Zone 22 Hazardous (Classified) Locations (IEC61241-10 Mod).*

FPN No. 2: Zone 20 classification includes locations inside dust containment systems; hoppers, silos, etc., cyclones and filters, dust transport systems, except some parts of belt and chain conveyors, etc.; blenders, mills, dryers, bagging equipment, etc.

(2) Zone 21 A Zone 21 location is a location

(a) In which ignitible concentrations of combustible dust or ignitible fibers or flyings are likely to exist occasionally under normal operating conditions; or

(b) In which ignitible concentrations of combustible dust or ignitible fibers or flyings may exist frequently because of repair or maintenance operations or because of leakage; or

(c) In which equipment is operated or processes are carried on, of such a nature that equipment breakdown or faulty operations could result in the release of ignitible concentrations of combustible dust, or ignitible fibers or flyings and also cause simultaneous failure of electrical equipment in a mode to cause the electrical equipment to become a source of ignition; or

(d) That is adjacent to a Zone 20 location from which ignitible concentrations of dust or ignitible fibers or flyings could be communicated, unless communication is prevented by adequate positive pressure ventilation from a source of clean air and effective safeguards against ventilation failure are provided.

FPN No. 1: As a guide to classification of Zone 21 locations, refer to ISA 12.10.05, *Electrical Apparatus for Use In Zone 20, Zone 21, and Zone 22 Hazardous (Classified) Locations — Classification of Zone 20, Zone 21, and Zone 22 Hazardous (Classified) Locations (IEC61241-10 Mod).*

FPN No. 2: This classification usually includes locations outside dust containment and in the immediate vicinity of access doors subject to frequent removal or opening for operation purposes when internal combustible mixtures are present; locations outside dust containment in the proximity of filling and emptying points, feed belts, sampling points, truck dump stations, belt dump over points, etc. where no measures are employed to prevent the formation of combustible mixtures; locations outside dust containment where dust accumulates and where due to process operations the dust layer is likely to be disturbed and form combustible mixtures; locations inside dust containment where explosive dust clouds are likely to occur (but neither continuously, nor for long periods, nor frequently) as, for example, silos (if filled and/or emptied only occasionally) and the dirty side of filters if large self-cleaning intervals are occurring.

(3) Zone 22 A Zone 22 location is a location

(a) In which ignitible concentrations of combustible dust or ignitible fibers or flyings are not likely to occur in normal operation, and if they do occur, will only persist for a short period; or

(b) In which combustible dust, or fibers, or flyings are handled, processed, or used but in which the dust, fibers, or flyings are normally confined within closed containers of closed systems from which they can escape only as a result of the abnormal operation of the equipment with which the dust, or fibers, or flyings are handled, processed, or used; or

(c) That is adjacent to a Zone 21 location, from which ignitible concentrations of dust or fibers or flyings could be communicated, unless such communication is prevented by adequate positive pressure ventilation from a source of clean air and effective safeguards against ventilation failure are provided.

FPN No. 1: As a guide to classification of Zone 22 locations, refer to ISA 12.10.05, *Electrical Apparatus for Use in Zone 20, Zone 21, and Zone 22 Hazardous (Classified) Locations — Classification of Zone 20, Zone 21, and Zone 22 Hazardous (Classified) Locations (IEC61241-10 Mod).*

FPN No. 2: Zone 22 locations usually include outlets from bag filter vents, because in the event of a malfunction there can be emission of combustible mixtures; locations near equipment that has to be opened at infrequent intervals or equipment that from experience can easily form leaks where, due to pressure above atmospheric, dust will blow out; pneumatic equipment, flexible connections that can become damaged, etc.; storage locations for bags containing dusty product, since failure of bags can occur during handling, causing dust leakage; and locations where controllable dust layers are formed that are likely to be raised into explosive dust/air mixtures. Only if the layer is removed by cleaning before hazardous dust–air mixtures can be formed is the area designated non-hazardous.

FPN No. 3: Locations that normally are classified as Zone 21 can fall into Zone 22 when measures are employed to prevent the formation of explosive dust–air mixtures. Such measures include exhaust ventilation. The measures should be used in the vicinity of (bag) filling and emptying points, feed belts, sampling points, truck dump stations, belt dump over points, etc.

506.6 Special Precaution Article 506 requires equipment construction and installation that ensures safe performance under conditions of proper use and maintenance.

FPN: It is important that inspection authorities and users exercise more than ordinary care with regard to the installation and maintenance of electrical equipment in hazardous (classified) locations.

(A) Implementation of Zone Classification System Classification of areas, engineering and design, selection of equipment and wiring methods, installation, and inspection shall be performed by qualified persons.

(B) Dual Classification In instances of areas within the same facility classified separately, Zone 22 locations shall be permitted to abut, but not overlap, Class II or Class III, Division 2 locations. Zone 20 or Zone 21 locations shall not abut Class II or Class III, Division 1 or Division 2 locations.

(C) Reclassification Permitted A Class II or Class III, Division 1 or Division 2 location shall be permitted to be reclassified as a Zone 20, Zone 21, or Zone 22 location, provided that all of the space that is classified because of a single combustible dust or ignitible fiber or flying source is reclassified under the requirements of this article.

(D) Simultaneous Presence of Flammable Gases and Combustible Dusts, Fibers, or Flyings Where flammable gases or combustible dusts, fibers, or flyings are or may be present at the same time, the simultaneous presence shall be considered during the selection and installation of the electrical equipment and the wiring methods, including the determination of the safe operating temperature of the electrical equipment.

506.8 Protection Techniques Acceptable protection techniques for electrical and electronic equipment in hazardous (classified) locations shall be as described in 506.8(A) through 506.8(F).

(A) Dust Ignitionproof This protection technique shall be permitted for equipment in Zone 20, Zone 21, and Zone 22 locations for which it is identified.

(B) Pressurized This protection technique shall be permitted for equipment in Zone 21, and Zone 22 locations for which it is identified.

(C) Intrinsic Safety This protection technique shall be permitted for equipment in Zone 20, Zone 21, and Zone 22 locations for which it is identified. Installation of intrinsically safe apparatus and wiring shall be in accordance with the requirements of Article 504.

(D) Dusttight This protection technique shall be permitted for equipment in Zone 22 locations for which it is identified.

(E) Nonincendive Circuit This protection technique shall be permitted for equipment in Zone 22 locations for which it is identified.

(F) Nonincendive Equipment This protection technique shall be permitted for equipment in Zone 22 locations for which it is identified.

506.9 Equipment Requirements

(A) Suitability Suitability of identified equipment shall be determined by one of the following:

(1) Equipment listing or labeling
(2) Evidence of equipment evaluation from a qualified testing laboratory or inspection agency concerned with product evaluation
(3) Evidence acceptable to the authority having jurisdiction such as a manufacturer's self-evaluation or an owner's engineering judgment

(B) Listing

(1) Equipment that is listed for Zone 20 shall be permitted in a Zone 21 or Zone 22 location of the same dust, or ignitible fiber, or flying. Equipment that is listed for Zone 21 may be used in a Zone 22 location of the same dust, fiber, or flying.
(2) Equipment shall be permitted to be listed for a specific dust, or ignitible fiber or flying, or any specific combination of dusts, fibers, or flyings.

(C) Marking Equipment identified for Class II, Division 1 or Class II, Division 2 shall, in addition to being marked in accordance with 500.8(B), be permitted to be marked with both of the following:

(1) Zone 20, 21, or 22 (as applicable)
(2) Temperature classification in accordance with 506.9(D)

(D) Temperature Classifications Equipment shall be marked to show the operating temperature referenced to a 40°C (104°F) ambient. Electrical equipment designed for use in the ambient temperature range between −20°C and +40°C shall require no additional ambient temperature marking. Electrical equipment that is designed for use in a range of ambient temperatures other than −20°C and +40°C is considered to be special; and the ambient temperature range shall then be marked on the equipment, including either the symbol "Ta" or "Tamb" together with the special range of ambient temperatures. As an example, such a marking might be "–30°C ≤ Ta ≤ +40°C." Electrical equipment suitable for ambient temperatures exceeding 40°C (104°F) shall be marked with both the maximum ambient temperature and the operating temperature at that ambient temperature.

Exception No. 1: Equipment of the non–heat-producing type, such as conduit fittings, shall not be required to have a marked operating temperature.

Exception No. 2: Equipment identified for Class II, Division 1 or Class II, Division 2 locations as permitted by 506.20(B) and 506.20(C) shall be permitted to be marked in accordance with 500.8(B) and Table 500.8(B).

(E) Threading All NPT threads referred to herein shall be threaded with a National (American) Standard Pipe Taper (NPT) thread that provides a taper of 1 in 16 (¾-in. taper per foot). Conduit and fittings shall be made wrenchtight to prevent sparking when the fault current flows through the conduit system, and to ensure the integrity of the conduit system. Equipment provided with threaded entries for field wiring connections shall be installed in accordance with 506.9(E)(1) or (E)(2).

(1) Equipment Provided with Threaded Entries for NPT Threaded Conduit or Fittings For equipment provided with threaded entries for NPT threaded conduit or fittings, listed conduit fittings, or cable fittings shall be used.

(2) Equipment Provided with Threaded Entries for Metric Threaded Conduit or Fittings For equipment with metric threaded entries, such entries shall be identified as being metric, or listed adapters to permit connection to conduit or NPT-threaded fittings shall be provided with the equipment. Adapters shall be used for connection to conduit or NPT-threaded fittings. Listed cable fittings that have metric threads shall be permitted to be used.

506.15 Wiring Methods Wiring methods shall maintain the integrity of the protection techniques and shall comply with 506.15(A), (B), or (C).

(A) Zone 20 In Zone 20 locations, the wiring methods in (1) through (5) shall be permitted.

(1) Threaded rigid metal conduit, or threaded steel intermediate metal conduit.
(2) Type MI cable with termination fittings listed for the location. Type MI cable shall be installed and supported in a manner to avoid tensile stress at the termination fittings.

Exception: MI cable and fittings listed for Class II, Division 1 locations are permitted to be used.

(3) In industrial establishments with limited public access, where the conditions of maintenance and supervision ensure that only qualified persons service the installation, Type MC cable, listed for continuous use in Zone 20 locations, with a gas/vaportight continuous corrugated metallic sheath, and overall jacket of suitable polymeric material, separate grounding conductors in accordance with 250.122, and provided with termination fittings listed for the application, shall be permitted.

Exception: MC cable and fittings listed for Class II, Division 1 locations are permitted to be used.

(4) Fittings and boxes shall be identified for use in Zone 20 locations.

Exception: Boxes and fittings listed for Class II, Division 1 locations are permitted to be used.

(5) Where necessary to employ flexible connections, liquidtight flexible metal conduit with listed fittings, liquidtight flexible nonmetallic conduit with listed fittings, or flexible cord listed for extra-hard usage and provided with listed fittings shall be used. Where flexible cords are used, they shall also comply with 506.17. Where flexible connections are subject to oil or other corrosive conditions, the insulation of the conductors shall be of a type listed for the condition or shall be protected by means of a suitable sheath.

Exception: Flexible conduit and flexible conduit and cord fittings listed for Class II, Division 1 locations are permitted to be used.

FPN: See 506.25 for grounding requirements where flexible conduit is used.

(B) Zone 21 In Zone 21 locations, the wiring methods in (B)(1) and (B)(2) shall be permitted.

(1) All wiring methods permitted in 506.15(A)
(2) Fittings and boxes that are dusttight, provided with threaded bosses for connection to conduit, in which taps, joints, or terminal connections are not made, and are not used in locations where metal dust is present, may be used

(C) Zone 22 In Zone 22 locations, the wiring methods in (1) through (8) shall be permitted.

(1) All wiring methods permitted in 506.15(B).
(2) Rigid metal conduit, intermediate metal conduit, electrical metallic tubing, dusttight wireways.
(3) Type MC or MI cable with listed termination fittings.
(4) Type PLTC in cable trays.
(5) Type ITC in cable trays.
(6) Type MC, MI, MV, or TC cable installed in ladder, ventilated trough, or ventilated channel cable trays in a single layer, with a space not less than the larger cable diameter between two adjacent cables, shall be the wiring method employed. Single conductor Type MV cables shall be shielded or metallic armored.
(7) Nonincendive field wiring shall be permitted using any of the wiring methods permitted for unclassified locations. Nonincendive field wiring systems shall be installed in accordance with the control drawing(s). Simple apparatus, not shown on the control drawing,

shall be permitted in a nonincendive field wiring circuit, provided the simple apparatus does not interconnect the nonincendive field wiring circuit to any other circuit.

FPN: *Simple apparatus* is defined in 504.2.

Separation of nonincendive field wiring circuits shall be in accordance with one of the following:
a. Be in separate cables
b. Be in multiconductor cables where the conductors of each circuit are within a grounded metal shield
c. Be in multiconductor cables where the conductors have insulation with a minimum thickness of 0.25 mm (0.01 in.)

(8) Boxes and fittings shall be dusttight.

506.16 Sealing Where necessary to protect the ingress of combustible dust, or ignitible fibers, or flyings, or to maintain the type of protection, seals shall be provided. The seal shall be identified as capable of preventing the ingress of combustible dust or ignitible fibers or flyings and maintaining the type of protection but need not be explosionproof or flameproof.

506.17 Flexible Cords Flexible cords used in Zone 20, Zone 21, and Zone 22 locations shall comply with all of the following:

(1) Be of a type listed for extra-hard usage
(2) Contain, in addition to the conductors of the circuit, a grounding conductor in complying with 400.23
(3) Be connected to terminals or to supply conductors in an approved manner
(4) Be supported by clamps or by other suitable means in such a manner to minimize tension on the terminal connections
(5) Be provided with suitable seals to prevent the entrance of combustible dust, or ignitible fibers, or flyings where the flexible cord enters boxes or fittings

506.20 Equipment Installation

(A) Zone 20 In Zone 20 locations, only equipment listed and marked as suitable for the location shall be permitted.

Exception: Intrinsically safe apparatus listed for use in Class II, Division 1 locations with a suitable temperature class shall be permitted.

(B) Zone 21 In Zone 21 locations, only equipment listed and marked as suitable for the location shall be permitted.

Exception No. 1: Apparatus listed for use in Class II, Division 1 locations with a suitable temperature class shall be permitted.

Exception No. 2: Pressurized equipment identified for Class II, Division 1 shall be permitted.

(C) Zone 22 In Zone 22 locations, only equipment listed and marked as suitable for the location shall be permitted.

Exception No. 1: Apparatus listed for use in Class II, Division 1 or Class II, Division 2 locations with a suitable temperature class shall be permitted.

Exception No. 2: Pressurized equipment identified for Class II, Division 1 or Division 2 shall be permitted.

(D) Manufacturer's Instructions Electrical equipment installed in hazardous (classified) locations shall be installed in accordance with the instructions (if any) provided by the manufacturer.

(E) Temperature The temperature marking specified in 506.9(C)(2)(5) shall comply with (E)(1) or (E)(2).

(1) For combustible dusts, less than the lower of either the layer or cloud ignition temperature of the specific combustible dust. For organic dusts that may dehydrate or carbonize, the temperature marking shall not exceed the lower of either the ignition temperature or 165°C (329°F).
(2) For ignitible fibers or flyings, less than 165°C (329°F) for equipment that is not subject to overloading, or 120°C (248°F) for equipment (such as motors or power transformers) that may be overloaded.

FPN: See NFPA 499-2004, *Recommended Practice for the Classification of Combustible Dusts and of Hazardous (Classified) Locations for Electrical Installations in Chemical Processing Areas*, for minimum ignition temperatures of specific dusts.

506.21 Multiwire Branch Circuits In Zone 20 and Zone 21 locations, a multiwire branch circuit shall not be permitted.

Exception: Where the disconnect device(s) for the circuit opens all ungrounded conductors of the multiwire circuit simultaneously.

506.25 Grounding and Bonding Grounding and bonding shall comply with Article 250 and the requirements in 506.25(A) and 506.25(B).

(A) Bonding The locknut-bushing and double-locknut types of contacts shall not be depended on for bonding purposes, but bonding jumpers with proper fittings or other approved means of bonding shall be used. Such means of bonding shall apply to all intervening raceways, fittings, boxes, enclosures, and so forth, between Zone 20, Zone 21, and Zone 22 locations and the point of grounding for service equipment or point of grounding of a separately derived system.

Exception: The specific bonding means shall be required only to the nearest point where the grounded circuit conductor and the grounding electrode conductor are connected together on the line side of the building or structure disconnecting means as specified in 250.32(A), (B), and (C), if the branch side overcurrent protection is located on the load side of the disconnecting means.

FPN: See 250.100 for additional bonding requirements in hazardous (classified) locations.

(B) Types of Equipment Grounding Conductors Where flexible conduit is used as permitted in 506.15, it shall be installed with internal or external bonding jumpers in parallel with each conduit and complying with 250.102.

Exception: In Zone 22 locations, the bonding jumper shall be permitted to be deleted where all of the following conditions are met:

(1) Listed liquidtight flexible metal conduit 1.8 m (6 ft) or less in length, with fittings listed for grounding, is used.
(2) Overcurrent protection in the circuit is limited to 10 amperes or less.
(3) The load is not a power utilization load.

ARTICLE 504
Intrinsically Safe Systems

Editor's Note: *The text of Section 504.2 is not included because it is included in Definitions.*

504.1 Scope. This article covers the installation of intrinsically safe (I.S.) apparatus, wiring, and systems for Class I, II, and III locations.

FPN: For further information, see ANSI/ISA RP 12.06.01-2002, *Wiring Methods for Hazardous (Classified) Locations Instrumentation — Part 1: Intrinsic Safety.*

504.3 Application of Other Articles. Except as modified by this article, all applicable articles of this *Code* shall apply.

504.4 Equipment. All intrinsically safe apparatus and associated apparatus shall be listed.

Exception: Simple apparatus, as described on the control drawing, shall not be required to be listed.

504.10 Equipment Installation.

(A) Control Drawing. Intrinsically safe apparatus, associated apparatus, and other equipment shall be installed in accordance with the control drawing(s).

Exception: A simple apparatus that does not interconnect intrinsically safe circuits.

FPN: The control drawing identification is marked on the apparatus.

(B) Location. Intrinsically safe apparatus shall be permitted to be installed in any hazardous (classified) location for which it has been identified. General-purpose enclosures shall be permitted for intrinsically safe apparatus.

Associated apparatus shall be permitted to be installed in any hazardous (classified) location for which it has been identified or, if protected by other means, permitted by Articles 501 through 503 and Article 505.

Simple apparatus shall be permitted to be installed in any hazardous (classified) location in which the maximum surface temperature of the simple apparatus does not exceed the ignition temperature of the flammable gases or vapors, flammable liquids, combustible dusts, or ignitible fibers or flyings present.

For simple apparatus, the maximum surface temperature can be determined from the values of the output power from the associated apparatus or apparatus to which it is connected to obtain the temperature class. The temperature class can be determined by:

(1) Reference to Table 504.10(B)
(2) Calculation using the formula:

$$P_o R_{th} + T_{amb}$$

Where:

T = is the surface temperature
P_o = is the output power marked on the associated apparatus or intrinsically safe apparatus
R_{th} = is the thermal resistance of the simple apparatus
T_{amb} = is the ambient temperature (normally 40°C) and reference Table 500.8(B)

In addition, components with a surface area smaller than 10 cm^2 (excluding lead wires) may be classified as T5 if their surface temperature does not exceed 150°C.

Table 504.10(B) Assessment for T4 Classification According to Component Size and Temperature

Total Surface Area Excluding Lead Wires	Requirement for T4 Classification (Based on 40°C Ambient Temperature)
$<20\ mm^2$	Surface temperature ≤275°C
$\geq 20\ mm^2 \leq 10\ cm^2$	Surface temperature ≤200°C
$\geq 20\ mm^2$	Power not exceeding 1.3 W*

*Reduce to 1.2 W with an ambient of 60°C or 1.0 W with 80°C ambient temperature.

FPN: The following apparatus are examples of simple apparatus:

(1) Passive components, for example, switches, junction boxes, resistance temperature devices, and simple semiconductor devices such as LEDs
(2) Sources of generated energy, for example, thermocouples and photocells, which do not generate more than 1.5 V, 100 mA, and 25 mW

504.20 Wiring Methods. Intrinsically safe apparatus and wiring shall be permitted to be installed using any of the wiring methods suitable for unclassified locations, including Chapter 7 and Chapter 8. Sealing shall be as provided in 504.70, and separation shall be as provided in 504.30.

504.30 Separation of Intrinsically Safe Conductors.

(A) From Nonintrinsically Safe Circuit Conductors.

(1) In Raceways, Cable Trays, and Cables. Conductors of intrinsically safe circuits shall not be placed in any raceway, cable tray, or cable with conductors of any nonintrinsically safe circuit.

Exception No. 1: Where conductors of intrinsically safe circuits are separated from conductors of nonintrinsically safe circuits by a distance of at least 50 mm (2 in.) and secured, or by a grounded metal partition or an approved insulating partition.

FPN: No. 20 gauge sheet metal partitions 0.91 mm (0.0359 in.) or thicker are generally considered acceptable.

Exception No. 2: Where either (1) all of the intrinsically safe circuit conductors or (2) all of the nonintrinsically safe circuit conductors are in grounded metal-sheathed or metal-clad cables where the sheathing or cladding is capable of carrying fault current to ground.

FPN: Cables meeting the requirements of Articles 330 and 332 are typical of those considered acceptable.

(2) Within Enclosures.

(1) Conductors of intrinsically safe circuits shall be separated at least 50 mm (2 in.) from conductors of any nonintrinsically safe circuits, or as specified in 504.30(A)(2).
(2) All conductors shall be secured so that any conductor that might come loose from a terminal cannot come in contact with another terminal.

FPN No. 1: The use of separate wiring compartments for the intrinsically safe and nonintrinsically safe terminals is the preferred method of complying with this requirement.

FPN No. 2: Physical barriers such as grounded metal partitions or approved insulating partitions or approved restricted access wiring ducts separated from other such ducts by at least 19 mm (¾ in.) can be used to help ensure the required separation of the wiring.

(3) Other (Not in Raceway or Cable Tray Systems). Conductors and cables of intrinsically safe circuits run in other than raceway or cable tray systems shall be separated by at least 50 mm (2 in.) and secured from conductors and cables of any nonintrinsically safe circuits.

Exception: Where either (1) all of the intrinsically safe circuit conductors are in Type MI or MC cables or (2) all of the nonintrinsically safe circuit conductors are in raceways or Type MI or MC cables where the sheathing or cladding is capable of carrying fault current to ground.

(B) From Different Intrinsically Safe Circuit Conductors. Different intrinsically safe circuits shall be in separate cables or shall be separated from each other by one of the following means:

(1) The conductors of each circuit are within a grounded metal shield.
(2) The conductors of each circuit have insulation with a minimum thickness of 0.25 mm (0.01 in.).

Exception: Unless otherwise identified.

(3) The clearance between two terminals for connection of field wiring of different intrinsically safe circuits shall be at least 6 mm (0.25 in.) unless this clearance is permitted to be reduced by the control drawing.

504.50 Grounding.

(A) Intrinsically Safe Apparatus, Associated Apparatus, and Raceways. Intrinsically safe apparatus, associated apparatus, cable shields, enclosures, and raceways, if of metal, shall be grounded.

FPN: Supplementary bonding to the grounding electrode may be needed for some associated apparatus, for example, zener diode barriers, if specified in the control drawing. See ANSI/ISA RP 12.06.01-2002, *Wiring Methods for Hazardous (Classified) Locations Instrumentation Part 1: Intrinsic Safety.*

(B) Connection to Grounding Electrodes. Where connection to a grounding electrode is required, the grounding electrode shall be as specified in 250.52(A)(1), (A)(2), (A)(3), and (A)(4) and shall comply with 250.30(A)(7). Section 250.52(A)(5), (A)(6), and (A)(7) shall not be used if electrodes specified in 250.52(A)(1), (A)(2), (A)(3), or (A)(4) are available.

(C) Shields. Where shielded conductors or cables are used, shields shall be grounded.

Exception: Where a shield is part of an intrinsically safe circuit.

504.60 Bonding.

(A) Hazardous Locations. In hazardous (classified) locations, intrinsically safe apparatus shall be bonded in the hazardous (classified) location in accordance with 250.100.

(B) Unclassified. In unclassified or nonhazardous locations, where metal raceways are used for intrinsically safe system wiring in hazardous (classified) locations, associated apparatus shall be bonded in accordance with 501.30(A), 502.30(A), 503.30(A), or 505.25, as applicable.

504.70 Sealing. Conduits and cables that are required to be sealed by 501.15, 502.15, and 505.16 shall be sealed to minimize the passage of gases, vapors, or dusts. Such seals shall not be required to be explosionproof or flameproof.

Exception: Seals shall not be required for enclosures that contain only intrinsically safe apparatus, except as required by 501.15(F)(3).

504.80 Identification. Labels required by this section shall be suitable for the environment where they are installed with consideration given to exposure to chemicals and sunlight.

(A) Terminals. Intrinsically safe circuits shall be identified at terminal and junction locations in a manner that will prevent unintentional interference with the circuits during testing and servicing.

(B) Wiring. Raceways, cable trays, and other wiring methods for intrinsically safe system wiring shall be identified with permanently affixed labels with the wording "Intrinsic Safety Wiring" or equivalent. The labels shall be located so as to be visible after installation and placed so that they may be readily traced through the entire length of the installation. Intrinsic safety circuit labels shall appear in every section of the wiring system that is separated by enclosures, walls, partitions, or floors. Spacing between labels shall not be more than 7.5 m (25 ft).

Exception: Circuits run underground shall be permitted to be identified where they become accessible after emergence from the ground.

FPN No. 1: Wiring methods permitted in unclassified locations may be used for intrinsically safe systems in hazardous (classified) locations. Without labels to identify the application of the wiring, enforcement authorities cannot determine that an installation is in compliance with this *Code*.

FPN No. 2: In unclassified locations, identification is necessary to ensure that nonintrinsically safe wire will not be inadvertently added to existing raceways at a later date.

(C) Color Coding. Color coding shall be permitted to identify intrinsically safe conductors where they are colored light blue and where no other conductors colored light blue are used. Likewise, color coding shall be permitted to identify raceways, cable trays, and junction boxes where they are colored light blue and contain only intrinsically safe wiring.

ARTICLE 510
Hazardous (Classified) Locations — Specific

510.1 Scope. Articles 511 through 517 cover occupancies or parts of occupancies that are or may be hazardous because of atmospheric concentrations of flammable liquids, gases, or vapors, or because of deposits or accumulations of materials that may be readily ignitible.

510.2 General. The general rules of this *Code* and the provisions of Articles 500 through 504 shall apply to electric wiring and equipment in occupancies within the scope of Articles 511 through 517, except as such rules are modified in Articles 511 through 517. Where unusual conditions exist in a specific occupancy, the authority having jurisdiction shall judge with respect to the application of specific rules.

ARTICLE 511
Commercial Garages, Repair and Storage

Editor's Note: *The applicable text of NFPA 30A-2003, Code for Motor Fuel Dispensing Facilities and Repair Garages, is in NFPA 30A, Section 7.4.5.4.*

FPN: Rules that are followed by a reference in brackets contain text that has been extracted from NFPA 30A-2003, *Code for Motor Fuel Dispensing Facilities and Repair Garages*. Only editorial changes were made to the extracted text to make it consistent with this *Code*.

511.1 Scope. These occupancies shall include locations used for service and repair operations in connection with self-propelled vehicles (including, but not limited to, passenger automobiles, buses, trucks, and tractors) in which volatile flammable liquids or flammable gases are used for fuel or power.

511.3 Classifications of Locations.

(A) Unclassified Locations.

(1) Parking and Repair Garages. Parking garages used for parking or storage shall be permitted to be unclassified. Repair garages shall be permitted to be unclassified when designed in accordance with 511.3(A)(2) through 511.3(A)(7).

FPN: For further information, see NFPA 88A-2002, *Standard for Parking Structures*, and NFPA 30A-2003, *Code for Motor Fuel Dispensing Facilities and Repair Garages.*

(2) Alcohol-Based Windshield Washer Fluid. The storage, handling, or dispensing into motor vehicles of alcohol-based windshield washer fluid in areas used for the service and repair operations of the vehicles shall not cause such areas to be classified as hazardous (classified) locations.

FPN: For further information, see 8.3.5, Exception, of NFPA 30A-2003, *Code for Motor Fuel Dispensing Facilities and Repair Garages.*

(3) Specific Areas Adjacent to Classified Locations. Areas adjacent to classified locations in which flammable vapors are not likely to be released, such as stock rooms, switchboard rooms, and other similar locations, shall not be classified where mechanically ventilated at a rate of four or more air changes per hour, or designed with positive air pressure, or where effectively cut off by walls or partitions.

(4) Pits in Lubrication or Service Room Where Class I Liquids Are Not Transferred. Any pit, belowgrade work area, or subfloor work area that is provided with exhaust ventilation at a rate of not less than 0.3 $m^3/min/m^2$ (1 cfm/ft^2) of floor area at all times that the building is occupied or when vehicles are parked in or over this area and where exhaust air is taken from a point within 300 mm (12 in.) of the floor of the pit, belowgrade work area, or subfloor work area is unclassified. [NFPA 30A:7.4.5.4 and Table 8.3.1]

(5) Up to a Level of 450 mm (18 in.) Above the Floor in Lubrication or Service Rooms Where Class I Liquids Are Transferred. For each floor, the entire area up to a level of 450 mm (18 in.) above the floor shall be considered unclassified where there is mechanical ventilation providing a minimum of four air changes per hour or one cubic foot per minute of exchanged air for each square foot of floor area. Ventilation shall provide for air exchange across the entire floor area, and exhaust air shall be taken at a point within 0.3 m (12 in.) of the floor.

(6) Flammable Liquids Having Flash Points Below 38°C (100°F). Where flammable liquids having a flash point below 38°C (100°F) (such as gasoline) or gaseous fuels (such as natural gas, hydrogen, or LPG) will not be transferred, such location shall be considered to be unclassified. unless the location is required to be classified in accordance with 511.3(B)(2) or (B)(4).

(7) Within 450 mm (18 in.) of the Ceiling. In major repair garages, where lighter-than-air gaseous fuels (such as natural gas or hydrogen) vehicles are repaired or stored, the area within 450 mm (18 in.) of the ceiling shall be considered unclassified where ventilation of at least 1 cfm/sq ft of ceiling area taken from a point within 450 mm (18 in.) of the highest point in the ceiling is provided.

FPN: For further information on the definition of *major repair garage*, see 3.3.12.1 of NFPA 30A-2003, *Code for Motor Fuel Dispensing Facilities and Repair Garages.*

(B) Classified Locations.

(1) Flammable Fuel Dispensing Areas. Areas in which flammable fuel is dispensed into vehicle fuel tanks shall conform to Article 514.

(2) Lubrication or Service Room Where Class I Liquids or Gaseous Fuels (Such as Natural Gas, Hydrogen, or LPG) Are Not Transferred. The following spaces that are not designed in accordance with 511.3(A)(4) shall be classified as Class I, Division 2:

(1) Entire area within any unventilated pit, belowgrade work area, or subfloor area.
(2) Area up to 450 mm (18 in.) above any such unventilated pit, belowgrade work area, or subfloor work area and extending a distance of 900 mm (3 ft) horizontally from the edge of any such pit, belowgrade work area, or subfloor work area.

(3) Lubrication or Service Room Where Class I Liquids or Gaseous Fuels (Such as Natural Gas, Hydrogen, or LPG) Are Transferred. The following spaces that are not designed in accordance with 511.3(A)(5) shall be classified as follows:

(1) Up to a Level of 450 mm (18 in.) Above the Floor. For each floor, the entire area up to a level of 450 mm (18 in.) above the floor shall be a Class I, Division 2 location.
(2) Any Unventilated Pit or Depression Below Floor Level. Any unventilated pit or depression below floor level shall be a Class I, Division 1 location and shall extend up to said floor level.
(3) Any Ventilated Pit or Depression Below Floor Level. Any ventilated pit or depression in which six air changes per hour are exhausted from a point within 300 mm (12 in.) of the floor level of the pit shall be a Class I, Division 2 location.
(4) Space Above an Unventilated Pit or Depression Below Floor Level. Above a pit, or depression below floor level, the space up to 450 mm (18 in.) above the floor or grade level and 900 mm (3 ft) horizontally from a lubrication pit shall be a Class I, Division 2 location.
(5) Dispenser for Class I Liquids, Other Than Fuels. Within 900 mm (3 ft) of any fill or dispensing point, extending in all directions shall be a Class I, Division 2 location. See also 511.3(B)(1).

(4) Within 450 mm (18 in.) of the Ceiling. In major repair garages where lighter-than-air gaseous fuel (such as natural gas or hydrogen) vehicles are repaired or stored, ceiling spaces that are not designed in accordance with 511.3(A)(7) shall be classified as Class I, Division 2.

FPN: For further information on the definition of *major repair garage*, see 3.3.12.1 of NFPA 30A, 2003, *Code for Motor Fuel Dispensing Facilities and Repair Garages*.

511.4 Wiring and Equipment in Class I Locations.

(A) Wiring Located in Class I Locations. Within Class I locations as classified in 511.3, wiring shall conform to applicable provisions of Article 501.

(B) Equipment Located in Class I Locations. Within Class I locations as defined in 511.3, equipment shall conform to applicable provisions of Article 501.

(1) Fuel-Dispensing Units. Where fuel-dispensing units (other than liquid petroleum gas, which is prohibited) are located within buildings, the requirements of Article 514 shall govern.

Where mechanical ventilation is provided in the dispensing area, the control shall be interlocked so that the dispenser cannot operate without ventilation, as prescribed in 500.5(B)(2).

(2) Portable Lighting Equipment. Portable lighting equipment shall be equipped with handle, lampholder, hook, and substantial guard attached to the lampholder or handle. All exterior surfaces that might come in contact with battery terminals, wiring terminals, or other objects shall be of non-conducting material or shall be effectively protected with insulation. Lampholders shall be of an unswitched type and shall not provide means for plug-in of attachment plugs. The outer shell shall be of molded composition or other suitable material. Unless the lamp and its cord are supported or arranged in such a manner that they cannot be used in the locations classified in 511.3, they shall be of a type identified for Class I, Division 1 locations.

511.7 Wiring and Equipment Installed Above Class I Locations.

(A) Wiring in Spaces Above Class I Locations.

(1) Fixed Wiring Above Class I Locations. All fixed wiring above Class I locations shall be in metal raceways, rigid nonmetallic conduit, electrical nonmetallic tubing, flexible metal conduit, liquidtight flexible metal conduit, or liquidtight flexible nonmetallic conduit, or shall be Type MC, AC, MI, manufactured wiring systems, or PLTC cable in accordance with Article 725, or Type TC cable or Type ITC cable in accordance with Article 727. Cellular metal floor raceways or cellular concrete floor raceways shall be permitted to be used only for supplying ceiling outlets or extensions to the area below the floor, but such raceways shall have no connections leading into or through any Class I location above the floor.

(2) Pendant. For pendants, flexible cord suitable for the type of service and listed for hard usage shall be used.

(B) Electrical Equipment Installed Above Class I Locations.

(1) Fixed Electrical Equipment. Electrical equipment in a fixed position shall be located above the level of any defined Class I location or shall be identified for the location.

(a) Arcing Equipment. Equipment that is less than 3.7 m (12 ft) above the floor level and that may produce arcs, sparks, or particles of hot metal, such as cutouts, switches, charging panels, generators, motors, or other equipment (excluding receptacles, lamps, and lampholders) having make-and-break or sliding contacts, shall be of the totally enclosed type or constructed so as to prevent the escape of sparks or hot metal particles.

(b) Fixed Lighting. Lamps and lampholders for fixed lighting that is located over lanes through which vehicles are commonly driven or that may otherwise be exposed to physical damage shall be located not less than 3.7 m (12 ft) above floor level, unless of the totally enclosed type or constructed so as to prevent escape of sparks or hot metal particles.

511.9 Sealing. Seals conforming to the requirements of 501.15 and 501.15(B)(2) shall be provided and shall apply to horizontal as well as vertical boundaries of the defined Class I locations.

511.10 Special Equipment.

(A) Battery Charging Equipment. Battery chargers and their control equipment, and batteries being charged, shall not be located within locations classified in 511.3.

(B) Electric Vehicle Charging Equipment.

(1) General. All electrical equipment and wiring shall be installed in accordance with Article 625, except as noted in 511.10(B)(2) and (B)(3). Flexible cords shall be of a type identified for extra-hard usage.

(2) Connector Location. No connector shall be located within a Class I location as defined in 511.3.

(3) Plug Connections to Vehicles. Where the cord is suspended from overhead, it shall be arranged so that the lowest point of sag is at least 150 mm (6 in.) above the floor. Where an automatic arrangement is provided to pull both cord and plug beyond the range of physical damage, no additional connector shall be required in the cable or at the outlet.

511.12 Ground-Fault Circuit-Interrupter Protection for Personnel. All 125-volt, single-phase, 15- and 20-ampere receptacles installed in areas where electrical diagnostic equipment, electrical hand tools, or portable lighting equipment are to be used shall have ground-fault circuit-interrupter protection for personnel.

511.16 Grounded and Grounding Requirements.

(A) General Grounding Requirements. All metal raceways, the metal armor or metallic sheath on cables, and all non–current-carrying metal parts of fixed or portable electrical equipment, regardless of voltage, shall be grounded as provided in Article 250.

(B) Supplying Circuits with Grounded and Grounding Conductors in Class I Locations. Grounding in Class I locations shall comply with 501.30.

(1) Circuits Supplying Portable Equipment or Pendants. Where a circuit supplies portables or pendants and includes a grounded conductor as provided in Article 200, receptacles, attachment plugs, connectors, and similar devices shall be of the grounding type, and the grounded conductor of the flexible cord shall be connected to the screw shell of any lampholder or to the grounded terminal of any utilization equipment supplied.

(2) Approved Means. Approved means shall be provided for maintaining continuity of the grounding conductor between the fixed wiring system and the non–current-carrying metal portions of pendant luminaires (fixtures), portable lamps, and portable utilization equipment.

ARTICLE 513
Aircraft Hangars

Editor's Note: *The applicable text from NFPA 409-2004, Standard on Aircraft Hangers, is in NFPA 409, Chapter 3.*

513.1 Scope. This article shall apply to buildings or structures in any part of which aircraft containing Class I (flammable) liquids or Class II (combustible) liquids whose temperatures are above their flash points are housed or stored and in which aircraft might undergo service, repairs, or alterations. It shall not apply to locations used exclusively for aircraft that have never contained fuel or unfueled aircraft.

FPN No. 1: For definitions of aircraft hangar and unfueled aircraft, see NFPA 409-2004, *Standard on Aircraft Hangars*.

FPN No. 2: For further information on fuel classification see NFPA 30-2003, *Flammable and Combustible Liquids Code*.

513.2 Definitions. For the purpose of this article, the following definitions shall apply.

Mobile Equipment. Equipment with electric components suitable to be moved only with mechanical aids or is provided with wheels for movement by person(s) or powered devices.

Portable Equipment. Equipment with electric components suitable to be moved by a single person without mechanical aids.

513.3 Classification of Locations.

(A) Below Floor Level. Any pit or depression below the level of the hangar floor shall be classified as a Class I, Division 1 or Zone 1 location that shall extend up to said floor level.

(B) Areas Not Cut Off or Ventilated. The entire area of the hangar, including any adjacent and communicating areas not suitably cut off from the hangar, shall be classified as a Class I, Division 2 or Zone 2 location up to a level 450 mm (18 in.) above the floor.

(C) Vicinity of Aircraft. The area within 1.5 m (5 ft) horizontally from aircraft power plants or aircraft fuel tanks shall be classified as a Class I, Division 2 or Zone 2 location that shall extend upward from the floor to a level 1.5 m (5 ft) above the upper surface of wings and of engine enclosures.

(D) Areas Suitably Cut Off and Ventilated. Adjacent areas in which flammable liquids or vapors are not likely to be released, such as stock rooms, electrical control rooms, and other similar locations, shall not be classified where adequately ventilated and where effectively cut off from the hangar itself by walls or partitions.

513.4 Wiring and Equipment in Class I Locations.

(A) General. All wiring and equipment that is or may be installed or operated within any of the Class I locations defined in 513.3 shall comply with the applicable provisions of Article 501 or Article 505 for the division or zone in which they are used.

Attachment plugs and receptacles in Class I locations shall be identified for Class I locations or shall be designed such that they cannot be energized while the connections are being made or broken.

(B) Stanchions, Rostrums, and Docks. Electric wiring, outlets, and equipment (including lamps) on or attached to stanchions, rostrums, or docks that are located or likely to be located in a Class I location, as defined in 513.3(C), shall comply with the applicable provisions of Article 501 or Article 505 for the division or zone in which they are used.

513.7 Wiring and Equipment Not Installed in Class I Locations.

(A) Fixed Wiring. All fixed wiring in a hangar but not installed in a Class I location as classified in 513.3 shall be installed in metal raceways or shall be Type MI, TC, or MC cable.

Exception: Wiring in unclassified locations, as described in 513.3(D), shall be permitted to be any suitable type wiring method recognized in Chapter 3.

(B) Pendants. For pendants, flexible cord suitable for the type of service and identified for hard usage or extra-hard usage shall be used. Each such cord shall include a separate equipment grounding conductor.

(C) Arcing Equipment. In locations above those described in 513.3, equipment that is less than 3.0 m (10 ft) above wings and engine enclosures of aircraft and that may produce arcs, sparks, or particles of hot metal, such as lamps and lampholders for fixed lighting, cutouts, switches, receptacles, charging panels, generators, motors, or other equipment having make-and-break or sliding contacts, shall be of the totally enclosed type or constructed so as to prevent the escape of sparks or hot metal particles.

Exception: Equipment in areas described in 513.3(D) shall be permitted to be of the general-purpose type.

(D) Lampholders. Lampholders of metal-shell, fiber-lined types shall not be used for fixed incandescent lighting.

(E) Stanchions, Rostrums, or Docks. Where stanchions, rostrums, or docks are not located or likely to be located in a Class I location, as defined in 513.3(C), wiring and equipment shall comply with 513.7, except that such wiring and equipment not more than 457 mm (18 in.) above the floor in any position shall comply with 513.4(B). Receptacles and attachment plugs shall be of a locking type that will not readily disconnect.

(F) Mobile Stanchions. Mobile stanchions with electric equipment complying with 513.7(E) shall carry at least one permanently affixed warning sign with the following words or equivalent:

WARNING
KEEP 5 FT CLEAR OF AIRCRAFT
ENGINES AND FUEL TANK AREAS

or

WARNING
KEEP 1.5 METERS CLEAR OF AIRCRAFT
ENGINES AND FUEL TANK AREAS

513.8 Underground Wiring.

(A) Wiring and Equipment Embedded, Under Slab, or Underground. All wiring installed in or under the hangar floor shall comply with the requirements for Class I, Division 1 locations. Where such wiring is located in vaults, pits, or ducts, adequate drainage shall be provided.

(B) Uninterrupted Raceways, Embedded, Under Slab, or Underground. Uninterrupted raceways that are embedded in a hangar floor or buried beneath the hangar floor shall be considered to be within the Class I location above the floor, regardless of the point at which the raceway descends below or rises above the floor.

513.9 Sealing. Seals shall be provided in accordance with 501.15 or 505.16, as applicable. Sealing requirements specified shall apply to horizontal as well as to vertical boundaries of the defined Class I locations.

513.10 Special Equipment.

(A) Aircraft Electrical Systems.

(1) De-energizing Aircraft Electrical Systems. Aircraft electrical systems shall be de-energized when the aircraft is stored in a hangar and, whenever possible, while the aircraft is undergoing maintenance.

(2) Aircraft Batteries. Aircraft batteries shall not be charged where installed in an aircraft located inside or partially inside a hangar.

(B) Aircraft Battery Charging and Equipment. Battery chargers and their control equipment shall not be located or operated within any of the Class I locations defined in 513.3 and shall preferably be located in a separate building or in an area such as defined in 513.3(D). Mobile chargers shall carry at least one permanently affixed warning sign with the following words or equivalent:

WARNING
KEEP 5 FT CLEAR OF AIRCRAFT
ENGINES AND FUEL TANK AREAS

or

WARNING
KEEP 1.5 METERS CLEAR OF AIRCRAFT
ENGINES AND FUEL TANK AREAS

Tables, racks, trays, and wiring shall not be located within a Class I location and, in addition, shall comply with Article 480.

(C) External Power Sources for Energizing Aircraft.

(1) Not Less Than 450 mm (18 in.) Above Floor. Aircraft energizers shall be designed and mounted such that all electric equipment and fixed wiring will be at least 450 mm (18 in.) above floor level and shall not be operated in a Class I location as defined in 513.3(C).

(2) Marking for Mobile Units. Mobile energizers shall carry at least one permanently affixed warning sign with the following words or equivalent:

WARNING
KEEP 5 FT CLEAR OF AIRCRAFT
ENGINES AND FUEL TANK AREAS

or

WARNING
KEEP 1.5 METERS CLEAR OF AIRCRAFT
ENGINES AND FUEL TANK AREAS

(3) Cords. Flexible cords for aircraft energizers and ground support equipment shall be identified for the type of service and extra-hard usage and shall include an equipment grounding conductor.

(D) Mobile Servicing Equipment with Electric Components.

(1) General. Mobile servicing equipment (such as vacuum cleaners, air compressors, air movers) having electric wiring and equipment not suitable for Class I, Division 2 or Zone 2 locations shall be so designed and mounted that all such fixed wiring and equipment will be at least 450 mm (18 in.) above the floor. Such mobile equipment shall not be operated within the Class I location defined in 513.3(C) and shall carry at least one permanently affixed warning sign with the following words or equivalent:

WARNING
KEEP 5 FT CLEAR OF AIRCRAFT ENGINES
AND FUEL TANK AREAS

or

WARNING
KEEP 1.5 METERS CLEAR OF AIRCRAFT ENGINES
AND FUEL TANK AREAS

(2) Cords and Connectors. Flexible cords for mobile equipment shall be suitable for the type of service and identified for extra-hard usage and shall include an equipment grounding conductor. Attachment plugs and receptacles shall be identified for the location in which they are installed and shall provide for connection of the equipment grounding conductor.

(3) Restricted Use. Equipment that is not identified as suitable for Class I, Division 2 locations shall not be operated in locations where maintenance operations likely to release flammable liquids or vapors are in progress.

(E) Portable Equipment.

(1) Portable Lighting Equipment. Portable lighting equipment that is used within a hangar shall be identified for the location in which they are used. For portable lamps, flexible cord suitable for the type of service and identified for extra-hard usage shall be used. Each such cord shall include a separate equipment grounding conductor.

(2) Portable Utilization Equipment. Portable utilization equipment that is or may be used within a hangar shall be of a type suitable for use in Class I, Division 2 or Zone 2 locations. For portable utilization equipment, flexible cord suitable for the type of service and approved for extra-hard usage shall be used. Each such cord shall include a separate equipment grounding conductor.

513.12 Ground-Fault Circuit-Interrupter Protection for Personnel. All 125-volt, 50/60 Hz, single phase, 15– and 20-ampere receptacles installed in areas where electrical diagnostic equipment, electrical hand tools, or portable lighting equipment are to be used shall have ground-fault circuit-interrupter protection for personnel.

513.16 Grounded and Grounding Requirements.

(A) General Grounding Requirements. All metal raceways, the metal armor or metallic sheath on cables, and all non–current-carrying metal parts of fixed or portable electrical equipment, regardless of voltage, shall be grounded as provided in Article 250. Grounding in Class I locations shall comply with 501.30 for Class I, Division 1 and 2 locations and 505.25 for Class I, Zone 0, 1, and 2 locations.

(B) Supplying Circuits with Grounded and Grounding Conductors in Class I Locations.

(1) Circuits Supplying Portable Equipment or Pendants. Where a circuit supplies portables or pendants and includes a grounded conductor as provided in Article 200, receptacles, attachment plugs, connectors, and similar devices shall be of the grounding type, and the grounded conductor of the flexible cord shall be connected to the screw shell of any lampholder or to the grounded terminal of any utilization equipment supplied.

(2) Approved Means. Approved means shall be provided for maintaining continuity of the grounding conductor between the fixed wiring system and the non–current-carrying metal portions of pendant luminaires (fixtures), portable lamps, and portable utilization equipment.

ARTICLE 514
Motor Fuel Dispensing Facilities

Editor's Note: *The applicable text of NFPA 30A-2003, Code for Motor Fuel Dispensing Facilities and Repair Garages, is in NFPA 30A, 3.3.1, 8.1, 8.3, 12.1, 12.4, 12.5, 6.7.1, and 6.7.2.*

FPN: Rules that are followed by a reference in brackets contain text that has been extracted from NFPA 30A-2003, *Code for Motor Fuel Dispensing Facilities and Repair Garages*. Only editorial changes were made to the extracted text to make it consistent with this *Code*.

514.1 Scope. This article shall apply to motor fuel dispensing facilities, marine/motor fuel dispensing facilities, motor fuel dispensing facilities located inside buildings, and fleet vehicle motor fuel dispensing facilities.

FPN: For further information regarding safeguards for motor fuel dispensing facilities, see NFPA 30A-2003, *Code for Motor Fuel Dispensing Facilities and Repair Garages.*

514.2 Definition.

Motor Fuel Dispensing Facility. That portion of a property where motor fuels are stored and dispensed from fixed equipment into the fuel tanks of motor vehicles or marine craft or into approved containers, including all equipment used in connection therewith. [NFPA 30A:3.3.11]

FPN: Refer to Articles 510 and 511 with respect to electric wiring and equipment for other areas used as lubritoriums, service rooms, repair rooms, offices, salesrooms, compressor rooms, and similar locations.

514.3 Classification of Locations.

(A) Unclassified Locations. Where the authority having jurisdiction can satisfactorily determine that flammable liquids having a flash point below 38°C (100°F), such as gasoline, will not be handled, such location shall not be required to be classified.

(B) Classified Locations.

(1) Class I Locations. Table 514.3(B)(1) shall be applied where Class I liquids are stored, handled, or dispensed and shall be used to delineate and classify motor fuel dispensing facilities and commercial garages as defined in Article 511. Table 515.3 shall be used for the purpose of delineating and classifying aboveground tanks. A Class I location shall not extend beyond an unpierced wall, roof, or other solid partition. [NFPA 30A:8.1, 8.3]

(2) Compressed Natural Gas, Liquefied Natural Gas, and Liquefied Petroleum Gas Areas. Table 514.3(B)(2) shall be used to delineate and classify areas where compressed natural gas (CNG), liquefied natural gas (LNG), or liquefied petroleum gas (LPG) are stored, handled, or dispensed. Where CNG or LNG dispensers are installed beneath a canopy or enclosure, either the canopy or the enclosure shall be designed to prevent accumulation or entrapment of ignitible vapors, or all electrical equipment installed beneath the canopy or enclosure shall be suitable for Class I, Division 2 hazardous (classified) locations. Dispensing devices for liquefied petroleum gas shall be located not less than 1.5 m (5 ft) from any dispensing device for Class I liquids. [NFPA 30A:12.1, 12.4, 12.5]

FPN No. 1: For information on area classification where liquefied petroleum gases are dispensed, see NFPA 58-2004, *Liquefied Petroleum Gas Code.*

FPN No. 2: For information on classified areas pertaining to LP-Gas systems other than residential or commercial, see NFPA 58-2004, *Liquefied Petroleum Gas Code,* and NFPA 59-2004, *Utility LP-Gas Plant Code.*

FPN No. 3: See 555.21 for motor fuel dispensing stations in marinas and boatyards.

514.4 Wiring and Equipment Installed in Class I Locations. All electrical equipment and wiring installed in Class I locations as classified in 514.3 shall comply with the applicable provisions of Article 501.

Exception: As permitted in 514.8.

FPN: For special requirements for conductor insulation, see 501.20.

514.7 Wiring and Equipment Above Class I Locations. Wiring and equipment above the Class I locations as classified in 514.3 shall comply with 511.7.

514.8 Underground Wiring. Underground wiring shall be installed in threaded rigid metal conduit or threaded steel intermediate metal conduit. Any portion of electrical wiring that is below the surface of a Class I, Division 1, or a Class I, Division 2, location [as classified in Table 514.3(B)(1) and Table 514.3(B)(2)] shall be sealed within 3.05 m (10 ft) of the point of emergence above grade. Except for listed explosionproof reducers at the conduit seal, there shall be no union, coupling, box, or fitting between the conduit seal and the point of emergence above grade. Refer to Table 300.5.

Exception No. 1: Type MI cable shall be permitted where it is installed in accordance with Article 332.

Exception No. 2: Rigid nonmetallic conduit shall be permitted where buried under not less than 600 mm (2 ft) of cover. Where rigid nonmetallic conduit is used, threaded rigid metal conduit or threaded steel intermediate metal conduit shall be used for the last 600 mm (2 ft) of the underground run to emergence or to the point of connection to the aboveground raceway, and an equipment grounding conductor shall be included to provide electrical continuity of the raceway system and for grounding of non–current-carrying metal parts.

514.9 Sealing.

(A) At Dispenser. A listed seal shall be provided in each conduit run entering or leaving a dispenser or any cavities

Table 514.3(B)(1) Class I Locations — Motor Fuel Dispensing Facilities

Location	Class I, Group D Division	Extent of Classified Location
Underground Tank		
Fill opening	1	Any pit or box below grade level, any part of which is within the Division 1 or Division 2, Zone 1 or Zone 2 classified location
	2	Up to 450 mm (18 in.) above grade level within a horizontal radius of 3.0 m (10 ft) from a loose fill connection and within a horizontal radius of 1.5 m (5 ft) from a tight fill connection
Vent — discharging upward	1	Within 900 mm (3 ft) of open end of vent, extending in all directions
	2	Space between 900 mm (3 ft) and 1.5 m (5 ft) of open end of vent, extending in all directions
Dispensing Device[2,5] (except overhead type)[3]		
Pits	1	Any pit or box below grade level, any part of which is within the Division 1 or Division 2, Zone 1 or Zone 2 classified location
Dispenser		FPN: Space classification inside the dispenser enclosure is covered in ANSI/UL 87-1995, *Power Operated Dispensing Devices for Petroleum Products.*
	2	Within 450 mm (18 in.) horizontally in all directions extending to grade from the dispenser enclosure or that portion of the dispenser enclosure containing liquid-handling components FPN: Space classification inside the dispenser enclosure is covered in ANSI/UL 87-1995, *Power Operated Dispensing Devices for Petroleum Products.*
Outdoor	2	Up to 450 mm (18 in.) above grade level within 6.0 m (20 ft) horizontally of any edge of enclosure.
Indoor with mechanical ventilation	2	Up to 450 mm (18 in.) above grade or floor level within 6.0 m (20 ft) horizontally of any edge of enclosure
with gravity ventilation	2	Up to 450 mm (18 in.) above grade or floor level within 7.5 m (25 ft) horizontally of any edge of enclosure
Dispensing Device[5]		
Overhead type[3]	1	The space within the dispenser enclosure, and all electrical equipment integral with the dispensing hose or nozzle
	2	A space extending 450 mm (18 in.) horizontally in all directions beyond the enclosure and extending to grade
	2	Up to 450 mm (18 in.) above grade level within 6.0 m (20 ft) horizontally measured from a point vertically below the edge of any dispenser enclosure
Remote Pump — Outdoor	1	Any pit or box below grade level if any part is within a horizontal distance of 3.0 m (10 ft) from any edge of pump
	2	Within 900 mm (3 ft) of any edge of pump, extending in all directions. Also up to 450 mm (18 in.) above grade level within 3.0 m (10 ft) horizontally from any edge of pump
Remote Pump — Indoor	1	Entire space within any pit
	2	Within 1.5 m (5 ft) of any edge of pump, extending in all directions. Also up to 900 mm (3 ft) above grade level within 7.5 m (25 ft) horizontally from any edge of pump

Table 514.3(B)(1) Continued

Location	Class I, Group D Division	Extent of Classified Location
Lubrication or Service Room — Without Dispensing	2	Entire area within any pit used for lubrication or similar services where Class I liquids may be released
	2	Area up to 450 mm (18 in.) above any such pit and extending a distance of 900 mm (3 ft) horizontally from any edge of the pit
	2	Entire unventilated area within any pit, belowgrade area, or subfloor area
	2	Area up to 450 mm (18 in.) above any such unventilated pit, belowgrade work area, or subfloor work area and extending a distance of 900 mm (3 ft) horizontally from the edge of any such pit, belowgrade work area, or subfloor work area
	Unclassified	Any pit, belowgrade work area, or subfloor work area that is provided with exhaust ventilation at a rate of not less than 0.3 $m^3/min/m^2$ (1 cfm/ft^2) of floor area at all times that the building is occupied or when vehicles are parked in or over this area and where exhaust air is taken from a point within 300 mm (12 in.) of the floor of the pit, belowgrade work area, or subfloor work area
Special Enclosure Inside Building[4]	1	Entire enclosure
Sales, Storage, and Rest Rooms	Unclassified	If there is any opening to these rooms within the extent of a Division 1 location, the entire room shall be classified as Division 1
Vapor Processing Systems Pits	1	Any pit or box below grade level, any part of which is within a Division 1 or Division 2 classified location or that houses any equipment used to transport or process vapors
Vapor Processing Equipment Located Within Protective Enclosures FPN: See 10.1.7 of NFPA 30A-2003, *Code for Motor Fuel Dispensing Facilities and Repair Garages.*	2	Within any protective enclosure housing vapor processing equipment
Vapor Processing Equipment Not Within Protective Enclosures (excluding piping and combustion devices)	2	The space within 450 mm (18 in.) in all directions of equipment containing flammable vapor or liquid extending to grade level. Up to 450 mm (18 in.) above grade level within 3.0 m (10 ft) horizontally of the vapor processing equipment
Equipment Enclosures	1	Any space within the enclosure where vapor or liquid is present under normal operating conditions
Vacuum-Assist Blowers	2	The space within 450 mm (18 in.) in all directions extending to grade level. Up to 450 mm (18 in.) above grade level within 3.0 m (10 ft) horizontally

[1]For marine application, *grade level* means the surface of a pier extending down to water level.
[2]Refer to Figure 514.3 for an illustration of classified location around dispensing devices.
[3]Ceiling mounted hose reel.
[4]FPN: See 4.3.9 of NFPA 30A-2003, *Code for Motor Fuel Dispensing Facilities and Repair Garages.*
[5]FPN: Area classification inside the dispenser enclosure is covered in ANSI/UL 87-1995, *Power-Operated Dispensing Devices for Petroleum Products.* [NFPA 30A:Table 8.3.1]

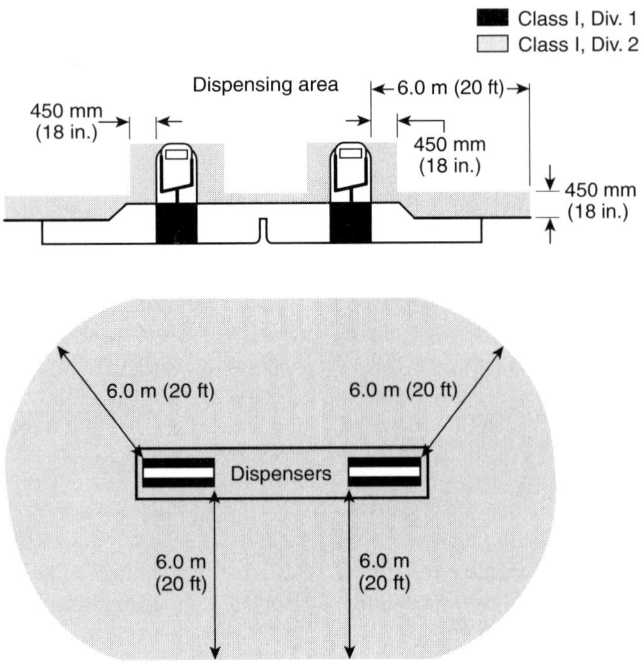

Figure 514.3 Classified Areas Adjacent to Dispensers as Detailed in Table 514.3(B)(1). [NFPA 30A:Figure 8.3.1]

or enclosures in direct communication therewith. The sealing fitting shall be the first fitting after the conduit emerges from the earth or concrete.

(B) At Boundary. Additional seals shall be provided in accordance with 501.15. Sections 501.15(A)(4) and (B)(2) shall apply to horizontal as well as to vertical boundaries of the defined Class I locations.

514.11 Circuit Disconnects.

(A) General. Each circuit leading to or through dispensing equipment, including equipment for remote pumping systems, shall be provided with a clearly identified and readily accessible switch or other acceptable means, located remote from the dispensing devices, to disconnect simultaneously from the source of supply, all conductors of the circuits, including the grounded conductor, if any.

Single-pole breakers utilizing handle ties shall not be permitted.

(B) Attended Self-Service Motor Fuel Dispensing Facilities. Emergency controls as specified in 514.11(A) shall be installed at a location acceptable to the authority having jurisdiction, but controls shall not be more than 30 m (100 ft) from dispensers. [NFPA 30A:6.7.1]

(C) Unattended Self-Service Motor Fuel Dispensing Facilities. Emergency controls as specified in 514.11(A) shall be installed at a location acceptable to the authority having jurisdiction, but the control shall be more than 6 m (20 ft) but less than 30 m (100 ft) from the dispensers. Additional emergency controls shall be installed on each group of dispensers or the outdoor equipment used to control the dispensers. Emergency controls shall shut off all power to all dispensing equipment at the station. Controls shall be manually reset only in a manner approved by the authority having jurisdiction. [NFPA 30A:6.7.2]

FPN: For additional information, see 6.7.1 and 6.7.2 of NFPA 30A-2003, *Code for Motor Fuel Dispensing Facilities and Repair Garages.*

514.13 Provisions for Maintenance and Service of Dispensing Equipment. Each dispensing device shall be provided with a means to remove all external voltage sources, including feedback, during periods of maintenance and service of the dispensing equipment. The location of this means shall be permitted to be other than inside or adjacent to the dispensing device. The means shall be capable of being locked in the open position.

Table 514.3(B)(2) Electrical Equipment Classified Areas for Dispensing Devices

	Extent of Classified Area	
Dispensing Device	**Class I, Division 1**	**Class I, Division 2**
Compressed natural gas	Entire space within the dispenser enclosure	1.5 m (5 ft) in all directions from dispenser enclosure
Liquefied natural gas	Entire space within the dispenser enclosure and 1.5 m (5 ft) in all directions from the dispenser enclosure	From 1.5 m to 3.0 m (5 ft to 10 ft) in all directions from the dispenser enclosure
Liquefied petroleum gas	Entire space within the dispenser enclosure; 450 mm (18 in.) from the exterior surface of the dispenser enclosure to an elevation of 1.2 m (4 ft) above the base of the dispenser; the entire pit or open space beneath the dispenser and within 6.0 m (20 ft) horizontally from any edge of the dispenser when the pit or trench is not mechanically ventilated.	Up to 450 mm (18 in.) aboveground and within 6.0 m (20 ft) horizontally from any edge of the dispenser enclosure, including pits or trenches within this area when provided with adequate mechanical ventilation

[NFPA 30A:Table 12.6.2]

514.16 Grounding. All metal raceways, the metal armor or metallic sheath on cables, and all non–current-carrying metal parts of fixed portable electrical equipment, regardless of voltage, shall be grounded as provided in Article 250. Grounding in Class I locations shall comply with 501.130.

ARTICLE 515
Bulk Storage Plants

Editor's Note: *The applicable text of NFPA 30-2003, Flammable and Combustible Liquids Code, is included in NFPA 30, 3.3.32.1, 8.1, 8.2.2, and Chapter 5.*

FPN: Rules that are followed by a reference in brackets contain text that has been extracted from NFPA 30-2003, *Flammable and Combustible Liquids Code*. Only editorial changes were made to the extracted text to make it consistent with this *Code*.

515.1 Scope. This article covers a property or portion of a property where flammable liquids are received by tank vessel, pipelines, tank car, or tank vehicle and are stored or blended in bulk for the purpose of distributing such liquids by tank vessel, pipeline, tank car, tank vehicle, portable tank, or container.

515.2 Definition.

Bulk Plant or Terminal. That portion of a property where liquids are received by tank vessel, pipelines, tank car, or tank vehicle and are stored or blended in bulk for the purpose of distributing such liquids by tank vessel, pipeline, tank car, tank vehicle, portable tank, or container. [NFPA 30:3.3.32.1]

FPN: For further information, see NFPA 30-2003, *Flammable and Combustible Liquids Code*.

515.3 Class I Locations. Table 515.3 shall be applied where Class I liquids are stored, handled, or dispensed and shall be used to delineate and classify bulk storage plants. The class location shall not extend beyond a floor, wall, roof, or other solid partition that has no communicating openings. [NFPA 30:8.1, 8.2.2]

FPN No. 1: The area classifications listed in Table 515.3 are based on the premise that the installation meets the applicable requirements of NFPA 30-2003, *Flammable and Combustible Liquids Code*, Chapter 5, in all respects. Should this not be the case, the authority having jurisdiction has the authority to classify the extent of the classified space.

FPN No. 2: See 555.21 for gasoline dispensing stations in marinas and boatyards.

515.4 Wiring and Equipment Located in Class I Locations. All electrical wiring and equipment within the Class I locations defined in 515.3 shall comply with the applicable provisions of Article 501 or Article 505 for the division or zone in which they are used.

Exception: As permitted in 515.8.

515.7 Wiring and Equipment Above Class I Locations.

(A) Fixed Wiring. All fixed wiring above Class I locations shall be in metal raceways or PVC Schedule 80 rigid nonmetallic conduit, or equivalent, or be Type MI, TC, or MC cable.

(B) Fixed Equipment. Fixed equipment that may produce arcs, sparks, or particles of hot metal, such as lamps and lampholders for fixed lighting, cutouts, switches, receptacles, motors, or other equipment having make-and-break or sliding contacts, shall be of the totally enclosed type or be constructed so as to prevent the escape of sparks or hot metal particles.

Table 515.3 Electrical Area Classifications

Location	NEC Class I Division	Zone	Extent of Classified Area
Indoor equipment installed in accordance with Section 5.3 of NFPA 30 where flammable vapor–air mixtures can exist under normal operation	1	0	The entire area associated with such equipment where flammable gases or vapors are present continuously or for long periods of time
	1	1	Area within 1.5 m (5 ft) of any edge of such equipment, extending in all directions
	2	2	Area between 1.5 m and 2.5 m (5 ft and 8 ft) of any edge of such equipment, extending in all directions; also, space up to 900 mm (3 ft) above floor or grade level within 1.5 m to 7.5 m (5 ft to 25 ft) horizontally from any edge of such equipment[1]

continues

Table 515.3 Continued

Location	NEC Class I Division	Zone	Extent of Classified Area
Outdoor equipment of the type covered in Section 5.3 of NFPA 30 where flammable vapor–air mixtures may exist under normal operation	1	0	The entire area associated with such equipment where flammable gases or vapors are present continuously or for long periods of time
	1	1	Area within 900 mm (3 ft) of any edge of such equipment, extending in all directions
	2	2	Area between 900 mm (3 ft) and 2.5 m (8 ft) of any edge of such equipment, extending in all directions; also, space up to 900 mm (3 ft) above floor or grade level within 900 mm to 3.0 m (3 ft to 10 ft) horizontally from any edge of such equipment
Tank storage installations inside buildings	1	1	All equipment located below grade level
	2	2	Any equipment located at or above grade level
Tank – aboveground	1	0	Inside fixed roof tank
	1	1	Area inside dike where dike height is greater than the distance from the tank to the dike for more than 50 percent of the tank circumference
Shell, ends, or roof and dike area	2	2	Within 3.0 m (10 ft) from shell, ends, or roof of tank; also, area inside dike to level of top of tank
Vent	1	0	Area inside of vent piping or opening
	1	1	Within 1.5 m (5 ft) of open end of vent, extending in all directions
	2	2	Area between 1.5 m and 3.0 m (5 ft and 10 ft) from open end of vent, extending in all directions
Floating roof with fixed outer roof	1	0	Area between the floating and fixed roof sections and within the shell
Floating roof with no fixed outer roof	1	1	Area above the floating roof and within the shell
Underground tank fill opening	1	1	Any pit, or space below grade level, if any part is within a Division 1 or 2, or Zone 1 or 2, classified location
	2	2	Up to 450 mm (18 in.) above grade level within a horizontal radius of 3.0 m (10 ft) from a loose fill connection, and within a horizontal radius of 1.5 m (5 ft) from a tight fill connection
Vent – discharging upward	1	0	Area inside of vent piping or opening
	1	1	Within 900 mm (3 ft) of open end of vent, extending in all directions
	2	2	Area between 900 mm and 1.5 m (3 ft and 5 ft) of open end of vent, extending in all directions
Drum and container filling – outdoors or indoors	1	0	Area inside the drum or container
	1	1	Within 900 mm (3 ft) of vent and fill openings, extending in all directions
	2	2	Area between 900 mm and 1.5 m (3 ft and 5 ft) from vent or fill opening, extending in all directions; also, up to 450 mm (18 in.) above floor or grade level within a horizontal radius of 3.0 m (10 ft) from vent or fill opening
Pumps, bleeders, withdrawal fittings, Indoors	2	2	Within 1.5 m (5 ft) of any edge of such devices, extending in all directions; also, up to 900 mm (3 ft) above floor or grade level within 7.5 m (25 ft) horizontally from any edge of such devices
Outdoors	2	2	Within 900 mm (3 ft) of any edge of such devices, extending in all directions. Also, up to 450 mm (18 in.) above grade level within 3.0 m (10 ft) horizontally from any edge of such devices

Table 515.3 Continued

Location	NEC Class I Division	Zone	Extent of Classified Area
Pits and sumps			
Without mechanical ventilation	1	1	Entire area within a pit or sump if any part is within a Division 1 or 2, or Zone 1 or 2, classified location
With adequate mechanical ventilation	2	2	Entire area within a pit or sump if any part is within a Division 1 or 2, or Zone 1 or 2, classified location
Containing valves, fittings, or piping, and not within a Division 1 or 2, or Zone 1 or 2, classified location	2	2	Entire pit or sump
Drainage ditches, separators, impounding basins			
Outdoors	2	2	Area up to 450 mm (18 in.) above ditch, separator, or basin; also, area up to 450 mm (18 in.) above grade within 4.5 m (15 ft) horizontally from any edge
Indoors			Same classified area as pits
Tank vehicle and tank car[2] loading through open dome	1	0	Area inside of the tank
	1	1	Within 900 mm (3 ft) of edge of dome, extending in all directions
	2	2	Area between 900 mm and 4.5 m (3 ft and 15 ft) from edge of dome, extending in all directions
Loading through bottom connections with atmospheric venting	1	0	Area inside of the tank
	1	1	Within 900 mm (3 ft) of point of venting to atmosphere, extending in all directions
	2	2	Area between 900 mm and 4.5 m (3 ft and 15 ft) from point of venting to atmosphere, extending in all directions; also, up to 450 mm (18 in.) above grade within a horizontal radius of 3.0 m (10 ft) from point of loading connection
Office and rest rooms	Ordinary		If there is any opening to these rooms within the extent of an indoor classified location, the room shall be classified the same as if the wall, curb, or partition did not exist.
Loading through closed dome with atmospheric venting	1	1	Within 900 mm (3 ft) of open end of vent, extending in all directions
	2	2	Area between 900 mm and 4.5 m (3 ft and 15 ft) from open end of vent, extending in all directions; also, within 900 mm (3 ft) of edge of dome, extending in all directions
Loading through closed dome with vapor control	2	2	Within 900 mm (3 ft) of point of connection of both fill and vapor lines extending in all directions
Bottom loading with vapor control or any bottom unloading	2	2	Within 900 mm (3 ft) of point of connections, extending in all directions; also up to 450 mm (18 in.) above grade within a horizontal radius of 3.0 m (10 ft) from point of connections
Storage and repair garage for tank vehicles	1	1	All pits or spaces below floor level
	2	2	Area up to 450 mm (18 in.) above floor or grade level for entire storage or repair garage
Garages for other than tank vehicles	Ordinary		If there is any opening to these rooms within the extent of an outdoor classified location, the entire room shall be classified the same as the area classification at the point of the opening.

continues

Table 515.3 Continued

Location	NEC Class I Division	Zone	Extent of Classified Area
Outdoor drum storage	Ordinary		
Inside rooms or storage lockers used for the storage of Class I liquids	2	2	Entire room
Indoor warehousing where there is no flammable liquid transfer	Ordinary		If there is any opening to these rooms within the extent of an indoor classified location, the room shall be classified the same as if the wall, curb, or partition did not exist.
Piers and wharves			See Figure 515.3.

[1]The release of Class I liquids may generate vapors to the extent that the entire building, and possibly an area surrounding it, should be considered a Class I, Division 2 or Zone 2 location.
[2]When classifying extent of area, consideration shall be given to fact that tank cars or tank vehicles may be spotted at varying points. Therefore, the extremities of the loading or unloading positions shall be used. [NFPA 30:Table 8.2.2]

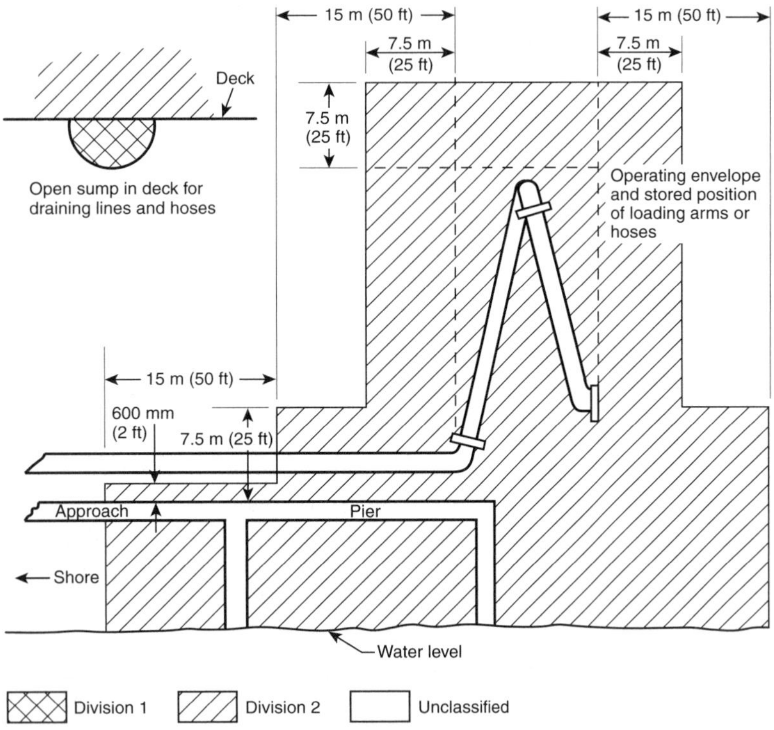

Notes:
(1) The "source of vapor" shall be the operating envelope and stored position of the outboard flange connection of the loading arm (or hose).
(2) The berth area adjacent to tanker and barge cargo tanks is to be Division 2 to the following extent:
a. 7.6 m (25 ft) horizontally in all directions on the pier side from that portion of the hull containing cargo tanks
b. From the water level to 7.6 m (25 ft) above the cargo tanks at their highest position
(3) Additional locations may have to be classified as required by the presence of other sources of flammable liquids on the berth, by Coast Guard, or other regulations.

Figure 515.3 Marine Terminal Handling Flammable Liquids. [NFPA 30:Figure 7.7.16]

(C) Portable Lamps or Other Utilization Equipment. Portable lamps or other utilization equipment and their flexible cords shall comply with the provisions of Article 501 or Article 505 for the class of location above which they are connected or used.

515.8 Underground Wiring.

(A) Wiring Method. Underground wiring shall be installed in threaded rigid metal conduit or threaded steel intermediate metal conduit or, where buried under not less than 600 mm (2 ft) of cover, shall be permitted in rigid nonmetallic conduit or a listed cable. Where rigid nonmetallic conduit is used, threaded rigid metal conduit or threaded steel intermediate metal conduit shall be used for the last 600 mm (2 ft) of the conduit run to emergence or to the point of connection to the aboveground raceway. Where cable is used, it shall be enclosed in threaded rigid metal conduit or threaded steel intermediate metal conduit from the point of lowest buried cable level to the point of connection to the aboveground raceway.

(B) Insulation. Conductor insulation shall comply with 501.20.

(C) Nonmetallic Wiring. Where rigid nonmetallic conduit or cable with a nonmetallic sheath is used, an equipment grounding conductor shall be included to provide for electrical continuity of the raceway system and for grounding of non–current-carrying metal parts.

515.9 Sealing. Sealing requirements shall apply to horizontal as well as to vertical boundaries of the defined Class I locations. Buried raceways and cables under defined Class I locations shall be considered to be within a Class I, Division 1 or Zone 1 location.

515.10 Special Equipment — Gasoline Dispensers. Where gasoline or other volatile flammable liquids or liquefied flammable gases are dispensed at bulk stations, the applicable provisions of Article 514 shall apply.

515.16 Grounding. All metal raceways, the metal armor or metallic sheath on cables, and all non–current-carrying metal parts of fixed or portable electrical equipment, regardless of voltage, shall be grounded as provided in Article 250. Grounding in Class I locations shall comply with 501.30 for Class I, Division 1 and 2 locations and 505.25 for Class I, Zone 0, 1, and 2 locations.

FPN: For information on grounding for static protection, see 4.5.3.4 and 4.5.3.5 of NFPA 30-2003, *Flammable and Combustible Liquids Code.*

ARTICLE 516
Spray Application, Dipping, and Coating Processes

***Editor's Note:** The applicable text from NFPA 33-2003, Standard for Spray Application Using Flammable and Combustible Materials; and NFPA 34-2003, Standard for Dipping and Coating Processes Using Flammable and Combustible Liquids, is included in NFPA 33, Chapters 11, 12, and 15; and NFPA 34, Sections 6.4 and 6.5.*

FPN: Rules that are followed by a reference in brackets contain text that has been extracted from NFPA 33-2003, *Standard for Spray Application Using Flammable and Combustible Materials*, or NFPA 34-2003, *Standard for Dipping and Coating Processes Using Flammable or Combustible Liquids*. Only editorial changes were made to the extracted text to make it consistent with this *Code*.

516.1 Scope. This article covers the regular or frequent application of flammable liquids, combustible liquids, and combustible powders by spray operations and the application of flammable liquids, or combustible liquids at temperatures above their flashpoint, by dipping, coating, or other means.

FPN: For further information regarding safeguards for these processes, such as fire protection, posting of warning signs, and maintenance, see NFPA 33-2003, *Standard for Spray Application Using Flammable and Combustible Materials*, and NFPA 34-2003, *Standard for Dipping and Coating Processes Using Flammable or Combustible Liquids*. For additional information regarding ventilation, see NFPA 91-2004, *Standard for Exhaust Systems for Air Conveying of Vapors, Gases, Mists, and Noncombustible Particulate Solids*.

516.2 Definitions. For the purpose of this article, the following definitions shall apply.

Spray Area. Normally, locations outside of buildings or localized operations within a larger room or space. Such are normally provided with some local vapor extraction/ventilation system. In automated operations, the area limits shall be the maximum area in the direct path of spray operations. In manual operations, the area limits shall be the maximum area of spray when aimed at 180 degrees to the application surface.

Spray Booth. An enclosure or insert within a larger room used for spray/coating/dipping applications. A spray booth may be fully enclosed or have open front or face and may include separate conveyor entrance and exit. The spray booth is provided with a dedicated ventilation exhaust but may draw supply air from the larger room or have a dedicated air supply.

Spray Room. A purposefully enclosed room built for spray/coating/dipping applications provided with dedicated ventilation supply and exhaust. Normally the room is configured to house the item to be painted, providing reasonable access around the item/process. Depending on the size of the item being painted, such rooms may actually be the entire building or the major portion thereof.

516.3 Classification of Locations. Classification is based on dangerous quantities of flammable vapors, combustible mists, residues, dusts, or deposits.

(A) Class I, Division 1 or Class I, Zone 0 Locations. The following spaces shall be considered Class I, Division 1, or Class I, Zone 0, as applicable:

(1) The interior of any open or closed container of a flammable liquid
(2) The interior of any dip tank or coating tank

FPN: For additional guidance and explanatory diagrams, see 4.3.5 of NFPA 33-2003, *Standard for Spray Application Using Flammable or Combustible Materials*, and Sections 4.2, 4.3, and 4.4 of NFPA 34-2003, *Standard for Dipping and Coating Processes Using Flammable or Combustible Liquids*.

(B) Class I or Class II, Division 1 Locations. The following spaces shall be considered Class I, Division 1, or Class I, Zone 1, or Class II, Division 1 locations, as applicable:

(1) The interior of spray booths and rooms except as specifically provided in 516.3(D).
(2) The interior of exhaust ducts.
(3) Any area in the direct path of spray operations.
(4) For open dipping and coating operations, all space within a 1.5-m (5-ft) radial distance from the vapor sources extending from these surfaces to the floor. The vapor source shall be the liquid exposed in the process and the drainboard, and any dipped or coated object from which it is possible to measure vapor concentrations exceeding 25 percent of the lower flammable limit at a distance of 300 mm (1 ft), in any direction, from the object.
(5) Sumps, pits, or belowgrade channels within 7.5 m (25 ft) horizontally of a vapor source. If the sump, pit, or channel extends beyond 7.5 m (25 ft) from the vapor source, it shall be provided with a vapor stop or it shall be classified as Class I, Division 1 for its entire length.
(6) All space in all directions outside of but within 900 mm (3 ft) of open containers, supply containers, spray gun cleaners, and solvent distillation units containing flammable liquids.

(C) Class I or Class II, Division 2 Locations. The following spaces shall be considered Class I, Division 2 or Class I, Zone 2, or Class II, Division 2 as applicable.

(1) Open Spraying. For open spraying, all space outside of but within 6 m (20 ft) horizontally and 3 m (10 ft) vertically of the Class I, Division 1 or Class I, Zone 1 location as defined in 516.3(A), and not separated from it by partitions. See Figure 516.3(B)(1). [NFPA 33:6.5.1]

(2) Closed-Top, Open-Face, and Open-Front Spraying. If spray application operations are conducted within a closed-top, open-face, or open-front booth or room, any electrical wiring or utilization equipment located outside of the booth or room but within the boundaries designated as Division 2 or Zone 2 in Figure 516.3(B)(2) shall be suitable for Class I, Division 2, Class I, Zone 2, or Class II, Division 2 locations, whichever is applicable. The Class I, Division 2, Class I, Zone 2, or Class II, Division 2 locations shown in Figure 516.3(B)(2) shall extend from the edges of the open face or open front of the booth or room in accordance with the following:

(a) If the exhaust ventilation system is interlocked with the spray application equipment, the Division 2 or Zone 2 location shall extend 1.5 m (5 ft) horizontally and 900 mm (3 ft) vertically from the open face or open front

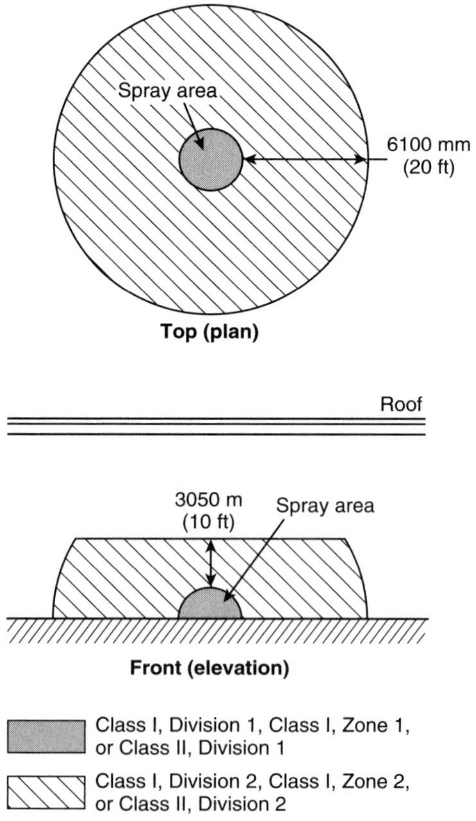

Figure 516.3(B)(1) Electrical Area Classification for Open Spray Areas. [NFPA 33:Figure 6.5.1]

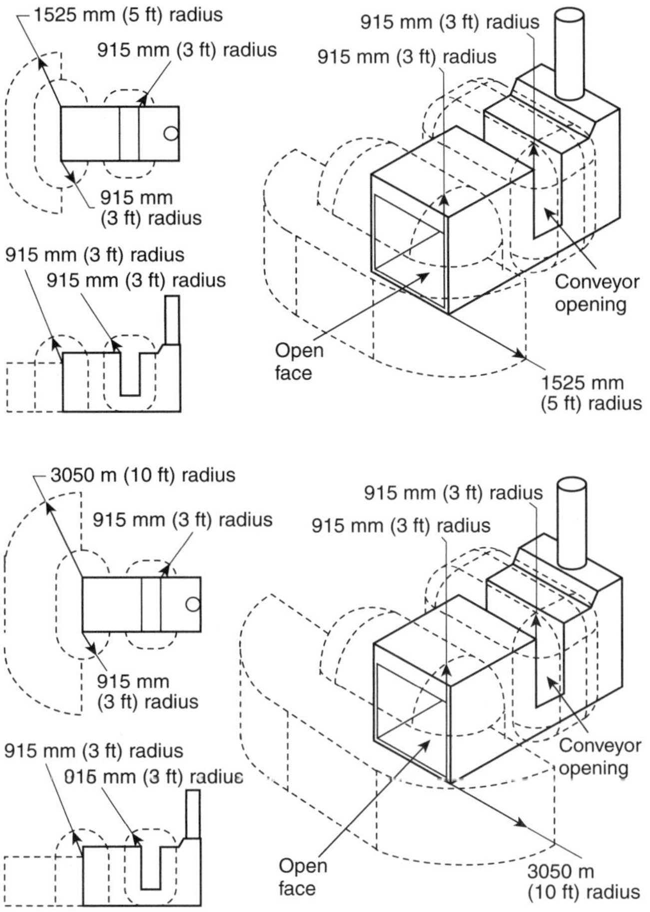

Figure 516.3(B)(2) Class I, Division 2, Class I, Zone 2, or Class II, Division 2 Locations Adjacent to a Closed Top, Open Face, or Open Front Spray Booth or Room. [NFPA 33:Figures 6.5.2(a) and 6.5.2(b)]

of the booth or room, as shown in Figure 516.3(B)(2), top.

(b) If the exhaust ventilation system is not interlocked with the spray application equipment, the Division 2 or Zone 2 location shall extend 3 m (10 ft) horizontally and 900 mm (3 ft) vertically from the open face or open front of the booth or room, as shown in Figure 516.3(B)(2), bottom.

For the purposes of this subsection, *interlocked* shall mean that the spray application equipment cannot be operated unless the exhaust ventilation system is operating and functioning properly and spray application is automatically stopped if the exhaust ventilation system fails. [NFPA 33:6.5.2]

(3) Open-Top Spraying. For spraying operations conducted within an open top spray booth, the space 900 mm (3 ft) vertically above the booth and within 900 mm (3 ft) of other booth openings shall be considered Class I, Division 2, Class I, Zone 2, or Class II, Division 2. [NFPA 33:6.5.3]

(4) Enclosed Booths and Rooms. For spraying operations confined to an enclosed spray booth or room, the space within 900 mm (3 ft) in all directions from any openings shall be considered Class I, Division 2, Class I, or Zone 2, or Class II, Division 2 as shown in Figure 516.3(B)(4). [NFPA 33:6.5.4]

(5) Dip Tanks and Drain Boards — Surrounding Space. For dip tanks and drain boards, the 914-mm (3-ft) space surrounding the Class I, Division 1 or Class I, Zone 1 location as defined in 516.3(A)(4) and as shown in Figure 516.3(B)(5). [NFPA 34:6.4.3]

(6) Dip Tanks and Drain Boards — Space Above Floor. For dip tanks and drain boards, the space 900 mm (3 ft) above the floor and extending 6 m (20 ft) horizontally in all directions from the Class I, Division 1 or Class I, Zone 1 location.

Exception: This space shall not be required to be considered a hazardous (classified) location where the vapor source area is 0.46 m^2 (5 ft^2) or less and where the contents of the open tank trough or container do not exceed 19 L (5 gal).

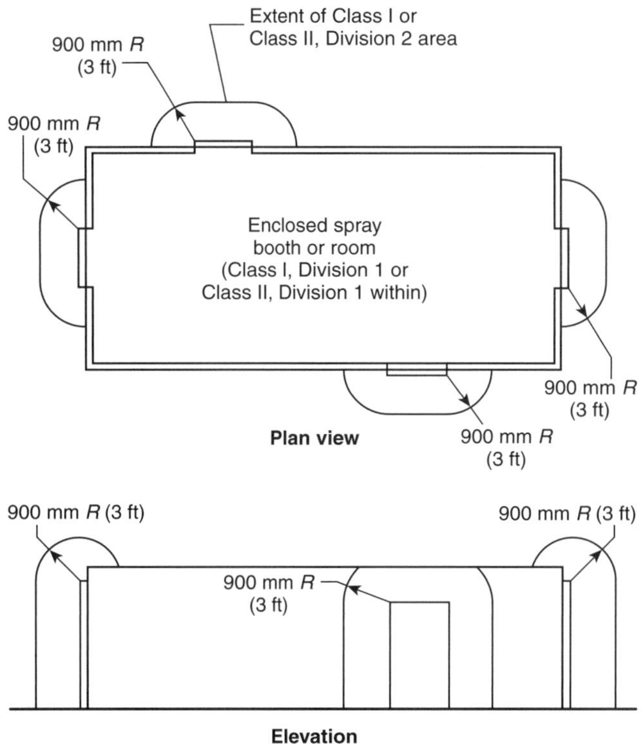

Figure 516.3(B)(4) Class I, Division 2, Class I, Zone 2, or Class II, Division 2 Locations Adjacent to an Enclosed Spray Booth or Spray Room. [NFPA 33:Figure 6.5.4]

In addition, the vapor concentration during operation and shutdown periods shall not exceed 25 percent of the lower flammable limit outside the Class I location specified in 516.3(A)(4). [NFPA 34:6.4.4]

(7) Open Containers. All space in all directions within 600 mm (2 ft) of the Division 1 or Zone 1 area surrounding open containers, supply containers, spray gun cleaners, and solvent distillation units containing flammable liquids, as well as the area extending 1.5 m (5 ft) beyond the Division 1 or Zone 1 area up to a height of 460 mm (18 in.) above the floor or grade level. [NFPA 33:6.5.5.1(2)]

(D) Enclosed Coating and Dipping Operations. The space adjacent to an enclosed dipping or coating process or apparatus shall be considered unclassified. [NFPA 34:6.5.2]

Exception: The space within 900 mm (3 ft) in all directions from any opening in the enclosures shall be classified as Class I, Division 2 or Class I, Zone 2, as applicable. [NFPA 34:6.5.3]

(E) Adjacent Locations. Adjacent locations that are cut off from the defined Class I or Class II locations by tight partitions without communicating openings, and within which flammable vapors or combustible powders are not likely to be released, shall be unclassified.

(F) Unclassified Locations. Locations using drying, curing, or fusion apparatus and provided with positive mechanical ventilation adequate to prevent accumulation of flammable concentrations of vapors, and provided with effective interlocks to de-energize all electrical equipment (other than equipment identified for Class I locations) in case the ventilating equipment is inoperative, shall be permitted to be unclassified where the authority having jurisdiction so judges.

FPN: For further information regarding safeguards, see NFPA 86-2003, *Standard for Ovens and Furnaces.*

516.4 Wiring and Equipment in Class I Locations.

(A) Wiring and Equipment — Vapors. All electric wiring and equipment within the Class I location (containing vapor only — not residues) defined in 516.3 shall comply with the applicable provisions of Article 501 or Article 505, as applicable.

(B) Wiring and Equipment — Vapors and Residues. Unless specifically listed for locations containing deposits of dangerous quantities of flammable or combustible vapors, mists, residues, dusts, or deposits (as applicable), there shall be no electrical equipment in any spray area as herein defined whereon deposits of combustible residue may readily accumulate, except wiring in rigid metal conduit, intermediate metal conduit, Type MI cable, or in metal boxes or fittings containing no taps, splices, or terminal connections. [NFPA 33:6.4.2]

(C) Illumination. Illumination of readily ignitible areas through panels of glass or other transparent or translucent material shall be permitted only if it complies with the following:

(1) Fixed lighting units are used as the source of illumination.

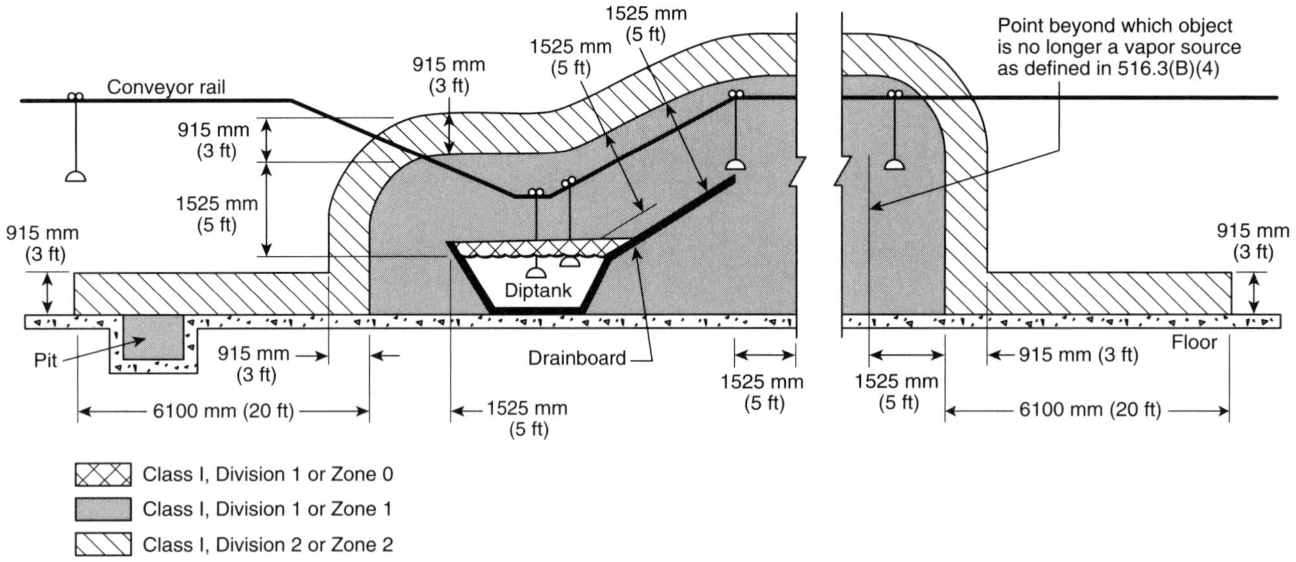

Figure 516.3(B)(5) Electrical Area Classification for Open Processes Without Vapor Containment or Ventilation. [NFPA 34:Figure 6.4(a)]

(2) The panel effectively isolates the Class I location from the area in which the lighting unit is located.
(3) The lighting unit is identified for its specific location.
(4) The panel is of a material or is protected so that breakage is unlikely.
(5) The arrangement is such that normal accumulations of hazardous residue on the surface of the panel will not be raised to a dangerous temperature by radiation or conduction from the source of illumination.

(D) Portable Equipment. Portable electric lamps or other utilization equipment shall not be used in a spray area during spray operations.

Exception No. 1: Where portable electric lamps are required for operations in spaces not readily illuminated by fixed lighting within the spraying area, they shall be of the type identified for Class I, Division 1 or Class 1, Zone 1 locations where readily ignitible residues may be present. [NFPA 33:6.9]

Exception No. 2: Where portable electric drying apparatus is used in automobile refinishing spray booths and the following requirements are met.

(a) The apparatus and its electrical connections are not located within the spray enclosure during spray operations.
(b) Electrical equipment within 450 mm (18 in.) of the floor is identified for Class I, Division 2 or Class I, Zone 2 locations.
(c) All metallic parts of the drying apparatus are electrically bonded and grounded.
(d) Interlocks are provided to prevent the operation of spray equipment while drying apparatus is within the spray enclosure, to allow for a 3-minute purge of the enclosure before energizing the drying apparatus and to shut off drying apparatus on failure of ventilation system.

(E) Electrostatic Equipment. Electrostatic spraying or detearing equipment shall be installed and used only as provided in 516.10.

FPN: For further information, see NFPA 33-2003, *Standard for Spray Application Using Flammable or Combustible Materials*.

516.7 Wiring and Equipment Not Within Class I and II Locations.

(A) Wiring. All fixed wiring above the Class I and II locations shall be in metal raceways, rigid nonmetallic conduit, or electrical nonmetallic tubing, or shall be Type MI, TC, or MC cable. Cellular metal floor raceways shall be permitted only for supplying ceiling outlets or extensions to the area below the floor of a Class I or II location, but such raceways shall have no connections leading into or through the Class I or II location above the floor unless suitable seals are provided.

(B) Equipment. Equipment that may produce arcs, sparks, or particles of hot metal, such as lamps and lampholders for fixed lighting, cutouts, switches, receptacles, motors, or other equipment having make-and-break or sliding contacts, where installed above a Class I or II location or above a location where freshly finished goods are handled, shall be of the totally enclosed type or be constructed so as to prevent the escape of sparks or hot metal particles.

516.10 Special Equipment.

(A) Fixed Electrostatic Equipment. This section shall apply to any equipment using electrostatically charged elements for the atomization, charging, and/or precipitation of hazardous materials for coatings on articles or for other similar purposes in which the charging or atomizing device is attached to a mechanical support or manipulator. This shall include robotic devices. This section shall not apply to devices that are held or manipulated by hand. Where robot or programming procedures involve manual manipulation of the robot arm while spraying with the high voltage on, the provisions of 516.10(B) shall apply. The installation of electrostatic spraying equipment shall comply with 516.10(A)(1) through (A)(10). Spray equipment shall be listed. All automatic electrostatic equipment systems shall comply with 516.4(A)(1) through (A)(9).

(1) Power and Control Equipment. Transformers, high-voltage supplies, control apparatus, and all other electric portions of the equipment shall be installed outside of the Class I location as defined in 516.3 or be of a type identified for the location.

Exception: High-voltage grids, electrodes, electrostatic atomizing heads, and their connections shall be permitted within the Class I location.

(2) Electrostatic Equipment. Electrodes and electrostatic atomizing heads shall be adequately supported in permanent locations and shall be effectively insulated from ground. Electrodes and electrostatic atomizing heads that are permanently attached to their bases, supports, reciprocators, or robots shall be deemed to comply with this section.

(3) High-Voltage Leads. High-voltage leads shall be properly insulated and protected from mechanical damage or exposure to destructive chemicals. Any exposed element at high voltage shall be effectively and permanently supported on suitable insulators and shall be effectively guarded against accidental contact or grounding.

(4) Support of Goods. Goods being coated using this process shall be supported on conveyors or hangers. The conveyors or hangers shall be arranged (1) to ensure that the parts being coated are electrically connected to ground with a resistance of 1 megohm or less and (2) to prevent parts from swinging.

(5) Automatic Controls. Electrostatic apparatus shall be equipped with automatic means that will rapidly de-energize the high-voltage elements under any of the following conditions:

(1) Stoppage of ventilating fans or failure of ventilating equipment from any cause
(2) Stoppage of the conveyor carrying goods through the high-voltage field unless stoppage is required by the spray process
(3) Occurrence of excessive current leakage at any point in the high-voltage system
(4) De-energizing the primary voltage input to the power supply

(6) Grounding. All electrically conductive objects in the spray area, except those objects required by the process to be at high voltage, shall be adequately grounded. This requirement shall apply to paint containers, wash cans, guards, hose connectors, brackets, and any other electrically conductive objects or devices in the area.

(7) Isolation. Safeguards such as adequate booths, fencing, railings, interlocks, or other means shall be placed about the equipment or incorporated therein so that they, either by their location, character, or both, ensure that a safe separation of the process is maintained.

(8) Signs. Signs shall be conspicuously posted to convey the following:

(1) Designate the process zone as dangerous with regard to fire and accident
(2) Identify the grounding requirements for all electrically conductive objects in the spray area
(3) Restrict access to qualified personnel only

(9) Insulators. All insulators shall be kept clean and dry.

(10) Other Than Nonincendive Equipment. Spray equipment that cannot be classified as nonincendive shall comply with (A)(10)(a) and (A)(10)(b).

(a) Conveyors or hangers shall be arranged so as to maintain a safe distance of at least twice the sparking distance between goods being painted and electrodes, electrostatic atomizing heads, or charged conductors. Warnings defining this safe distance shall be posted.
(b) The equipment shall provide an automatic means of rapidly de-energizing the high-voltage elements in the event the distance between the goods being painted and the electrodes or electrostatic atomizing heads falls below that specified in (a). [NFPA 33:Chapter 11]

(B) Electrostatic Hand-Spraying Equipment. This section shall apply to any equipment using electrostatically charged elements for the atomization, charging, and/or precipitation of materials for coatings on articles, or for other similar purposes in which the atomizing device is hand-held or manipulated during the spraying operation. Electrostatic hand-spraying equipment and devices used in connection with paint-spraying operations shall be of listed types and shall comply with 516.10(B)(1) through (B)(5).

(1) General. The high-voltage circuits shall be designed so as not to produce a spark of sufficient intensity to ignite the most readily ignitible of those vapor–air mixtures likely to be encountered, or result in appreciable shock hazard upon coming in contact with a grounded object under all normal operating conditions. The electrostatically charged exposed elements of the handgun shall be capable of being energized only by an actuator that also controls the coating material supply.

(2) Power Equipment. Transformers, power packs, control apparatus, and all other electric portions of the equipment shall be located outside of the Class I location or be identified for the location.

Exception: The handgun itself and its connections to the power supply shall be permitted within the Class I location.

(3) Handle. The handle of the spraying gun shall be electrically connected to ground by a metallic connection and be constructed so that the operator in normal operating position is in intimate electrical contact with the grounded handle to prevent buildup of a static charge on the operator's body. Signs indicating the necessity for grounding other persons entering the spray area shall be conspicuously posted.

(4) Electrostatic Equipment. All electrically conductive objects in the spraying area shall be adequately grounded. This requirement shall apply to paint containers, wash cans, and any other electrical conductive objects or devices in the area. The equipment shall carry a prominent, permanently installed warning regarding the necessity for this grounding feature.

(5) Support of Objects. Objects being painted shall be maintained in metallic contact with the conveyor or other grounded support. Hooks shall be regularly cleaned to ensure adequate grounding of 1 megohm or less. Areas of contact shall be sharp points or knife edges where possible. Points of support of the object shall be concealed from random spray where feasible; and, where the objects being sprayed are supported from a conveyor, the point of attachment to the conveyor shall be located so as to not collect spray material during normal operation. [NFPA 33:Chapter 12]

(C) Powder Coating. This section shall apply to processes in which combustible dry powders are applied. The hazards associated with combustible dusts are present in such a process to a degree, depending on the chemical composition of the material, particle size, shape, and distribution.

(1) Electric Equipment and Sources of Ignition. Electric equipment and other sources of ignition shall comply with the requirements of Article 502. Portable electric lamps and other utilization equipment shall not be used within a Class II location during operation of the finishing processes. Where such lamps or utilization equipment are used during cleaning or repairing operations, they shall be of a type identified for Class II, Division 1 locations, and all exposed metal parts shall be effectively grounded.

Exception: Where portable electric lamps are required for operations in spaces not readily illuminated by fixed lighting within the spraying area, they shall be of the type listed for Class II, Division 1 locations where readily ignitible residues may be present.

(2) Fixed Electrostatic Spraying Equipment. The provisions of 516.10(A) and 516.10(C)(1) shall apply to fixed electrostatic spraying equipment.

(3) Electrostatic Hand-Spraying Equipment. The provisions of 516.10(B) and 516.10(C)(1) shall apply to electrostatic hand-spraying equipment.

(4) Electrostatic Fluidized Beds. Electrostatic fluidized beds and associated equipment shall be of identified types. The high-voltage circuits shall be designed so that any discharge produced when the charging electrodes of the bed are approached or contacted by a grounded object shall not be of sufficient intensity to ignite any powder–air mixture likely to be encountered or to result in an appreciable shock hazard.

(a) Transformers, power packs, control apparatus, and all other electric portions of the equipment shall be located outside the powder-coating area or shall otherwise comply with the requirements of 516.10(C)(1).

Exception: The charging electrodes and their connections to the power supply shall be permitted within the powder-coating area.

(b) All electrically conductive objects within the powder-coating area shall be adequately grounded. The powder-coating equipment shall carry a prominent, permanently installed warning regarding the necessity for grounding these objects.
(c) Objects being coated shall be maintained in electrical contact (less than 1 megohm) with the conveyor or other support in order to ensure proper grounding. Hangers shall be regularly cleaned to ensure effective electrical contact. Areas of electrical contact shall be sharp points or knife edges where possible.
(d) The electric equipment and compressed air supplies shall be interlocked with a ventilation system so that the equipment cannot be operated unless the ventilating fans are in operation. [NFPA 33:Chapter 15]

516.16 Grounding. All metal raceways, the metal armors or metallic sheath on cables, and all non–current-carrying metal parts of fixed or portable electrical equipment, regardless of voltage, shall be grounded as provided in Article 250. Grounding shall comply with 501.30, 502.30, or 505.25, as applicable.

ARTICLE 517
Health Care Facilities

517.1 Scope. The provisions of this article shall apply to electrical construction and installation criteria in health care facilities that provide services to human beings.

The requirements in Parts II and III not only apply to single-function buildings but are also intended to be individually applied to their respective forms of occupancy within a multifunction building (e.g., a doctor's examining room located within a limited care facility would be required to meet the provisions of 517.10).

FPN: For information concerning performance, maintenance, and testing criteria, refer to the appropriate health care facilities documents.

IV. Inhalation Anesthetizing Locations

517.60 Anesthetizing Location Classification.

(A) Hazardous (Classified) Location.

(1) Use Location. In a location where flammable anesthetics are employed, the entire area shall be considered to be a Class I, Division 1 location that extends upward to a level 1.52 m (5 ft) above the floor. The remaining volume up to the structural ceiling is considered to be above a hazardous (classified) location. [NFPA 99:Annex E, E.1, and E.2]

(2) Storage Location. Any room or location in which flammable anesthetics or volatile flammable disinfecting agents are stored shall be considered to be a Class I, Division 1 location from floor to ceiling.

517.61 Wiring and Equipment.

(A) Within Hazardous (Classified) Anesthetizing Locations.

(1) Isolation. Except as permitted in 517.160, each power circuit within, or partially within, a flammable anesthetizing location as referred to in 517.60 shall be isolated from any distribution system by the use of an isolated power system. [NFPA 99:Annex E, E.6.6.2]

(2) Design and Installation. Where an isolated power system is utilized, the isolated power equipment shall be listed as isolated power equipment, and the isolated power system shall be designed and installed in accordance with 517.160.

(3) Equipment Operating at More Than 10 Volts. In hazardous (classified) locations referred to in 517.60, all fixed wiring and equipment and all portable equipment, including lamps and other utilization equipment, operating at more than 10 volts between conductors shall comply with the requirements of 501.1 through 501.25, and 501.100 through 501.150, and 501.30(A) and 501.30(B) for Class I, Division 1 locations. All such equipment shall be specifically approved for the hazardous atmospheres involved. [NFPA 99:Annex E, E.2.1, E.4.5, E.4.6, and E.4.7]

(4) Extent of Location. Where a box, fitting, or enclosure is partially, but not entirely, within a hazardous (classified) location(s), the hazardous (classified) location(s) shall be considered to be extended to include the entire box, fitting, or enclosure.

(5) Receptacles and Attachment Plugs. Receptacles and attachment plugs in a hazardous (classified) location(s) shall be listed for use in Class I, Group C hazardous (classified) locations and shall have provision for the connection of a grounding conductor.

(6) Flexible Cord Type. Flexible cords used in hazardous (classified) locations for connection to portable utilization equipment, including lamps operating at more than 8 volts between conductors, shall be of a type approved for extra-hard usage in accordance with Table 400.4 and shall include an additional conductor for grounding.

(7) Flexible Cord Storage. A storage device for the flexible cord shall be provided and shall not subject the cord to bending at a radius of less than 75 mm (3 in.).

(B) Above Hazardous (Classified) Anesthetizing Locations.

(1) Wiring Methods. Wiring above a hazardous (classified) location referred to in 517.60 shall be installed in rigid metal conduit, electrical metallic tubing, intermediate metal conduit, Type MI cable, or Type MC cable that employs a continuous, gas/vaportight metal sheath.

(2) Equipment Enclosure. Installed equipment that may produce arcs, sparks, or particles of hot metal, such as lamps and lampholders for fixed lighting, cutouts, switches, generators, motors, or other equipment having make-and-break or sliding contacts, shall be of the totally enclosed type or be constructed so as to prevent escape of sparks or hot metal particles.

Exception: Wall-mounted receptacles installed above the hazardous (classified) location in flammable anesthetizing locations shall not be required to be totally enclosed or have openings guarded or screened to prevent dispersion of particles.

(3) Luminaires (Lighting Fixtures). Surgical and other luminaires (lighting fixtures) shall conform to 501.130(B).

Exception No. 1: The surface temperature limitations set forth in 501.130(B)(1) shall not apply.

Exception No. 2: Integral or pendant switches that are located above and cannot be lowered into the hazardous (classified) location(s) shall not be required to be explosionproof.

(4) Seals. Approved seals shall be provided in conformance with 501.15, and 501.15(A)(4) shall apply to horizontal as well as to vertical boundaries of the defined hazardous (classified) locations.

(5) Receptacles and Attachment Plugs. Receptacles and attachment plugs located above hazardous (classified) anesthetizing locations shall be listed for hospital use for services of prescribed voltage, frequency, rating, and number of conductors with provision for the connection of the grounding conductor. This requirement shall apply to attachment plugs and receptacles of the 2-pole, 3-wire grounding type for single-phase, 120-volt, nominal, ac service.

(6) 250-Volt Receptacles and Attachment Plugs Rated 50 and 60 Amperes. Receptacles and attachment plugs rated 250 volts, for connection of 50-ampere and 60-ampere ac medical equipment for use above hazardous (classified) locations, shall be arranged so that the 60-ampere receptacle will accept either the 50-ampere or the 60-ampere plug. Fifty-ampere receptacles shall be designed so as not to accept the 60-ampere attachment plug. The attachment plugs shall be of the 2-pole, 3-wire design with a third contact connecting to the insulated (green or green with yellow stripe) equipment grounding conductor of the electrical system.

PART II REQUIREMENTS FROM OTHER NFPA DOCUMENTS

Editor's Note: *As with the NEC sections in Part I, this part contains only sections of NFPA documents that directly address electrical installations in hazardous locations (although a few brief documents are shown in their entirety, including the annexes). Since the requirements of any code or standard are represented only by the complete document, other elements of the following documents will apply for context and for non-electrical rules.*

NFPA 30, *Flammable and Combustible Liquids Code*, 2003

1.1 Scope.

1.7.3* Classification of Liquids. Any liquid within the scope of this code and subject to the requirements of this code shall be known generally as either a flammable liquid or a combustible liquid and shall be defined and classified in accordance with this subsection.

A.1.7.3 The classification of liquids is based on flash points that have been corrected to sea level, in accordance with the relevant ASTM test procedures. At high altitudes, the actual flash points will be significantly lower than those either observed at sea level or corrected to atmospheric pressure at sea level. Allowances could be necessary for this difference in order to appropriately assess the risk.

Table A.1.7.3 presents a comparison of the definitions and classification of flammable and combustible liquids, as set forth in Section 1.7 of this code, with similar definitions and classification systems used by other regulatory bodies.

The Hazardous Materials Regulations of the U.S. Department of Transportation (DOT), as set forth in the 49 CFR 173.120(b)(2) and 173.150(f), provide an exception

Table A.1.7.3 Comparative Classification of Liquids

		Agency Flash Point		NFPA Flash Point			
Agency	Agency Classification	°F	°C	NFPA Definition	NFPA Classification	°F	°C
ANSI/CMA Z129.1-1994	Flammable	<141	<60.5	Flammable	Class I	<100	<37.8
				Combustible	Class II	≥100 to <140	≥37.8 to <60
					Class IIIA	≥140 to <200	≥60 to <93
	Combustible	≥141 to <200	≥60.5 to <93	Combustible	Class IIIA	≥140 to <200	≥60 to <93
DOT	Flammable	<141	<60.5	Flammable	Class I	<100	<37.8
				Combustible	Class II	≥100 to <140	≥37.8 to <60
					Class IIIA	≥140 to <200	≥60 to <93
	Combustible	≥141 to <200	≥60.5 to <93	Combustible	Class IIIA	≥140 to <200	≥60 to <93
DOT HM-181 Domestic Exemption[1]	Flammable	<100	<37.8	Flammable	Class I	<100	<37.8
	Combustible	≥100 to <200	≥37.8 to <93	Combustible	Class II	≥100 to <140	≥37.8 to <60
					Class IIIA	≥140 to <200	≥60 to <93
UN	Flammable	<141	<60.5	Flammable	Class I	<100	< 37.8
				Combustible	Class II	≥100 to <140	≥37.8 to <60
					Class IIIA	≥140 to <200	≥60 to <93
	Combustible	≥141 to <200	≥60.5 to <93	Combustible	Class II	≥100 to <140	≥37.8 to <60
					Class IIIA	≥140 to <200	≥60 to <93
OSHA	Flammable	<100	<37.8	Flammable	Class I	<100	<37.8
	Combustible[2]	≥100	≥37.8	Combustible	Class II	≥100 to <140	≥37.8 to <60
					Class IIIA	≥140 to <200	≥60 to <93
					Class IIIB[2]	≥200	≥93

[1]See A.1.7.3.
[2]See 29 CFR 1910.106 for Class IIIB liquid exemptions.

whereby a flammable liquid that has a flash point between 37.8°C (100°F) and 60.5°C (141°F) and does not also meet the definition of any other DOT hazard class can be reclassified as a combustible liquid [i.e., one having a flash point above 60.5°C (141°F)] for shipment by road or rail within the United States.

1.7.3.1 Combustible Liquid. Any liquid that has a closed-cup flash point at or above 100°F (37.8°C), as determined by the test procedures and apparatus set forth in 1.7.4. Combustible liquids are classified as Class II or Class III as follows: (1) *Class II Liquid* — any liquid that has a flash point at or above 100°F (37.8°C) and below 140°F (60°C); (2) *Class IIIA* — any liquid that has a flash point at or above 140°F (60°C), but below 200°F (93°C); (3) *Class IIIB* — any liquid that has a flash point at or above 200°F (93°C).

1.7.3.2* Flammable Liquid. Any liquid that has a closed-cup flash point below 100°F (37.8°C), as determined by the test procedures and apparatus set forth in 1.7.4. Flammable liquids are classified as Class I as follows: *Class I Liquid* — any liquid that has a closed-cup flash point below 100°F (37.8°C) and a Reid vapor pressure not exceeding 40 psia (2068.6 mm Hg) at 100°F (37.8°C), as determined by ASTM D 323, *Standard Method of Test for Vapor Pressure of Petroleum Products (Reid Method).* Class I liquids are further classified as follows: (1) *Class IA liquids* — those liquids that have flash points below 73°F (22.8°C) and boiling points below 100°F (37.8°C); (2) *Class IB liquids* — those liquids that have flash points below 73°F (22.8°C) and boiling points at or above 100°F (37.8°C); (3) *Class IC liquids* — those liquids that have flash points at or above 73°F (22.8°C), but below 100°F (37.8°C).

A.1.7.3.2 For the purposes of this code, a material with a Reid Vapor Pressure greater than 2068 mm Hg absolute (40 psia) is considered to be a gas and is, therefore, not within the scope of NFPA 30. See NFPA 58, *Liquefied Petroleum Gas Code.*

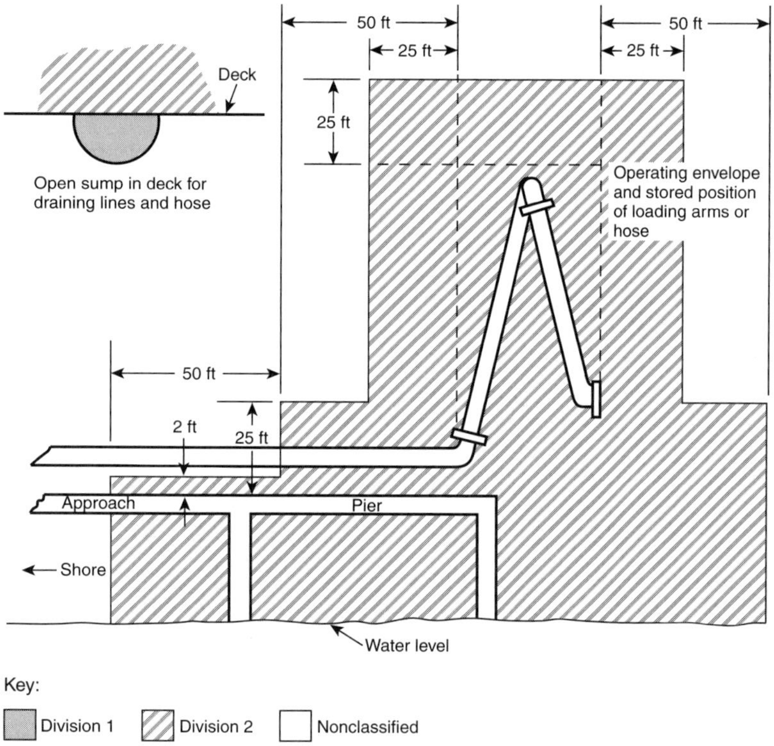

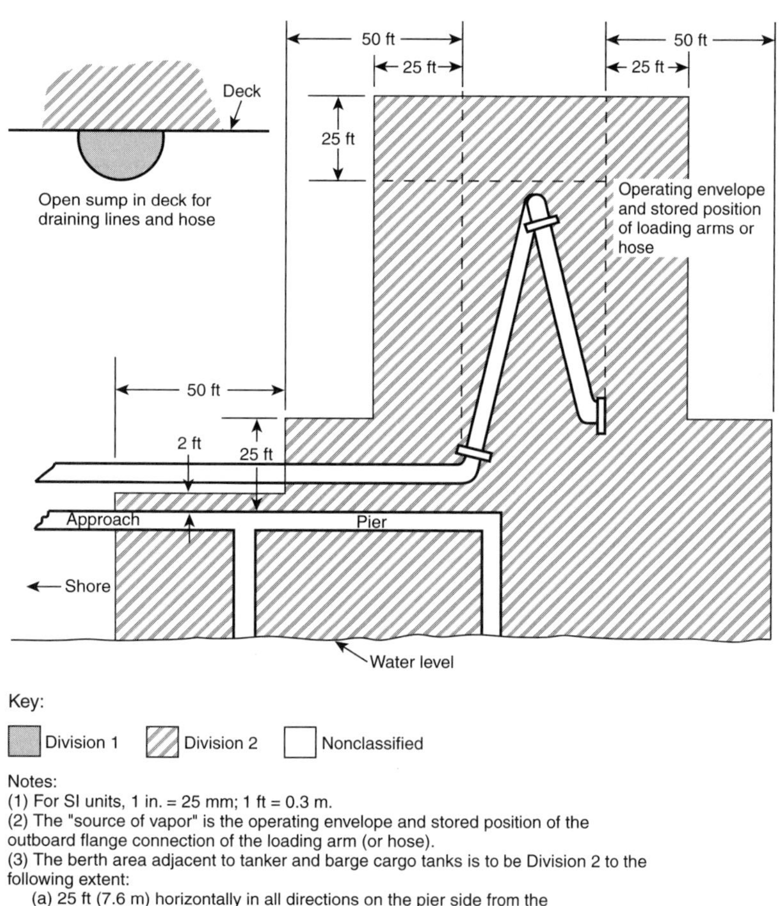

Notes:
(1) For SI units, 1 in. = 25 mm; 1 ft = 0.3 m.
(2) The "source of vapor" is the operating envelope and stored position of the outboard flange connection of the loading arm (or hose).
(3) The berth area adjacent to tanker and barge cargo tanks is to be Division 2 to the following extent:
(a) 25 ft (7.6 m) horizontally in all directions on the pier side from the portion of the hull containing cargo tanks.
(b) From the water level to 25 ft (7.6 m) above the cargo tanks at their highest position.
(4) Additional locations can be classified as required by the presence of other sources of flammable liquids on the berth, or by Coast Guard or other regulations.

FIGURE 7.7.16 Marine Terminal Handling Flammable Liquids.

7.7 Wharves.

7.7.16 For marine terminals handling flammable liquids, Figure 7.7.16 shall be used to determine the extent of classified areas for the purpose of installation of electrical equipment.

7.9 Control of Ignition Sources.

7.9.5 Electrical Installations. Electrical equipment and wiring installations shall be in accordance with Chapter 8.

Chapter 8 Electrical Equipment and Installations

8.1 Scope.

This chapter shall apply to areas where Class I liquids are stored or handled and to areas where Class II or Class III liquids are stored or handled at or above their flash points.

8.2 General.

Any electrical equipment provided shall not constitute a source of ignition for the flammable vapor that might be present under normal operation or during a spill. Compliance with 8.2.1 through 8.2.4 shall be deemed as meeting the requirements of Section 8.2.

8.2.1 All electrical equipment and wiring shall be of a type specified by and installed in accordance with NFPA 70, *National Electrical Code* (*NEC*).

8.2.2* So far as it applies, Table 8.2.2 shall be used to delineate and classify areas for the purpose of installation of electrical equipment under normal conditions. In the application of classified areas, a classified area shall not extend beyond a floor, wall, roof, or other solid partition that has no openings within the classified area. The designation of classes, divisions, and zones shall be as defined in NFPA 70, *National Electrical Code*, Chapter 5, Article 500.

A.8.2.2 For additional information, see NFPA 497, *Recommended Practice for the Classification of Flammable Liquids, Gases, or Vapors and of Hazardous (Classified) Locations for Electrical Installations in Chemical Process Areas.*

8.2.3 The area classifications listed in Table 8.2.2 are based on the premise that the installation shall meet the applicable requirements of this code in all respects. Should this not be the case, the authority having jurisdiction shall have the authority to classify the extent of the area.

8.2.4* Where the provisions of 8.2.1 through 8.2.4 require the installation of electrical equipment suitable for Class I,

Table 8.2.2 Electrical Area Classifications

Location	*NEC* Class I Division	*NEC* Class I Zone	Extent of Classified Area
Indoor equipment installed in accordance with Section 7.3 where flammable vapor–air mixtures can exist under normal operation	1	0	The entire area associated with such equipment where flammable gases or vapors are present continuously or for long periods of time
	1	1	Area within 5 ft of any edge of such equipment, extending in all directions
	2	2	Area between 5 ft and 8 ft of any edge of such equipment, extending in all directions; also, space up to 3 ft above floor or grade level within 5 ft to 25 ft horizontally from any edge of such equipment[1]
Outdoor equipment of the type covered in Section 7.3 where flammable vapor–air mixtures can exist under normal operation	1	0	The entire area associated with such equipment where flammable gases or vapors are present continuously or for long periods of time
	1	1	Area within 3 ft of any edge of such equipment, extending in all directions
	2	2	Area between 3 ft and 8 ft of any edge of such equipment, extending in all directions; also, space up to 3 ft above floor or grade level within 3 ft to 10 ft horizontally from any edge of such equipment
Tank storage installations inside buildings	1	1	All equipment located below grade level

(continues)

Table 8.2.2 Continued

Location	*NEC* Class I Division	*NEC* Class I Zone	Extent of Classified Area
Tank — aboveground	2	2	Any equipment located at or above grade level
	1	0	Inside fixed-roof tank
	1	1	Area inside dike where dike height is greater than the distance from the tank to the dike for more than 50 percent of the tank circumference
Shell, ends, or roof and dike area	2	2	Within 10 ft from shell, ends, or roof of tank; also, area inside dike up to top of dike wall
Vent	1	0	Area inside of vent piping or opening
	1	1	Within 5 ft of open end of vent, extending in all directions
	2	2	Area between 5 ft and 10 ft from open end of vent, extending in all directions
Floating roof			
With fixed outer roof	1	0	Area between the floating and fixed-roof sections and within the shell
With no fixed outer roof	1	1	Area above the floating roof and within the shell
Underground tank fill opening	1	1	Any pit, box, or space below grade level, if any part is within a Division 1 or 2 or Zone 1 or 2 classified location
	2	2	Up to 18 in. above grade level within a horizontal radius of 10 ft from a loose fill connection and within a horizontal radius of 5 ft from a tight fill connection
Vent — discharging upward	1	0	Area inside of vent piping or opening
	1	1	Within 3 ft of open end of vent, extending in all directions
	2	2	Area between 3 ft and 5 ft of open end of vent, extending in all directions
Tank Vault — Interior	1	1	Entire interior volume, if Class I liquids are stored within
Drum and container filling — outdoors or indoors	1	0	Area inside the drum or container
	1	1	Within 3 ft of vent and fill openings, extending in all directions
	2	2	Area between 3 ft and 5 ft from vent or fill opening, extending in all directions; also, up to 18 in. above floor or grade level within a horizontal radius of 10 ft from vent or fill opening
Pumps, bleeders, withdrawal fittings			
Indoor	2	2	Within 5 ft of any edge of such devices, extending in all directions; also, up to 3 ft above floor or grade level within 25 ft horizontally from any edge of such devices
Outdoor	2	2	Within 3 ft of any edge of such devices, extending in all directions; also, up to 18 in. above grade level within 10 ft horizontally from any edge of such devices
Pits and sumps			
Without mechanical ventilation	1	1	Entire area within a pit or sump if any part is within a Division 1 or 2 or Zone 1 or 2 classified location
With adequate mechanical ventilation	2	2	Entire area within a pit or sump if any part is within a Division 1 or 2 or Zone 1 or 2 classified location
Containing valves, fittings, or piping, and not within a Division 1 or 2 or Zone 1 or 2 classified location	2	2	Entire pit or sump
Drainage ditches, separators, impounding basins			
Outdoor	2	2	Area up to 18 in. above ditch, separator, or basin; also, area up to 18 in. above grade within 15 ft horizontally from any edge
Indoor			Same classified area as pits

Table 8.2.2 Continued

Location	*NEC* Class I Division	*NEC* Class I Zone	Extent of Classified Area
Tank vehicle and tank car[2]			
Loading through open dome	1	0	Area inside of the tank
	1	1	Within 3 ft of edge of dome, extending in all directions
	2	2	Area between 3 ft and 15 ft from edge of dome, extending in all directions
Loading through bottom connections with atmospheric venting	1	0	Area inside of the tank
	1	1	Within 3 ft of point of venting to atmosphere, extending in all directions
	2	2	Area between 3 ft and 15 ft from point of venting to atmosphere, extending in all directions; also, up to 18 in. above grade within a horizontal radius of 10 ft from point of loading connection
Office and rest rooms	Ordinary		If there is any opening to these rooms within the extent of an indoor classified location, the room shall be classified the same as if the wall, curb, or partition did not exist
Loading through closed dome with atmospheric venting	1	1	Within 3 ft of open end of vent, extending in all directions
	2	2	Area between 3 ft and 15 ft from open end of vent, extending in all directions; also, within 3 ft of edge of dome, extending in all directions
Loading through closed dome with vapor control	2	2	Within 3 ft of point of connection of both fill and vapor lines, extending in all directions
Bottom loading with vapor control or any bottom unloading	2	2	Within 3 ft of point of connections, extending in all directions; also, up to 18 in. above grade within a horizontal radius of 10 ft from point of connections
Storage and repair garage for tank vehicles	1	1	All pits or spaces below floor level
	2	2	Area up to 18 in. above floor or grade level for entire storage or repair garage
Garages for other than tank vehicles	Ordinary		If there is any opening to these rooms within the extent of an outdoor classified location, the entire room shall be classified the same as the area classification at the point of the opening
Outdoor drum storage	Ordinary		
Inside rooms or storage lockers used for the storage of Class I liquids	2	2	Entire room
Indoor warehousing where there is no flammable liquid transfer	Ordinary		If there is any opening to these rooms within the extent of an indoor classified location, the room shall be classified the same as if the wall, curb, or partition did not exist
Piers and wharves			See Figure 7.7.16

Note: For SI units, 1 in. = 25 mm; 1 ft = 0.3 m.

[1]The release of Class I liquids can generate vapors to the extent that the entire building, and possibly an area surrounding it, should be considered a Class I, Division 2, or Zone 2 location.

[2]When classifying extent of area, consideration shall be given to the fact that tank cars or tank vehicles can be spotted at varying points. Therefore, the extremities of the loading or unloading positions shall be used.

Division 1 or 2 or Zone 1 or 2 locations, ordinary electrical equipment, including switchgear, shall be permitted to be used if installed in a room or enclosure that is maintained under positive pressure with respect to the classified area. Ventilation make-up air shall not be contaminated.

A.8.2.4 NFPA 496, *Standard for Purged and Pressurized Enclosures for Electrical Equipment*, provides details for these types of installations.

NFPA 30A, *Code for Motor Fuel Dispensing Facilities and Repair Garages,* 2003

1.1* Scope.

A.1.1 This code, known as the *Code for Motor Fuel Dispensing Facilities and Repair Garages,* is recommended for use as the basis for legal regulations. Its provisions are intended to reduce the hazards of motor fuels to a degree consistent with reasonable public safety, without undue interference with public convenience and necessity. Thus, compliance with this code does not eliminate all hazards in the use of these fuels.

See the *Flammable and Combustible Liquids Code Handbook* for additional explanatory information.

1.1.1 This code shall apply to motor fuel dispensing facilities; marine/motor fuel dispensing facilities; and motor fuel dispensing facilities located inside buildings, at fleet vehicle motor fuel facilities, and at farms and isolated construction sites. This code shall also apply to motor vehicle repair garages.

1.1.2* This code shall not apply to those motor fuel dispensing facilities where only liquefied petroleum gas (LP-Gas), liquefied natural gas (LNG), or compressed natural gas (CNG) is dispensed as motor fuel.

A.1.1.2 See NFPA 52, NFPA 57, and NFPA 58 for requirements for facilities where only these fuels are dispensed.

1.7* Classification of Liquids.

Any liquid within the scope of this code and subject to the requirements of this code shall be known generally as either a flammable liquid or a combustible liquid and shall be defined and classified in accordance with this section. [**30:**1.7.3]

A.1.7 The classification of liquids is based on flash points that have been corrected to sea level, in accordance with the relevant ASTM test procedures. At high altitudes, the actual flash points will be significantly lower than those either observed at sea level or corrected to atmospheric pressure at sea level. Allowances could be necessary for this difference in order to appropriately assess the risk.

Table A.1.7 presents a comparison of the definitions and classification of flammable and combustible liquids, as set forth in Section 1.7 of this code, with similar definitions and classification systems used by other regulatory bodies.

The Hazardous Materials Regulations of the U.S. Department of Transportation (DOT), as set forth in the 49 CFR 173.120(b)(2) and 173.150(f), provide an exception whereby a flammable liquid that has a flash point between 37.8°C (100°F) and 60.5°C (141°F) and does not also meet the definition of any other DOT hazard class, can be reclassified as a combustible liquid [i.e., one having a flash point above 60.5°C (141°F)] for shipment by road or rail within the United States

1.7.1 Combustible Liquid. Any liquid that has a closed-cup flash point at or above 100°F (37.8°C), as determined by the test procedures and apparatus set forth in 1.7.4 of NFPA 30. Combustible liquids are classified as Class II or Class III as follows: (1) *Class II Liquid* — any liquid that has a flash point at or above 100°F (37.8°C) and below 140°F (60°C); (2) *Class IIIA* — any liquid that has a flash point at or above 140°F (60°C), but below 200°F (93°C); (3) *Class IIIB* — any liquid that has a flash point at or above 200°F (93°C). [**30:**1.7.3.1]

1.7.2* Flammable Liquid. Any liquid that has a closed-cup flash point below 100°F (37.8°C), as determined by the test procedures and apparatus set forth in 1.7.4 of NFPA 30. Flammable liquids are classified as Class I as follows: *Class I Liquid* — any liquid that has a closed-cup flash point below 100°F (37.8°C) and a Reid vapor pressure not exceeding 40 psia (2068.6 mm Hg) at 100°F (37.8°C), as determined by ASTM D 323, *Standard Method of Test for Vapor Pressure of Petroleum Products (Reid Method).* Class I liquids are further classified as follows: (1) *Class IA liquids* — those liquids that have flash points below 73°F (22.8°C) and boiling points below 100°F (37.8°C); (2) *Class IB liquids* — those liquids that have flash points below 73°F (22.8°C) and boiling points at or above 100°F (37.8°C); (3) *Class IC liquids* — those liquids that have flash points at or above 73°F (22.8°C), but below 100°F (37.8°C). [**30:**1.7.3.2]

A.1.7.2 For the purposes of this code, a material with a Reid Vapor Pressure greater than 2068 mm Hg absolute (40 psia) is considered to be a gas and is, therefore, not within the scope of NFPA 30A. See NFPA 58, *Liquefied Petroleum Gas Code.*

3.3 General Definitions.

3.3.2 Bulk Plant or Terminal. That portion of a property where liquids are received by tank vessel, pipeline, tank car,

Table A.1.7 Comparative Classification of Liquids

Agency	Agency Classification	Agency Flash Point		NFPA Flash Point	NFPA Definition	NFPA Classification	
		°F	°C			°F	°C
ANSI/CMA Z129.1-1994	Flammable	<141	<60.5	Flammable	Class I	<100	<37.8
				Combustible	Class II	≥100 to <140	≥37.8 to <60
					Class IIIA	≥140 to <200	≥60 to <93
	Combustible	≥141 to <200	≥60.5 to <93	Combustible	Class IIIA	≥140 to <200	≥60 to <93
DOT	Flammable	<141	<60.5	Flammable	Class I	<100	<37.8
				Combustible	Class II	≥100 to <140	≥37.8 to <60
					Class IIIA	≥140 to <200	≥60 to <93
	Combustible	≥141 to <200	≥60.5 to <93	Combustible	Class IIIA	≥140 to <200	≥60 to <93
DOT HM-181 Domestic Exemption[1]	Flammable	<100	<37.8	Flammable	Class I	<100	<37.8
	Combustible	≥100 to <200	≥37.8 to <93	Combustible	Class II	≥100 to <140	≥37.8 to <60
					Class IIIA	≥140 to <200	≥60 to <93
UN	Flammable	<141	<60.5	Flammable	Class I	<100	< 37.8
				Combustible	Class II	≥100 to <140	≥37.8 to <60
					Class IIIA	≥140 to <200	≥60 to <93
	Combustible	≥141 to <200	≥60.5 to <93	Combustible	Class II	≥100 to <140	≥37.8 to <60
					Class IIIA	≥140 to <200	≥60 to <93
OSHA	Flammable	<100	<37.8	Flammable	Class I	<100	<37.8
	Combustible[2]	≥100	≥37.8	Combustible	Class II	≥100 to <140	≥37.8 to <60
					Class IIIA	≥140 to <200	≥60 to <93
					Class IIIB2	≥200	≥93

[1]See A.1.7.3 of NFPA 30.
[2]See 29 CFR 1910.106 for Class IIIB liquid exemptions.

or tank vehicle and are stored or blended in bulk for the purpose of distributing such liquids by tank vessel, pipeline, tank car, tank vehicle, portable tank, or container.

3.3.6* Dispensing Device, Overhead Type. A dispensing device that consists of one or more individual units intended for installation in conjunction with each other, mounted above a dispensing area typically within the service station canopy structure, and characterized by the use of an overhead hose reel.

A.3.3.6 Dispensing Device, Overhead Type. This definition applies to an overhead dispenser that uses a retractable hose on an overhead reel, as distinguished from the now-common dispensing device that has one or more hose outlets located in a canopy at the top of the dispensing device. The latter, also called *high-hose units* or *multi-product dispensers*, are treated by NFPA 30A as conventional dispensing devices.

3.3.11 Motor Fuel Dispensing Facility. That portion of a property where motor fuels are stored and dispensed from fixed equipment into the fuel tanks of motor vehicles or marine craft or into approved containers, including all equipment used in connection therewith.

Chapter 8 Electrical Installations

8.1 Scope.

This chapter shall apply to the installation of electrical wiring and electrical utilization equipment in areas where liquids are stored, handled, or dispensed.

8.2 General Requirements.

Electrical wiring and electrical utilization equipment shall be of a type specified by and shall be installed in accordance with NFPA 70. Electrical wiring and electrical utilization equipment shall be approved for the locations in which they are installed.

8.2.1* In major repair garages where CNG vehicles are repaired or stored, the area within 455 mm (18 in.) of the ceiling shall be designated a Class I, Division 2 hazardous (classified) location.

Exception: In major repair garages, where ventilation equal to not less than four air changes per hour is provided, this requirement shall not apply.

A.8.2.1 The intent is that the electrical utilization equipment be placed below a volume located at the highest area of the building that is equal to 150 percent of the released volume of the largest CNG tank.

8.3 Installation in Classified Locations.

8.3.1* Table 8.3.1 shall be used to delineate and classify areas for the purposes of installing electrical wiring and electrical utilization equipment. *(See also Figure 8.3.1.)*

A.8.3.1 The designation of classes and divisions of classified locations is defined in Chapter 5, Article 500, of NFPA 70.

Exception: The extent of the classified area around a vacuum-assist blower shall be permitted to be reduced if the blower is specifically listed for such reduced distances.

8.3.2 A designated classified area, as specified in Table 8.3.1, shall not extend beyond a floor, wall, roof, or other solid partition that has no openings.

8.3.3 The area classifications given in Table 8.3.1 shall be based on the premise that the installation meets the applicable requirements of this code in all respects. Should this not be the case, the authority having jurisdiction shall be permitted to determine the extent of the classified area.

8.3.4 All electrical wiring and electrical utilization equipment that is integral with the dispensing hose or dispensing nozzle shall be approved for use in Class I, Division 1 classified locations.

8.3.5 Where Class I liquids are stored, handled, or dispensed, electrical wiring and electrical utilization equipment shall be designed and installed in accordance with the requirements for Class I, Division 1 or Division 2 classified locations, as set forth in Table 8.3.1 and in NFPA 70.

Exception: The storage, handling, and dispensing of methyl alcohol–based windshield washer fluids shall not cause an area to be designated as a hazardous (classified) location.

8.3.6 The storage, handling, and dispensing of Class II or Class III liquids shall not cause an area to be designated as a hazardous (classified) location.

8.4 Emergency Electrical Disconnects.

Emergency electrical disconnects shall be installed at the locations required by Section 6.7.

8.5 Specific Requirements for Marine Fuel Facilities.

8.5.1 Where excessive stray currents are encountered, piping handling Class I and Class II liquids shall be electrically isolated from the shore piping.

8.5.2* Pipelines on piers shall be bonded and grounded. Bonding and grounding connections on all pipelines shall be located on the pier side of hose riser insulating flanges, if used, and shall be accessible for inspection.

A.8.5.2 NFPA 77 contains information on this subject.

8.5.3 The fuel delivery nozzle shall be put into contact with the vessel fill pipe before the flow of fuel commences, and this bonding contact shall be continuously maintained until fuel flow has stopped to avoid possibility of electrostatic discharge.

Chapter 12 Additional Requirements for CNG, LNG, and LPG

12.6 Electrical Equipment.

12.6.1 All electrical wiring and electrical utilization equipment shall be of a type specified by, and shall be installed in accordance with, NFPA 70.

12.6.2* Table 12.6.2 shall be used to delineate and classify areas for the purpose of installation of electrical wiring and electrical utilization equipment.

A.12.6.2 The designation of classes and divisions of classified locations is defined in Chapter 5, Article 500, of NFPA 70.

NFPA 30B, *Code for the Manufacture and Storage of Aerosol Products*, 2002

1.1 Scope.

1.1.1 This code shall apply to the manufacture, storage, and display of aerosol products as herein defined.

1.1.2* This code shall not apply to the manufacture, storage, and display of aerosol products that contain only a nonflammable base product and a nonflammable propellant.

Table 8.3.1 Electrical Equipment Classified Areas — Motor Fuel Dispensing Facilities

Location	*NEC* Class I, Group D Division	Extent of Classified Area[1]
Dispensing device[2,3]	(except Overhead Type)	(see Figure 8.3.1)
Pits	1	Any pit or box below grade level, any part of which is within a Division 1 or 2 classified area
Dispenser	2	Within 46 cm (18 in.) horizontally in all directions extending to grade from the dispenser enclosure or that portion of the dispenser enclosure containing liquid handling components[3]
Outdoor	2	Up to 46 cm (18 in.) above grade level within 6 m (20 ft) horizontally of any edge of enclosure
Indoor		
With mechanical ventilation	2	Up to 46 cm (18 in.) above grade or floor level within 6 m (20 ft) horizontally of any edge of enclosure
With gravity ventilation	2	Up to 46 cm (18 in.) above grade or floor level within 7.6 m (25 ft) horizontally of any edge of enclosure
Dispensing device — overhead[3,4]		The area within the dispenser enclosure and all electrical equipment integral with the dispensing hose or nozzle
	2	An area extending 46 cm (18 in.) horizontally in all directions beyond the enclosure and extending to grade
	2	Up to 46 cm (18 in.) above grade level within 6 m (20 ft) horizontally measured from a point vertically below the edge of any dispenser enclosure
Remote pump — outdoor	1	Any pit or box below grade level if any part is within a horizontal distance of 3 m (10 ft) from any edge of pump
	2	Within 3 ft of any edge of pump, extending in all directions; also up to 46 cm (18 in.) above grade level within 3 m (10 ft) horizontally from any edge of pump
Remote pump — indoor	1	Entire area within any pit
	2	Within 1.5 (5 ft) of any edge of pump, extending in all directions; also up to 3 ft above floor or grade level within 7.6 m (25 ft) horizontally from any edge of pump
Lubrication or service room where Class I liquids are dispensed *(see 8.3.5)*	1	Any pit within any unventilated area
	2	Any pit with ventilation
	2	Area up to 46 cm (18 in.) above floor or grade level and 3 ft horizontally from a lubrication pit
Dispenser for Class I liquids[3]	2	Within 3 ft of any fill or dispensing point, extending in all directions
Lubrication or service room where Class I liquids are not dispensed *(see 8.3.5)*	2	Entire area within any pit used for lubrication or similar services where Class I liquids can be released
	2	Area up to 46 cm (18 in.) above any such pit and extending a distance of 3 ft horizontally from any edge of the pit
	2	Entire unventilated area within any pit, belowgrade area, or subfloor area
	2	Area up to 46 cm (18 in.) above any such unventilated pit, belowgrade work area, or subfloor work area and extending a distance of 3 ft horizontally from the edge of any such pit, belowgrade work area, or subfloor work area

(continues)

Table 8.3.1 Continued

Location	*NEC* Class I, Group D Division	Extent of Classified Area[1]
	Nonclassified	Any pit, belowgrade work area, or subfloor work area that is ventilated in accordance with 7.4.5.4
Interior of special enclosure or vault	1	Entire interior volume, if Class I liquids are stored within
Sales, storage, and rest rooms	Nonclassified	If there is any opening to these rooms within the extent of a Division 1 area, the entire room is classified as Division 1
Tank, aboveground	1	Area inside dike where dike height is greater than the distance from the tank to the dike for more than 50 percent of the tank circumference
Shell, ends, or roof and dike area	2	Within 3 m (10 ft) of shell, ends, or roof of tank; area within dike to level of top of dike
Vent	1	Within 1.5 m (5 ft) of open end of vent, extending in all directions
	2	Between 1.5 (5 ft) and 3 m (10 ft) from open end of vent, extending in all directions
Underground tank fill opening	1	Any pit or box below grade level, any part of which is within a Division 1 or 2 classified area
	2	Up to 46 cm (18 in.) above grade level within a horizontal radius of 3 m (10 ft) from a loose fill connection and within a horizontal radius of 1.5 m (5 ft) from a tight fill connection
Vapor processing systems pits	1	Any pit or box below grade level, any part of which is within a Division 1 or 2 classified area or that houses any equipment used to transport or process vapors
Vapor processing equipment located within protective enclosures *(see 10.1.7)*	2	Within any protective enclosure housing vapor processing equipment
Vapor processing equipment not within protective enclosures (excluding piping and combustion devices)	2	The space within 46 cm (18 in.) in all directions of equipment containing flammable vapors or liquid extending to grade level; up to 46 cm (18 in.) above grade level within 3 m (10 ft) horizontally of the vapor processing equipment
Equipment enclosures	1	Any area within the enclosure where vapor or liquid is present under normal operating conditions
	2	Entire area within the enclosure other than Division 1
Vacuum-assist blowers	2	The space within 46 cm (18 in.) in all directions extending to grade level; up to 46 cm (18 in.) above grade level within 3 m (10 ft) horizontally
Vault	1	Entire interior volume if Class I liquids are stored within
Vent discharging upward	1	Within 3 ft of open end of vent, extending in all directions
	2	Area between 3 ft and 1.5 m (5 ft) of open end of vent, extending in all directions

[1]For marine application, *grade level* means the surface of a pier, extending down to water level.
[2]Refer to Figure 8.3.1 for an illustration of classified areas around dispensing devices.
[3]Area classification inside the dispenser enclosure is covered in ANSI/UL 87, *Standard for Power-Operated Dispensing Devices for Petroleum Products.*
[4]Ceiling-mounted hose reel.

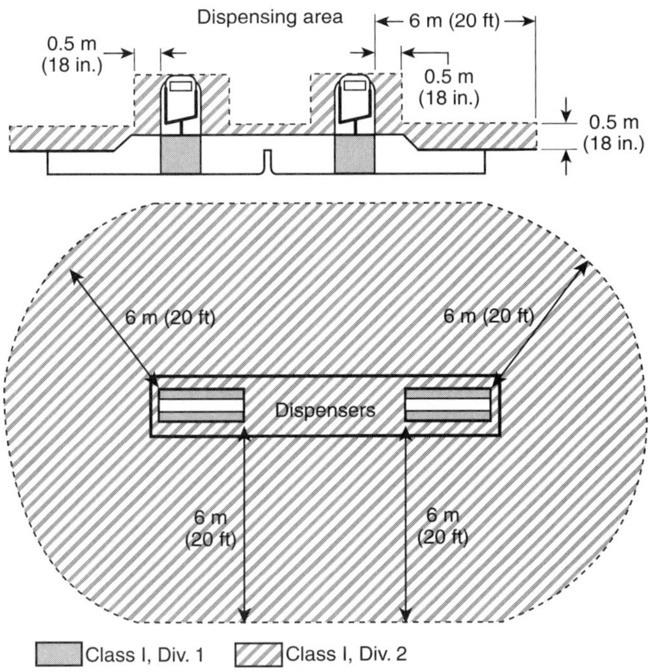

FIGURE 8.3.1 Classified Areas Adjacent to Dispensers as Detailed in Table 8.3.1.

A.1.1.2 An example of an aerosol product that is not flammable and, therefore, not covered by this code is whipped cream: the base product is a water-based material and the propellant is nitrous oxide, which is nonflammable.

1.1.3* This code shall not apply to the storage and display of containers whose contents are comprised entirely of LP-Gas products.

A.1.1.3 See NFPA 58, *Liquefied Petroleum Gas Code.*

1.1.4 This code shall not apply to post-consumer processing of aerosol containers.

Chapter 4 Basic Requirements

4.3 Electrical Installations.

4.3.1 All electrical equipment and wiring, including heating equipment, shall be installed in accordance with NFPA 70, *National Electrical Code®*.

4.3.1.1 Electrical equipment and wiring in areas where flammable liquids or flammable gases are handled shall meet the additional requirements of Articles 500 and 501 of NFPA 70, *National Electrical Code*.

4.3.2 Aerosol product storage and display areas shall be considered unclassified for purposes of electrical installation.

Table 12.6.2 Electrical Equipment Classified Areas for Dispensing Devices

Dispensing Device	Extent of Classified Area: Class I, Division 1	Extent of Classified Area: Class I, Division 2
Compressed natural gas	Entire space within the dispenser enclosure	1.5 m (5 ft) in all directions from dispenser enclosure
Liquefied natural gas	Entire space within the dispenser enclosure and1.5 m (5 ft) in all directions from the dispenser enclosure	From 1.5 m (5 ft) to3 m (10 ft) in all directions from the dispenser enclosure
Liquefied petroleum gas	Entire space with the dispenser enclosure; 46 cm (18 in.) from the exterior surface of the dispenser enclosure to an elevation of 4 ft above the base of the dispenser; the entire pit or open space beneath the dispenser and within 6 m (20 ft) horizontally from any edge of the dispenser when the pit or trench is not mechanically ventilated	Up to 46 cm (18 in.) above ground and within 6 m (20 ft) horizontally from any edge of the dispenser enclosure, including pits or trenches within this area when provided with adequate mechanical ventilation

Chapter 5 Manufacturing Facilities

5.5 Electrical Equipment.

Electrical equipment and wiring in flammable propellant charging and pump rooms shall be suitable for Class I, Division 1 or Class I, Zone 1 locations in accordance with Articles 500, 501, 504, and 505 of NFPA 70, *National Electrical Code.*

5.5.1 If the vacuum pumps for propellant charging are installed remotely (i.e., not in the charging room), the area within 1.5 m (5 ft) of the extremities of the pumps shall be classified as a Class I, Division 2 or Class I, Zone 2 location.

5.5.2* Electrical equipment and wiring in areas where flammable liquids are handled shall be suitable for the classification of the area, as defined in Chapter 6 of NFPA 30, *Flammable and Combustible Liquids Code.*

A.5.5.2 See also NFPA 497, *Recommended Practice for the Classification of Flammable Liquids, Gases, or Vapors and of Hazardous (Classified) Locations for Electrical Installations in Chemical Process Areas.*

5.5.3 The area enclosed by the test bath shall be classified as a Class I, Division 1 or Class I, Zone 1 location.

5.5.4 The area within 1.5 m (5 ft) in all directions of the hot tank shall be classified as a Class I, Division 2 or Class I, Zone 2 location.

NFPA 32, *Standard for Drycleaning Plants,* 2004

1.1 Scope.

This standard shall apply to establishments hereinafter defined as drycleaning plants.

Chapter 3 Definitions

3.3.16 Solvents.
3.3.16.2 ***Class II Solvents.*** Liquids having a flash point at or above 38°C (100°F) and below 60°C (140°F).

3.3.18 Systems.
3.3.18.2 ***Type II.*** Systems employing Class II solvents and complying with the requirements of Chapters 4, 5, 6, and 7 (e.g., Stoddard solvent).

Chapter 7 Type II Drycleaning Plants

7.3.2 Electrical Installations. Electrical equipment and wiring in a Type II drycleaning room shall comply with the provisions of NFPA 70, *National Electrical Code*, for use in Class I, Division 2 hazardous locations.

NFPA 33, *Standard for Spray Application Using Flammable or Combustible Materials*, 2003

1.1* Scope.

A.1.1 The risk to life and property because of the fire and explosion hazards of spray application of flammable and combustible materials will vary depending on the arrangement and operation of the particular process and on the nature of the material being sprayed. The principal hazards addressed in this standard are those of the materials being sprayed: flammable and combustible liquids and combustible powders, as well as their vapors, mists, and dusts, and the highly combustible deposits and residues that result from their use. Properly designed, constructed, and ventilated spray areas are able to confine and control combustible residues, dusts, or deposits and to remove vapors and mists from the spray area and discharge them to a safe location, thus reducing the likelihood of fire or explosion. Likewise, accumulations of overspray residues, some of which are not only highly combustible but also subject to spontaneous ignition, can be controlled.

The control of sources of ignition in spray areas and in areas where flammable and combustible liquids or powders are handled, together with constant supervision and maintenance, is essential to safe spray application operations. The human element requires careful consideration of the location of spray application operations and the installation of fire extinguishing systems so that the potential for spread of fire to other property and damage to property by extinguishing agent discharge is reduced.

1.1.1 This standard shall apply to the spray application of flammable or combustible materials, as herein defined, either continuously or intermittently, by any of the following methods:

(1) Compressed air atomization
(2) Airless or hydraulic atomization
(3) Electrostatic application methods
(4) Other means of atomized application

1.1.2 This standard shall also apply to the application of flammable or combustible materials, as herein defined, either continuously or intermittently, by any of the following methods:

(1) Fluidized bed application methods
(2) Electrostatic fluidized bed application methods
(3) Other means of fluidized application

1.1.3 This standard shall also apply to spray application of water-borne, water-based, and water-reducible materials that contain flammable or combustible liquids or that produce combustible deposits or residues.

1.1.4* This standard shall not apply to spray application processes or operations that are conducted outdoors.

A.1.1.4 This standard does not cover spray application operations that are conducted outdoors on buildings, bridges, tanks, or similar structures. These situations occur only occasionally for any given structure and overspray deposits are not likely to present a hazardous condition. Also, the space where there might be an ignitable vapor–air or dust–air mixture is very limited due to atmospheric dilution.

1.1.5* This standard shall not apply to the use of small portable spraying equipment or aerosol products that are not used repeatedly in the same location.

A.1.1.5 The occasional use of small portable spray equipment or aerosol spray containers is not likely to result in hazardous accumulations of overspray. Therefore, such operations are not within the scope of this standard. The following safeguards, however, should be observed:

(1) Adequate ventilation should be provided at all times, particularly when spray application is conducted in relatively small rooms or enclosures.
(2) Spray application should not be conducted in the vicinity of open flames or other sources of ignition. Either the spray operation should be relocated or the source of ignition should be removed or turned off.
(3) Containers of coating materials, thinners, or other hazardous materials should be kept tightly closed when not actually being used.
(4) Oily or coating-laden rags or waste should be disposed of promptly and in a safe manner at the end of each day's operations, due to the potential for spontaneous ignition.
(5) The same fundamental rules for area cleanliness and housekeeping that are required for industrial spray application operations should be observed.

1.1.6 This standard shall not apply to the spray application of noncombustible materials.

1.1.7 This standard shall not apply to the hazards of toxicity or industrial health and hygiene. *(See 1.2.2.)*

Chapter 3 Definitions

3.3.3 Combustible Powder. Any finely divided solid coating material that is capable of being ignited.

3.3.5 Fluidized Bed. A chamber holding powder coating material that is aerated from below to form an air-supported, expanded cloud of the powder. The object or material being coated is preheated, then immersed into the cloud.

3.3.5.1 Electrostatic Fluidized Bed. A chamber holding powder coating material that is aerated from below to form an air-supported, expanded cloud of the powder. The powder is electrically charged with a charge opposite to that of the object or material being coated.

3.3.12* Spray Booth. A power-ventilated enclosure for a spray application operation or process that confines and limits the escape of the material being sprayed, including vapors, mists, dusts, and residues that are produced by the spraying operation and conducts or directs these materials to an exhaust system.

A.3.3.12 Spray Booth. Spray booths are manufactured in a variety of forms, including automotive refinishing, downdraft, open-face, traveling, tunnel, and updraft booths. This definition is not intended to limit the term spray booth to any particular design. The entire spray booth is part of the spray area. A spray booth is not a spray room.

3.3.13* Spray Room. A power-ventilated fully enclosed room used exclusively for open spraying of flammable or combustible materials.

A.3.3.13 Spray Room. The entire spray room is considered part of the spray area. A spray booth is not a spray room.

3.3.15 Workstation.

3.3.15.1 Limited Finishing Workstation. An apparatus that is capable of confining the vapors, mists, residues, dusts, or deposits that are generated by a spray application process and that meets the requirements of Section 14.3, but does not meet the requirements of a spray booth or spray room, as herein defined.

3.3.15.2 Preparation Workstation. An enclosed, partially enclosed, or unenclosed power-ventilated apparatus that is used to control the dusts and residues generated by surface preparation activities, such as sanding. A preparation workstation is not a limited finishing workstation, spray booth, or spray room, as herein defined.

Chapter 11 Automated Electrostatic Spray Equipment

11.1 Scope.

This chapter shall apply to any equipment using electrostatically charged elements for the atomization, charging, and/or precipitation of flammable and combustible materials for coatings on articles or for other purposes in which the charging or atomizing device is attached to a mechanical support or manipulator, including robotic devices. This chapter shall not apply to devices that are held or manipulated by hand.

11.2 General.

11.2.1 The installation and use of automated electrostatic spray application apparatus shall comply with the requirements of this chapter and also shall comply with the applicable requirements of all other chapters.

11.2.2 Where robot programming procedures involve manual manipulation of the robot arm while spraying with the high voltage components energized, the provisions of Section 12.5 also shall apply.

11.3 Automated Electrostatic Systems.

All automated electrostatic equipment systems shall comply with the requirements of 11.3.1 through 11.3.9.

11.3.1 Transformers, high voltage supplies, control apparatus, and all other electrical portions of the equipment shall be located outside of the spray area, as defined in 3.3.1.3, or shall otherwise meet the requirements of Chapter 6 of this standard.

Exception: High voltage grids, electrodes, electrostatic atomizing heads, integral power supplies, and their connections shall not be required to meet this requirement.

11.3.2 Electrodes and electrostatic atomizing heads shall be insulated from ground. Electrodes and electrostatic atomizing heads that are permanently attached to their bases, supports, reciprocators, or robots shall be deemed to comply with Section 11.3.

11.3.3 High voltage leads shall be insulated and protected from mechanical damage or exposure to destructive chemicals. Any exposed element at high voltage shall be effectively and permanently supported on insulators and shall be effectively guarded against accidental contact or grounding.

11.3.4* All electrically conductive objects in the spray area, except those objects required by the process to be at high voltage, shall be electrically connected to ground with a resistance of not more than 1 megohm. This requirement shall apply to containers of coating material, wash cans, guards, hose connectors, brackets, and any other electrically conductive objects or devices in the area. This requirement shall also apply to any personnel who enter the spray area.

A.11.3.4 Ground circuit resistance should be measured using a test instrument that applies at least 500 volts to the objects being evaluated. A megohm meter should be used for this test, rather than a volt-ohm meter.

11.3.4.1 Containers for any liquids in the spray area also shall be electrically conductive.

11.3.5 Objects or material being coated shall be electrically connected to ground with a resistance of not more than 1 megohm. Areas of contact shall be sharp points or knife edges, where possible, and those areas of contact shall be protected from overspray, where practical.

11.3.5.1 Objects or material transported by a conveyor shall be maintained in electrical contact with the conveyor or other grounding contacts. Hooks and hangers shall be cleaned regularly to ensure grounding.

11.3.6 Electrostatic apparatus shall be equipped with automatic means that rapidly de-energizes the high voltage elements under any one of the following conditions:

(1) Shutdown of ventilating fans or failure of ventilating equipment from any cause
(2) Stopping of the conveyor carrying objects or material through the high voltage field unless stopping is required by the spray process
(3) De-energizing the primary voltage input to the power supply
(4) Occurrence of excessive current leakage at any point on the high voltage system

11.3.7 Safeguards such as booths, fencing, railings, interlocks, or other means shall be placed about the equipment or incorporated therein so that they, either by their location or character or both, ensure that a safe separation of the process is maintained.

11.3.8 Signs shall be conspicuously posted for the following purposes:

(1) To designate the process zone as dangerous with regard to fire and accident
(2) To identify the grounding requirements for all electrically conductive objects in the spray area, including persons
(3) To restrict access to qualified personnel only

11.3.9 All insulators shall be kept clean and dry.

11.4 Incendive Equipment.

Spray equipment that is not considered to be nonincendive shall comply with 11.4.1 and 11.4.2.

11.4.1 Conveyors, hangers, and application equipment shall be arranged so that a minimum separation of at least twice the sparking distance is maintained between the workpiece or material being sprayed and the electrodes, electrostatic atomizing heads, or charged conductors. Warnings defining this safe distance shall be provided.

11.4.2 The high voltage elements shall be automatically deenergized rapidly enough to prevent an arc in the event the clearance between the objects or material being coated and the electrodes or electrostatic atomizing heads falls below that specified in 11.4.1.

11.5 Listing and Approval of Equipment.

Spray equipment shall be listed.

Exception No. 1: Spray equipment that was installed prior to December 31, 1997 shall be listed or approved.

Exception No. 2: This requirement shall not apply to automatic electrostatic spray equipment protected in accordance with Section 9.7 and Section 9.8.

Chapter 12 Handheld Electrostatic Spray Equipment

12.1 Scope.

This chapter shall apply to any equipment using electrostatically charged elements for the atomization, charging, and/or precipitation of flammable and combustible materials for coatings on articles or for other purposes in which the charging or atomizing device is handheld and manipulated during the spraying operation.

12.2 General.

The installation and use of handheld electrostatic spray application apparatus shall comply with the requirements of this chapter and also shall comply with the applicable requirements of all other chapters.

12.3 Handheld Apparatus.

Handheld electrostatic spray apparatus and devices shall be listed. The high voltage circuits shall be designed so that they cannot produce a spark that is capable of igniting the most hazardous vapor–air mixture or powder–air mixture likely to be encountered and so that they cannot result in an ignition hazard upon coming in contact with a grounded object under all normal operating conditions.

12.3.1 The electrostatically charged exposed elements of the hand gun shall be capable of being energized only by an actuator that also controls the coating material supply.

12.3.2 Where the liquid coating material is electrically energized, precautions shall be taken to prevent electric shock.

12.4 Electrical Components.

Transformers, high voltage supplies, control apparatus, and all other electrical portions of the equipment, with the exception of the hand gun itself and its connections to the power supply, shall be located outside of the spray area or shall otherwise meet the requirements of Chapter 6.

12.5 Grounding.

12.5.1* The handle of the spray gun shall be electrically connected to ground by a conductive material. It shall be constructed so that the operator, in normal operating position, is in electrical contact with the grounded handle by a resistance of not more than 1 megohm to prevent buildup of a static charge on the operator's body. Signs indicating the necessity for grounding persons entering the spray area shall be conspicuously posted.

A.12.5.1 This requirement can be met with the use of suitable gloves.

12.5.2 All electrically conductive objects in the spray area, except those objects required by the process to be at high voltage, shall be electrically connected to ground with a resistance of not more than 1 megohm. This requirement shall apply to containers of coating material, wash cans, guards, hose connectors, brackets, and any other electrically conductive objects or devices in the area. This requirement also shall apply to any personnel who enter the area.

12.5.3 Objects or material being coated shall be electrically connected to ground with a resistance of not more than 1 megohm. Areas of contact shall be sharp points or knife edges, where possible, and those areas of contact shall be protected from overspray, where practical.

12.5.4 Objects or material transported by a conveyor shall be maintained in electrical contact with the conveyor or other grounding contacts. Hooks and hangers shall be cleaned to ensure grounding.

Chapter 14 Miscellaneous Spray Operations

14.3 Limited Finishing Workstations.

A limited finishing workstation shall be designed and operated in accordance with the requirements of 14.3.1 through 14.3.9.

14.3.5 The area inside the curtains or partitions shall be considered a Class I, Division 1; Class I, Zone 1; or Class II, Division 1 hazardous (classified) location, as defined by NFPA 70.

Chapter 15 Powder Coating

15.6 Electrical and Other Sources of Ignition.

15.6.1 Electrical utilization equipment and other sources of ignition shall meet both the requirements of Chapter 6 of this standard and Articles 500, 502, 504, and 516 of NFPA 70, as applicable.

15.10 Automated Electrostatic Powder Spraying Equipment.

The provisions of Chapter 11 and other sections of Chapter 15 shall apply to fixed electrostatic equipment, except that electrical equipment not covered therein shall conform to Section 15.6.

15.11 Handheld Electrostatic Powder Spraying Equipment.

The provisions of Chapter 12 and other sections of Chapter 15 shall apply to electrostatic hand guns where used in powder coating, except that the high voltage circuits shall be designed so as not to produce a spark capable of igniting any powder–air mixtures likely to be encountered instead of the vapor–air mixtures referred to, and except that electrical equipment not covered therein shall conform to Section 15.6.

15.12 Electrostatic Fluidized Beds.

15.12.1 The high voltage circuits shall be designed so that any discharge produced when the charging electrodes of the bed are approached or contacted by a grounded object cannot produce a spark that is capable of igniting the most hazardous powder–air mixture likely to be encountered and so that they cannot result in a shock hazard.

15.12.2 Transformers, power packs, control apparatus, and all other electrical portions of the equipment, with the exception of the charging electrodes and their connections to the power supply, shall be located outside of the area classified as hazardous or otherwise shall conform to the requirements of Section 15.6.

15.12.3 All electrically conductive objects within the powder coating area, except those objects required by the process to be at high voltage, shall be electrically connected to ground with a resistance of not more than 10^6 ohms (1 megohm). This requirement shall also apply to any personnel who might enter the area. The powder coating equipment shall carry a prominent, permanently installed warning regarding the necessity for grounding these objects.

15.12.4 Objects or material being coated shall be maintained in electrical contact [less than 10^6 ohms (1 megohm)] with the conveyor or other support in order to ensure grounding. Hangers shall be cleaned to ensure effective contact. Areas of contact shall be sharp points or knife edges where possible.

15.12.5 The electrical equipment and compressed air supplies shall be interlocked with the ventilation system so that the equipment cannot be operated unless the ventilation fans are in operation.

Chapter 17 Styrene Cross-Linked Composites Manufacturing (Glass Fiber Reinforced Plastics)

17.1* Scope.

This chapter shall apply to manufacturing processes involving spray application of styrene cross-linked thermoset resins (commonly known as glass fiber reinforced plastics) for hand lay-up or spray fabrication methods, that is, resin application areas, and where the processes do not produce vapors that exceed 25 percent of the lower flammable limit.

A.17.1 The reinforced styrene-polyester composites industry uses a variety of fabrication techniques to manufacture a wide range of useful products. Most of these products are fabricated with polyester- or vinyl ester-based resins and a fiber reinforcement, most commonly glass fiber. The resins contain a monomer, usually styrene, and are mixed with a catalyst to initiate curing. Other volatile organic chemicals used include the organic peroxide formulations, such as methyl ethyl ketone peroxide (MEKP), used to cure the resin, and various dyes and admixtures.

Open molding is the predominant molding method, with mold sizes ranging from less than 0.1 m^2 (1 ft^2) to very large structures, such as boat hulls over 30 m (100 ft) in length. The two most widely used application methods are hand lay-up and spray-up. In the hand lay-up fabrication method, a glass fiber mat is saturated with the resin by direct spray application or by manual application of the liquid resin. The spray-up fabrication method employs a "chopper gun" that simultaneously applies catalyzed resin and chopped glass fiber to a mold. In addition, many operations use a spray-applied polyester resin gelcoat, as for in-mold coating. Products produced by this industry include boats, bathtubs and shower enclosures, sinks and lavatories, underground storage tanks, auto and truck bodies, recreational vehicles, pollution control equipment, piping, and other specialized parts.

17.5 Electrical and Other Hazards.

17.5.1 Electrical wiring and utilization equipment located in resin application areas that is not subject to deposits of combustible residues shall be installed in accordance with the requirements of NFPA 70 for ordinary hazard locations.

17.5.2 Electrical wiring and utilization equipment located in resin application areas that is subject to deposits of combustible residues shall be listed for such exposure and shall be suitable for Class I, Division 1; Class I, Zone 1; or Class II, Division 1 locations, whichever is applicable. Such wiring and utilization equipment shall be installed in accordance with the requirements of NFPA 70 for the hazardous (classified) location involved.

NFPA 34, *Standard for Dipping and Coating Processes Using Flammable or Combustible Liquids,* 2003

1.1 Scope.

1.1.1 This standard shall apply to processes in which articles or materials are passed through tanks, vats, containers, or process equipment that contain flammable or combustible liquids, including but not limited to dipping, roll coating, flow coating, curtain coating, and cleaning.

1.1.2 This standard shall also apply to the use of water-borne, water-based, and water-reducible materials that contain flammable or combustible liquids or that produce combustible deposits or residues.

1.1.3 This standard shall not apply to processes involving noncombustible liquids.

1.1.4 This standard shall not apply to the use of a liquid that does not have a fire point when tested in accordance with ASTM D 92, *Standard Test Method for Flash and Fire Points by Cleveland Open Cup,* up to the boiling point of the liquid or up to a temperature at which the sample being tested shows an obvious physical change.

Chapter 3 Definitions

3.3.2 Closed Container. A container that is sealed by means of a lid or other device so that neither liquid nor vapor can escape from it at ordinary temperatures.

3.3.3 Coating Processes.
3.3.3.1 Curtain Coating. A coating process by which an object or material is coated by passing it through a vertically flowing film of liquid.

3.3.3.2 Flow Coating. A coating process by which the coating liquid is discharged in an un-atomized state from nozzles, slots, or other openings onto the object or material to be coated.

3.3.3.3 Roll Coating. The process of applying or impregnating objects or materials by bringing them into contact with a roller that is coated with a liquid.

3.3.4 Detearing. A process for removing excess wet coating material from the bottom edge of a dipped or coated object or material by passing it through an electrostatic field.

3.3.5 Dip Tank. A tank, vat, or container of flammable or combustible liquid into which objects or materials are immersed for the purpose of coating, cleaning, or similar processes.

3.3.9* Vapor Area. Any area in the vicinity of: (1) a dipping or coating process and its drainboard, or (2) associated drying or conveying equipment, or (3) the interior of any exhaust plenum or any exhaust duct leading from the process, or (4) other associated equipment that might contain a flammable vapor concentration exceeding 25 percent of the lower flammable limit (LFL) during operation or shut-down periods. *(See Chapter 5.)*

A.3.3.9 Vapor Area. A vapor area is created by the exposed surface of a liquid when the temperature of the liquid is equal to or above its flash point. Hence, a liquid with a flash point of 37.8°C (100°F) (closed cup) might create a vapor area without the application of heat when used in a very warm atmosphere. When heat is applied to a liquid, automatic arrangements to properly limit the liquid temperature will assist in preventing the formation of a vapor area.

When unenclosed dipping operations involve highly volatile liquids or large exposed surfaces, either in an open tank or on dipped materials, the vapor area might extend to all portions of the room where the process is located. When, however, operations are provided with adequate continuous ventilation, the vapor area might extend only a limited distance. *(See Chapter 7.)*

The information in Chapter 7 of this publication and Annex D of NFPA 86, *Standard for Ovens and Furnaces,* can be of assistance in determining the adequacy of ventilation necessary to prevent the formation or limit the extent of a vapor area under the many variable conditions encountered in the dipping and coating processes.

Any vapor concentration exceeding 25 percent of that required to produce a lower flammable limit mixture is considered dangerous and susceptible to fire or explosion. An approved combustible gas indicator should be used to establish the extent of a vapor area. In many cases a further reduction in vapor concentration is needed to prevent a toxic effect on personnel.

In any given situation, the authority having jurisdiction

can determine the extent of the vapor area, taking into consideration the characteristics of the liquid, the degree of sustained ventilation, and the nature of the operations.

3.3.10 Vapor Source. The liquid exposed in the process and on the drainboard. Also, any dipped or coated object from which it is possible to measure vapor concentrations exceeding 25 percent of the lower flammable limit at a distance of 305 mm (1 ft) in any direction from the object.

Chapter 6 Electrical and Other Sources of Ignition

6.1 Scope.

Chapter 6 shall apply to electrical wiring and electrical utilization equipment that is used in, or on, or is attached to the dipping and coating equipment or is in the vicinity of the dipping and coating equipment. This chapter shall also apply to other sources of ignition.

6.2 General.

6.2.1 Dipping and coating process areas where Class I liquids are used, or where Class II or Class III liquids are used at temperatures at or above their flash points, shall meet the requirements of 6.2.1.1 and 6.2.1.2.

6.2.1.1 The extent of hazardous (classified) locations around dipping and coating processes shall be determined in accordance with Sections 6.2, 6.4, and 6.5 of this standard and with Article 500 of NFPA 70, *National Electrical Code.*

6.2.1.2 Electrical wiring and electrical utilization equipment shall be suitable for the location in which they are installed and shall be installed in accordance with the applicable requirements of this chapter and with the applicable requirements of Articles 500, 501, 505, and 516 of NFPA 70, *National Electrical Code.*

6.2.1.3 For the purposes of this standard, the Zone classification system shall be applied as follows. For purposes of area classification, Division classification and Zone classification shall not be intermixed for a given source of release.

(1) Class I, Zone 0 locations are inside open or closed flammable liquid containers.
(2) For all other conditions, a Class I, Division 1 area is permitted to alternatively be classified as a Class I, Zone 1 area.
(3) A Class I, Division 2 is permitted to be alternatively classified as a Class I, Zone 2 area.

6.2.1.4 In instances of areas within the same facility classified separately, Class I, Zone 2 locations shall be permitted to abut, but not overlap, Class I, Division 2 locations. Class I, Zone 0 or Zone 1 locations shall not abut Class I, Division 1 or Division 2 locations. [**70:**505.7(B)]

6.2.2* Open flames, spark-producing equipment or processes, and equipment whose exposed surfaces exceed the autoignition temperature of the dipping or coating liquid shall not be located in the dipping or coating process area or in surrounding areas that are classified as Division 2 or Zone 2.

A.6.2.2 There should be no open flames, hot surfaces, or spark-producing equipment in any dipping or coating process area. Open flames, hot surfaces, or spark-producing equipment should not be located where they can be exposed to deposits of combustible residues. Some residues can ignite at low temperatures, such as those produced by steam pipes, light fixtures, power tools, and so forth.

6.2.3* Any electrical utilization equipment or apparatus that is capable of producing sparks or particles of hot metal, and is located above or adjacent to either the dipping or coating process area or the surrounding Division 2 or Zone 2 areas, shall be of the totally enclosed type or shall be constructed to prevent the escape of sparks or particles of hot metal.

A.6.2.3 Equipment known to produce flames, sparks, or particles of hot metal, including light fixtures, that is located adjacent to areas that are safe under normal operating conditions but which might become dangerous due to accident or careless operation, should not be installed in those areas unless the equipment is totally enclosed or separated from the area by partitions to prevent sparks or particles of hot metal from entering that

6.2.4* Electrical wiring and electrical utilization equipment that is located in the process area and is not subject to deposits of combustible residues shall be suitable for Class I, Division 1 or Class I, Zone 1 locations, whichever is applicable.

A.6.2.4 See NFPA 70, *National Electrical Code.*

6.2.5* Electrical wiring and electrical utilization equipment that is located in the process area, and is subject to deposits of combustible residues, shall be listed for such exposure and shall be suitable for Class I, Division 1 or Class I, Zone 1 locations, whichever is applicable.

Exception: Electrostatic detearing apparatus shall meet the requirements of Chapter 11.

A.6.2.5 See NFPA 70, *National Electrical Code*.

6.3 Electrical Area Classification.

6.3.1* Class I Locations. A Class I location shall be any location where a flammable gas or vapor is present or might be present in the air in quantities sufficient to produce an explosive or ignitable mixture.

A.6.3.1 See NFPA 70, *National Electrical Code*.

6.3.1.1* Class I, Division 1 Locations. A Class I, Division 1 location shall be any location where one of the following conditions exist:

(1) Where an ignitable concentration of flammable gas or vapor can exist under normal operating conditions.
(2) Where an ignitable concentration of flammable gas or vapor can exist frequently because of repair or maintenance operations or because of leakage.
(3) Where breakdown or faulty operation of equipment or processes might release an ignitable concentration of flammable gas or vapor and might also cause simultaneous failure of electrical equipment in such a way as to directly cause the electrical equipment to become a source of ignition.

A.6.3.1.1 See also 500.5(B)(1) of NFPA 70, *National Electrical Code*, for further information and descriptions.

6.3.1.2* Class I, Division 2 Locations. A Class I, Division 2 location shall be any location where one of the following conditions exist:

(1) Where a flammable gas or a volatile flammable liquid is handled, processed, or used, but any flammable gas, vapor, or liquid is confined within a closed container or a closed system from which it can escape only in case of accidental rupture or breakdown of the container or system or in case of abnormal operation of the equipment.
(2) Where an ignitable concentration of flammable gas or vapor is normally prevented by positive mechanical ventilation, but which might exist because of failure or abnormal operation of the ventilating equipment.
(3) Where an ignitable concentration of flammable gas or vapor might occasionally be transmitted from an adjacent Class I, Division 1 location, unless such transmission is prevented by positive pressure ventilation from a source of clean air and effective safeguards against ventilation failure are provided.

A.6.3.1.2 See also 500.5(B)(2) of NFPA 70, *National Electrical Code*, for further information and descriptions.

6.3.1.3* Class I, Zone 0 Locations. A Class I, Zone 0 location shall be any location where an ignitable concentration of flammable gas or vapor is present either continuously or for long periods of time.

A.6.3.1.3 See also 505.5(B)(1) of NFPA 70, *National Electrical Code*, for further information and descriptions.

6.3.1.4* Class I, Zone 1 Locations. A Class I, Zone 1 location shall be any location where one of the following conditions exist:

(1) Where an ignitable concentration of flammable gas or vapor is likely to exist under normal operating conditions.
(2) Where an ignitable concentration of flammable gas or vapor can exist frequently because of repair or maintenance operations or because of leakage.
(3) Where breakdown or faulty operation of equipment or processes might release an ignitable concentration of flammable gas or vapor and might also cause simultaneous failure of electrical equipment in such a way as to directly cause the electrical equipment to become a source of ignition.
(4) Where an ignitable concentration of flammable gas or vapor might occasionally be transmitted from an adjacent Class I, Zone 0 location, unless such transmission is prevented by positive pressure ventilation from a source of clean air and effective safeguards against ventilation failure are provided.

A.6.3.1.4 See also 505.5(B)(2) of NFPA 70, *National Electrical Code*, for further information and descriptions.

6.3.1.5* Class I, Zone 2 Locations. A Class I, Zone 2 location shall be any location where one of the following conditions exist:

(1) Where an ignitable concentration of a flammable gas or vapor is not likely to exist under normal operating conditions and, if an ignitable concentration does exist, will exist only for a short period of time.
(2) Where a flammable gas or a volatile flammable liquid is handled, processed, or used, but any flammable gas, vapor, or liquid is confined within a closed container or a closed system from which it can escape only in case of accidental rupture or breakdown of the container or system or in case of abnormal operation of the equipment.
(3) Where an ignitable concentration of flammable gas or vapor is normally prevented by positive mechanical ventilation, but which might exist because of failure or abnormal operation of the ventilating equipment.
(4) Where an ignitable concentration of flammable gas or

vapor might occasionally be transmitted from an adjacent Class I, Zone 1 location, unless such transmission is prevented by positive pressure ventilation from a source of clean air and effective safeguards against ventilation failure are provided.

A.6.3.1.5 See also 505.5(B)(3) of NFPA 70, *National Electrical Code*, for further information and descriptions.

6.3.2 Class II Locations. A Class II location shall be any location that might be hazardous because of the presence of a combustible dust.

6.3.2.1* Class II, Division 1 Locations. A Class II, Division 1 location shall be any location where one of the following conditions exist:

(1) Where combustible dust is in the air in quantities sufficient to produce explosive or ignitable mixtures under normal operating conditions.
(2) Where mechanical failure or abnormal operation of machinery or equipment might cause an explosive or ignitable mixture of combustible dust in air and might also provide a source of ignition through simultaneous failure of electrical equipment, operation of protection devices, or from other causes.
(3) Where combustible dust of an electrically conductive nature might be present in hazardous quantities.

A.6.3.2.1 See also 500.5(C)(1) of NFPA 70, *National Electrical Code*, for further information and descriptions.

6.3.2.2* Class II, Division 2 Locations. A Class II, Division 2 location shall be any location where one of the following conditions exist:

(1) Where combustible dust is not normally in the air in quantities sufficient to produce explosive or ignitable mixtures and accumulations of dust are normally insufficient to interfere with the normal operation of electrical equipment or other apparatus, but combustible dust might be in suspension in air as a result of infrequent malfunctioning of handling or processing equipment.
(2) Where accumulations of combustible dust on, in, or in the vicinity of the electrical equipment might be sufficient to interfere with the safe dissipation of heat from electrical equipment or might be ignited by abnormal operation or failure of the electrical equipment.

A.6.3.2.2 See also 505.5(C)(2) of NFPA 70, *National Electrical Code*, for further information and descriptions.

6.4 Areas Adjacent to Open Processes.

Electrical wiring and electrical utilization equipment located adjacent to open dipping and coating processes shall meet the requirements of 6.4.1 through 6.4.4 and Figure 6.4(a) or Figure 6.4(b), whichever is applicable.

6.4.1 Electrical wiring and electrical utilization equipment located in any sump, pit, or belowgrade channel that is within 7620 mm (25 ft) horizontally of a vapor source, as defined by this standard, shall be suitable for Class I, Division 1 or Class I, Zone 1 locations. If the sump, pit, or channel extends

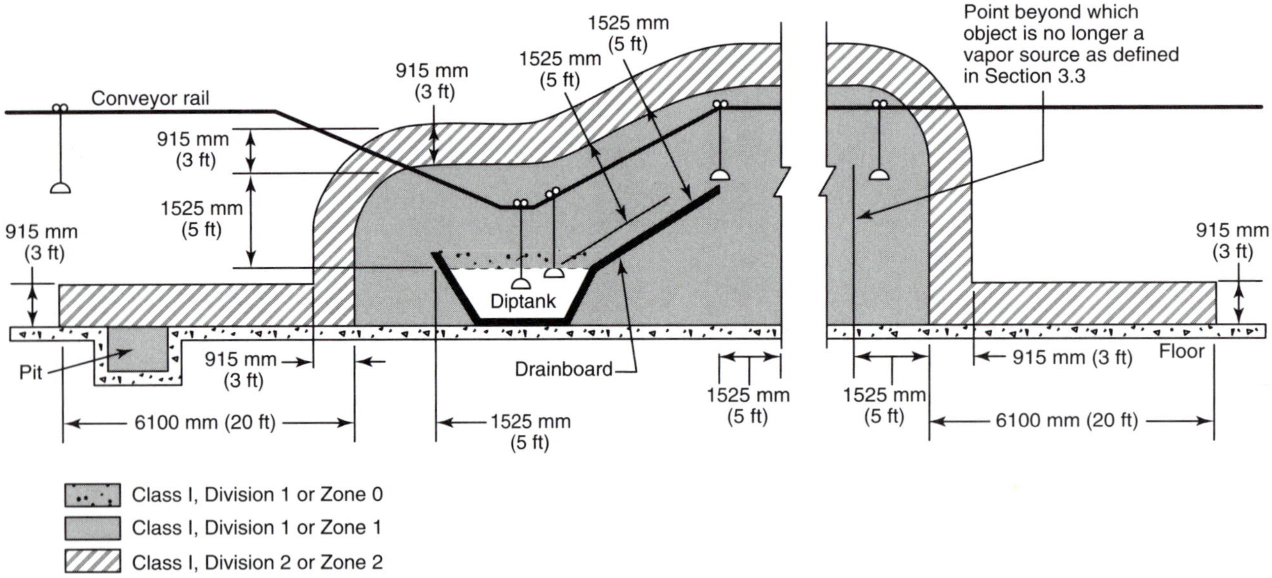

FIGURE 6.4(a) Electrical Area Classification for Open Dipping and Coating Processes Without Vapor Containment or Ventilation.

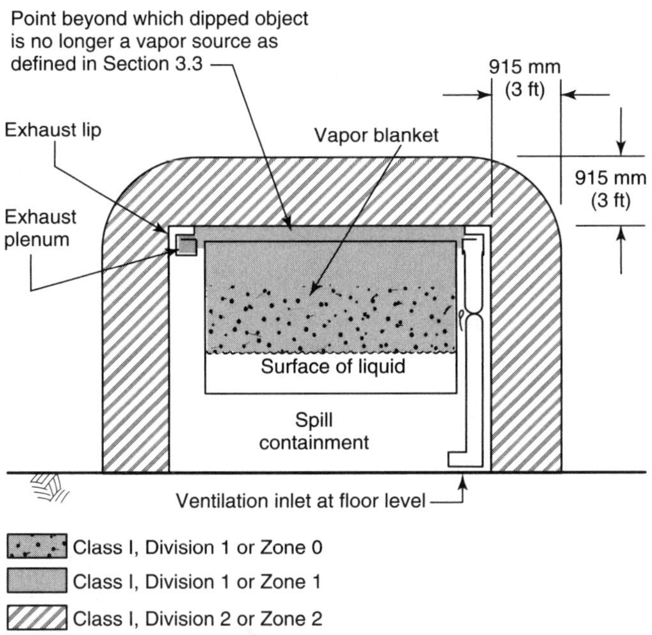

FIGURE 6.4(b) Electrical Area Classification for Open Dipping and Coating Processes with Peripheral Vapor Containment and Ventilation.

beyond 7620 mm (25 ft) of the vapor source, it shall be provided with a vapor stop, or it shall be classified as Class I, Division 1 or Class I, Zone 1 for its entire length.

6.4.2 Electrical wiring and electrical utilization equipment located within 1525 mm (5 ft) of a vapor source shall be suitable for Class I, Division 1 or Class I, Zone 1 locations. The area inside the dip tank shall be classified as Class I, Division 1 or Class I, Zone 0, whichever is applicable.

6.4.3 Electrical wiring and electrical utilization equipment located within 915 mm (3 ft) of the Class I, Division 1 or Class I, Zone 1 location described in 6.4.2 shall be suitable for Class I, Division 2 or Class I, Zone 2 locations, whichever is applicable.

6.4.4 The space 915 mm (3 ft) above the floor and extending 6100 mm (20 ft) horizontally in all directions from the Class I, Division 1 or Class I, Zone 1 location described in 6.4.3 shall be classified as Class I, Division 2 or Class I, Zone 2 and electrical wiring and electrical utilization equipment located within this space shall be suitable for Class I, Division 2 or Class I, Zone 2 locations, whichever is applicable.

Exception: This space shall be permitted to be nonclassified for purposes of electrical installations if the surface area of the vapor source does not exceed 0.5 m² (5 ft²) and the contents of the tank do not exceed 19 L (5 gal), and the vapor concentration during operating and shutdown periods does not exceed 25 percent of the lower flammable limit.

6.5 Areas Adjacent to Enclosed Processes.

Areas adjacent to enclosed dipping and coating processes shall be classified in accordance with 6.5.1, 6.5.2, and Figure 6.5.

6.5.1 The interior of any enclosed dipping or coating process or apparatus shall be a Class I, Division 1 or Class I, Zone 1 location, and electrical wiring and electrical utilization equipment located within this space shall be suitable for Class I, Division 1 or Class I, Zone 1 locations, whichever is applicable. The area inside the dip tank shall be classified as Class I, Division 1 or Class I, Zone 0, whichever is applicable.

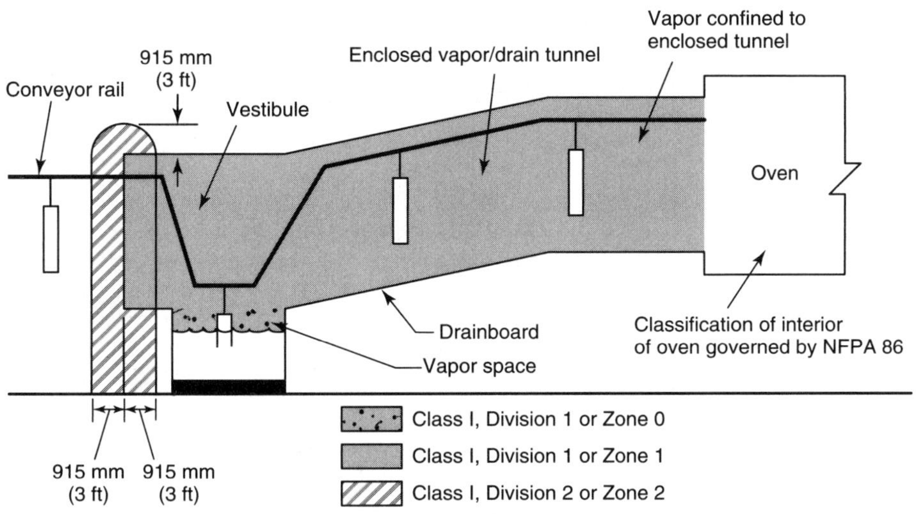

FIGURE 6.5 Electrical Area Classification Around Enclosed Dipping and Coating Processes.

6.5.2 The space within 915 mm (3 ft) in all directions from any opening in the enclosure and extending to the floor or grade level shall be classified as Class I, Division 2 or Class I, Zone 2, and electrical wiring and electrical utilization equipment located within this space shall be suitable for Class I, Division 2 locations or Class I, Zone 2, whichever is applicable.

6.5.3 All other spaces adjacent to an enclosed dipping or coating process or apparatus shall be classified as nonhazardous for purposes of electrical installations.

6.6 Equipment and Containers in Ventilated Areas.

Where dipping or coating equipment or supply containers are located in an area that is provided with ventilation as defined in this standard and are adjacent to a process area, but outside of a storage room or mixing room, the area within 915 mm (3 ft) in all directions from any open container or equipment and extending to the floor or grade level shall be classified as Class I, Division 1 or Class I, Zone 1 as shown in Figure 6.6. The area extending 610 mm (2 ft) beyond the Division 1 or Zone 1 location shall be classified as Class I, Division 2 or Class I, Zone 2, whichever is applicable. In addition, the area within 3050 mm (10 ft) horizontally of the perimeter of such open container or equipment, up to a height of 460 mm (18 in.) above the floor or grade level, shall be classified as Class I, Division 2 or Class I, Zone 2, as shown in Figure 6.6. The area inside the tank or supply container shall be classified as Class I, Division 1, or Class I, Zone 0, whichever is applicable. Electrical wiring and electrical utilization equipment installed in these areas shall be suitable for the location.

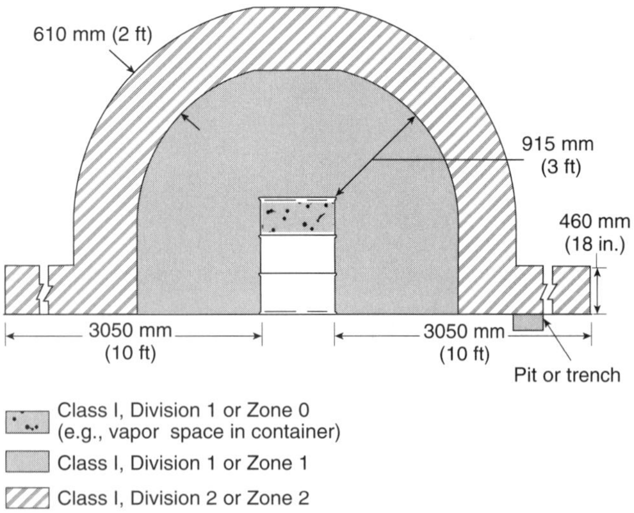

FIGURE 6.6 Electrical Area Classification Around an Open Container in a Ventilated Area.

6.7 Light Fixtures.

6.7.1 Light fixtures that are attached to the walls or ceilings of a process enclosure, but are outside of any classified area and are separated from the process area by glass panels that meet the requirements of 5.2.2 shall be suitable for use in ordinary hazard (general purpose) locations. Such fixtures shall be serviced from outside the enclosure.

6.7.2 Light fixtures that are attached to the walls or ceilings of a process enclosure, are located within the Class I, Division 2 or Class I, Zone 2 location, and are separated from the process area by glass panels that meet the requirements of 5.2.2 shall be suitable for use in that location. Such fixtures shall be serviced from outside the enclosure.

6.8* Static Electric Discharges.

In order to prevent discharges or sparks from the accumulation of static electricity, all persons and all electrically conductive objects, including any metal parts of the process equipment or apparatus, containers of material, exhaust ducts, and piping systems that convey flammable or combustible liquids, shall be electrically grounded.

A.6.8 NFPA 77, *Recommended Practice on Static Electricity,* contains information about grounding for static electric charge.

Chapter 11 Electrostatic Detearing Apparatus

11.1 Scope.

Chapter 11 shall apply to any dipping or coating process that incorporates electrostatic detearing systems to remove excess coating material.

11.2 General.

11.2.1 Electrostatic detearing equipment shall meet the requirements of Chapter 1 through Chapter 10, except as hereinafter modified, and shall also meet the requirements of this chapter.

11.2.2 Electrostatic apparatus and devices used in connection with paint detearing operations shall be listed or approved.

11.3 Requirements for Electrical System and Components.

11.3.1 Transformers, high voltage supplies, control apparatus, and all other electrical portions of the equipment, with the exception of high voltage grids and their connections, shall be located outside the vapor area defined in 3.3.9 or shall meet the requirements of Chapter 6.

11.3.2 Electrodes shall be supported and shall be insulated from ground.

11.3.3 High voltage leads to electrodes shall be supported on insulators and shall be guarded against accidental contact or grounding. Insulators shall be kept clean and dry.

11.3.4 A separation of at least twice the sparking distance shall be maintained between the object or material being deteared and the electrodes or conductors. A sign indicating this separation shall be conspicuously posted near the assembly.

11.4 Support of Workpieces.

11.4.1 Workpieces or material being deteared shall be supported on conveyors or hangers.

11.4.2* The conveyor shall be arranged to ensure that the workpieces or material being deteared are electrically connected to ground with a resistance of not more than 10^6 ohms (1 megohm), and that the distance required by 11.3.4 is maintained between the workpiece or material and the electrodes at all times.

A.11.4.2 Ungrounded parts can, if they are near high voltage electrodes, become electrically charged. In this condition, they constitute an energy source capable of producing an ignition-capable spark when approached by a grounded object or person. This condition can be avoided if the electrical resistance between the part and ground is 10^6 ohms (1 megohm) or less. Further detailed information on this subject can be found in NFPA 77, *Recommended Practice on Static Electricity.*

11.4.3 Workpieces or material being deteared shall be supported to prevent swinging or movement that would reduce the distance to less than that required.

11.5 Manual Operations.

Electrostatic detearing shall not be used where the workpieces or material being deteared are manipulated by hand.

11.6 Electrical Safety Requirements.

Electrostatic apparatus shall be equipped with automatic means that will de-energize and ground the high voltage elements and signal the operator under any of the following conditions:

(1) Stopping of ventilating fans or failure of ventilating equipment from any cause
(2) Stopping of the conveyor carrying the objects or material through the high voltage field
(3) Occurrence of a fault to ground or excessive current leakage at any point on the high voltage system
(4) Reduction of clearances to below that specified in 11.3.4
(5) De-energizing of the primary side of the power supply

11.7 Personnel Safety.

Safeguards such as enclosures, fencing, railings, or other means shall be placed about the equipment or incorporated therein so that they, either by their location or character or both, ensure that isolation of the process is maintained from plant storage or personnel.

11.8* Grounding Requirements.

All electrically conductive objects in the process area, except those objects required by the process to be at high voltage, shall be electrically connected to ground with a resistance of not more than 10^6 ohms (1 megohm). This requirement shall apply to paint containers, wash cans, guards, and any other electrically conductive objects or devices in the area. This requirement also shall apply to any personnel in the process area. The equipment shall carry a prominent, permanently installed warning regarding the necessity for this grounding feature.

A.11.8 The grounding requirements for parts being deteared *(see A.9.4)* apply, for the same reasons, to all other conductive objects (including personnel) that are in the vicinity of the high voltage electrodes.

11.9 Signs.

Signs designating the process zone as dangerous in regard to fire and accident shall be conspicuously posted.

11.10 Drip Plates and Screens.

Drip plates and screens subject to deposits of coating material shall be removable and shall be taken to a safe place for cleaning.

NFPA 35, *Standard for the Manufacture of Organic Coatings*, 1999

1-1 Scope.

1-1.1 This standard shall apply to facilities that use flammable and combustible liquids, as herein defined, to manufacture organic coatings for automotive, industrial, institutional, household, marine, printing, transportation, and other applications.

1-1.2 This standard shall not apply to the following:

(1)* Operations involving the use or application of coating materials
(2)* Storage of organic coatings in locations other than the manufacturing facility

Chapter 3 Building Construction

3-6 Electrical Equipment.

3-6.2* Where Class I liquids are exposed to the air or where Class II or Class III liquids are exposed to the air at temperatures at or above their flash points, the equipment used in the building and the ventilation of the building shall be designed so that flammable vapor-air mixtures are confined, under normal operating conditions, to the inside of the equipment and to a zone that extends not more than 5 ft (1.5 m) from the equipment. Table 5-9.5.3 of NFPA 30, *Flammable and Combustible Liquids Code*, shall be used to determine the extent of hazardous (classified) locations for purposes of installation of electrical equipment and wiring. In establishing the extent of hazardous (classified) locations, such locations shall not extend beyond floors, walls, roofs, or other solid partitions that have no communicating openings into the hazardous (classified) locations.

A-3-6.2 Examples of such equipment are dispensing stations, sand mills, open centrifuges, plate-and-frame filter presses, open vacuum filters, and the surfaces of open equipment. The classifications listed in Table 5-9.5.3 of NFPA 30, *Flammable and Combustible Liquids Code*, are based on the premise that the installation meets all applicable requirements of NFPA 30 and NFPA 70, *National Electrical Code*. Should this not be the case, the authority having jurisdiction has the authority to determine the extent of hazardous (classified) locations.

For additional information, see NFPA 497, *Recommended Practice for the Classification of Flammable Liquids, Gases, or Vapors and of Hazardous (Classified) Locations for Electrical Installations in Chemical Process Areas*.

3-6.3 Where the provisions of 3-6.2 require the installation of Class I, Division 1, or Class I, Division 2 electrical equipment, ordinary electrical equipment, including switch gear, shall be permitted to be used if installed in a room or enclosure that is maintained under positive pressure with respect to the classified area in accordance with NFPA 496, *Standard for Purged and Pressurized Enclosures for Electrical Equipment*. Ventilation makeup air shall not be contaminated.

NFPA 36, *Standard for Solvent Extraction Plants*, 2004

1.1 Scope.

1.1.1* This standard shall apply to the commercial scale extraction processing of animal and vegetable oils and fats by the use of Class I flammable hydrocarbon liquids, hereinafter referred to as "solvents."

A.1.1.1 Extraction processes that use flammable liquids but are not within the scope of NFPA 36 might be within the scope of NFPA 30, *Flammable and Combustible Liquids Code*, and the user is referred to that document for guidance. *(See Chapter 3 for definitions of terms, including "extraction process" and "solvent.")*

1.1.2 This standard shall also apply to any equipment and buildings that are located within 30 m (100 ft) of the extraction process.

1.1.3 This standard shall also apply to the unloading, storage, and handling of solvents, regardless of distance from the extraction process.

1.1.4 This standard shall also apply to the means by which material to be extracted is conveyed from the preparation process to the extraction process.

1.1.5 This standard shall also apply to the means by which extracted desolventized solids and oils are conveyed from the extraction process.

1.1.6 This standard shall also apply to preparation and meal finishing processes that are connected by conveyor to the extraction process, regardless of intervening distance.

1.1.7* This standard shall not apply to the storage of raw materials or finished products.

A.1.1.7 See NFPA 61, *Standard for the Prevention of Fires and Dust Explosions in Agricultural and Food Processing Facilities*.

1.1.8 This standard shall not apply to extraction processes that use liquids that are miscible with water.

1.1.9 This standard shall not apply to extraction processes that use flammable gases, liquefied petroleum gases, or nonflammable gases.

1.1.10 This standard shall prohibit the use of processes that employ oxygen-active compounds that are heat or shock sensitive, such as certain organic peroxides, within the area defined in 1.1.2.

Chapter 3 Definitions

3.3 General Definitions.

3.3.1 Condensate. Any material that has been condensed from the vapor state to the liquid state.

Chapter 4 General Requirements

4.3 Sources of Ignition.

4.3.1 Electrical installations shall meet all applicable requirements of NFPA 70, *National Electrical Code*®.

Chapter 5 Bulk Solvent Unloading and Storage

5.5 Sources of Ignition.

5.5.1 Electrical Equipment. All electrical utilization equipment and electrical wiring shall be suitable for Class I, Division 1 or 2, Group D hazardous (classified) locations and shall be designed and installed in accordance with NFPA 70, *National Electrical Code.*

5.5.1.1 Where enclosures that house solvent-handling equipment such as solvent pumps or valves are provided or where solvents are transferred to individual containers, these enclosures shall be considered to be Division 1 locations.

5.5.1.2* In outdoor locations, areas adjacent to loading racks or platforms or to aboveground tanks shall be considered to be Division 2 locations. The Division 2 locations shall extend 7.5 m (25 ft) horizontally from such racks or tanks, and 4.5 m (15 ft) upward from adjacent ground level.

A.5.5.1.2 See Chapter 6 of NFPA 30, *Flammable and Combustible Liquids Code*, and NFPA 70, *National Electrical Code*.

Chapter 6 Preparation and Meal Finishing Processes

6.1 Scope.

6.1.1 This chapter shall apply to processes that are used to prepare raw material for the extraction process and are connected by conveyor or other materials-handling equipment to the extraction process, regardless of intervening distance.

6.1.2 This chapter shall also apply to processes used to prepare extracted and desolventized product for storage.

6.1.3 Where the processing operations do not involve liberation of combustible dusts, the requirements of 6.2.2, 6.2.3, and 6.3.1 shall not apply.

6.3 Electrical Systems.

6.3.1* In areas where combustible dust presents a hazard, all electrical wiring and equipment shall conform to the requirements for Class II, Division 1, Group G locations.

A.6.3.1 See NFPA 70, *National Electrical Code*.

6.3.2* Static protection shall be provided in equipment located in areas where combustible dust presents a hazard.

A.6.3.2 See NFPA 77, *Recommended Practice on Static Electricity.*

Chapter 7 Extraction Process

7.7 Electrical Systems.

7.7.1* Class I, Division 1 Locations. Electrical wiring and electrical utilization equipment of the extraction process shall be installed in accordance with the requirements for Class I, Division 1 locations as specified by NFPA 70, *National Electrical Code*. The Class I, Division 1 location shall extend outward from the extraction process and into the restricted area for a horizontal distance of not less than 4.5 m (15 ft) and a vertical distance of not less than 1.5 m (5 ft) above the highest vent, vessel, or equipment containing solvent, as shown in Figure 7.7.1.

A.7.7.1 Electrical equipment and wiring systems that are installed in areas classified as hazardous (classified) locations in accordance with this standard and NFPA 70, *National Electrical Code*, should meet certain specific requirements to ensure that they will not provide a means of ignition for any ignitible atmosphere that might be present. Usually, this is accomplished by using explosionproof electrical equipment and wiring methods that are listed for use in such locations. Installation of such equipment and wiring should meet the requirements of Chapter 5 of NFPA 70. Due to their nature, explosionproof electrical devices and wiring methods are costly. Other means of providing equivalent safety are available.

One alternate method is to use purged or pressurized enclosures. Purged and pressurized enclosures are built to be relatively tight and are supplied with clean air from a compressed air system or from a fan taking suction from an uncontaminated source. The air supply is arranged to maintain a slight positive pressure inside the enclosure. Clean air leaks out, but contaminated air cannot leak in. General purpose electrical equipment that otherwise would not be allowed in the hazardous location can be installed in these enclosures, and the slight positive pressure that is maintained

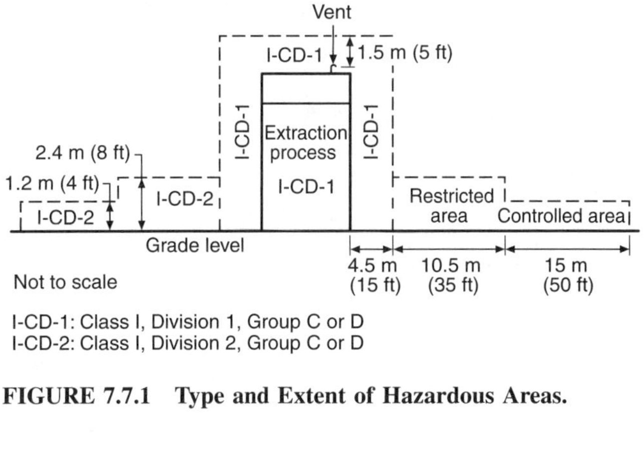

FIGURE 7.7.1 Type and Extent of Hazardous Areas.

in the enclosure prevents ignitible atmospheres from entering the enclosure and reaching a source of ignition. NFPA 496, *Standard for Purged and Pressurized Enclosures for Electrical Equipment*, provides design and performance requirements for such systems. Note that purging and pressurization can be used for small enclosures, large enclosures, and even control rooms.

For low-energy process control systems, intrinsically safe and nonincendive devices can be used. These electrical devices are typically of such low voltage, amperage, and capacitance that they cannot release ignition-capable energy. Nonincendive devices can only be used in Division 2 locations, while intrinsically safe devices can be used in either Division 1 or Division 2 locations.

Finally, moving some electrical devices and equipment to other areas of the plant that are not classified is always an option.

7.7.2 Electrical wiring and electrical utilization equipment within the restricted area beyond the 4.5-m (15-ft) distance specified in 7.7.1 and to a height of 2.4 m (8 ft) above the extraction process grade level shall be installed in accordance with the requirements of Class I, Division 2 locations, as specified in NFPA 70, *National Electrical Code*, and as shown in Figure 7.7.1.

7.7.3 Electrical wiring and electrical utilization equipment within the controlled area and to within a height of 1.2 m (4 ft) above grade level shall be installed in accordance with the requirements of Class I, Division 2 locations, as shown in Figure 7.7.1.

Exception: This requirement shall not apply to the preparations process. (See 7.2.8.)

7.7.4 Permanent luminaires (lighting fixtures) shall be installed where needed.

7.7.5 Flashlights approved for Class I, Group D locations shall be provided.

NFPA 40, *Standard for the Storage and Handling of Cellulose Nitrate Film*, 2001

Chapter 1 Administration

1.1* Scope.

A.1.1 Although past experience in the storage and handling of cellulose nitrate film has resulted in a good safety record, fire tests conducted prior to 1967 indicated the desirability of a modification of existing standards. The requirements of this standard, therefore, apply strictly to long term storage of cellulose nitrate film.

1.1.1* This standard shall apply to all facilities that are involved with the storage and handling of cellulose nitrate–based film.

A.1.1.1 Cellulose nitrate–based film includes, but is not limited to, original negative, duplicate negative, interpositive (fine grain), color separation master (YCM), successive exposure master (SEN), optical soundtrack negative or master, mattes, title bands, and release prints.

1.1.2 This standard shall not apply to the storage and handling of film having a base other than cellulose nitrate.

Chapter 4 Construction Requirements and Arrangements of Buildings

4.6.1 All electrical wiring and equipment shall comply with NFPA 70, *National Electrical Code®*, for Class I, Group D, Division 2 locations. The temperature rating of electrical equipment shall be Class T6.

NFPA 45, *Standard on Fire Protection for Laboratories Using Chemicals*, 2004

1.1 Scope.

1.1.1 This standard shall apply to laboratory buildings, laboratory units, and laboratory work areas whether located above or below grade in which chemicals, as defined, are handled or stored.

1.1.2 This standard shall not apply to the following:

(1)* If conditions (a) and (b) exist, this standard shall not apply:
  (a) Laboratory units that contain less than or equal to 4 L (1 gal) of flammable or combustible liquid
  (b) Laboratory units that contain less than 2.2 standard m^3 (75 scf) of flammable gas, not including piped-in low-pressure utility gas installed in accordance with NFPA 54, *National Fuel Gas Code*

(2) Laboratories that are pilot plants

(3) Laboratories that handle only chemicals with a hazard rating of zero or one, as defined by NFPA 704, *Standard System for the Identification of the Hazards of Materials for Emergency Response*, for all of the following: health, flammability, and instability

(4) Laboratories that are primarily manufacturing plants

(5) Incidental testing facilities

(6) Physical, electronic, instrument, laser, or similar laboratories that use chemicals only for incidental purposes, such as cleaning

(7) Laboratories that work only with radioactive materials, as covered by NFPA 801, *Standard for Fire Protection for Facilities Handling Radioactive Materials*

(8) Laboratories that work only with explosive material, as covered by NFPA 495, *Explosive Materials Code*

Chapter 5 Laboratory Unit Design and Construction

5.6 Electrical Installation.

5.6.2* Laboratory work areas, laboratory units, and chemical fume hood interiors shall be considered as unclassified electrically with respect to Article 500 of NFPA 70, *National Electrical Code.*

Exception: Under some conditions of hazard, it could be necessary to classify a laboratory work area, or a part thereof, as a hazardous location, for the purpose of designating the electrical installations. [See 10.5.5 (electric motors) and 12.2.2.2 (refrigerators).]

A.5.6.2 Table A.5.6.2 and the sample calculations were developed to explain the committee's decision to allow unclassified electrical equipment inside the hood. The requirement in 5.6.2 identifies laboratory units, laboratory work areas,

Table A.5.6.2 Properties of Commonly Used Flammable Liquid Loading in Chemical Fume Hoods

Flammable Liquid	Specific Gravity (Water = 1.0)	Vapor Density (Air = 1.0)	Flash Point		Lower Flam. Limit (Vol%)	Vapor Volume		Ratio of the Vapor Volume to Hood Exhaust Volume		Quantity of Liquid Vaporized to Reach Lower Flam. Limit	
			°C	°F		m^3/L	ft^3/gal	$0.472\ m^3/sec$	$1000\ ft^3/min$	L	gal
Acetone	0.8	2.0	20	4	2.5	0.33	44.4	0.7	0.044	2.2	0.57
Acetonitrile	0.8	1.4	5.5	42	3.0	0.48	63.4	1	0.063	1.8	0.48
Amyl Ether	0.8	5.5	57	135	2.3	0.12	16.2	0.26	0.016	5.5	1.44
Amyl Acetate	0.9	4.4	15.5	60	1.1	0.17	22.2	0.35	0.022	1.9	0.50
Aniline	1.1	3.2	70	158	1.3	0.29	38.2	0.60	0.038	1.3	0.34
Benzene	0.9	2.6	11	12	1.2	0.27	35.7	0.57	0.036	1.2	0.33
Butyl Alcohol	0.8	2.5	37	98	1.4	0.26	34.2	0.54	0.034	1.6	0.41
Chloro-benzene	1.1	3.9	28	82	1.3	0.23	31.3	0.50	0.031	1.6	0.42
Cumene	0.9	4.1	35.5	96	0.9	0.18	24.4	0.39	0.024	1.4	0.38
Decane	0.7	4.9	46	115	0.8	0.12	15.9	0.25	0.016	1.9	0.50
Ethyl Alcohol	0.8	1.5	13	55	3.3	0.42	55.5	0.88	0.055	2.3	0.60
Ethyl Ether	0.7	2.5	45	49	1.9	0.22	29.9	0.47	0.030	2.4	0.63
Gasoline	0.8	3.4	43	45	1.4	0.19	25.4	0.40	0.025	2.1	0.56
Heptane	0.7	3.4	4	25	1.1	0.17	22.2	0.35	0.022	1.9	0.50
Hexane	0.7	2.9	22	7	1.1	0.19	25.9	0.41	0.026	1.6	0.42
Methyl Alcohol	0.8	1.1	11	52	6.0	0.60	80.7	1.28	0.081	2.8	0.74
Methylcyclohexane	0.8	3.4	4	25	1.2	0.20	26.1	0.4	0.026	1.7	0.46
Methylene Chloride	1.3	2.9	N/A	N/A	13.0	0.37	49.8	0.79	0.050	9.8	2.60
Methyl Ethyl Ketone	0.8	2.4	9	16	1.4	0.27	35.5	0.56	0.035	1.5	0.40
Pentane	0.6	2.5	40	40	1.5	0.20	26.6	0.42	0.027	2.1	0.56
Propyl Alcohol	0.8	2.0	23	74	2.2	0.32	42.3	0.67	0.042	2.0	0.52
Propylene Oxide	0.83	2.0	37	35	2.3	0.35	46.1	0.73	0.046	1.9	0.50
Propyl Ether	0.75	3.5	21	70	1.3	0.18	23.8	0.38	0.024	2.0	0.54
Pyridine	1.0	2.7	20	68	1.8	0.31	41.1	0.65	0.041	1.7	0.44
Tetrahydro-furan	0.9	2.4	14	6	2.0	0.30	39.9	0.63	0.040	1.9	0.50
Toluene	0.9	3.1	4	40	1.1	0.24	32.2	0.51	0.032	1.3	0.34
m-Xylene	0.9	3.6	27	81	1.1	0.20	27.0	0.43	0.027	1.6	0.41

Notes:
(1) Air volume and liquid quantity normalized at 0.472 m^3/sec (1000 ft^3/min) and 1 L (1 gal), respectively, to enable extrapolation to other conditions.
(2) Columns 1–4 are constants.
(3) Column 5 is quantity of vapor generated from 1 L (1 gal) of liquid.
(4) Column 6 is the vapor volume in 0.472 m^3/sec (1000 ft^3/min) air from 1 L (1 gal) of liquid that is assumed to vaporize over 1 minute with perfect mixing.

and the interior of chemical fume hoods as electrically unclassified areas. The ensuing discussion about the classification of chemical fume hoods will support the classification of chemical fume hoods as general purpose when the combination of the airflow rate and flammable vapor concentration will not permit a stoichiometric vapor–air mixture to form (which could result in an explosion).

$$\text{Vapor volume} = \frac{8.33\ \text{lb/gal}\ (H_2O) \times \text{specific gravity (solvent)}}{0.075\ \text{lb/ft}^3 \times \text{vapor density (relative to air)}}$$

$$\text{Vapor volume} = \frac{0.998\ \text{kg/L}\ (H_2O) \times \text{specific gravity (solvent)}}{1.2\ \text{kg/m}^3 \times \text{vapor density (relative to air)}}$$

The effective ventilation of laboratory work areas where toxic or flammable chemicals are handled or stored is critical to safe operations for two important reasons.

First, to prevent personnel exposure to toxic or odiferous materials during the course of conducting normal laboratory activities, the resulting vapors or gases are usually exhausted with nonrecirculated, conditioned air through hoods and local exhaust arrangements. This is one reason why it is very important to assure that laboratory ventilation systems are operating properly while activities are being carried out in the work area. Depending on the nature of the chemicals and the type of operations, it might be possible to reduce ventilation rates during off-hours when work areas are unoccupied.

Second, when activities involve the use of toxic or flammable materials in quantities that cannot be safely handled in work areas, the activity should be done in a well-ventilated area or hood. Typical, modern hoods operate under negative pressure compared to the work area with hood face velocities that range from 0.41 m/sec to 0.61 m/sec (80 linear ft/min to 120 linear ft/min), with 0.5 m/sec (100 linear ft/min) being the most common. These ventilation rates mix, dilute, and exhaust toxic or flammable materials to a safe location.

Many chemical fume hoods contain a wide variety of electrical equipment ranging from vaporproof lighting, pumps, motors, heated surfaces, and electronic instruments. The electrical classification of the hoods is dependent on the ventilation rate, the quantity of flammable materials, and the type of chemical process or activity (Schram and Earley, 1988). The challenge is to determine how flammable vapors or gases can be safely used in a hood with unclassified electrical equipment and heated surfaces.

When the quantity of flammable vapor is minimized and ventilation rates are maximized, a determination (as shown in Table A.5.6.2) of the volume of vapor generated by the vaporization of flammable liquid can help to approximate the maximum quantity of liquid that can be used to operate below the lower flammable limit (LFL). The calculations are made assuming that liquid vaporizes within 1 minute and mixes perfectly with air moving through the hood with a hood size of 4 ft wide and 2.5 ft high as shown in the sample calculations. The exhaust air volume is a critical parameter necessary to calculate the maximum quantity that can be vaporized in 1 minute before reaching LFL. Under these ventilation rates and controlled quantities of flammable liquids, the hood is considered to be electrically unclassified, since the LFL of the specific vapor will not be reached. Ventilation is assumed to be constant, and its failure simultaneously with an unexpected development of flammable vapors above the LFL is highly unlikely. The sample calculations for acetone illustrate how the table is derived.

Sample Calculations (Metric Units)

Given: Acetone

Specific gravity	0.8	Water	1.0
Vapor density	2.0	Air	1.0
Lower flammable limit	2.5 % by volume		

Fume hood

Face, 4 ft (1.22 m) wide

Face velocity, 100 ft/min (30.48 m/min)

2.5 ft (0.76 m) high

Face opening, 10 ft^2 (0.93 m^2)

Assume: 1 L/min

$$\text{Vapor volume} = \frac{0.998\ \text{kg/L} \times 0.8}{1.2\ \text{kg/m}^3 \times 2.0} = 0.33\ \text{m}^3/\text{L}$$

Fume hood dilution: 0.93 m^2 × 30.48 m/min = 28.35 m^3/min

$$\frac{0.33\ \text{m}^3/\text{L}}{28.35\ \text{m}^3/\text{min}} = 0.01164\ \text{min/L} \times 1\ \text{L/min} = 0.01164$$

$$0.01164(60) = 0.70\ \text{per sec or}\ 1.164\%$$

$$\text{LFL} = \frac{2.5\%}{1.164\%} = \frac{X}{1\ \text{L}}$$

X = 2.15 L in 1 minute to reach LFL

Note: One common control used is 25% of LFL or 0.54 L.

Sample Calculations (English Units)

Given: Acetone

Specific gravity	0.8	Water	1.0
Vapor density	2.0	Air	1.0
Lower flammable limit	2.5 % by volume		

Fume hood

Face, 4 ft wide 2.5 ft high

Face opening, 10 ft^2

Face velocity, 100 ft/min

Assume: 1 gal/min

$$\text{Vapor Volume} = \frac{8.33\ \text{lb/gal} \times 0.8}{0.075\ \text{lb/ft}^3 \times 2.0} = 44.4\ \text{ft}^3\text{/gal}$$

Fume hood dilution: $10\ \text{ft}^2 \times 100\ \text{ft/min} = 1000\ \text{ft}^3\text{/min}$

$$\frac{44.4\ \text{ft}^3\text{/gal}}{1000\ \text{ft}^3\text{/min}} = 0.0444\ \text{min/gal}$$

$0.0444\ \text{min/gal} \times 1\ \text{gal/min} = 0.0444$ or 4.4%

$$\text{LFL} = \frac{2.5\%}{4.4\%} = \frac{X}{1\ \text{gal}}$$

$X = 0.57$ gal in 1 minute to reach LFL

Note: One common control used is 25% of LFL or 0.1425 gal.

Before undertaking any laboratory activity involving toxic or flammable materials, a hazard assessment should be completed. This is especially important when unfamiliar chemicals, apparatus, and reactions are used or when operating parameters are altered. Generally, because a chemical fume hood is used to conduct various operations involving larger quantities of toxic or flammables compared to work done on the laboratory bench, the following items should be reviewed and considered before attempting to assemble apparatus in a hood:

(1) The minimum quantity of flammable or combustible liquids or flammable gases that are necessary to achieve the desired results considering the airflow capabilities of the hood should be determined.
(2) Possible ignition sources, including hot surfaces, electrical equipment, static charge, and so forth, should be identified.
(3) If the activity is conducted when personnel are not in attendance, the equipment has to be designed to shut down in a fail-safe mode in the event of ventilation failure, fire, over-temperature, and so forth. Audible and visible alarms should also be installed.
(4) Switches that control power to equipment should be outside the hood and within 15 m (50 ft) so apparatus can be de-energized in an emergency.
(5) When a risk assessment is performed, a list of possible hazards should be prepared and engineering controls should be used to reduce the probability of the event occurring.
(6) If the operation is designed for unattended use, a measuring device for hood airflow equipped with electrical contacts can be used to warn of defective hood performance and to provide a means to initiate an automatic safe shutdown of equipment.
(7) When flammable or combustible liquids are pressurized, the equipment layout should be arranged so that potential leaks from pump seals, piping components, glassware, and so forth, cannot occur near ignition sources. In some extra-hazardous operations, a partition between ignition sources and flammable and combustible liquid or flammable gas supplies might be required. If partitions are used, a smoke test of the airflow patterns should be conducted to assure good distribution across the hood opening.
(8) Spill containment should be provided to prevent spilled liquids from escaping from the hood.
(9) Hoods with once-through flow that are equipped with a bypass airfoil or mechanical stops on the sash(es) that assure good ventilation rates even with the sash(es) in the closed position should be used.
(10) All flammable and combustible materials that are not being used should be removed.
(11) Unauthorized use or operation of the hood should be prevented.
(12) Risk of accidental spill of liquids should be minimized by using nonglass apparatus where possible.
(13) Hood sash(es) should be kept closed when operations do not require them to be open.
(14) An emergency plan and off-hour instructions that can be followed during an emergency should be developed.

Chapter 12 Laboratory Operations and Apparatus

12.2.2 Refrigeration and Cooling Equipment.

12.2.2.1* Each refrigerator, freezer, or cooler shall be prominently marked to indicate whether or not it meets the requirements for safe storage of flammable liquids.

A.12.2.2.1 Figure A.12.2.2.1 gives examples of labels that can be used on laboratory refrigerators.

12.2.2.2* Refrigerators, freezers, and other cooling equipment used to store or cool flammable liquids shall be listed

Do not store flammable solvents in this refrigerator.

Label used for unmodified domestic models

Notice: This is not an explosionproof refrigerator, but it has been designed to permit safe storage of materials producing flammable vapors. Containers should be well-stoppered or tightly closed.

Label for laboratory-safe or modified domestic models

FIGURE A.12.2.2.1 Labels to Be Used in Laboratory Refrigerators.

special purpose units for use in laboratories in commercial occupancies or units listed for Class I, Division 1 locations, as described in Article 501 of NFPA 70, *National Electrical Code*.

A.12.2.2.2 Protection against the ignition of flammable vapors in refrigerated equipment is available through three types of laboratory refrigerators: explosionproof, "laboratory-safe" (or "explosion-safe"), and modified domestic models.

Explosionproof refrigeration equipment is designed to protect against ignition of flammable vapors both inside and outside the refrigerated storage compartment. This type is intended and recommended for environments such as pilot plants or laboratory work areas where all electrical equipment is required to meet the requirements of Article 501 of NFPA 70, *National Electrical Code*.

The design concepts of the "explosion-safe" or "laboratory-safe" type of refrigerator are based on the typical laboratory environment. The primary intent is to eliminate ignition of vapors inside the storage compartment by sources also within the compartment. In addition, commercially available "laboratory-safe" refrigerators incorporate such design features as thresholds, self-closing doors, friction latches or magnetic door gaskets, and special materials for the inner shell. All of these features are intended to control or limit the damage should an exothermic reaction occur within the storage compartment. Finally, the compressor and its circuits and controls are located at the top of the unit to further reduce the potential for ignition of floor-level vapors. In general, the design features of a commercially available "laboratory-safe" refrigerator are such that they provide important safeguards not easily available through modification of domestic models.

12.2.2.2.1* Domestic refrigerators, freezers, and other cooling equipment shall be permitted to store or cool flammable liquids if modified as follows:

(1) Any electrical equipment located within the outer shell, within the storage compartment, on the door, or on the door frame shall meet the requirements for Class I, Division 1 locations, as described in Article 501 of NFPA 70, *National Electrical Code*.
(2) Electrical equipment mounted on the outside of the storage compartment shall be installed in one of the following ways:
    (a) To meet the requirements for Class I, Division 2 locations
    (b) To be located above the storage compartment
    (c) To be located on the outside surface of the equipment where exposure to hazardous concentrations of vapors will be minimal

A.12.2.2.2.1 The use of domestic refrigerators for the storage of typical laboratory solvents presents a significant hazard to the laboratory work area. Refrigerator temperatures are almost universally higher than the flash points of the flammable liquids most often stored in them. In addition to vapor accumulation, a domestic refrigerator contains readily available ignition sources, such as thermostats, light switches, and heater strips, all within or exposed to the refrigerated storage compartment. Furthermore, the compressor and its circuits are typically located at the bottom of the unit, where vapors from flammable liquid spills or leaks could easily accumulate.

Although not considered optimum protection, it is possible to modify domestic refrigerators to achieve some degree of protection. However, the modification process can be applied only to manual defrost refrigerators; the self-defrosting models cannot be successfully modified to provide even minimum safeguards against vapor ignition. The minimum procedures for modification include the following:

(1) Relocation of manual temperature controls to the exterior of the storage compartment, sealing all points where capillary tubing or wiring formerly entered the storage compartment
(2) Removal of light switches and light assemblies and sealing of all resulting openings
(3) Replacement of positive mechanical door latches with magnetic door gaskets

Regardless of the approach used (explosionproof, "laboratory-safe," modified domestic, or unmodified domestic), every laboratory refrigerator should be clearly marked to indicate whether or not it is safe for storage of flammable materials. Internal laboratory procedures should ensure that laboratory refrigerators are being properly used.

12.2.2.3 Refrigerators, freezers, and cooling equipment located in a laboratory work area designated as a Class I location, as specified in the Exception to 5.6.2, shall be approved for Class I, Division 1 or 2 locations and shall be installed in accordance with Article 501 of NFPA 70, *National Electrical Code*.

NFPA 50A, *Standard for Gaseous Hydrogen Systems at Consumer Sites*, 1999

Editor's Note: *NFPA 50A has been incorporated into NFPA 55, Standard for the Storage, Use, and Handling of Compressed Gases and Cryogenic Fluids in Portable and Stationary Containers, Cylinders, and Tanks, 2005 edition.*

1-1 Scope.

This standard covers the requirements for the installation of gaseous hydrogen systems on consumer premises where the hydrogen supply to the consumer premises originates outside the consumer premises and is delivered by mobile equipment.

4-1 Outdoor Locations.

4-1.2 Electrical equipment within 15 ft (4.6 m) shall be in accordance with Article 501 of NFPA 70, *National Electrical Code®*, for Class I, Division 2 locations.

4-2 Separate Buildings.

4-2.5 Electrical equipment shall be in accordance with Article 501 of NFPA 70, *National Electrical Code*, for Class I, Division 2 locations.

4-3 Special Rooms.

4-3.5 Electrical equipment shall be in accordance with Article 501 of NFPA 70, *National Electrical Code*, for Class I, Division 2 locations.

NFPA 50B, *Standard for Liquefied Hydrogen Systems at Consumer Sites*, 1999

Editor's Note: *NFPA 50B has been incorporated into NFPA 55, Standard for the Storage, Use, and Handling of Compressed Gases and Cryogenic Fluids in Portable and Stationary Containers, Cylinders, and Tanks, 2005 edition.*

1-1 Scope.

1-1.1 Application. This standard covers the requirements for the installation of liquefied hydrogen systems on consumer premises where the liquid hydrogen supply to the consumer's premises originates outside the consumer's premises and is delivered by mobile equipment.

1-1.2* Nonapplication. This standard does not apply to the following:

(a) Portable containers having a total liquefied hydrogen content of less than 39.7 gal (150 L)

(b) Liquefied hydrogen manufacturing plants or other establishments operated by the hydrogen supplier or his or her agent for the sole purpose of storing liquefied hydrogen and refilling portable containers, trailers, mobile supply trucks, or tank cars

A-1-1.2 For information on gaseous hydrogen systems, see NFPA 50A, *Standard for Gaseous Hydrogen Systems at Consumer Sites*.

Table 2-7 Electrical Area Classification

Location	Division	Extent of Classified Area
Points where connections are regularly made and disconnected	1	Within 3 ft (1 m) of connection
	2	Between 3 ft (1 m) and 25 ft (7.6 m) of connection

Chapter 2 Design of Liquefied Hydrogen Systems

2-7 Electrical Systems.

Electrical wiring and equipment shall be in accordance with Table 2-7 and Article 501 of NFPA 70, *National Electrical Code®*.

Exception No. 1: Where equipment approved for Class I, Group B atmospheres is not commercially available, the equipment used shall meet the following requirements:

(a) Purged or ventilated in accordance with NFPA 496, Standard for Purged and Pressurized Enclosures for Electrical Equipment
(b) Intrinsically safe
(c) Approved for Class I, Group C atmospheres

Exception No. 2: Electrical equipment installed on mobile supply trucks or tank cars from which the storage container is filled.

NFPA 51, *Standard for the Design and Installation of Oxygen–Fuel Gas Systems for Welding, Cutting, and Allied Processes*, 2002

1.1 Scope.

1.1.1 This standard applies to the following:

(1) Design and installation of oxygen–fuel gas welding and cutting systems and allied processes *(see 3.3.2)*, except for systems meeting the criteria in 1.1.5
(2) Utilization of gaseous fuels generated from flammable liquids under pressure when such fuels are used with oxygen
(3) Storage, on the site of a welding and cutting system installation, of the following:

(a) Gases to be used with such systems where more than one cylinder each of oxygen and fuel gas are stored in any single storage area [This category includes storage of more than one cylinder each in any single storage area even though all such stored cylinders may be intended for use in systems of the kind described in 1.1.5(1).]

(b) Calcium carbide

1.1.2 Unless specifically indicated otherwise, the term *welding and cutting systems* shall be considered to include *allied processes* in this standard.

1.1.3 When only a portion of a fuel gas system is to be used for welding, cutting, or allied processes, only that portion of the system need comply with this standard.

1.1.4 When only a portion of an oxygen system is to be used with fuel gas for welding, cutting, or allied processes, only that portion of the system need comply with this standard.

1.1.5 This standard shall not apply to the following:

(1)* Systems consisting of a single cylinder not exceeding 120 ft^3 of oxygen and a single cylinder not exceeding 120 ft^3 of fuel gas used for welding and cutting
(2) Systems in which fuel gases are not to be used with oxygen, as described in NFPA 54, *National Fuel Gas Code*, and NFPA 58, *Liquefied Petroleum Gas Code*
(3) The manufacture of gases and the filling of cylinders
(4) Storage of empty cylinders
(5) Compressed air–fuel gas systems

Chapter 8 Acetylene Generators

8.4.4 Acetylene Compressor or Booster Pump Equipment.

8.4.4.2 Wiring and electrical equipment in compressor or booster pump rooms or enclosures shall conform to the provisions of Article 501 of NFPA 70, *National Electrical Code®*, for Class I, Division 2 or Class I, Zone 2 locations.

8.5.3 Heating Systems, Electrical Equipment, and Sources of Ignition.

8.5.3.2 Electrical equipment and wiring in outside generator houses or inside generator rooms shall conform to the provisions of Article 501 of NFPA 70, *National Electrical Code*, for Class I, Division 2 or Class I, Zone 2 locations.

Chapter 10 Mobile Acetylene Trailer Systems

10.2 Discharge Stations at Consumer Sites.

10.2.8 Electrical equipment within 4.6 m (15 ft) of outdoor installations shall be in accordance with Article 501 of NFPA 70, *National Electrical Code*, for Class I, Division 2 or Class I, Zone 2 locations.

10.3 Indoor Discharge Stations.

10.3.4 Electrical equipment shall be in accordance with Article 501 of NFPA 70, *National Electrical Code*, for Class I, Division 2 or Class I, Zone 2 locations.

NFPA 51A, *Standard for Acetylene Cylinder Charging Plants,* 2001

1.2 Scope.

1.2.1 This standard shall apply to plants that are engaged in the generation and compression of acetylene and in the charging of acetylene cylinders, either as their sole operation or in conjunction with facilities for charging other compressed gas cylinders.

1.2.2 An existing plant that is not in strict compliance with the provisions of this standard shall be permitted to continue operations where such use does not constitute a distinct hazard to life or adjoining property.

1.2.3 This standard shall not apply to plants that only produce and compress acetylene for chemical operations or to plants that only produce and compress acetylene below a gauge pressure of 15 psi (103 kPa). (Refer to NFPA 51, *Standard for the Design and Installation of Oxygen-Fuel Gas Systems for Welding, Cutting, and Allied Processes,* for acetylene generating plants where the acetylene is used with oxygen for welding, cutting, heating, and heat-treating operations.)

Chapter 2 Plant Location, Arrangement, Construction, and Utilities

2.6 Electrical Equipment.

2.6.2 Electrical equipment and wiring in rooms housing acetylene operations, except rooms used exclusively for calcium carbide storage, shall conform to NFPA 70, *National Electrical Code®*, Article 501, for Class I, Division 2 locations.

NFPA 52, *Compressed Natural Gas (CNG) Vehicular Fuel Systems Code,* 2002

1.1* Scope.

This code shall apply to the design and installation of compressed natural gas (CNG) engine fuel systems on vehicles of all types, including the following:

(1) Original equipment manufacturers
(2) Vehicle converters
(3) Vehicle fueling (dispensing) systems

A.1.1 *Properties of CNG.* Natural gas is a flammable gas. It is colorless, tasteless, and nontoxic. It is a light gas, weighing about two-thirds as much as air. As used in the systems covered by this standard, it tends to rise and diffuses rapidly in air when it escapes from the system.

Natural gas burns in air with a luminous flame. At atmospheric pressure, the ignition temperature of natural gas–air mixtures has been reported to be as low as 900°F (482°C). The flammable limits of natural gas–air mixtures at atmospheric pressure are about 5 percent to 15 percent by volume natural gas.

Natural gas is nontoxic but can cause anoxia (asphyxiation) when it displaces the normal 21 percent oxygen in air in a confined area without adequate ventilation.

1.1.1 Vehicles and fuel supply containers complying with Federal Motor Vehicle Safety Standards covering the installation of CNG fuel systems on vehicles and certified by the respective manufacturer as meeting these standards shall not be required to comply with Section 4.4, 4.8.4, and Chapter 5 (except Section 5.11, 5.12.4, Section 5.13, and Section 5.14.).

Chapter 6 CNG Compression, Storage, and Dispensing Systems

6.4.3.8 Buildings and rooms used for compression, storage, and dispensing shall be classified in accordance with Table 6.4.3.8 for installations of electrical equipment.

6.12* Installation of Electrical Equipment.

Fixed electrical equipment and wiring within areas specified in Table 6.4.3.8 shall comply with Table 6.4.3.8 and shall be installed in accordance with NFPA 70, *National Electrical Code®*.

Exception: Electrical equipment on internal combustion engines installed in accordance with NFPA 37, Standard for the Installation and Use of Stationary Combustion Engines and Gas Turbines.

Table 6.4.3.8 Electrical Installations

Location	Division or Zone	Extent of Classified Area
Containers (other than mounted fuel supply containers)	2	Within 10 ft (3 m) of container
Area containing compression and ancillary equipment	2	Up to 15 ft (4.6 m) from equipment
Dispensing equipment		
Outdoors	1	Inside the dispenser enclosure
Outdoors	2	From 0 to 5 ft (1.5 m) from the dispenser
Indoors	1	Inside the dispenser enclosure
Indoors	2	Entire room, with adequate ventilation *(see 6.4.3)*
Discharge from relief valves or vents		
Outdoors	1	5 ft (1.5 m) in all directions from the point source
Outdoors	2	Beyond 5 ft (1.5 m) but within 15 ft (4.6 m) in all directions from point of discharge
Valves, flanges of screwed fittings	None	Unclassified
Discharge from relief valves within 15 degrees of the line of discharge	1	15 ft (4.6 m)

A.6.12 See Figure A.6.12 for an illustration of classified areas in and around dispensers.

The electrical classification specified in Table 6.4.3.8 can be permitted to be reduced, or hazardous areas limited or eliminated, by adequate positive pressure ventilation from a source of clean air or inert gas in conjunction with effective safeguards against ventilator failure by purging methods recognized in NFPA 496, *Standard for Purged and Pressurized Enclosures for Electrical Equipment.* Such changes should be subject to approval by the authority having jurisdiction.

6.12.1 With the approval of the authority having jurisdiction, the classified areas specified in Table 6.4.3.8 shall be permitted to be reduced or eliminated by positive pressure ventilation from a source of clean air or inert gas in conjunction with effective safeguards against ventilator failure by purging methods recognized in NFPA 496, *Standard for Purged and Pressurized Enclosures for Electrical Equipment.*

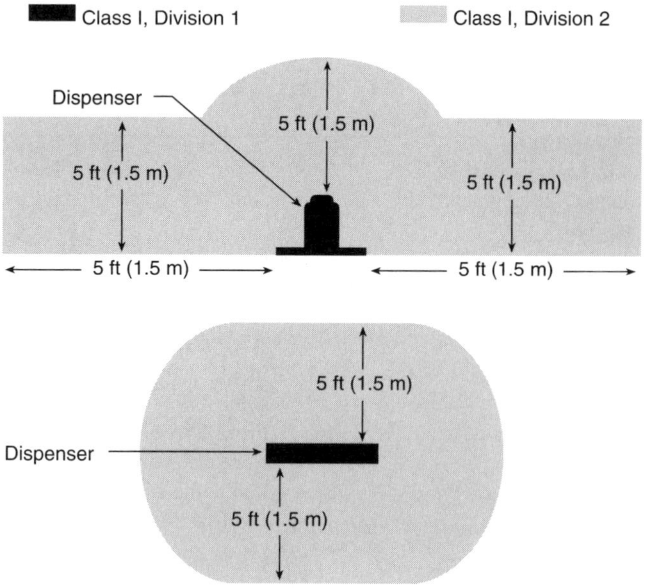

FIGURE A.6.12 Classified Areas in and Around Dispensers as Detailed in Table 6.4.3.8.

6.12.2 Classified areas shall not extend beyond an unpierced wall, roof, or vaportight partition. Space around welded pipe and equipment without flanges, valves, or fittings shall be a nonhazardous location.

Exception: Listed dispensers shall be permitted to be installed using classified areas in accordance with the terms of the listing.

Chapter 7 Residential Fueling Facility

7.3.3 Vehicles shall be considered as unclassified electrically with respect to NFPA 70, *National Electrical Code*, Article 500.

Exception: Vehicles containing fuel-fired equipment (e.g., recreational vehicles) shall be considered a source of ignition unless this equipment is shut off completely before entering an area in which ignition sources shall not be permitted.

Chapter 8 Commercial Marine Vessels and Pleasure Craft

8.1 Scope.

This chapter shall apply to all commercial marine vessels and pleasure craft operating on CNG, including new and retrofit construction.

8.2 Modifications to NFPA 52, *Compressed Natural Gas (CNG) Vehicular Fuel Systems Code.*

8.2.1 Chapters 3, 4, 6, and 7 of this code shall apply to commercial marine vessels and pleasure craft operating on compressed natural gas with the exception of subsections 6.14.5, 6.14.6, and 6.14.10.

NFPA 53, *Recommended Practice on Materials, Equipment, and Systems Used in Oxygen-Enriched Atmospheres,* 2004

1.1 Scope.

This document establishes recommended minimum criteria for the safe use of oxygen (liquid/gaseous) and the design of systems for use in oxygen and oxygen-enriched atmospheres (OEAs).

Chapter 3 Definitions

3.3 General Definitions.

3.3.25 Oxygen-Enriched Atmosphere (OEA). An atmosphere in which the concentration of oxygen exceeds 21 percent by volume or its partial pressure exceeds 21.3 kPa (160 torr).

Chapter 7 System Design

7.7 Electrical Equipment.

7.7.2* Fixed electrical equipment within an OEA should comply with the requirements of Article 500, Class I, Division 1, NFPA 70, *National Electrical Code® (NEC®)*, and, in addition, equipment installed therein should be approved for use in the specific hazardous atmospheres at the maximum proposed pressure and oxygen concentration.

A.7.7.2 Because there are no flexible cords available with noncombustible insulation, it is essential for safe operation that portable equipment be used in OEA only if required for life safety and under rigorously controlled conditions.

7.7.3 All electrical wiring installed in a system or chamber should comply with the requirements of the *NEC*, Article 500, Class I, Division 1.

NFPA 54, *National Fuel Gas Code,* 2002

1.1 Scope.

1.1.1 Applicability.

1.1.1.1 This code is a safety code that shall apply to the installation of fuel gas piping systems, fuel gas utilization equipment, and related accessories as shown in 1.1.1.1(A) through 1.1.1.1(D).

(A) Coverage of piping systems shall extend from the point of delivery to the connections with each gas utilization device. For other than undiluted liquefied petroleum gas systems, the point of delivery shall be considered the outlet of the service meter assembly or the outlet of the service regulator or service shutoff valve where no meter is provided. For undiluted liquefied petroleum gas systems, the point of delivery shall be considered the outlet of the final pressure regulator, exclusive of line gas regulators, in the system.

(B) The maximum operating pressure shall be 125 psi (862 kPa).

Exception No. 1: Piping systems for gas–air mixtures within the flammable range are limited to a maximum pressure of 10 psi (69 kPa).

Exception No. 2: LP-Gas piping systems are limited to 20 psi (140 kPa), except as provided in 5.5.2.

(C) Piping systems requirements shall include design, materials, components, fabrication, assembly, installation, testing, inspection, operation, and maintenance.

(D) Requirements for gas utilization equipment and related accessories shall include installation, combustion, and ventilation air and venting.

1.1.1.2 This code shall not apply to the following items (reference standards for some of which appear in Annex L):

(1) Portable LP-Gas equipment of all types that are not connected to a fixed fuel piping system
(2) Installation of farm equipment such as brooders, dehydrators, dryers, and irrigation equipment
(3) Raw material (feedstock) applications except for piping to special atmosphere generators
(4) Oxygen–fuel gas cutting and welding systems
(5) Industrial gas applications using such gases as acetylene and acetylenic compounds, hydrogen, ammonia, carbon monoxide, oxygen, and nitrogen
(6) Petroleum refineries, pipeline compressor or pumping stations, loading terminals, compounding plants, refinery tank farms, and natural gas processing plants
(7) Large integrated chemical plants or portions of such plants where flammable or combustible liquids or gases are produced by chemical reactions or used in chemical reactions
(8) LP-Gas installations at utility gas plants
(9) Liquefied natural gas (LNG) installations
(10) Fuel gas piping in power and atomic energy plants
(11) Proprietary items of equipment, apparatus, or instruments such as gas generating sets, compressors, and calorimeters
(12) LP-Gas equipment for vaporization, gas mixing, and gas manufacturing
(13) LP-Gas piping for buildings under construction or renovations that is not to become part of the permanent building piping system — that is, temporary fixed piping for building heat
(14) Installation of LP-Gas systems for railroad switch heating
(15) Installation of LP-Gas and compressed natural gas systems on vehicles
(16) Gas piping, meters, gas pressure regulators, and other appurtenances used by the serving gas supplier in distribution of gas, other than undiluted LP-Gas
(17) Building design and construction, except as specified herein
(18) Fuel gas systems on recreational vehicles manufactured in accordance with NFPA 1192, *Standard on Recreational Vehicles*
(19) Fuel gas systems using hydrogen as a fuel

1.1.2 Other Standards. In applying this code, reference shall also be made to the manufacturers' instructions and the serving gas supplier regulations.

Chapter 6 Gas Piping Installation

6.12 Systems Containing Flammable Gas–Air Mixtures.

6.12.5 Installation of Gas-Mixing Machines.

6.12.5.2 Electrical Requirements. Where gas-mixing machines are installed in well-ventilated areas, the type of electrical equipment shall be in accordance with NFPA 70, *National Electrical Code®*, for general service conditions unless other hazards in the area prevail. Where gas-mixing machines are installed in small detached buildings or cutoff rooms, the electrical equipment and wiring shall be installed in accordance with NFPA 70 for hazardous locations (Articles 500 and 501, Class I, Division 2).

NFPA 55, *Standard for the Storage, Use, and Handling of Compressed Gases and Cryogenic Fluids in Portable and Stationary Containers, Cylinders, and Tanks,* 2003

1.1 Scope.

1.1.1 Applicability. This standard shall apply to the installation, storage, use, and handling of compressed gases and cryogenic fluids in portable and stationary containers, cylinders, and tanks in all occupancies.

1.1.2 Exemptions. This standard shall not apply to the following:

(1)* Off-site transportation of materials covered by this standard

A.1.1.2(1) For regulations on the transportation of gases, see 49 CFR 100 to 179 (Transportation) and *Transportation of Dangerous Goods Regulations* of Transport Canada.

(2) Storage, use, and handling of radioactive gases in accordance with NFPA 801, *Standard for Fire Protection for Facilities Handling Radioactive Materials*
(3) Storage, use, and handling of medical compressed gases at health care facilities in accordance with NFPA 99, *Standard for Health Care Facilities*
(4) Systems consisting of cylinders of oxygen and cylinders of fuel gas used for welding and cutting in accordance with NFPA 51, *Standard for the Design and Installation of Oxygen–Fuel Gas Systems for Welding, Cutting, and Allied Processes*
(5) Acetylene cylinders in acetylene cylinder charging plants in accordance with NFPA 51A, *Standard for Acetylene Cylinder Charging Plants*
(6) Ethylene oxide containers stored, handled, or used for sterilization and fumigation in accordance with NFPA 560, *Standard for the Storage, Handling, and Use of Ethylene Oxide for Sterilization and Fumigation*
(7)* Flammable gases used as a vehicle fuel when stored on a vehicle

A.1.1.2(7) For information, see NFPA 52, *Compressed Natural Gas (CNG) Vehicular Fuel Systems Code*, or NFPA 58, *Liquefied Petroleum Gas Code.*

(8)* Storage, use, and handling of liquefied and nonliquefied compressed gases, in laboratory work areas that are in accordance with NFPA 45, *Standard on Fire Protection for Laboratories Using Chemicals*

A.1.1.2(8) The storage of gases outside of laboratory work areas is covered by this standard.

(9) Storage, use, and handling of liquefied petroleum gases in accordance with NFPA 58, *Liquefied Petroleum Gas Code*
(10) Storage, use and handling of gases within closed-cycle refrigeration systems complying with the mechanical code
(11) LNG storage at utility plants under NFPA 59A, *Standard for the Production, Storage, and Handling of Liquefied Natural Gas (LNG)*)
(12) LNG handled as a vehicle fuel under NFPA 57, *Liquefied Natural Gas (LNG) Vehicular Fuel Systems Code*
(13) CNG handled as a vehicle fuel under NFPA 52, *Compressed Natural Gas (CNG) Vehicular Fuel Systems Code*

Chapter 7 Compressed Gases

7.6 Flammable Gases.

7.6.4 Electrical. Areas in which the storage or use of compressed gases exceeds the quantity thresholds for gases requiring special provisions shall be in accordance with NFPA 70, *National Electrical Code®*.

NFPA 57, *Liquefied Natural Gas (LNG) Vehicular Fuel Systems Code,* 2002

Editor's Note: *NFPA 57 was incorporated into NFPA 52, 2006.*

1.1* Scope.

This code shall apply to the design, installation, operation, and maintenance of liquefied natural gas (LNG) engine fuel systems on vehicles of all types, to their associated fueling (dispensing) facilities, and to LNG to CNG facilities with LNG storage in ASME containers of 70,000 gal (265 m^3) or less.

A.1.1 For information on on-site storage of LNG in ASME tanks larger than 70,000 gal (265 m^3) and in tanks built to API or other standards, see NFPA 59A, *Standard for the Production, Storage, and Handling of Liquefied Natural Gas (LNG).*

At the time this code was developed, the use of LNG as an aviation fuel, fueling site liquefaction facilities, and the use of residential LNG fueling facilities were not being considered actively. The committee intends to provide coverage for these applications at the appropriate time.

1.1.1 This code shall include marine, highway, rail, off-road, and industrial vehicles.

1.1.2 Vehicles that are required to comply with federal motor vehicle safety standards covering the installation of LNG fuel systems on vehicles and that are certified by the manufacturer as meeting these standards shall not be required to comply with Chapter 4, except 4.12.8.

Chapter 5 LNG Fueling Facilities

3.2 NFPA Official Definitions.

5.12 Electrical Equipment.

5.12.1 Electrical equipment and wiring shall be as specified by and shall be installed in accordance with NFPA 70, *National Electrical Code®*, and shall meet the requirements of Class I, Group D, Division or Zone as specified in Table 5.12.1.

Table 5.12.1 LNG Fueling Facility Electrical Area Classification

Part	Location	Class I, Group D Division or Zone[1]	Extent of Classified Area[2]
A	**LNG Fueling Facility Container Area**		
	Indoors	1	Entire room
	Outdoor, aboveground containers (other than portable)	1	Open area between a high-type dike and container wall where dike wall height exceeds distances between dike and container walls
		2	Within 15 ft (4.6 m) in all directions from container, plus area inside a low-type diked or impounding area up to the height of the dike impoundment wall
	Outdoor, belowground containers	1	Within any open space between container walls and surrounding grade or dike
		2	Within 15 ft (4.6 m) in all directions from roof and sides above grade
B	**Nonfired LNG Process Areas Containing Pumps, Compressors, Heat Exchangers, Piping, Connections Vessels, etc.**		
	Indoors with adequate ventilation	2	Entire room and any adjacent room not separated by a gastight partition, and 15 ft (4.6 m) beyond any ventilation discharge vent or lower
	Outdoors in open air at or above grade	2	Within 15 ft (4.6 m) in all directions from this equipment
C	**Pits, Trenches, or Sumps Located in or Adjacent to Division 1 or 2 Areas**	1	Entire pit, trench, or sump
D	**Discharge from Relief Valves, Drains**	1	Within 5 ft (1.5 m) from point of discharge
		2	Beyond 5 ft (1.5 m) but within 15 ft (4.6 m) in all directions from point of discharge
E	**Vehicle/Cargo Transfer Area**		
	Indoors with adequate ventilation[3]	1	Within 5 ft (1.5 m) in all directions from point of transfer
		2	Beyond 5 ft (1.5 m) of entire room and 15 ft (4.6 m) beyond ventilation vent
	Outdoors in open air at or above grade	1	Within 5 ft (1.5 m) in all directions from point of transfer
		2	Beyond 5 ft (1.5 m) but within 15 ft (4.6 m) in all directions from the point of transfer

[1]See Article 500 — "Hazardous (Classified) Locations" in NFPA 70, *National Electrical Code®*, for definitions of classes, groups, and divisions.
[2]The classified area shall not extend beyond an unpierced wall, roof, or solid vaportight partition.
[3]Ventilation is considered adequate when provided in accordance with the provisions of this code.

Exception: Electrical equipment on internal combustion engines installed in accordance with NFPA 37, Standard for the Installation and Use of Stationary Combustion Engines and Gas Turbines.

5.12.2 Each interface between a flammable fluid system and an electrical conduit or wiring system, including process instrumentation connections, integral valve operators, foundation heating coils, canned pumps, and blowers, shall be sealed or isolated to prevent the passage of flammable fluids to another portion of the electrical installation.

5.12.3 Each seal, barrier, or other means used to comply with 5.12.2 shall be designed to prevent the passage of flammable fluids through the conduit, stranded conductors, and cables.

5.12.4* A primary seal shall be provided between the flammable fluid system and the electrical conduit wiring system. If the failure of the primary seal would allow the passage of flammable fluids to another portion of the conduit or wiring system, an additional approved seal, barrier, or other means shall be provided to prevent the passage of the flammable fluid beyond the additional device or means in the event that the primary seal fails. [**59A:**7-6.5]

A.5.12.4 Examples of such other means might include a physical interruption of the conduit run and of the stranded conductor(s) through the use of an adequately vented junction box containing terminal strip or busbar connections; an exposed section of MI cable using suitable fittings; or an exposed section of single conductor(s) that are incapable of transmitting gases or vapors. *[See Sections 501-5(a), (b), (c), and (d) of NFPA 70, National Electrical Code.]*

5.12.5 Each primary seal shall be designed to withstand the service conditions to which it is expected to be exposed. Each additional seal or barrier and interconnecting enclosure shall meet the pressure and temperature requirements of the condition to which it could be exposed in the event of failure of the primary seal, unless other approved means are provided to accomplish this purpose.

5.12.6 Unless specifically designed and approved for the purpose, the seals specified in 5.12.2 through 5.12.4 are not intended to replace the conduit seals required by Sections 501-5(a), (b), (c), and (d) of NFPA 70, *National Electrical Code.*

5.12.7 Where primary seals are installed, drains, vents, or other devices shall be provided for monitoring purposes to detect flammable fluids and leakage.

5.12.8 Static protection shall not be required when cargo transport vehicles or marine equipment are loaded or unloaded by conductive or nonconductive hose, flexible metallic tubing, or pipe connections through or from tight (top or bottom) outlets where both halves of metallic couplings are in contact. [**59A:**7-7.2]

NFPA 58, *Liquefied Petroleum Gas Code,* 2004

1.1* Scope.

This code applies to the storage, handling, transportation, and use of LP-Gas.

A.1.1 *General Properties of LP-Gas.* Liquefied petroleum gases (LP-Gases), as defined in this code *(see 3.3.37)*, are gases at normal room temperature and atmospheric pressure. They liquefy under moderate pressure and readily vaporize upon release of the pressure. It is this property that permits transportation and storage of LP-Gases in concentrated liquid form, although they normally are used in vapor form.

For additional information on other properties of LP-Gases, see Annex B.

Federal Regulations. Regulations of the U.S. Department of Transportation (DOT) are referenced throughout this code. Prior to April 1, 1967, these regulations were promulgated by the Interstate Commerce Commission (ICC). The Federal Hazardous Substances Act (15 U.S.C. 1261) requires cautionary labeling of refillable cylinders of liquefied petroleum gases distributed for consumer use. They are typically 40 lb (13 kg) and less and are used with outdoor cooking appliances, portable lamps, camp stoves, and heaters. The Federal Hazardous Substances Act is administered by the U.S. Consumer Product Safety Commission under regulations codified at 16 CFR 1500, Commercial Practices, Chapter 11, "Consumer Product Safety Commission."

1.3 Application.

1.3.1 Application of the Code. This code shall apply to the operation of all LP-Gas systems including the following:

(1) Containers, piping, and associated equipment, when delivering LP-Gas to a building for use as a fuel gas.
(2) Highway transportation of LP-Gas.
(3) The design, construction, installation, and operation of marine terminals whose primary purpose is the receipt of LP-Gas for delivery to transporters, distributors, or users except for marine terminals associated with refineries, petrochemicals, gas plants, and marine terminals whose purpose is the delivery of LP-Gas to marine vessels.
(4)* The design, construction, installation, and operation of pipeline terminals that receive LP-Gas from pipelines under the jurisdiction of the U.S. Department of Transportation, whose primary purpose is the receipt of LP-Gas for delivery to transporters, distributors, or users. Coverage shall begin downstream of the last pipeline valve or tank manifold inlet.

A.1.3.1(4) For further information on the storage and handling of LP-Gas at natural gas processing plants, refineries, and petrochemical plants, see API 2510, *Design and Construction of LP-Gas Installations.*

1.3.2 Nonapplication of Code. This code shall not apply to the following:

(1) Frozen ground containers and underground storage in caverns including associated piping and appurtenances used for the storage of LP-Gas.
(2) Natural gas processing plants, refineries, and petrochemical plants.
(3) LP-Gas (including refrigerated storage) at utility gas plants *(see NFPA 59, Utility LP-Gas Plant Code).*
(4) Chemical plants where specific approval of construction and installation plans, based on substantially similar requirements, is obtained from the authority having jurisdiction.
(5)* LP-Gas used with oxygen.

A.1.3.2(5) For information on the use of LP-Gas with oxygen, see NFPA 51, *Standard for the Design and Installation of Oxygen–Fuel Gas Systems for Welding, Cutting, and Allied Processes*, and ANSI Z49.1, *Safety in Welding, Cutting, and Allied Processes.*

(6)* The portions of LP-Gas systems covered by NFPA 54 (ANSI Z223.1), *National Fuel Gas Code*, where NFPA 54 (ANSI Z223.1) is adopted, used, or enforced.

A.1.3.2(6) Several types of LP-Gas systems are not covered by NFPA 54, *National Fuel Gas Code*, as noted. These include, but are not restricted to, most portable applications; many farm installations; vaporization, mixing, and gas manufacturing; temporary systems, for example, in construction; and systems on vehicles.

(7) Transportation by air (including use in hot air balloons), rail, or water under the jurisdiction of the U.S. Department of Transportation (DOT).
(8)* Marine fire protection.

A.1.3.2(8) For information on the use of LP-Gas in vessels, see NFPA 302, *Fire Protection Standard for Pleasure and Commercial Motor Craft.*

(9) Refrigeration cycle equipment and LP-Gas used as a refrigerant in a closed cycle.
(10) The manufacturing requirements for recreational vehicle LP-Gas systems that are addressed by NFPA 1192, *Standard on Recreational Vehicles.*
(11) Propane dispensers located at multiple fuel refueling stations shall comply with NFPA 30A, *Code for Motor Fuel Dispensing Facilities and Repair Garages.*

Chapter 6 Installation of LP-Gas Systems

6.20 Ignition Source Control

6.20.2 Electrical Equipment.

6.20.2.1 Electrical equipment and wiring installed in unclassified areas shall be in accordance with NFPA 70, *National Electrical Code*, for nonclassified locations.

6.20.2.2* Fixed electrical equipment and wiring installed within a classified area specified in Table 6.20.2.2 shall be

Table 6.20.2.2 Electrical Area Classification

Part	Location	Extent of Classified Area[a]	Equipment Shall Be Approved for *National Electrical Code,* Class I[a], Group D[b]
A	Unrefrigerated containers other than cylinders and ASME vertical containers of less than 1000 lb (454 kg) water capacity	Within 15 ft (4.6 m) in all directions from connections, except connections otherwise covered in Table 6.20.2.2	Division 2
B	Refrigerated storage containers	Within 15 ft (4.6 m) in all directions from connections otherwise covered in Table 6.20.2.2	Division 2
		Area inside dike to the level of the top of the dike	Division 2
C[c]	Tank vehicle and tank car loading and unloading	Within 5 ft (1.5 m) in all directions from connections regularly made or disconnected for product transfer	Division 1
		Beyond 5 ft (1.5 m) but within 15 ft (4.6 m) in all directions from a point where connections are regularly made or disconnected and within the cylindrical volume between the horizontal equator of the sphere and grade	Division 2
D	Gauge vent openings other than those on cylinders and ASME vertical containers of less than 1000 lb (454 kg) water capacity	Within 5 ft (1.5 m) in all directions from point of discharge	Division 1
		Beyond 5 ft (1.5 m) but within 15 ft (4.6 m) in all directions from point of discharge	Division 2

(continues)

Table 6.20.2.2 Continued

Part	Location	Extent of Classified Area[a]	Equipment Shall Be Approved for *National Electrical Code,* Class I[a], Group D[b]
E	Relief device discharge other than those on cylinders and ASME vertical containers of less than 1000 lb (454 kg) water capacity and vaporizers	Within direct path of discharge	Note: Fixed electrical equipment should preferably not be installed.
F[c]	Pumps, vapor compressors, gas–air mixers and vaporizers (other than direct-fired or indirect-fired with an attached or adjacent gas-fired heat source)		
	Indoors without ventilation	Entire room and any adjacent room not separated by a gastight partition	Division 1
		Within 15 ft (4.6 m) of the exterior side of any exterior wall or roof that is not vaportight or within 15 ft (4.6 m) of any exterior opening	Division 2
	Indoors with ventilation	Entire room and any adjacent room not separated by a gastight partition	Division 2
	Outdoors in open air at or above grade	Within 15 ft (4.6 m) in all directions from this equipment and within the cylindrical volume between the horizontal equator of the sphere and grade	Division 2
G	Vehicle fuel dispenser	Entire space within dispenser enclosure, and 18 in. (460 mm) horizontally from enclosure exterior up to an elevation 4 ft (1.2 m) above dispenser base; entire pit or open space beneath dispenser	Division 1
		Up to 18 in. (460 mm) above ground within 20 ft (6.1 m) horizontally from any edge of enclosure (Note: For pits within this area, see part H of this table.)	Division 2
H	Pits or trenches containing or located beneath LP-Gas valves, pumps, vapor compressors, regulators, and similar equipment		
	Without mechanical ventilation	Entire pit or trench	Division 1
		Entire room and any adjacent room not separated by a gastight partition	Division 2
		Within 15 ft (4.6 m) in all directions from pit or trench when located outdoors	Division 2
	With mechanical ventilation	Entire pit or trench	Division 2
		Entire room and any adjacent room not separated by a gastight partition	Division 2
		Within 15 ft (4.6 m) in all directions from pit or trench when located outdoors	Division 2
I	Special buildings or rooms for storage of cylinders	Entire room	Division 2

Table 6.20.2.2 Continued

Part	Location	Extent of Classified Area[a]	Equipment Shall Be Approved for *National Electrical Code,* Class I[a], Group D[b]
J	Pipelines and connections containing operational bleeds, drips, vents, or drains	Within 5 ft (1.5 m) in all directions from point of discharge	Division 1
		Beyond 5 ft (1.5 m) from point of discharge, same as part F of this table	
K[c]	Cylinder filling		
	Indoors with ventilation	Within 5 ft (1.5 m) in all directions from a point of transfer	Division 1
		Beyond 5 ft (1.5 m) and entire room	Division 2
	Outdoors in open air	Within 5 ft (1.5 m) in all directions from a point of transfer	Division 1
		Beyond 5 ft (1.5 m) but within 15 ft (4.6 m) in all directions from point of transfer and within the cylindrical volume between the horizontal equator of the sphere and grade	Division 2
L	Piers and wharves	Within 5 ft (1.5 m) in all directions from connections regularly made or disconnected for product transfer	Division 1
		Beyond 5 ft (1.5 m) but within 15 ft (4.6 m) in all directions from a point where connections are regularly made or disconnected and within the cylindrical volume between the horizontal equator of the sphere and the vessel deck	Division 2

[a]The classified area shall not extend beyond an unpierced wall, roof, or solid vaportight partition.
[b] See Article 500 Hazardous (Classified) Locations, in NFPA 70, *National Electrical Code*, for definitions of classes, groups, and divisions.
[c]See A.6.20.2.2.

installed in accordance with NFPA 70, *National Electrical Code*.

A.6.20.2.2 When classifying the extent of hazardous area, consideration should be given to possible variations in the spotting of railroad tank cars and cargo tank vehicles at the unloading points and the effect these variations of actual spotting point can have on the point of connection.

Where specified for the prevention of fire or explosion during normal operation, ventilation is considered adequate where provided in accordance with the provisions of this code.

6.20.2.3* The provisions of 6.20.2.2 shall apply to vehicle fuel operations.

A.6.20.2.3 See Figure A.6.20.2.3.

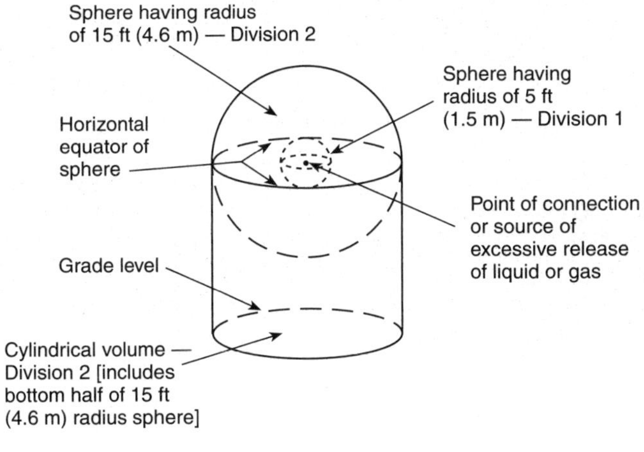

FIGURE A.6.20.2.3 Extent of Electrically Classified Area. ***(See Table 6.20.2.2.)***

6.20.2.4 The provisions of 6.20.2.2 shall not apply to fixed electrical equipment at residential or commercial installations of LP-Gas systems or to systems covered by Section 6.21.

6.20.2.5 Fired vaporizers, calorimeters with open flames, and other areas where open flames are present either intermittently or constantly shall not be considered electrically classified areas.

6.20.2.6 Electrical equipment installed on LP-Gas cargo tank vehicles shall comply with Section 9.2.

Chapter 10 Buildings or Structures Housing LP-Gas Distribution Facilities

10.2.3 Structure or Building Heating. Heating shall be by steam or hot water radiation or other heating transfer medium, with the heat source located outside of the building or structure *(see Section 6.20),* or by electrical appliances listed for Class I, Group D, Division 2 locations, in accordance with NFPA 70, *National Electrical Code.*

NFPA 59, *Utility LP-Gas Plant Code,* 2004

1.1 Scope.

1.1.1* This code shall apply to the design, construction, location, installation, operation, and maintenance of refrigerated and nonrefrigerated utility gas plants. Coverage of liquefied petroleum gas systems at utility gas plants shall extend to the point where LP-Gas or a mixture of LP-Gas and air is introduced into the utility distribution system.

A.1.1.1 Those portions of LP-Gas systems downstream of the point where LP-Gas or a mixture of LP-Gas and air is introduced into the utility distribution system are covered in the United States by the U.S. Department of Transportation, 49 CFR 192. *(See Figure A.1.1.1.)*

1.1.2 When operations that involve the liquid transfer of LP-Gas from the utility gas plant storage into cylinders or portable tanks (as defined by NFPA 58, *Liquefied Petroleum Gas Code*) are carried out in the utility gas plant, these operations shall conform to NFPA 58.

1.1.3 Installations that have an aggregate water capacity of 4000 gal (15.14 m^3) or less shall conform to NFPA 58, *Liquefied Petroleum Gas Code.*

Chapter 4 General Requirements

4.5 Ignition Source Control.

[**58:**6.20]

4.5.1 Scope. [**58:**6.20.1]

4.5.2 Electrical Equipment. [**58:**6.20.2]

4.5.2.1 Electrical equipment and wiring installed in unclassified areas shall be in accordance with NFPA 70, *National Electrical Code®*, for nonclassified locations. [**58:**6.20.2.1]

4.5.2.2* Fixed electrical equipment and wiring installed within a classified area specified in Table 4.5.2.2 shall be

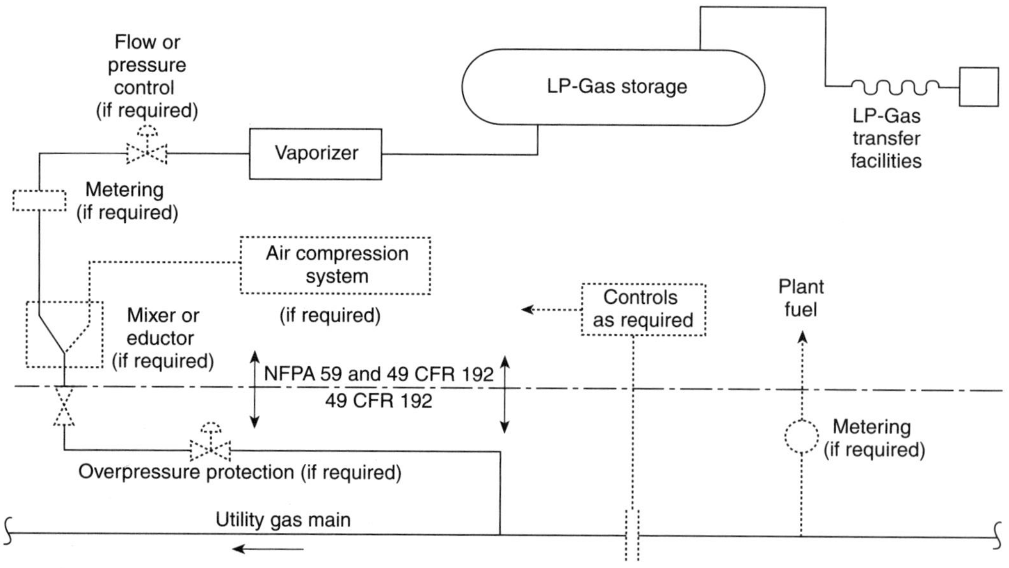

FIGURE A.1.1.1 Typical Installation of an LP-Gas Air Base or Peaking Facility, LP-Gas Vapor Base Load, or Enrichment Facility.

Table 4.5.2.2 Electrical Area Classification

Part	Location	Extent of Classified Area[a]	Equipment Shall Be Approved for *National Electrical Code,* Class I[a], Group D[b]
A	Unrefrigerated containers other than cylinders and ASME vertical containers of less than 1000 lb (454 kg) water capacity	Within 15 ft (4.6 m) in all directions from connections, except connections otherwise covered in Table 4.5.2.2	Division 2
B	Refrigerated storage containers	Within 15 ft (4.6 m) in all directions from connections otherwise covered in Table 4.5.2.2	Division 2
		Area inside dike to the level of the top of the dike	Division 2
C[c]	Tank vehicle and tank car loading and unloading	Within 5 ft (1.5 m) in all directions from connections regularly made or disconnected for product transfer	Division 1
		Beyond 5 ft (1.5 m) but within 15 ft (4.6 m) in all directions from a point where connections are regularly made or disconnected and within the cylindrical volume between the horizontal equator of the sphere and grade *(see Figure A.4.5.2.2)*	Division 2
D	Gauge vent openings other than those on cylinders and ASME vertical containers of less than 1000 lb (454 kg) water capacity	Within 5 ft (1.5 m) in all directions from point of discharge	Division 1
		Beyond 5 ft (1.5 m) but within 15 ft (4.6 m) in all directions from point of discharge	Division 2
E	Relief device discharge other than those on cylinders and ASME vertical containers of less than 1000 lb (454 kg) water capacity and vaporizers	Within direct path of discharge	Note: Fixed electrical equipment should preferably not be installed.
F[c]	Pumps, vapor compressors, gas–air mixers and vaporizers (other than direct-fired or indirect-fired with an attached or adjacent gas-fired heat source)		
	Indoors without ventilation	Entire room and any adjacent room not separated by a gastight partition	Division 1
		Within 15 ft (4.6 m) of the exterior side of any exterior wall or roof that is not vaportight or within 15 ft (4.6 m) of any exterior opening	Division 2
	Indoors with ventilation	Entire room and any adjacent room not separated by a gastight partition	Division 2
	Outdoors in open air at or above grade	Within 15 ft (4.6 m) in all directions from this equipment and within the cylindrical volume between the horizontal equator of the sphere and grade *(see Figure A.4.5.2.2)*	Division 2
G	Vehicle fuel dispenser	Entire space within dispenser enclosure, and 18 in. (460 mm) horizontally from enclosure exterior up to an elevation 4 ft (1.2 m) above dispenser base; entire pit or open space beneath dispenser	Division 1
		Up to 18 in. (460 mm) above ground within 20 ft (6.1 m) horizontally from any edge of enclosure (Note: For pits within this area, see part H of this table.)	Division 2

(continues)

Table 4.5.2.2 Continued

Part	Location	Extent of Classified Area[a]	Equipment Shall Be Approved for *National Electrical Code,* Class I[a], Group D[b]
H	Pits or trenches containing or located beneath LP-Gas valves, pumps, vapor compressors, regulators, and similar equipment		
	Without mechanical ventilation	Entire pit or trench	Division 1
		Entire room and any adjacent room not separated by a gastight partition	Division 2
		Within 15 ft (4.6 m) in all directions from pit or trench when located outdoors	Division 2
	With mechanical ventilation	Entire pit or trench	Division 2
		Entire room and any adjacent room not separated by a gastight partition	Division 2
		Within 15 ft (4.6 m) in all directions from pit or trench when located outdoors	Division 2
I	Special buildings or rooms for storage of cylinders	Entire room	Division 2
J	Pipelines and connections containing operational bleeds, drips, vents, or drains	Within 5 ft (1.5 m) in all directions from point of discharge	Division 1
		Beyond 5 ft (1.5 m) from point of discharge, same as part F of this table	
K[c]	Cylinder filling		
	Indoors with ventilation	Within 5 ft (1.5 m) in all directions from a point of transfer	Division 1
		Beyond 5 ft (1.5 m) and entire room	Division 2
	Outdoors in open air	Within 5 ft (1.5 m) in all directions from a point of transfer	Division 1
		Beyond 5 ft (1.5 m) but within 15 ft (4.6 m) in all directions from point of transfer and within the cylindrical volume between the horizontal equator of the sphere and grade *(see Figure A.4.5.2.2)*	Division 2
L	Piers and wharves	Within 5 ft (1.5 m) in all directions from connections regularly made or disconnected for product transfer	Division 1
		Beyond 5 ft (1.5 m) but within 15 ft (4.6 m) in all directions from a point where connections are regularly made or disconnected and within the cylindrical volume between the horizontal equator of the sphere and the vessel deck *(see Figure A.4.5.2.2)*	Division 2

[a]The classified area shall not extend beyond an unpierced wall, roof, or solid vaportight partition.
[b] See Article 500 Hazardous (Classified) Locations, in NFPA 70, *National Electrical Code*, for definitions of classes, groups, and divisions.
[c]See A.6.20.2.2 [of NFPA 58]. [**58:** Table 6.20.2.2]

installed in accordance with NFPA 70, *National Electrical Code*. [**58:**6.20.2.2]

A.4.5.2.2 When classifying the extent of hazardous area, consideration should be given to possible variations in the spotting of railroad tank cars and cargo tank vehicles at the unloading points and the effect these variations of actual spotting point can have on the point of connection.

Where specified for the prevention of fire or explosion during normal operation, ventilation is considered adequate where provided in accordance with the provisions of this code. [**58:**6.20.2.2]

See Figure A.4.5.2.2. [**58:** Figure A.6.20.2.3]

4.5.2.3 The provisions of 4.5.2.2 shall apply to vehicle fuel operations. [**58:**6.20.2.3]

4.5.2.4 The provisions of 4.5.2.2 shall not apply to fixed electrical equipment at residential or commercial installations of LP-Gas systems or to systems covered by Section 6.21 [of NFPA 58]. [**58:**6.20.2.4]

4.5.2.5 Fired vaporizers, calorimeters with open flames, and other areas where open flames are present either intermittently or constantly shall not be considered electrically classified areas. [**58:**6.20.2.5]

4.5.2.6 Electrical equipment installed on LP-Gas cargo tank vehicles shall comply with Section 9.2 [of NFPA 58]. [**58:**6.20.2.6]

4.7 Fixed Electrical Equipment in Classified Areas.

4.7.1 Fixed electrical equipment and wiring installed within the classified areas specified in Table 4.5.2.2 shall comply with Table 4.5.2.2 and shall be installed in accordance with NFPA 70, *National Electrical Code*®, for hazardous locations.

4.7.2 Direct-fired vaporizers, calorimeters with open flames, and other areas where open flames are present, either intermittently or constantly, shall not be considered electrically classified areas.

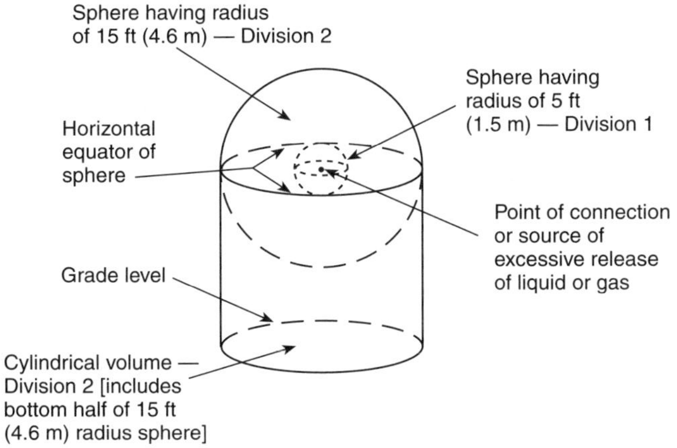

FIGURE A.4.5.2.2 Extent of Electrically Classified Area. *(See Table 4.5.2.2.)* **[58:Figure A.6.20.2.3]**

NFPA 59A, *Standard for the Production, Storage, and Handling of Liquefied Natural Gas (LNG)*, 2001

1.1* Scope.

A.1.1 This standard establishes essential requirements and standards for the design, installation, and safe operation of liquefied natural gas (LNG) facilities. It provides guidance to all persons concerned with the construction and operation equipment for the production, storage, and handling of LNG. It is not a design handbook, and competent engineering judgment is necessary for its proper use.

At sufficiently low temperatures, natural gas liquefies. At atmospheric pressure, natural gas can be liquefied by reducing its temperature to approximately −260°F (−162°C).

Upon release from the container to the atmosphere, LNG will vaporize and release gas that, at ambient temperature, has about 600 times the volume of the liquid vaporized. Generally, at temperatures below approximately −170°F (−112°C), this gas is heavier than ambient air at 60°F (15.6°C). However, as its temperature rises, it becomes lighter than air.

NOTE: The −260°F (−162°C) temperature value is for methane. If the constituents are present, see 1.7.18, Liquefied Natural Gas (LNG).

NOTE: For information on the use of LNG as a vehicle fuel, see NFPA 57, *Liquefied Natural Gas (LNG) Vehicular Fuel Systems Code*.

1.1.1 This standard shall apply to the following:

(1) Design
(2) Location
(3) Construction
(4) Operation
(5) Maintenance of facilities at any location for the liquefaction of natural gas and the storage, vaporization, transfer, handling, and truck transport of liquefied natural gas (LNG), as well as the personnel training

1.1.2 This standard shall apply to all containers for the storage of LNG, including those with insulation systems applying a vacuum.

1.1.3 This standard shall not apply to frozen ground containers.

Chapter 7 Instrumentation and Electrical Services

7.6 Electrical Equipment.

7.6.1 Electrical equipment and wiring shall be of the type specified by and shall be installed in accordance with NFPA 70, *National Electrical Code®*, or CSA C 22.1, *Canadian Electrical Code*, for hazardous locations.

7.6.2 Fixed electrical equipment and wiring installed within the classified areas specified in Table 7.6.2 shall comply with Table 7.6.2 and Figures 7.6.2(a) through 7.6.2(d) and shall be installed in accordance with NFPA 70, *National Electrical Code*, for hazardous locations.

Exception: For the purpose of designing electrical equipment, the interior of an LNG container shall be permitted to be unclassified where the following conditions are met:

(a) Electrical equipment shall be deenergized and locked out until the container is purged of air.

(b) Electrical equipment is deenergized and locked out prior to allowing air into the container.

(c) The electrical system is designed and operated to deenergize the equipment automatically when the pressure in the container is reduced to atmospheric pressure.

7.6.3 Each interface between a flammable fluid system and an electrical conduit or wiring system, including process instrumentation connections, integral valve operators, foundation heating coils, canned pumps, and blowers, shall be sealed or isolated to prevent the passage of flammable fluids to another portion of the electrical installation.

7.6.3.1 Each seal, barrier, or other means used to comply with 7.6.3 shall be designed to prevent the passage of flammable fluids through the conduit, stranded conductors, and cables.

7.6.3.2 A primary seal shall be provided between the flammable fluid system and the electrical conduit wiring system. If the failure of the primary seal allows the passage of flammable fluids to another portion of the conduit or wiring system, an additional approved seal, barrier, or other means shall be provided to prevent the passage of the flammable

Table 7.6.2 Electrical Area Classification

Part	Location	Group D, Division[1]	Extent of Classified Area[2]
A	**LNG storage containers with vacuum breakers**		
	Inside of containers	2	Entire container interior.
B	**LNG storage container area**		
	Indoors	1	Entire room.
	Outdoor, aboveground containers (other than small containers)[3]	1	Open area between a high-type dike and container wall where dike wall height exceeds distance between dike and container walls. *[See Figure 7.6.2(c).]*
		2	Within 15 ft (4.5 m) in all directions from container walls and roof, plus area inside a low-type diked or impounding area up to the height of the dike impoundment wall. *[See Figure 7.6.2(b).]*
	Outdoor, belowground containers	1	Within any open space between container walls and surrounding grade or dike. *[See Figure 7.6.2(d).]*
		2	Within 15 ft (4.5 m) in all directions from roof and sides. *[See Figure 7.6.2(d).]*
C	**Nonfired LNG process areas containing pumps, compressors, heat exchangers, pipelines, connections, small containers, etc.**		

Table 7.6.2 Continued

Part	Location	Group D, Division[1]	Extent of Classified Area[2]
	Indoors with adequate ventilation[4]	2	Entire room and any adjacent room not separated by a gastight partition, and 15 ft (4.5 m) beyond any wall or roof ventilation discharge vent or louver.
	Outdoors in open air at or above grade	2	Within 15 ft (4.5 m) in all directions from this equipment, and within the cylindrical volume between the horizontal equator of the sphere and grade. *[See Figure 7.6.2(a).]*
D	**Pits, trenches, or sumps located in or adjacent to Division 1 or 2 Areas**	1	Entire pit, trench, or sump.
E	**Discharge from relief valves**	1	Within direct path of relief valve discharge.
F	**Operational bleeds, drips, vents, or drains**		
	Indoors with adequate ventilation[4]	1	Within 5 ft (1.5 m) in all directions from point of discharge.
		2	Beyond 5 ft (1.5 m) and entire room and 15 ft (4.5 m) beyond any wall or roof ventilation discharge vent or louver.
	Outdoors in open air at or above grade	1	Within 5 ft (1.5 m) in all directions from point of discharge.
		2	Beyond 5 ft (1.5 m) but within 15 ft (4.5 m) in all directions from point of discharge.
G	**Tank car, tank vehicle, and container loading and unloading[5]**		
	Indoors with adequate ventilation[4]	1	Within 5 ft (1.5 m) in all directions from connections regularly made or disconnected for product transfer.
		2	Beyond 5 ft (1.5 m) and entire room and 15 ft (4.5 m) beyond any wall or roof ventilation discharge vent or louver.
	Outdoors in open air at or above grade	1	Within 5 ft (1.5 m) in all directions from connections regularly made or disconnected for product transfer.
		2	Beyond 5 ft (1.5 m) but within 15 ft (4.5 m) in all directions from a point where connections are regularly made or disconnected, and within the cylindrical volume between the horizontal equator of the sphere and grade. *[See Figure 7.6.2(a).]*
H	**Electrical seals and vents specified in 7.6.3, 7.6.4, and 7.6.5**	2	Within 15 ft (4.5 m) in all directions from the equipment and within the cylindrical volume between the horizontal equator of the sphere and grade.

[1] See Article 500, "Hazardous (Classified) Locations" in NFPA 70, *National Electrical Code®*, for definitions of classes, groups, and divisions. Most of the flammable vapors and gases found within the facilities covered by this standard are classified as Group D. Ethylene is classified as Group C. Much available electrical equipment for hazardous locations is suitable for both groups.
[2] The classified area shall not extend beyond an unpierced wall, roof, or solid vaportight partition.
[3] Small containers are those that are portable and of less than 200-gal (760-L) capacity.
[4] Ventilation is considered adequate where provided in accordance with the provisions of this standard.
[5] Where classifying the extent of the hazardous area, consideration shall be given to possible variations in the spotting of tank cars and tank vehicles at the unloading points and the effect these variations might have on the point of connection.

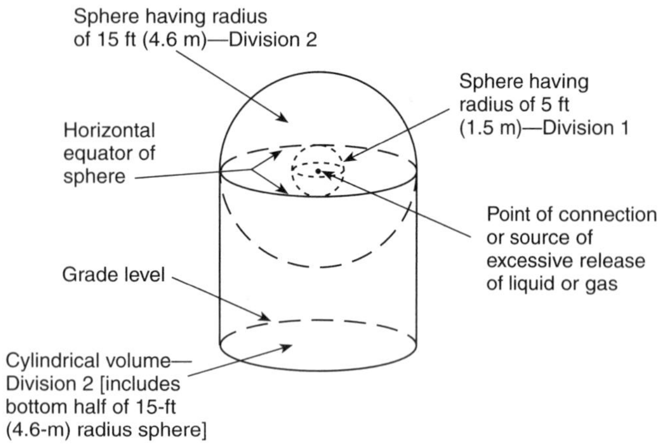

Figure 7.6.2(a) Extent of Classified Area Around Containers.

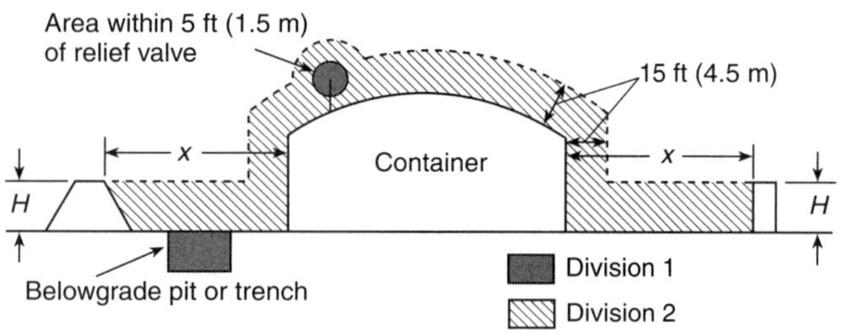

Figure 7.6.2(b) Dike Height Less than Distance from Container to Dike (*H* Less than *x*).

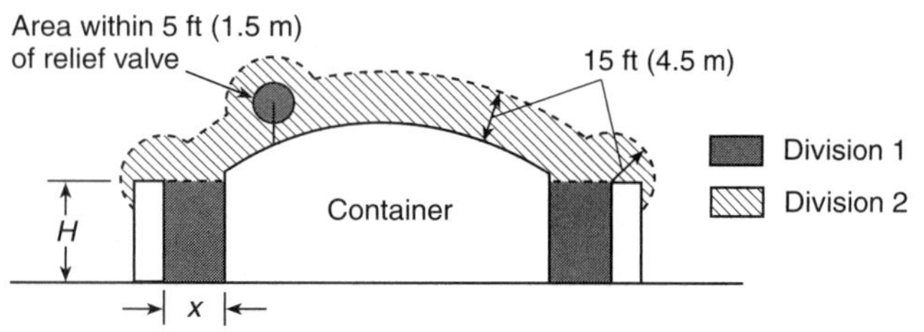

Figure 7.6.2(c) Dike Height Less than Distance from Container to Dike (*H* Greater than *x*).

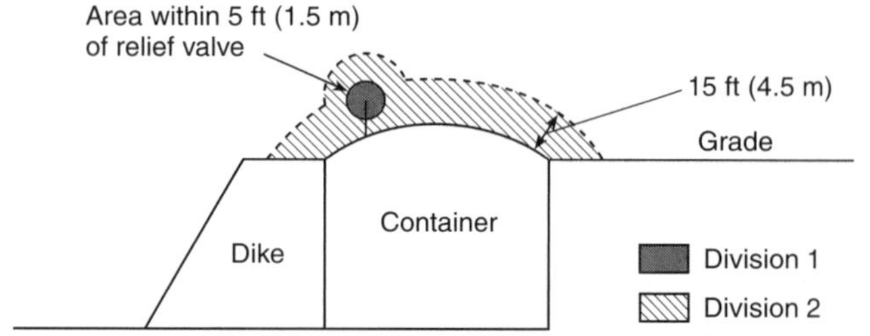

Figure 7.6.2(d) Container with Liquid Level Below Grade or Top of Dike.

fluid beyond the additional device or means if the primary seal fails.

7.6.3.3 Each primary seal shall be designed to withstand the service conditions to which it can be exposed. Each additional seal or barrier and interconnecting enclosure shall be designed to meet the pressure and temperature requirements of the condition to which it could be exposed in the event of failure of the primary seal unless other approved means are provided to accomplish the purpose.

7.6.3.4 Where secondary seals are used, the space between the primary and secondary seals shall be continuously vented to the atmosphere. Similar provisions shall be made on double-integrity primary sealant systems of the type used for submerged motor pumps.

7.6.3.5 The seals specified in 7.6.3, 7.6.4, and 7.6.5 shall not be used to meet the sealing requirements of NFPA 70, *National Electrical Code*, or CSA C 22.1, *Canadian Electrical Code*.

7.6.4 Where primary seals are installed, drains, vents, or other devices shall be provided for monitoring purposes to detect flammable fluids and leakage.

7.6.5 The venting of a conduit system shall be done in a manner that minimizes the possibility of damage to personnel and equipment, considering the properties of the liquid or gas and the potential for ignition.

NFPA 61, *Standard for the Prevention of Fires and Dust Explosions in Agricultural and Food Processing Facilities,* 2002

1.1 Scope.

1.1.1* This standard shall apply to all of the following:

(1) All facilities that receive, handle, process, dry, blend, use, mill, package, store, or ship dry agricultural bulk materials, their by-products, or dusts that include grains, oilseeds, agricultural seeds, legumes, sugar, flour, spices, feeds, and other related materials
(2) All facilities designed for manufacturing and handling starch, including drying, grinding, conveying, processing, packaging, and storing dry or modified starch, and dry products and dusts generated from these processes
(3) Those seed preparation and meal-handling systems of oilseed processing plants not covered by NFPA 36, *Standard for Solvent Extraction Plants*

A.1.1.1 Examples of facilities covered by this standard include, but are not limited to, bakeries, grain elevators, feed mills, flour mills, milling, corn milling (dry and wet), rice milling, dry milk products, mix plants, soybean and other oilseed preparation operations, cereal processing, snack food processing, tortilla plants, chocolate processing, pet food processing, cake mix processing, sugar refining and processing, and seed plants.

1.1.2 This standard shall not apply to oilseed extraction plants that are covered by NFPA 36, *Standard for Solvent Extraction Plants.*

Chapter 4 Construction Requirements

4.1.3 Electrical wiring and power equipment shall meet all applicable requirements of NFPA 70, *National Electrical Code®.*

4.1.4* Masonry shall not be used for the construction of exterior walls or roofs of areas classified as Class II, Group G, Division 1 in NFPA 70, *National Electrical Code.*

Exception: Masonry walls that are designed for explosion resistance to preclude failure of these walls before the explosion pressure can be vented safely to the outside.

A.4.1.4 For the purpose of this standard, masonry construction refers to stone, brick, gypsum, hollow tile, concrete block, and cinder block building units, or other similar building units or materials laid up unit by unit and set in mortar.

Chapter 10 Dust Control

10.2 Removal of Layered Agricultural Dust.

10.2.3* Portable electric vacuum cleaners, if used, shall be listed for use in Class II, Group G, Division 1 atmospheres as defined in NFPA 70, *National Electrical Code.*

A.10.2.3 Vacuum cleaning systems are preferred for removal of static dust on surfaces in order to prevent resuspension of the dust in ambient air, as is often caused by brushing down with brooms or using compressed air. Where the surfaces are inaccessible or create a hazard to employees working from stepladders or in hazardous positions while handling vacuum hoses and tools, alternative means should be followed under the conditions specified in 10.2.3. (*See also 10.3.1.*)

Chapter 11 Pneumatic Conveying

11.2.3 Electrical wiring and power equipment shall meet all applicable requirements of NFPA 70, *National Electrical Code.*

11.4 Air-Material Separators.

11.4.2 Cyclones with a 0.76-m (30-in.) diameter or less used as air-material separators shall be allowed to be placed inside buildings without explosion protection when the following conditions are present:

(1) The room, building, or other enclosure is not a Class I, Division 1 or 2 or Class II, Division 1 area as defined by Article 500 of NFPA 70, *National Electrical Code.*
(2) The material being processed has a minimum ignition energy of >10 mJ.
(3) The system is a closed process, excluding cleaning vacuum systems.
(4) The material being processed has a K_{St} of less than 200 bar-m/sec.

NFPA 70B, *Recommended Practice for Electrical Equipment Maintenance,* 2002

1.1 Scope.

1.1.1 This recommended practice applies to preventive maintenance for electrical, electronic, and communication systems and equipment and is not intended to duplicate or supersede instructions that manufacturers normally provide. Systems and equipment covered are typical of those installed in industrial plants, institutional and commercial buildings, and large multifamily residential complexes.

1.1.2 Consumer appliances and equipment intended primarily for use in the home are not included.

Chapter 22 Hazardous (Classified) Location Electrical Equipment

22.1 Types of Equipment.

Hazardous location electrical equipment is used in areas that commonly or infrequently contain ignitable vapors or dusts. Designs of hazardous location electrical equipment include explosionproof, dust-ignition-proof, dusttight, purged pressurized, intrinsically safe, nonincendive, oil immersion, hermetically sealed, and other types. Maintenance of each type of equipment requires attention to specific items.

22.2 Maintenance of Electrical Equipment for Use in Hazardous (Classified) Locations.

22.2.1 Electrical equipment designed for use in hazardous (classified) locations should be maintained through periodic inspections, tests, and servicing as recommended by the manufacturer. Electrical preventive maintenance documentation should define the classified area (the class, group, and division specification, and the extent of the classified area) and the equipment maintenance required. Electrical preventive maintenance documentation should identify who is authorized to work on this equipment, where this maintenance is to be performed, and what precautions are necessary. Although repairs to certain equipment should be done by the manufacturer or authorized representatives, inspection and servicing that can be performed in-house should be clearly identified.

22.2.2 Maintenance should be performed only by qualified personnel who are trained in safe maintenance practices and the special considerations necessary to maintain electrical equipment for use in hazardous (classified) locations. These individuals should be familiar with requirements for obtaining safe electrical installations. They should be trained to evaluate and eliminate ignition sources, including high surface temperatures, stored electrical energy, and the buildup of static charges, and to identify the need for special tools, equipment, tests, and protective clothing.

22.2.3 Where possible, repairs and maintenance should be performed outside the hazardous (classified) area. For maintenance involving permanent electrical installations, an acceptable method of compliance can include de-energizing the electrical equipment and removing the hazardous atmosphere for the duration of the maintenance period. All sources of hazardous vapors, gases, and dusts should be removed, and enclosed, trapped atmospheres should be cleared.

22.2.4 Electrical power should be disconnected and all other ignition sources abated before disassembling any electrical equipment in a hazardous (classified) location. Time should be allowed for parts to cool and electrical charges to dissipate, and other electrical maintenance precautions followed.

22.2.5 Electrical equipment designed for use in hazardous (classified) locations should be fully reassembled with original components or approved replacement before the hazardous atmosphere is reintroduced and before restoring power. Special attention should be given to joints and other openings in the enclosure. Cover(s) should not be interchanged unless identified for the purpose. Foreign objects, including burrs, pinched gaskets, pieces of insulation, and wiring, will prevent the proper closure of mating joints designed to prevent the propagation of flame upon explosion.

22.2.6 An approved system of conduit and equipment seals conforming to the requirements of NFPA 70, *National Electrical Code*, and manufacturer's specifications should be maintained. Corrective action should be taken upon maintenance actions that damage or discover damage to a seal. Damage to factory-installed seals within equipment can necessitate replacing the equipment.

22.2.7 Wherever electrical equipment cover bolts or screws require torquing to meet operating specifications, these bolts or screws should be maintained with the proper torque as specified by the manufacturer. Electrical equipment should not be energized when any such bolts or screws are missing. All bolts and screws should be replaced with original components or approved replacements.

22.2.8 Special care should be used in handling electrical devices and components approved for use in hazardous (classified) locations. Rough handling, and the use of tools that pry, impact, or abrade components, can dent, scratch, nick, or otherwise mar close-tolerance, precision-machined joints and make them unsafe.

22.2.8.1 Grease, paint, and dirt should be cleaned from machined joints using a bristle (not wire) brush, an acceptable noncorrosive solvent, or other methods recommended by the manufacturer.

22.2.8.2 Prior to replacing a cover on an enclosure designed to prevent flame propagation upon an explosion, mating surfaces should be cleaned and lubricated in accordance with the manufacturer's instructions.

22.2.9 Field modifications of equipment and parts replacement should be limited to those changes acceptable to the manufacturer and approved by the authority having jurisdiction. Normally, modifications to equipment will void any listing by nationally recognized testing laboratories.

22.2.10 The requirements of NFPA 70, *National Electrical Code*, should be followed.

22.2.10.1 Explosionproof enclosures, dust-ignition-proof enclosures, dusttight enclosures, raceway seals, vents, barriers, and other protective features are required for electrical equipment in certain occupancies. Equipment and facilities should be maintained in a way that will not compromise equipment performance or safety.

22.2.10.2 Intrinsically safe equipment and wiring is permitted in locations for which specific systems are approved. Such wiring should be separate from the wiring of other circuits. NFPA 70, *National Electrical Code, Article 504*, Intrinsically Safe Systems, describes control drawings, grounding, and other features involved in maintenance programs.

22.2.10.3 Purged and pressurized enclosures can be used in hazardous (classified) areas. NFPA 496, *Standard for Purged* *and Pressurized Enclosures for Electrical Equipment*, provides guidance useful to maintenance personnel.

NFPA 70E, *Standard for Electrical Safety in the Workplace,* 2004

90.1 Scope.

(A) Covered. This standard addresses those electrical safety requirements for employee workplaces that are necessary for the practical safeguarding of employees in their pursuit of gainful employment. This standard covers the installation of electric conductors, electric equipment, signaling and communications conductors and equipment, and raceways for the following:

(1) Public and private premises, including buildings, structures, mobile homes, recreational vehicles, and floating buildings
(2) Yards, lots, parking lots, carnivals, and industrial substations

FPN: For additional information concerning such installations in an industrial or multibuilding complex, see ANSI C2-2002, *National Electrical Safety Code.*

(3) Installations of conductors and equipment that connect to the supply of electricity
(4) Installations used by the electric utility, such as office buildings, warehouses, garages, machine shops, and recreational buildings, that are not an integral part of a generating plant, substation, or control center

(B) Not Covered. This standard does not cover the following:

(1) Installations in ships, watercraft other than floating buildings, railway rolling stock, aircraft, or automotive vehicles other than mobile homes and recreational vehicles

(2) Installations underground in mines and self-propelled mobile surface mining machinery and its attendant electrical trailing cable

(3) Installations of railways for generation, transformation, transmission, or distribution of power used exclusively for operation of rolling stock or installations used exclusively for signaling and communications purposes

(4) Installations of communications equipment under the exclusive control of communications utilities located outdoors or in building spaces used exclusively for such installations

(5) Installations under the exclusive control of an electric utility where such installations:

a. Consist of service drops or service laterals, and associated metering, or

b. Are located in legally established easements, rights-of-way, or by other agreements either designated by or recognized by public service commissions, utility commissions, or other regulatory agencies having jurisdiction for such installations, or

c. Are on property owned or leased by the electric utility for the purpose of communications, metering, generation, control, transformation, transmission, or distribution of electric energy.

ARTICLE 235 Hazardous (Classified) Locations

235.1 Scope. This article covers maintenance requirements in those areas identified as hazardous (classified) locations in accordance with Article 440 of this standard.

FPN: These locations require special types of equipment and installation that ensure safe performance under conditions of proper use and maintenance. It is important that inspection authorities and users exercise more than ordinary care with regard to installation and maintenance. The maintenance requirements for specific equipment and materials covered elsewhere in Chapter 2 are applicable to hazardous (classified) locations. Other maintenance is required to ensure that the form of construction and of installation that makes the equipment and materials suitable for the particular location are not nullified.

The maintenance required for specific hazardous (classified) locations requires that the classification of the specific location be known. The design principles and equipment characteristics — for example, use of positive pressure ventilation, explosionproof, nonincendive, intrinsically safe, and purged and pressurized equipment — that were applied in the installation to meet the requirements of the area classification must also be known. With this information, the employer and the inspection authority are able to determine whether the installation as maintained has retained the condition necessary for a safe workplace.

235.2 Maintenance Requirements for Hazardous (Classified) Locations. Equipment and installations in these locations shall be maintained such that the following apply:

(1) No energized parts are exposed.

Exception to (1): Intrinsically safe and nonincendive circuits.

(2) There are no breaks in conduit systems, fittings, or enclosures from damage, corrosion, or other causes.
(3) All bonding jumpers are securely fastened and intact.
(4) All fittings, boxes, and enclosures with bolted covers have all bolts installed and bolted tight.

(5) All threaded conduit shall be wrenchtight and enclosure covers shall be tightened in accordance with the manufacturer's instructions.
(6) There are no open entries into fittings, boxes, or enclosures that would compromise the protection characteristics.
(7) All close-up plugs, breathers, seals, and drains are securely in place.
(8) Marking of luminaires (lighting fixtures) for maximum lamp wattage and temperature rating is legible and not exceeded.
(9) Required markings are secure and legible.

ARTICLE 440 Hazardous (Classified) Locations, Class I, II, and III, Divisions 1 and 2 and Class I, Zones 0, 1, and 2

440.1 Scope. This article shall apply to the requirements for electric equipment and wiring in locations that are classified depending on the properties of the flammable vapors, liquids, or gases, or combustible dusts or fibers that might be present therein and the likelihood that a flammable or combustible concentration or quantity is present. Hazardous (classified) locations can be found in occupancies such as, but not limited to, aircraft hangars, gasoline dispensing and service stations, bulk storage plants for gasoline or other volatile flammable liquids, paint-finishing process plants, health care facilities, agricultural or other facilities where excessive combustible dusts might be present, marinas, boat yards, and petroleum and chemical processing plants. Each room, section, or area shall be considered individually in determining its classification.

440.2 Definition: Intrinsically Safe Equipment. Apparatus in which the circuits are not necessarily intrinsically safe themselves but that affect the energy in the intrinsically safe circuits and are relied on to maintain intrinsic safety. Associated apparatus may be either of the following:

(1) Electrical apparatus that has an alternative-type protection for use in the appropriate hazardous (classified) location
(2) Electrical apparatus not so protected that shall not be used within a hazardous (classified) location

440.3 General.

(A) Documentation. All areas designated as hazardous (classified) locations shall be properly documented. This documentation shall be available to those authorized to design, install, inspect, maintain, or operate electrical equipment at the location.

(B) Approval for Class and Properties.

(1) Equipment Identification. Equipment shall be identified not only for the class of location but also for the explosive, combustible, or ignitible properties of the specific gas, vapor, dust, fiber, or flyings that will be present. Class I equipment shall not have any exposed surface that operates at a temperature in excess of the ignition temperature of the specific gas or vapor.

FPN: Luminaires (lighting fixtures) and other heat-producing apparatus, switches, circuit breakers, and plugs and receptacles are potential sources of ignition and are investigated for suitability in classified locations. Such type of equipment, as well as cable terminations for entry into explosionproof enclosures, are available as listed for Class I, Division 2 locations. Fixed wiring, however, might utilize wiring methods that are not evaluated with respect to classified locations. Wiring products such as cable, raceways, boxes, and fittings, therefore, are not marked as being suitable for Class I, Division 2 locations.

Suitability of identified equipment shall be determined by any of the following:

(1) Equipment listing or labeling
(2) Evidence of equipment evaluation from a qualified testing laboratory or inspection agency concerned with product evaluation
(3) Evidence acceptable to the authority having jurisdiction such as manufacturer's self-evaluation or an owner's engineering judgment

(2) Division Location. Equipment has been identified for a Division 1 location shall be permitted in a Division 2 location of the same class and group.

(3) General Purpose Location. Where specifically permitted, general-purpose equipment or equipment in general-purpose enclosures shall be permitted to be installed in Division 2 locations if the equipment does not constitute a source of ignition under normal operating conditions.

(4) Equipment Requiring Sealing Means. Equipment, regardless of the classification of the location in which it is installed, that depends on a single compression seal, diaphragm, or tube to prevent flammable or combustible fluids from entering the equipment shall be identified for a Class I, Division 2 location. Equipment installed in a Class I, Division 1 location shall be identified for the Class I, Division 1 location.

(5) Normal Operating Conditions. Unless otherwise specified, normal operating conditions for motors shall be assumed to be rated full-load steady conditions.

(6) Flammable Gases and Dusts. Where flammable gases or combustible dusts are or may be present at the same time, the simultaneous presence of both shall be considered when determining the safe operation temperature of the electrical equipment.

FPN: The characteristics of various atmospheric mixtures of gases, vapors, and dusts depend on the specific material involved.

(C) Conduits. All threaded conduit or fittings shall be made wrenchtight to prevent sparking when fault current flows through the conduit system and to ensure the explosionproof and flameproof integrity of the conduit system where applicable. Equipment provided with threaded entries for field wiring connections shall be installed as applicable.

(D) Marking. All equipment shall be marked to show the class, group, and operating temperature or temperature class referenced to a 40°C (104°F) ambient.

Exception No. 1: Equipment of the non–heat-producing type, such as junction boxes, conduit, and fittings, and equipment of the heat-producing type and having a maximum temperature not more than 100°C (212°F) shall not be required to have a marked operating temperature or temperature class.

Exception No. 2: Fixed luminaires (lighting fixtures) marked for use in Class I, Division 2 or Class II, Division 2 locations only shall not be required to be marked to indicate the group.

Exception No. 3: Fixed general-purpose equipment in Class I locations, other than fixed luminaires (lighting fixtures), that is acceptable for use in Class I, Division 2 locations shall not be required to be marked with the class, group, division, or operating temperature.

Exception No. 4: Fixed dusttight equipment other than fixed luminaires (lighting fixtures) that is acceptable for use in Class II, Division 2 and Class III locations shall not be required to be marked with class, group, division, or operating temperature.

Exception No. 5: Electric equipment suitable for ambient temperatures exceeding 40°C (104°F) shall be marked with both the maximum ambient temperature and the operating temperature or temperature class at that ambient temperature.

440.4 Class I, Zone 0, 1, and 2 Locations.

(A) Scope. This article covers the requirements for the zone classification system as an alternative to the division classification system for electrical and electronic equipment and wiring for all voltage in Class I, Zone 0, Zone 1, and Zone 2 hazardous (classified) locations where fire or explosion hazards may exist due to flammable gases, vapors, or liquids.

FPN: Requirements for electric and electronic equipment and wiring for all voltages in Class I, Division 1 or Division 2; Class II, Division 1 or Division 2; and Class III, Division 1 or Division 2 hazardous (classified) locations where fire or explosion hazards may exist due to flammable gases or vapors, flammable liquids, or combustible dusts or fibers, are contained in Articles 500 through 504 of NFPA 70, *National Electrical Code,* 2002.

(B) Threading. All threaded conduit or fittings referred to herein shall be threaded with a National (American) Standard Pipe Taper (NPT) standard conduit cutting die that provides a taper of 1 in 16 (¾ in. taper per foot). Such conduit shall be made wrenchtight to prevent sparking when fault current flows through the conduit system and to ensure the explosionproof or flameproof integrity of the conduit system where applicable. Equipment provided with threaded entries for field wiring connection shall be installed in accordance with 440.4(B)(1) or 440.4(B)(2).

(1) Equipment Provided with Threaded Entries for NPT-Threaded Conduit or Fittings. For equipment provided with threaded entries for NPT-threaded conduit or fittings, listed conduit, conduit fittings, or cable fittings shall be used.

(2) Equipment Provided with Threaded Entries for Metric Threaded Conduit or Fittings. For equipment with metric threaded entries, such entries shall be identified as being metric, or listed adapters to permit connection to conduit or NPT-threaded fittings shall be provided with the equipment. Adapters shall be used for connection to conduit or NPT-threaded fittings. Listed cable fittings that have metric threads shall be permitted to be used.

FPN: Threading specifications for metric threaded entries are located in ISO 965/1-1980, *Metric Screw Threads,* and ISO 965/3-1980, *Metric Screw Threads.*

(C) Special Precaution. Article 440 requires equipment construction and installation that will ensure safe performance under conditions of proper use and maintenance.

FPN No. 1: It is important that inspection authorities and users exercise more than ordinary care with regard to the installation and maintenance of electrical equipment in hazardous (classified) locations.

FPN No. 2: Low ambient conditions require special consideration. Electrical equipment depending on the protection techniques may not be suitable for use at temperatures lower than −20°C (−4°F) unless they are approved for use at lower temperatures. However, at low ambient temperatures, flammable concentrations of vapors may exist in a location classified Class I, Zones 0, 1, or 2 at normal ambient temperature.

(1) Supervision of Work. Classification of areas and selection of equipment and wiring methods shall be under the supervision of a qualified Registered Professional Engineer.

(2) Dual Classification. In instances of areas within the same facility classified separately, Class I, Zone 2 locations shall be permitted to abut, but not overlap, Class I, Division 2 locations. Class I, Zone 0 or Zone 1 locations shall not abut Class I, Division 1 or Division 2 locations.

(3) Reclassification Permitted. A Class I, Division 1 or Division 2 location shall be permitted to be reclassified as a Class I, Zone 0, Zone 1, or Zone 2 location, provided all of the space that is classified because of a single flammable gas or vapor source is reclassified under the requirements of the section.

(D) Class I Temperature. The temperature marking shall not exceed the ignition temperature of the specific gas or vapor to be encountered.

FPN: For information regarding ignition temperatures of gases and vapors, see NFPA 497-1997, *Recommended Practice for the Classification of Flammable Liquids, Gases, or Vapors and of Hazardous (Classified) Locations for Electrical Installations in Chemical Process Areas,* and IEC 79-20-1996, *Electrical Apparatus for Explosive Gas Atmospheres, Data for Flammable Gases and Vapours, Relating to the Use of Electrical Apparatus.*

(E) Equipment.

(1) Suitability. Suitability of identified equipment shall be determined by one of the following:

(1) Equipment listing or labeling
(2) Evidence of equipment evaluation for a qualified testing laboratory or inspection agency concerned with product evaluation
(3) Evidence acceptable to the authority having jurisdiction such as a manufacturer's self-evaluation or an owner's engineering judgment

(2) Listing.

(a) Equipment that is listed for a Zone 0 location shall be permitted in a Zone 1 or Zone 2 location of the same gas or vapor. Equipment that is listed for a Zone 1 location shall be permitted in a Zone 2 location of the same gas or vapor.

(b) Equipment shall be permitted to be listed for a specific gas or vapor, specific mixtures of gases or vapors, or any specific combination of gases or vapors.

(3) Marking. Equipment shall be marked in accordance with 440.3(E)(3)(a) or 440.3(E)(3)(b).

(a) Division Equipment. Equipment approved for Class I, Division 1 or Class I, Division 2 shall, in addition to being marked, be permitted to be marked with all of the following:

(1) Class I, Zone 1 or Class I, Zone 2 (as applicable)
(2) Applicable gas classification group(s)
(3) Temperature classification

(b) Zone Equipment. Equipment meeting one or more of the protection techniques shall be marked with the following in the order shown:

(1) Class
(2) Zone
(3) Symbol "AEx"
(4) Protection technique(s)
(5) Applicable gas classification group(s)
(6) Temperature classification

Exception: Intrinsically safe associated apparatus shall be required to be marked only with items (4), (5) and (6).

(c) Group and Zone Markings for Zone Equipment. Electric equipment of types of protection "e," "m," "p," or "q" shall be marked Group II. Electric equipment of types of protection "d," "ia," "ib," "[ia]," or "[ib]" shall be marked Group IIA, or IIB, or IIC, or for a specific gas or vapor. Electric equipment of types of protection "n" shall be marked Group II unless it contains enclosed-break devices, nonincendive components, or energy-limited equipment or circuits, in which case it shall be marked Group IIA, IIB, or IIC, or a specific gas or vapor. Electrical equipment of other types of protection shall be marked Group II unless the type of protection utilized by the equipment requires that it be marked Group IIA, IIB, or IIC, or a specific gas or vapor. *[See Figure 440.4(E)(3)(c).]*

FPN: An example of such a required marking is "Class I, Zone 0, AEx ia IIC T6."

(F) Documentation for Industrial Occupancies. All areas in industrial occupancies designated as hazardous (classified) locations shall be properly documented. This documentation shall be available to those authorized to design, install, inspect, maintain, or operate electrical equipment at the location.

(G) Grounding and Bonding. Grounding and bonding shall comply with the following and other applicable requirements.

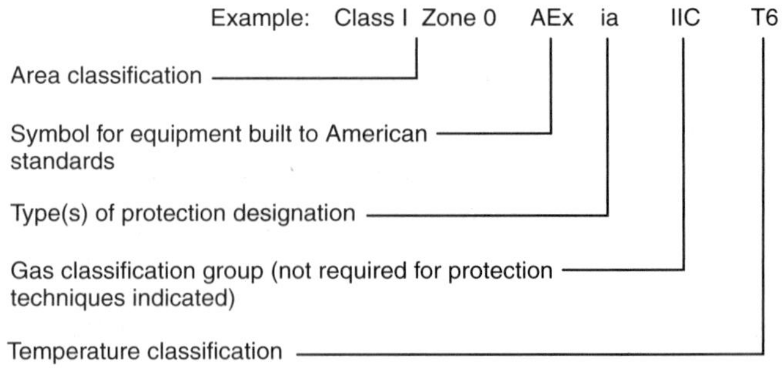

Figure 440.4(E)(3)(c) Class I, Zone 0, AEx ia IIC T6.

(1) Bonding. The locknut-bushing and double-locknut type of contacts shall not be depended on for bonding purposes, but bonding jumpers with proper fittings or other approved means of bonding shall be used. Such means of bonding shall apply to all intervening raceways, fittings, boxes, enclosures, and so forth, between Class I locations and the point of grounding for service equipment or point of grounding of a separately derived system.

Exception: The specific bonding means shall be required only to the nearest point where the grounded circuit conductor and the grounding electrode are connected together on the line side of the building or structure disconnecting means, provided the branch-circuit overcurrent protection is located on the line side of the disconnecting means.

(2) Flexible Metal Conduit. Where flexible metal conduit or liquidtight flexible metal conduit is used and is to be relied on to complete a sole equipment grounding path, it shall be installed with internal or external bonding jumpers in parallel with each conduit.

Exception: In Class I, Zone 2 locations, the bonding jumper shall be permitted to be deleted where all the following conditions are met:

(a) Listed liquidtight flexible metal conduit 1.8 m (6 ft) or less in length, with fittings listed for grounding, is used.
(b) Overcurrent protection in the circuit is limited to 10 amperes or less.
(c) The load is not a power utilization load.

NFPA 77, *Recommended Practice on Static Electricity,* 2000

***Editor's Note:** This document is reprinted here in its entirety, including its appendixes.*

Information on referenced publications can be found in Chapter 10 and Appendix E.

Chapter 1 Administration

1.1 Scope.

1.1.1 This recommended practice applies to the identification, assessment, and control of static electricity for purposes of preventing fires and explosions.

1.1.2* This recommended practice does not apply directly to shock hazards from static electricity. However, application of the principles set forth in this recommended practice can reduce such shock hazards to personnel.

1.1.3* This recommended practice does not apply to the prevention and control of static electricity in hospital operating rooms or in areas where flammable anesthetics are administered or handled.

1.1.4* This recommended practice does not apply to lightning.

1.1.5* This recommended practice does not apply to stray electrical currents or to induced currents from radio frequency (RF) energy.

1.1.6* This recommended practice does not apply to fueling of motor vehicles, marine craft, or aircraft.

1.1.7* This recommended practice does not apply to cleanrooms.

1.1.8 This recommended practice does not apply to control of static electricity and static electricity hazards involved with electronic components, which have their own requirements.

1.2 Purpose.

The purpose of this recommended practice is to assist the user in controlling the hazards associated with the generation, accumulation, and discharge of static electricity by providing the following:

(1) A basic understanding of the nature of static electricity
(2) Guidelines for identifying and assessing the hazards of static electricity
(3) Techniques for controlling the hazards of static electricity
(4) Guidelines for controlling static electricity in selected industrial applications

1.3 Applicability. [Reserved]

1.4 Equivalency.

Nothing in this recommended practice is intended to prevent the use of systems, methods, or devices of equivalent or superior quality, strength, fire resistance, effectiveness, durability, or safety over those prescribed by this recommended practice, provided technical documentation is submitted to the authority having jurisdiction to demonstrate equivalency and the system, method, or device is approved for the purpose.

Chapter 2 Referenced Publications [Reserved]

Chapter 3 Definitions, Units, and Symbols

3.1 Definitions.

The following terms are defined in this chapter for the purposes of this recommended practice.

3.1.1 Antistatic. Capable of dissipating a static electric charge at an acceptable rate.

3.1.2 Bonding. The process of connecting two or more conductive objects together by means of a conductor so that they are at the same electrical potential, but not necessarily at the same potential as the earth.

3.1.3 Breakdown Strength. The minimum voltage, measured in volts per meter of thickness, necessary to cause a spark through a solid material that is held between electrodes that produce a uniform electric field under specified test conditions.

3.1.4 Breakdown Voltage. The minimum voltage, measured in volts, necessary to cause a spark through a gas mixture between electrodes that produce a uniform electric field under specified test conditions.

3.1.5* Capacitance. The amount of charge, measured in coulombs, that must be stored on a specified body or material to raise the potential difference by 1 volt. Capacitance is measured in coulombs per volt or farads.

3.1.6 Combustible. Capable of burning.

3.1.7 Combustible Dust. Any finely divided solid material, 420 micrometers (microns, μm) or less in diameter (i.e., material that will pass through a U.S. No. 40 standard sieve), that presents a fire or explosion hazard when dispersed and ignited in air or other gaseous oxidizer.

3.1.8 Conductive. The ability to allow the flow of an electric charge; possessing a conductivity greater than 10^4 pS/m or a resistivity less than 10^8 Ω-m.

3.1.9 Conductor. A material or object that allows an electric charge to flow easily through it.

3.1.10 Grounding. The process of bonding one or more conductive objects to the ground, so that all objects are at zero (0) electrical potential; also referred to as "earthing."

3.1.11 Ignitible Mixture. A gas-air, vapor-air, mist-air, or dust-air mixture, or combinations of such mixtures, that can be ignited by a sufficiently strong source of energy, such as a static electric discharge.

3.1.12 Inert Gas. A nonflammable, nonreactive gas that renders the combustible material in a system incapable of supporting combustion.

3.1.13 Nonconductive. The ability to resist the flow of an electric charge.

3.1.14 Nonconductor. A material or object that resists the flow of an electric charge through it.

3.1.15 Semiconductive. Possessing a conductivity between 10^2 pS/m and 10^4 pS/m or a resistivity between 10^8 Ω-m and 10^{10} Ω-m.

3.1.16 Static Electric Discharge. A release of static electricity in the form of a spark, corona discharge, brush discharge, or propagating brush discharge that might be capable of causing ignition under appropriate circumstances.

3.1.17 Static Electricity. An electric charge that is significant only for the effects of its electrical field component and that manifests no significant magnetic field component.

3.2 Symbols and Units.

The following symbols are used throughout this recommended practice and are defined as follows.

A = ampere (coulombs per second). Electric current; the quantity of charge passing per second through a given point.
C = capacitance (farads)
d = diameter (meters)
e = 2.718 [base of Napierian (natural) logarithms]
E = V/m = electric field strength (volts per meter)
ϵ = dielectric constant of a material
ϵ_0 = 8.845×10^{-12} farads per meter (electrical permittivity of a vacuum)
$\epsilon\epsilon_0$ = electrical permittivity of a material (farads per meter)
I_s = streaming current (amperes)
K = liquid conductivity (picosiemens per meter)
μ = ion mobility (square meters per volt-second)
μm = micrometers (microns) = 10^{-6} meters
Ω-m = electrical resistivity (ohm-meter)
P = pressure (millimeters of mercury)
Q = quantity of electrical charge (coulombs)
R = electrical resistance (ohms)
ρ = volume resistivity (ohm-meters)

S = electrical conductance (siemens)
t = elapsed time (seconds)
τ = charge relaxation time constant (seconds)
v = flow velocity (meters per second)
V = electrical potential difference (volts)
W = energy or work done (joules)

Chapter 4 Fundamentals of Static Electricity

4.1 General.

4.1.1 The most common experiences of static electricity are the crackling and clinging of fabrics as they are removed from a clothes dryer or the electric shock felt as one touches a metal object after walking across a carpeted floor or stepping out of an automobile. Nearly everyone recognizes that these phenomena occur mainly when the atmosphere is very dry, particularly in winter. To most people, they are simply an annoyance. In many industries, particularly those where combustible materials are handled, static electricity can cause fires or explosions.

4.1.2 The word *electricity* is derived from the ancient Greek word for amber, *elektron*. The phenomenon of electrification was first noticed when pieces of amber were rubbed briskly. For centuries, the word *electricity* had no meaning other than the ability of some substances to attract or repel lightweight objects after being rubbed with a material like silk or wool. Stronger electrification accompanied by luminous effects and small sparks was first observed about 300 years ago by von Guericke.

In comparatively recent times, when the properties of flowing (current) electricity were discovered, the term *static* came into use as a means of distinguishing a charge that was at rest from one that was in motion. However, today the term is used to describe phenomena that originate from an electric charge, regardless of whether the charge is at rest or in motion.

4.1.3 All materials, whether solid or fluid, are composed of various arrangements of atoms. The atoms are composed of positively charged nuclear components that give them mass, surrounded by negatively charged electrons. Atoms can be considered electrically neutral in their normal state, meaning that there are equal amounts of positive and negative charge present. They can become charged when there is an excess or a deficiency of electrons relative to the neutral state. Electrons are mobile and of insignificant mass and are the charge carriers most associated with static electricity.

4.1.4 In materials that are conductors of electricity, such as metals, electrons can move freely. In materials that are insulators, electrons are more tightly bound to the nuclei of the atoms and are not free to move. Examples of insulators include the following:

(1) Nonconductive glass
(2) Rubber
(3) Plastic resins
(4) Dry gases
(5) Paper
(6) Petroleum fluids

The mobility of electrons in materials known as semiconductors is freer than in insulators but is still less than in conductors. Semiconductive materials are commonly characterized by their high electrical resistance, which can be measured with a megohmmeter.

4.1.5 In otherwise insulating fluids, an electron can separate from one atom and move freely or attach to another atom to form a negative ion. The atom losing the electron then becomes a positive ion. Ions are charged atoms and molecules.

4.1.6 Unlike charges attract each other and the attractive force can draw the charges together, if the charges are mobile. The energy stored is the result of the work done to keep the charges separated by a finite distance.

4.1.7 Separation of charge cannot be prevented absolutely, because the origin of the charge lies at the interface of materials. When materials are placed in contact, some electrons move from one material to the other until a balance (equilibrium condition) in energy is reached. This charge separation is most noticeable in liquids that are in contact with solid surfaces and in solids in contact with other solids. The flow of clean gas over a solid surface produces negligible charging.

4.1.8 The enhanced charging that results from rubbing materials together (triboelectric charging) is the result of exposing surface electrons to a broad variety of energies in an adjacent material, so that charge separation is more likely to take place. The breakup of liquids by splashing and misting results in a similar charge separation. It is only necessary to transfer about one electron for each 500,000 atoms to produce a condition that can lead to a static electric discharge. Surface contaminants at very low concentrations can play a significant role in charge separation at the interface of materials. *[See Figures 4.1.8(a) and (b).]*

4.1.9 Conductive materials can become charged when brought near a highly charged surface. Electrons in the conductive material are either drawn toward or forced away from the region of closest approach to the charged surface, depending on the nature of the charge on that surface. If the conductive material is then touched to ground or to a third object, additional electrons can pass to or from ground or the object. If contact is then broken and the conductive

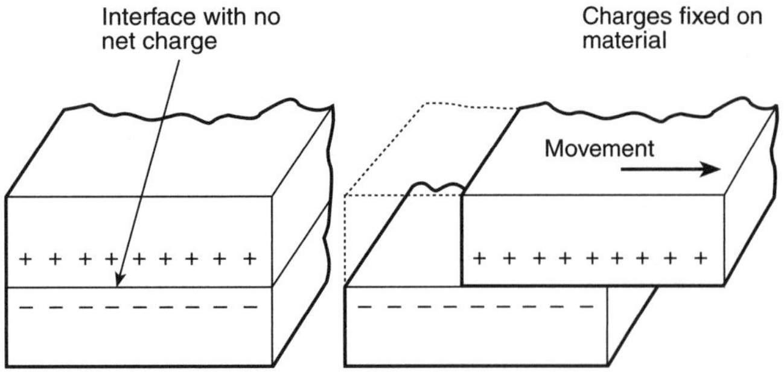

Figure 4.1.8(a) Typical Charge Generation by Bulk Motion of Insulating Materials. *(Walmsley, 1992, p. 19.)*

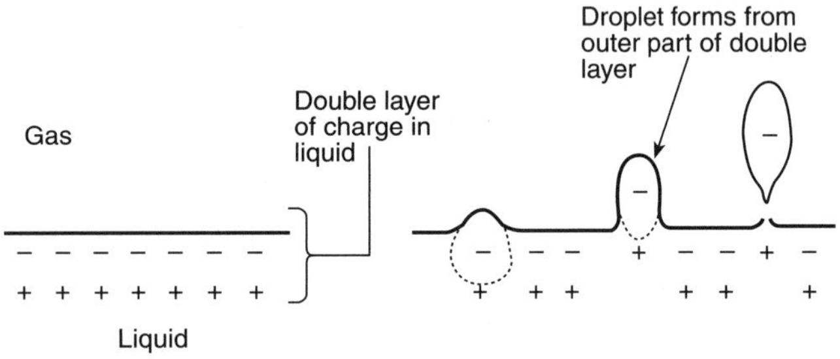

Figure 4.1.8(b) Typical Charge Generation by Atomization. *(Walmsley, 1992, p. 19.)*

material and charged surface are separated, the charge on the isolated conductive object changes. The net charge that is transferred is called *induced charge. [See Figures 4.1.9(a), (b), (c), and (d).]*

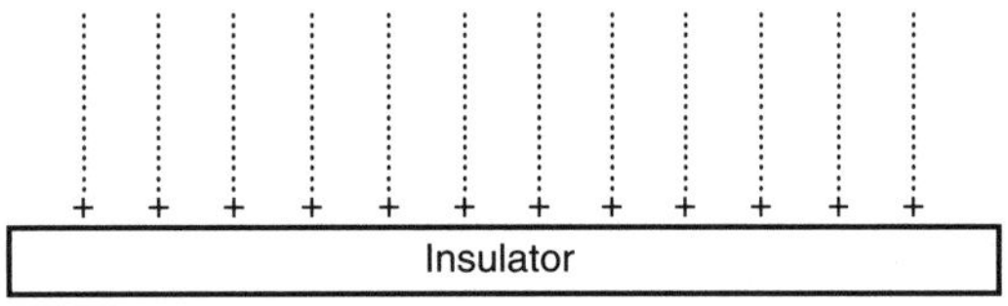

Figure 4.1.9(a) Charged Insulator with Field Lines Shown. *(Pratt, 1997, p. 29.)*

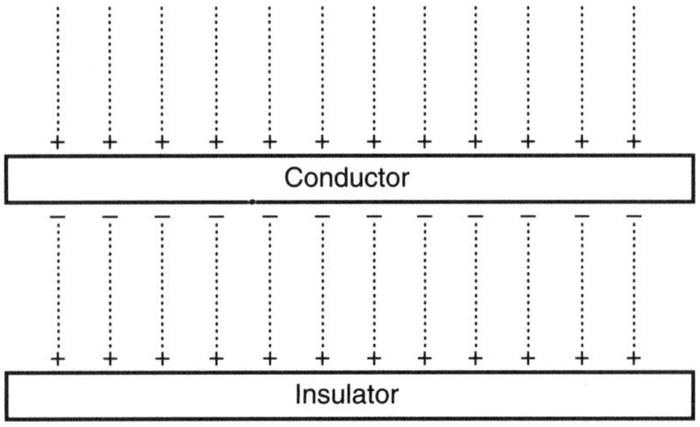

Figure 4.1.9(b) Induced Charge on Conductor; the Charge Remains on the Conductor as the Conductor is Removed from Contact with the Insulator. *(Pratt, 1997, p. 29.)*

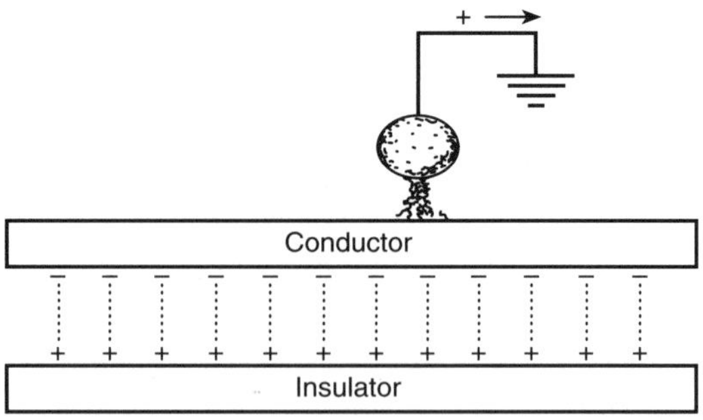

Figure 4.1.9(c) Discharge of Free Charge from a Conductor. *(Pratt, 1997, p. 29.)*

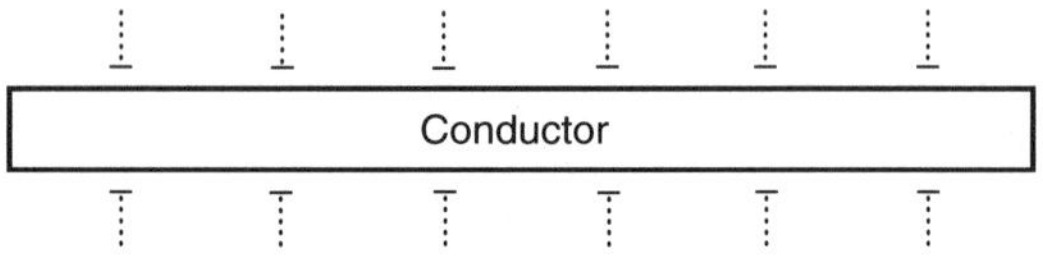

Figure 4.1.9(d) Conductor, Removed from an Insulator, Carrying a Charge.

4.1.10 The separation of charge on a neutral isolated conductor and its distribution near a charged insulating surface will produce electrical stresses near the point of closest approach. Sharp edges on the conductor can yield a localized electrical breakdown of the gas stream, known as *corona,* or an electric spark across the gap. Either of these events can transfer charge between the materials, leaving the isolated conductor charged. Such a transfer occurs, for example, when a person wearing nonconductive shoes receives a static electric shock by contacting the metal frame supporting a highly charged web. As a result, the person acquires a net static electric charge and can receive a second shock after leaving the area by touching a grounded metal object, thereby allowing the acquired charge to flow to earth.

4.1.11 Charge can also be imparted to a surface or into the bulk of a material by directing a stream of electrons or ions against it. If the surface is not conductive or is conductive but isolated from ground, the charge delivered by the bombarding stream will remain after the stream stops.

4.1.12 Charge can also be injected into a stream of nonconductive fluid by submerging within the stream a pointed electrode on which a high voltage has been impressed.

4.2 Accumulation and Dissipation of Charge.

4.2.1 A static electric charge will accumulate whenever the rate at which charges separate exceeds the rate at which charges recombine. Work must be done to separate charges, and there is a tendency for the charges to return to a neutral state. The potential difference, that is, the voltage, between any two points is the work per unit charge that would have to be done to move the charges from one point to the other. This work depends on the physical geometry (i.e., shape, size, nature of materials, and location of objects) of the particular system and can be expressed by the following equation:

$$C = \frac{Q}{V}$$

Where:

C = capacitance (farads)
Q = charge that has been separated (coulombs)
V = potential difference (volts)

Typical examples of accumulation are illustrated in Figure 4.2.1.

4.2.2 Separation of electric charge might not, in itself, be a potential fire or explosion hazard. There must be a discharge or sudden recombination of the separated charges to pose an ignition hazard. One of the best protections from static electric discharge is a conductive or semiconductive path that allows the controlled recombination of the charges.

4.2.3 In static electric phenomena, charge is generally separated by a resistive barrier, such as an air gap or insulation between the conductors or by the insulating property of the materials being handled or processed. In many applications, particularly those where the materials being processed are

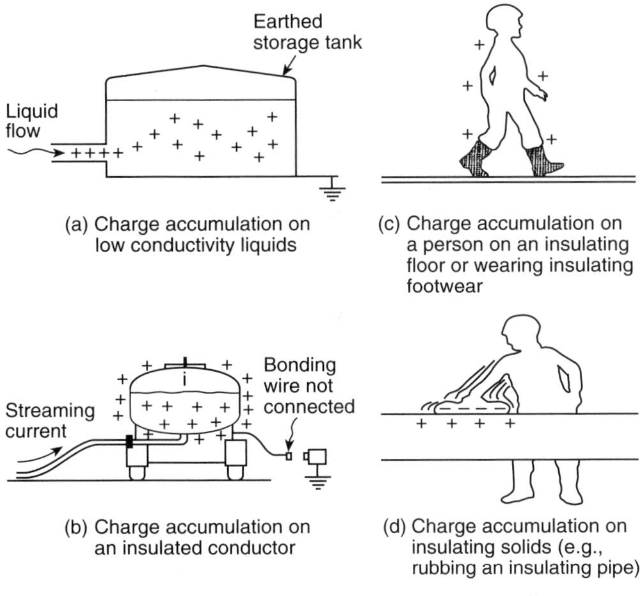

Figure 4.2.1 Typical Examples of Charge Accumulation. *(Walmsley, 1992, p. 37.)*

charged insulators (nonconductors), it is not easy to measure the charges or their potential differences.

4.2.4 When recombining of charges occurs through a path that has electrical resistance, the process proceeds at a finite rate, t/τ, and is described by the *relaxation time* or *charge decay time*, τ. This relaxation process is typically exponential and is expressed by the following equation:

$$Q_t = Q_0 e^{-t/\tau}$$

Where:

Q_t = charge remaining at time t (coulombs)
Q_0 = charge originally separated (coulombs)
e = 2.718 (base of natural logarithms)
t = elapsed time (seconds)
τ = time constant (seconds)

The rate of charge recombination depends on the capacitance of the material and its resistance and is expressed as follows:

$$\tau = RC$$

Where:

τ = time constant (seconds)
R = resistance (ohms)
C = capacitance (farads)

For bulk materials, the relaxation time is often expressed in terms of the volume electrical resistivity of the material and its electrical permittivity as follows:

$$\tau = \rho \epsilon \epsilon_0$$

Where:

τ = time constant (seconds)
ρ = resistivity (ohm-meters)
$\epsilon\epsilon_0$ = electrical permittivity (farads per meter)

4.2.5 The exponential decay model described in 4.2.4 is helpful in explaining the recombination process, but is not necessarily applicable to all situations. In particular, nonexponential decay is observed when the materials supporting the charge are certain low conductivity liquids or powders composed of combinations of insulating, semiconductive, and conductive materials. The decay in these cases is faster than the exponential model predicts.

4.2.6 Dissipation of static electric charges can be effected by modifying the volume or surface resistivity of insulating materials with antistatic additives, by grounding isolated conductors, or by ionizing the air near insulating materials or isolated conductors. Air ionization involves introducing mobile electric charges (positive, negative, or both) into the air around the charged objects. These ions are attracted to the charged objects until they become electrically neutral.

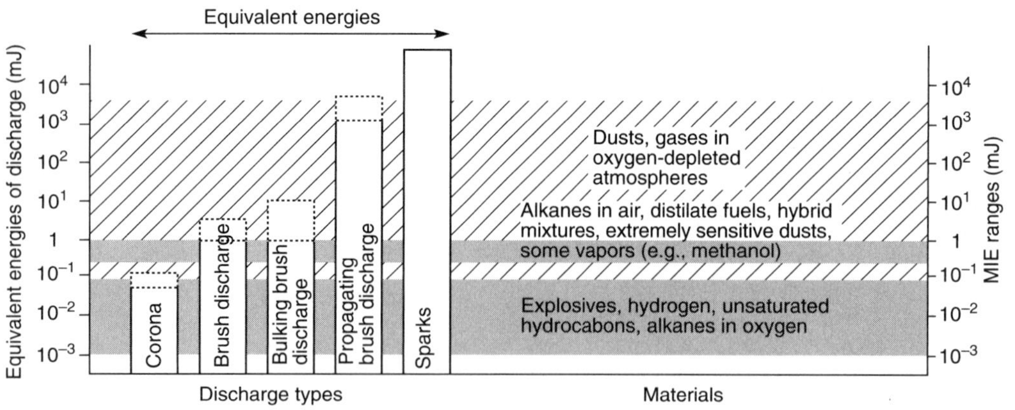

Figure 4.3.1 Approximate Energies of Types of Discharges Compared with Minimum Ignition Energies (MIE) of Typical Combustible Materials. ***(Adapted from Walmsley, 1992, p. 26.)***

The ion current in the air serves as the mechanism that brings the neutralizing charge to the otherwise bound or isolated charge.

4.3 Discharge of Static Electricity and Ignition Mechanisms.

4.3.1 General. As electric charge accumulates through separation, there will be an increase in the electrical forces trying to restore a neutral condition by reuniting those charges in the form of a static electric discharge. Many types of discharges can occur, and these are illustrated broadly in Figure 4.3.1. For a static electric discharge to be a source of ignition, the following four conditions must be met:

(1) An effective means of separating charge must be present.
(2) A means of accumulating the separated charges and maintaining a difference of electrical potential must be available.
(3) A discharge of the static electricity of adequate energy must occur.
(4) The discharge must occur in an ignitible mixture.

4.3.2 Corona Discharge. Corona is an electrical discharge in the microampere range that results from a localized electrical needle-shaped breakdown of gases by charges on surfaces such as sharp edges, points, and wires. The charges can arise on conductors at high voltages or on grounded conductors that are situated near a charged surface. Corona is accompanied by a faint luminosity. *(See Figure 4.3.2.)*

4.3.2.1 In most cases, the energy density of corona discharge is very low. Consequently, the hazard from corona discharge is small. Where corona discharge is more intense, prebreakdown streamers called *brush discharges* occur. These appear as random filaments of light that make faint hissing or frying sounds. Brush discharges that originate on needle-like tips with radii smaller than 1 mm do not, in general, lead to ignition. Discharges from blades, however, can ignite mixtures that have very low ignition energies, such as hydrogen-air or carbon disulfide-air mixtures. Gas-air and vapor-air mixtures can be ignited if brush discharges originate from elements with edge diameters greater than 5 mm, or from a rod with a hemispherical end, such as a human finger. *(See Figure 4.3.2.1.)*

4.3.2.2 Sharp edges, corners, and projections (e.g., those with an edge diameter of 5 mm or less) that point toward charged surfaces should be identified. These sharp regions concentrate the charge, providing intense, localized stresses that can lead to electrical corona and sparks.

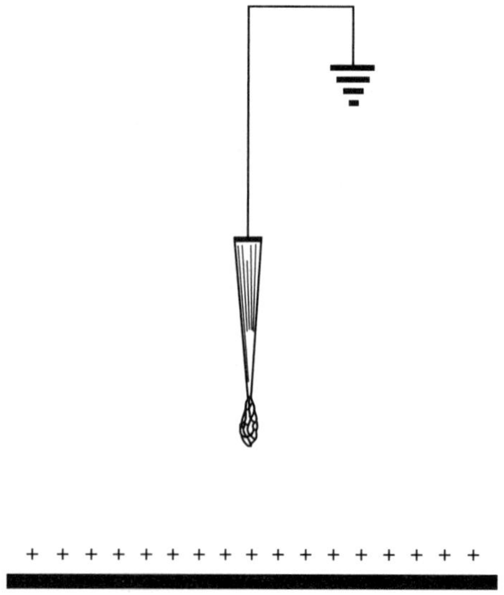

Figure 4.3.2 Corona Discharge. ***(Pratt, 1997, p. 32.)***

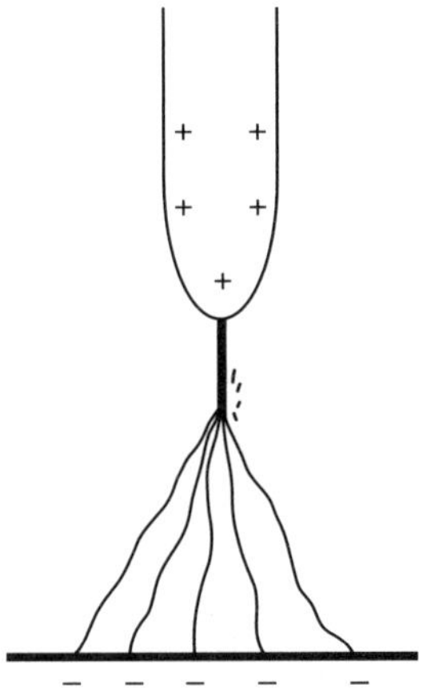

Figure 4.3.2.1 Brush Discharge. ***(Adapted from Walmsley, 1992, p. 27.)***

4.3.3 Sparks Between Conductors.

4.3.3.1 Sparks from ungrounded charged conductors, including the human body, are responsible for most fires and explosions ignited by static electricity. Sparks are typically intense capacitive discharges that occur in the gap between two charged conducting bodies, usually metal. The energy of a spark discharge is highly concentrated in space and in time.

4.3.3.2 The ability of a spark to produce ignition is governed largely by its energy, which will be some fraction of the total energy stored in the system.

4.3.3.3 The energy of a spark can be determined from the capacitance of the conductive system and the electrical potential or from the quantity of charge separated from the conductors and is expressed by the following equations:

$$W = \frac{1}{2}CV^2$$

$$W = \frac{1}{2}\frac{Q^2}{C}$$

Where:

W = energy (joules)
C = capacitance (farads)
V = potential difference (volts)
Q = charge (coulombs)

The relationships are shown in Figure 4.3.3.3.

4.3.3.4* To be capable of causing ignition, the energy released in the discharge must be at least equal to the minimum ignition energy (MIE) of the ignitible mixture. Other factors, such as the shape of the charged electrodes and the form of

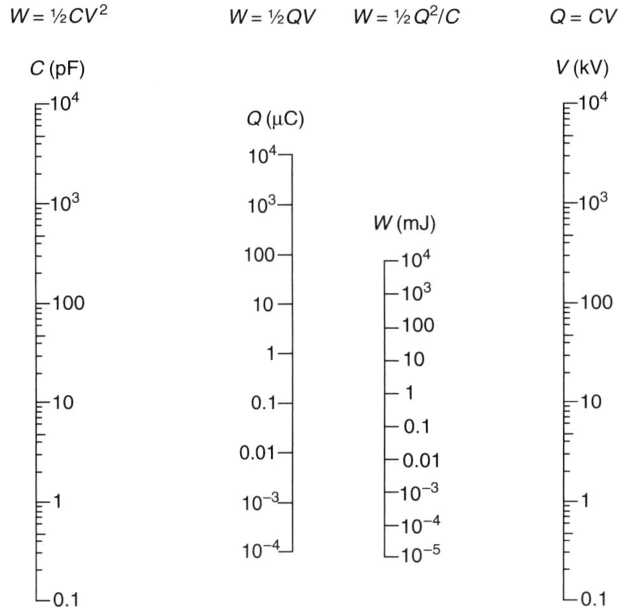

Figure 4.3.3.3 Nomograph for Estimating Energy in a Capacitive Spark Discharge. ***(Pratt, 1997, p. 113.)***

discharge, influence conditions for the static electric discharge and its likelihood of causing ignition.

4.3.3.5 Most gases and vapors of saturated hydrocarbons require about 0.25 mJ of energy for spark discharge ignition, assuming optimum mixtures with air. Unsaturated hydrocarbons can have lower MIEs. Discussion of MIEs for specific materials can be found in Chapters 7 through 9.

4.3.3.6 Mists, dusts, and fibers usually require MIEs that are one or two orders of magnitude greater than those for gases and vapors. Note that for any given particulate material, the MIE diminishes rapidly with decreasing particle size.

4.3.3.7 The ignition energies for gases, vapors, and dusts are reduced by an increase in the oxygen concentration relative to that for air. Likewise, ignition energies are increased by a decrease in oxygen concentration.

4.3.4 Hybrid Mixtures. Where two or more flammable materials of different phases (e.g., a dust and a vapor) are present in the same mixture, the mixture is known as a *hybrid*. Tests have shown that adding a flammable gas to a dust suspension can greatly lower the ignition energy of the dust. This phenomenon is especially true when the gas is present at a concentration below its lower flammable limit or the dust is below its minimum explosable concentration. Such hybrid mixtures can sometimes be ignited even if both components are below their respective lower limits. A hybrid mixture can be formed by:

(1) Vapor desorption from particulates (such as in resin product receivers)
(2) Reaction of particulates with atmospheric moisture that produces a flammable gas
(3) Introduction of a dust into a flammable vapor atmosphere (such as adding a dust or powder to a flammable liquid)

In such instances, the hybrid mixture can be ignited at the MIE of the most easily ignited component.

4.3.5 Static Electric Discharge from the Human Body.

4.3.5.1 The human body is a good electrical conductor and numerous incidents have resulted from static electric discharges from people.

4.3.5.2 If a person is insulated from ground, that person can accumulate a significant charge by walking on an insulating surface, by touching a charged object, by brushing surfaces while wearing nonconductive clothing, or by momentarily touching a grounded object in the presence of charges in the environment. During normal activity, the potential of the human body can reach 10 kV to 15 kV, and the energy of a possible spark can reach 20 mJ to 30 mJ. By comparing these values to the MIEs of gases or vapors, the hazard is readily apparent.

4.3.6 Discharges Between Conductors and Insulators.

4.3.6.1 Sparks often occur between conductors and insulators. Examples of such occurrences include situations in which plastic parts and structures, insulating films and webs, liquids, and particulate material are handled. The charging of these materials can result in surface discharges and sparks, depending on the accumulated charge and the shape of nearby conductive surfaces. The variable (both in magnitude and polarity) charge density observed on insulating surfaces is the effect of these discharges spreading over a limited part of the insulating surface.

4.3.6.2 Even with the use of static electricity neutralizers, some charges will remain in certain areas, but these will typically not be hazardous if there is no mechanism by which they can accumulate. However, a dangerous (i.e., ignition-capable) static electric charge can result due to concentration of individual charges. Examples are stacking or nesting of empty plastic containers, winding film onto a roll or drum, and filling a vessel with a nonconductive liquid or powder.

4.3.7 Discharge on the Surface of an Insulator Backed by a Conductor. A surface coated with a thin (less than 8 mm) layer of an insulating material will act as a capacitor

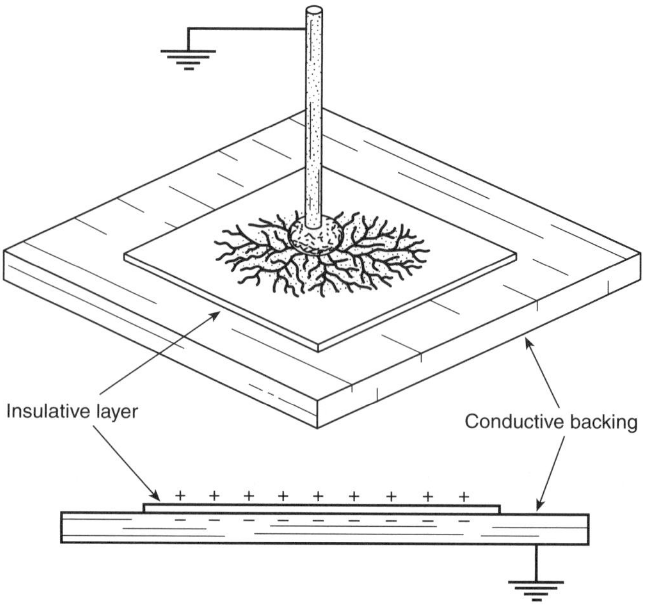

Figure 4.3.7 Propagating Brush Discharge. *(Pratt, 1997, p. 35.)*

to store charge. At sufficiently high charge levels (i.e., greater than 250 $\mu C/m^2$), a branching discharge will be observed on the surface of the coating. This branching discharge is called a *propagating brush discharge.* Alternatively, an electrical breakdown through the layer can occur. The energy stored in the coating can be as high as several joules per square meter, so the energy of the discharge, however distributed in space, can be sufficient to ignite gas-air, vapor-air, and dust-air mixtures. *(See Figure 4.3.7.)*

4.3.8* Discharges During Filling Operations. During filling of large silos with powders, granules, and pellets, surface flashes up to a meter in length have been observed. These discharges, called *bulking brush (cone) discharges,* are accompanied by a crackling sound capable of being heard above the noise of the material transfer. Bulking brush discharges have a maximum effective energy of 10 mJ to 25 mJ and are believed responsible for dust explosions in grounded silos. Similar discharges are observed during the filling of tank vehicles with nonconductive liquids. In this case, the phenomenon is known as a *surface streamer* or *go-devil. (See Figure 4.3.8.)*

Chapter 5 Evaluating the Hazard of Static Electricity

5.1 General.

5.1.1 The following are two basic steps in evaluating static electricity hazards:

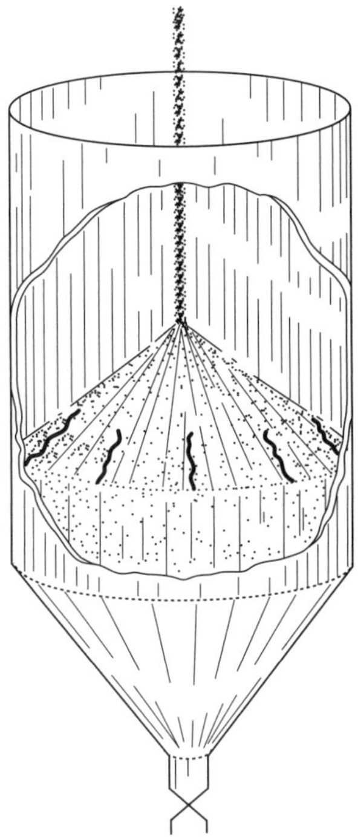

Figure 4.3.8 Bulking Brush Discharge During Filling of Silos with a Bulk Powder.

(1) Identifying locations where charge separates and accumulates
(2) Assessing the ignition hazards at these locations

The basic process is outlined in Figure 5.1.1.

5.1.2 On-site evaluation or survey of the process should be made to identify any ungrounded conductive objects, including personnel, and any materials that could serve as electrical insulators and could interfere with proper bonding and grounding. The survey should identify those locations that might pose a static electricity hazard, even if there is no evidence of accumulation of charge at the time of the evaluation.

5.1.2.1 Special attention should be given to insulating materials that are handled or processed. Each process operation should be considered separately and attention should be given to the likely range of exposure of the materials. For example, changes in temperature and relative humidity can significantly influence the bulk conductivity and surface conductivity of materials.

5.1.2.2 It is often helpful to complete a design review of the operation, process, or machine and a visual survey of the area first. An on-site instrumented evaluation should then be conducted during actual operating conditions to determine the nature and magnitude of any static electricity hazards present.

WARNING

During an audit, precautions should be taken that are consistent with the equipment and the materials in the area where measurements are to be taken. The primary ignition hazard comes from introducing a grounded electrode, such as the housing of a field meter *(see Section 5.4)*, into the vicinity of a charged surface, thus providing a route for a static electric discharge. The surface being measured should always be approached slowly while observing the meter's response. Extreme care should be taken so that neither instruments nor testing techniques cause ignition of flammable atmospheres. Appropriate safe work practices should be employed when taking measurements in and around physical hazards such as moving belts, webs, and pulleys.

5.2 Measuring a Static Electric Charge.

A meaningful evaluation requires using an appropriate instrument, using the instrument according to the manufacturer's instructions, maintaining calibration of the instrument, and interpreting the measurements according to the manufacturer's recommendations.

5.3 Measuring the Charge on a Conductor.

5.3.1 The voltage on a conductor is proportional to the charge it supports and is expressed by the following equation:

$$V = \frac{Q}{C}$$

Where:

V = potential difference (volts)
Q = charge supported by the conductor (coulombs)
C = capacitance of the conductor (farads)

5.3.2 The voltage on a conductor can be measured by direct contact using a voltmeter, provided the impedance of the voltmeter is high enough so that it does not discharge the conductor and the capacitance is small enough so that it does not collect a significant charge from the conductor. An electrostatic voltmeter with input impedance greater than 10^{12} ohms can be used for measuring voltages on most ungrounded conductors. Since conductors have the same voltage at every point on their surface, it is not important where the test probe of the voltmeter touches the surface of the conductor.

5.4 Measuring the Charge on a Nonconductor.

5.4.1 The charge on a nonconductor cannot be measured using a direct contact electrostatic voltmeter. A noncontact electrostatic voltmeter, or field meter, must be used. A noncontact electrostatic voltmeter senses the strength of the

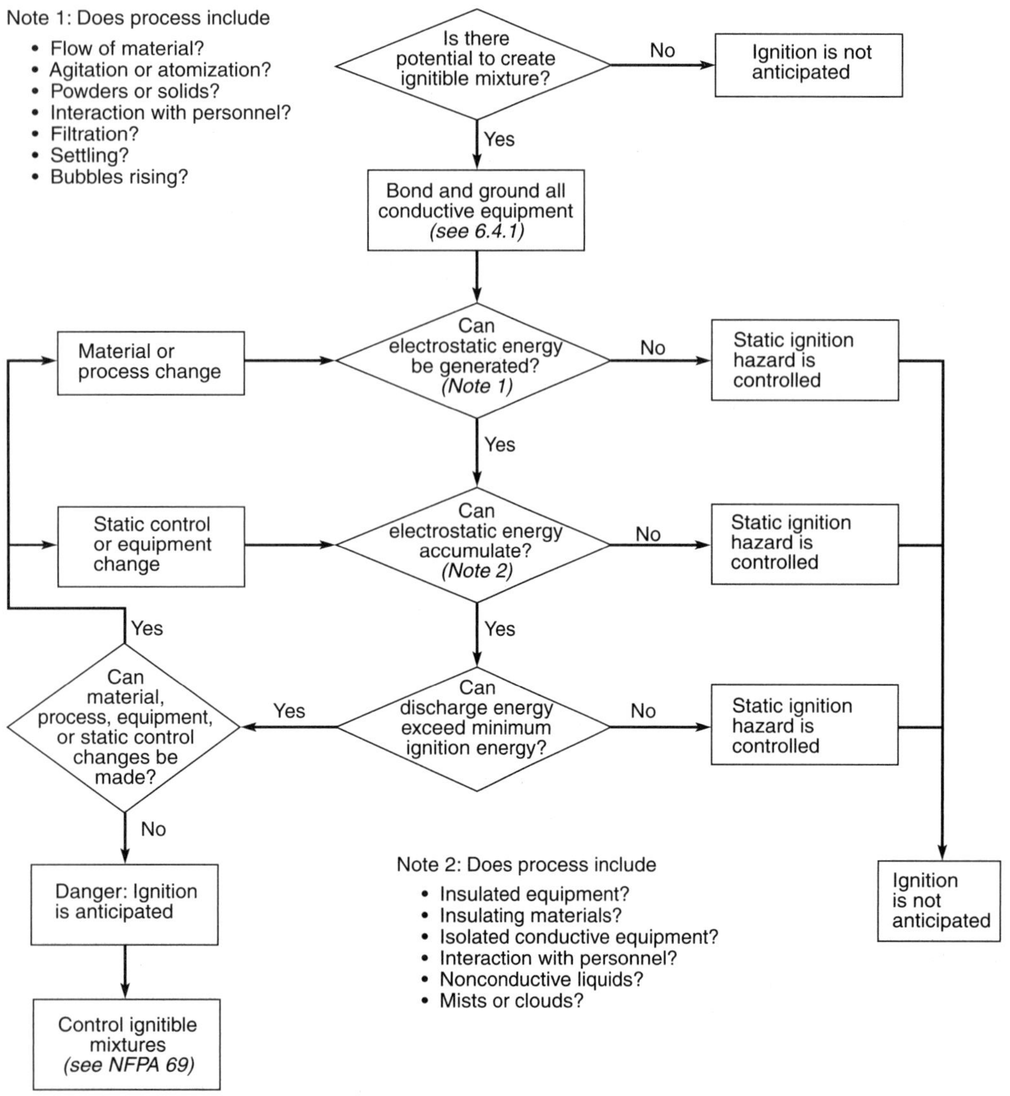

Figure 5.1.1 Flow Chart for Determining Static Electric Ignition Hazard.

static electric field from the net charge on or in the nonconductor. The field strength is proportional to the static electric force per unit charge and it describes the electric forces present near a charged object. For practical purposes, an electric field is the force that one experiences or measures around a charged object.

5.4.2 Field meters are calibrated to measure the electric field in units of volts per unit distance, typically in kilovolts per meter. In most cases, the measurements are proportional to the net static electric charge on the object being measured. Field meters are called *field mills* or *charge locators*, depending on their principle of operation and level of sophistication. Since the charge density on or in a nonconductor is typically not uniform, measurements should be taken at several locations.

5.4.3 Charged nonconductors exist in many forms, such as sheets, films, webs, powders, liquids, process rolls, and extrusions. Charges on these materials and objects will produce electric fields that will be influenced by the instrument, by the observer, and by other nearby conductive, semiconductive, or insulating materials. As a result, the electric field measured by the instrument will generally be different from the electric field present before the instrument was introduced. This phenomenon is a result of a change in capacitance.

5.4.4 The forces between electric charges exhibit themselves at a distance. For this reason, the effects of accumulated charge can be observed as the charged objects are approached. For example, the hands and arms of a person who approaches a highly charged object will tingle and might even draw a spark as the surface of the skin and the hair become "charged." Sometimes these observations and sparks provide the first indication that a potentially hazardous condition exists. This charging of the human body can happen even if the person is well grounded.

5.5 General Practices.

5.5.1 The primary instrument for locating a charge on either a conductor or a nonconductor is the noncontact voltmeter or field meter. As its name implies, the instrument does not contact the charged surface directly. Rather, it senses the magnitude and the polarity of the electric field that exists *at its sensing aperture*. As stated in 5.4.3, the instrument and its sensing aperture disturb the electric field around the charge to be measured, so the meter reading does not accurately indicate the actual magnitude of the field when the meter is not present. Aside from this deficiency, the field meter is an inexpensive and valuable tool for locating a static electric charge.

5.5.2 In some cases, primarily cases involving flowing liquids and flowing bulk solids, it is easier to collect a sample of the charged material in an isolated vessel or cup, referred to as a *Faraday cup*, and to use an electrometer to measure the streaming current or net charge flowing to the cup receiving the charged material.

5.5.3 The proper use of instruments to evaluate the magnitude of charge accumulation in specific applications is further discussed in Chapters 7 through 9.

5.6 Measuring the Accumulation and Relaxation of Charge.

5.6.1 Measuring the rate of accumulation and relaxation of static electric charge involves measuring changing potential differences or currents.

5.6.2 Field meters and dedicated charge decay monitors can be used to observe charge relaxation under conditions of prescribed initial voltages on conductors and nonconductors.

5.6.3 Leakage currents down to about 10^{-13} A can be measured from isolated conductors using commercially available electrometers. The isolated conductor can be a Faraday cup containing a bulk solid or fluid.

5.7 Measuring the Resistivity of Materials.

Electrical resistivity of materials often consists of volume (bulk) and surface components. In electrostatic processes, the approximate ranges of resistivities that define materials as insulating, semiconductive (antistatic), or conductive are summarized in Figure 5.7.

5.7.1 Electric charges can be conducted from a solid, liquid, or powder, either across the surface or through the material.

5.7.2 The volumetric resistivity of a material can be determined by applying a potential difference across a sample of known cross section and monitoring the current through the cross section.

5.7.3 Adsorbed material, particularly water vapor, and compaction of materials are known to lower the resistivity of materials. The resistivity of many materials also has been found to vary with the applied potential difference and with the duration of the test. Various designs of cells used to measure resistivity have been developed into standard test configurations that are applicable to specific kinds of samples. Appropriate test procedures include the following:

(1) ASTM D 257, *Standard Test Methods for DC Resistance or Conductance of Insulating Materials*, 1999
(2) JIS B 9915, *Measuring Methods for Dust Resistivity (with Parallel Electrodes)*
(3) CENELEC EN 61241 2 2, *Electrical Apparatus for Use in the Presence of Combustible Dust — Part 2: Test Methods; Section 2: Method for Determining the Electrical Resistivity of Dust in Layers*, 1995

5.8 Assessment of Bonding and Grounding.

5.8.1 As defined in Chapter 3, bonding is a process whereby two or more conductive objects are connected together by means of a conductor, so that they are at the same electrical potential; that is, the voltage difference between the objects

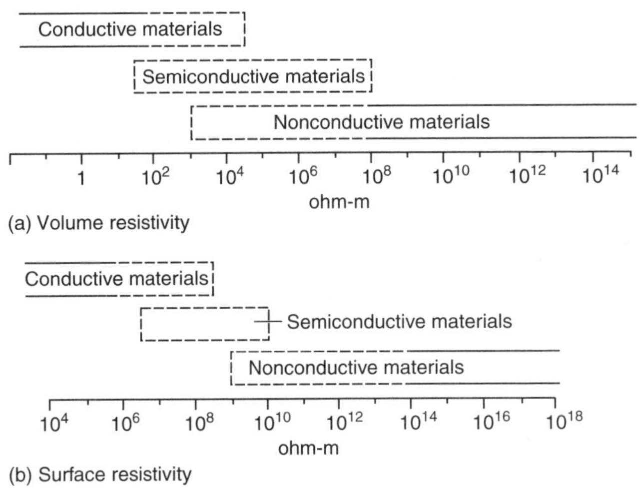

Figure 5.7 Ranges of Volume and Surface Resistivities. ***(Walmsley, 1992, p. 138.)***

is zero (0). The objects might or might not be at the same potential as the earth. In fact, a considerable potential difference between the objects and the earth could exist.

Grounding is the process of connecting the conductive objects to the earth, so that they are all at zero (0) electrical potential.

In both cases, the intent is to eliminate the occurrence of a static electric spark.

5.8.2 Identification of conductive equipment and objects within a process is critical for successful bonding and grounding. Periodic inspection and testing of bonding and grounding systems is equally important. Proper inspection and testing will ensure that the chances for a static electric charge to accumulate are minimized.

5.8.3 The resistance to ground of the bonding or grounding path is important, not only to ensure relaxation of the static electric charge, but also to maintain worker safety and satisfy other purposes, such as those for lightning protection and electrical system shock protection. Practices that constitute proper resistance to ground will vary from application to application. Chapters 6 through 9 provide examples of acceptable grounding practices.

5.8.4 The resistance to ground is measured with a common ohmmeter or megohmmeter.

5.9 Measuring Spark Energies.

The discharge energy for conductors is determined from the voltage on the conductor and its capacitance and is expressed by the following equation, which is also given in 4.3.3.3:

$$W = \frac{1}{2}CV^2$$

$$W = \frac{1}{2}\frac{Q^2}{C}$$

Where:

W = energy (joules)
C = capacitance (farads)
V = potential difference (volts)
Q = charge (coulombs)

A capacitance meter can often be used to measure electrostatic charge storage capacity when the charge is stored on a conductive element.

5.10 Measuring Ignition Energies.

Any combustible solid (dust), liquid (vapor), or gas should be considered for its potential as an ignitible atmosphere in the presence of discharges of static electricity. This requires determining the MIE of the material. Some data on MIE can be found in Appendix B. Standardized test equipment and procedures have been developed for measuring MIEs of particulate and gaseous materials. The equipment is highly specialized and requires trained technicians for its operation. Typically, the equipment is operated and maintained by specialized testing firms.

Chapter 6 Control of Static Electricity Hazards

6.1 General.

6.1.1 The objective of controlling a static electricity hazard is to provide a means whereby charges, separated by whatever cause, can recombine harmlessly before discharges can occur.

6.1.2 Ignition hazards from static electricity can be controlled by the following methods:

(1) Removing the ignitible mixture from the area where static electricity could cause an ignition-capable discharge
(2) Reducing charge generation, charge accumulation, or both by means of process or product modifications
(3) Neutralizing the charges

Grounding isolated conductors and air ionization are primary methods of neutralizing charges.

6.2 Control of Ignitible Mixtures by Inerting Equipment, by Ventilating, or by Relocating Equipment.

6.2.1 General. Despite efforts to prevent accumulation of static electric charges, which should be the primary aim of good design, many operations involving the handling of nonconductive materials or nonconductive equipment do not lend themselves to engineered solutions. It then becomes desirable or essential, depending on the nature of the materials involved, to provide other measures, such as inerting the equipment, ventilating the equipment or the area in which it is located, or relocating the equipment to a safer area.

6.2.2 Inerting. Where an ignitible mixture is contained, such as in a processing vessel, the atmosphere can be made oxygen deficient by introducing enough inert gas (e.g., nitrogen or combustion flue gas) to make the mixture nonignitible. This technique is known as *inerting*. When operations are normally conducted in an atmosphere containing a mixture above the upper flammable limit, it might be practical to introduce the inert gas only during those periods when the mixture passes through its flammable range. NFPA 69, *Standard on Explosion Prevention Systems*, contains requirements for inerting systems.

6.2.3 Ventilating. Mechanical ventilation can be used to dilute the concentration of a combustible material to a point well below its lower flammable limit (LFL) in the case of a gas or vapor or below its minimum explosible concentration (MEC) in the case of a dust. Usually, this means dilution to a concentration at or below 25 percent of the lower limit. Also, by properly directing the air movement, it might be practical to prevent the material from approaching an area of operation where an otherwise uncontrollable static electricity hazard exists.

6.2.4 Relocating Equipment. Where equipment that can accumulate a static electric charge is unnecessarily located in a hazardous area, it might be possible to relocate it to a safe location rather than to rely on other means of hazard control.

6.3 Control of Static Charge Generation.

Electric charges separate when materials are placed in contact and are pulled apart. Reducing process speeds and flow rates will reduce the rate of charge generation. Examples of such charge separation are found where plastic parts and structures, insulating films and webs, liquids, and particulate material are handled. If the material flows at a slow enough rate, a hazardous level of excess charge will not normally accumulate. This means of static electricity control might not be practical due to processing requirements. *(See Chapters 7 through 9 for recommended practices in specific applications.)*

6.4 Charge Dissipation.

6.4.1 Bonding and Grounding. Bonding is used to minimize the potential difference between conductive objects, even when the resulting system is not grounded. Grounding (i.e., earthing), on the other hand, equalizes the potential difference between the objects and the earth. The relationship between bonding and grounding is illustrated in Figure 6.4.1.

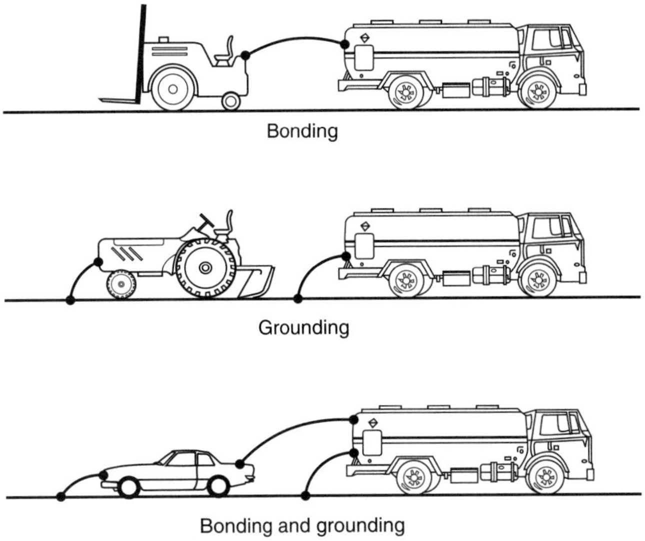

Figure 6.4.1 Bonding and Grounding.

6.4.1.1 A conductive object can be grounded by a direct conductive path to the earth or by bonding it to another conductive object that is already connected to the ground. Some objects are inherently bonded or inherently grounded because of their contact with the ground. Examples of inherently grounded objects are underground metal piping or large metal storage tanks resting on the ground.

6.4.1.2 The total resistance between a grounded object and the soil is the sum of the individual resistances of the ground wire, its connectors, other conductive materials along the intended grounding path, and the resistance of the ground electrode (i.e., ground rod) to the soil. Most of the resistance in a ground connection exists between the ground electrode and the soil. This ground resistance is quite variable, since it depends on the area of contact, the resistivity of the soil, and the amount of moisture present in the soil.

6.4.1.3 To prevent the accumulation of static electricity in conductive equipment, the total resistance of the ground path to earth should be sufficient to dissipate charges that are otherwise likely to be present. A resistance of 1 megohm (10^6 ohms) or less is generally considered adequate. Where the bonding/grounding system is all metal, resistance in continuous ground paths will typically be less than 10 ohms. Such systems include multiple component systems. Greater resistance usually indicates the metal path is not continuous, usually because of loose connections or corrosion. A grounding system that is acceptable for power circuits or for lightning protection is more than adequate for a static electricity grounding system.

Appendix C contains diagrams of various grounding devices, connections, and equipment.

6.4.1.4 Where wire conductors are used, the minimum size of the bonding or grounding wire is dictated by mechanical strength, not by its current-carrying capacity. Stranded or braided wires should be used for bonding wires that will be connected and disconnected frequently. *(See Appendix C for additional information.)*

6.4.1.5 Grounding conductors can be insulated (e.g., a jacketed or plastic-coated cable) or uninsulated (i.e., bare conductors). Uninsulated conductors are recommended, because it is easier to detect defects in them.

6.4.1.6 Permanent bonding or grounding connections can be made by brazing or welding. Temporary connections can be made using bolts, pressure-type ground clamps, or other special clamps. Pressure-type clamps should have sufficient pressure to penetrate any protective coating, rust, or spilled material to ensure contact with the base metal.

6.4.1.7 Workers should only be grounded through a resistance that limits the current to ground to less than 3 mA for the range of voltages experienced in the area. This method is called *soft grounding* and is used to prevent injury from an electric shock from line voltages or stray currents.

6.4.2 Humidification.

6.4.2.1 The surface resistivity of many materials can be controlled by the humidity of the surroundings. At humidities of 65 percent and higher, the surface of most materials will adsorb enough moisture to ensure a surface conductivity that is sufficient to prevent accumulation of static electricity. When the humidity falls below about 30 percent, these same materials could become good insulators, in which case accumulation of charge will increase.

6.4.2.2 While humidification does increase the surface conductivity of the material, the charge will only dissipate if there is a conductive path to ground.

6.4.2.3 Humidification is a not a cure-all for all static electricity problems. Some insulators do not adsorb moisture from the air and high humidity will not noticeably decrease their surface resistivity. Examples of such insulators are uncontaminated surfaces of some polymeric materials, such as plastic piping, containers, and films, and the surface of petroleum liquids. These surfaces are capable of accumulating a static electric charge even when the atmosphere has a humidity of 100 percent.

6.4.3 Charge Relaxation and Antistatic Treatments.

6.4.3.1 Based on their properties, liquid and solid materials carrying a static electric charge need time to dissipate, or "relax," this charge. In some cases, the materials can be allowed sufficient time for the charges to relax before being introduced into a hazardous area or process.

6.4.3.2 Charge relaxation can only occur if a path to ground for conduction of the charge is available. Increasing the conductivity of the material will not eliminate hazards if the material remains isolated from ground.

6.4.3.3 A nonconductive material can often be made sufficiently conductive to dissipate static electric charge, either by adding conductive ingredients to its composition or by applying hygroscopic agents to its surface to attract atmospheric moisture. *(See Chapters 7 through 9 for additional information.)*

6.4.3.4 Carbon black can be added to some plastics or rubbers to increase conductivity. Carbon-filled plastics and rubber articles are sometimes sufficiently conductive to be grounded like metal objects. Antistatic additives can also be mixed with liquid and particulate streams to foster charge relaxation.

6.4.3.5 In some cases, particularly with plastic films or sheeting, a material is added to the plastic to attract atmospheric moisture to the surface of the film, thus increasing its surface conductivity. Care should be taken when using antistatic plastic film or sheeting in low humidity situations. In environments with less than 30 percent humidity, film or sheeting can become nonconductive and accumulate static electric charge.

6.4.3.6 Topical hygroscopic coatings will attract atmospheric moisture and make the surface of the coated material conductive. However, these coatings could be easily washed away or rubbed off or could lose effectiveness over time. This type of coating should only be considered a temporary measure to reduce accumulation of static electric charge.

6.4.3.7 Conductive polymers, laminates with conductive elements, and metallized films have been developed for improved static dissipation.

6.5 Charge Neutralization.

6.5.1 General. Air can be made to contain mobile ions that will be attracted to and will eliminate unbalanced static electric charge from surfaces. In the use of air ionizers, one must consider certain factors that can influence their effectiveness, such as environmental conditions (e.g., dust and temperature) and positioning of the device in relation to the material processed, machine parts, and personnel. It is important to note that these control devices do not prevent the generation of static electric charge. They provide ions of opposite polarity to neutralize the generated static electric charge.

6.5.2 Inductive Neutralizers.

6.5.2.1 Inductive neutralizers include the following:

(1) Needle bars, which are metal bars equipped with a series of needle-like emitters
(2) Metal tubes wrapped with metal tinsel
(3) Conductive string
(4) Brushes made with metal fibers or conductive fibers

Each design is based on or consists of sharply pointed

elements arranged for placement in the static electric field near the charged surfaces.

6.5.2.2 A charge drawn from ground to the needle-like tips of an inductive neutralizer produces a concentrated electric field at the tips. If the tips are sharply pointed, the electrical field will be sufficient (i.e., greater than 3 kV/mm) to produce a localized electrical breakdown of the air. This electrical breakdown, known as corona, will inject ions into the air that are free to move to distant charges of opposite polarity. The flow of ions produced in corona constitute a neutralizing current. *(See Figure 6.5.2.2.)*

6.5.2.3 Although inexpensive and easy to install, inductive neutralizers require a minimum potential difference between the object and the needle tip to initiate corona and the neutralizing process. In the absence of this minimum charge, neutralization will not occur, and a residual potential of a few thousand volts will be left on the material when sharp inductive points are approximately within about 12 mm of the surface.

6.5.2.4 It is critically important that inductive neutralizers are connected to a secure ground. If the inductive neutralizer is not grounded, sparks from the induction bar can occur.

6.5.3 Active Electric Static Neutralizers.

6.5.3.1 Electric static neutralizers use a high voltage power supply to produce corona from sharp electrodes. The charge on any object near the device attracts charge from the corona to achieve neutralization. The use of a high voltage power supply eliminates the limitation of inductive neutralizers in control of charges having fields below the corona onset threshold.

6.5.3.2 Electric static neutralizers suitable for hazardous (classified) locations use a high voltage power supply to energize the corona electrodes at 50 Hz to 60 Hz. The use of an alternating field stresses the electrodes to produce both positive and negative ions for use in the neutralizing process. Current from the power supply is capacitive-coupled to each or to several of the sharp electrodes to limit spark energy in the event of a short circuit. These alternating current (AC)-powered static neutralizers must be approved for use in hazardous (classified) locations.

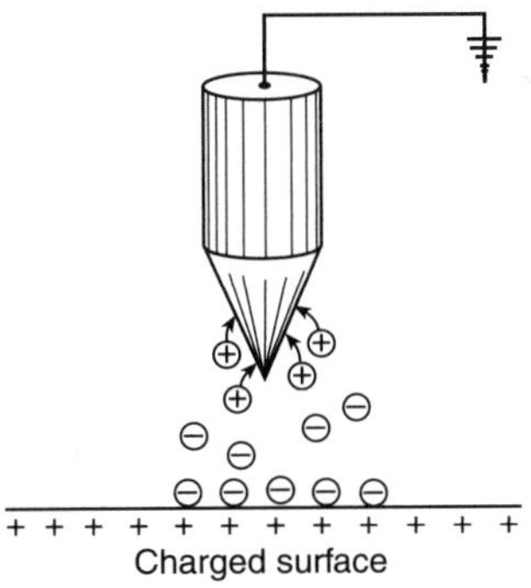

Figure 6.5.2.2 Example of an Induction Needle.

6.5.3.3 Electric static neutralizers that use pulsed or steady-state double-polarity direct current (DC) use a pulsed or steady field to stress the electrodes to produce ions for use in the neutralizing process. The spark energy, in the event of a short circuit, is controlled by current-limiting resistors. Pulsed or double-polarity DC ionizers that are used in hazardous (classified) locations should be listed for such use.

6.5.4 Active Radioactive Static Neutralizers. Radioactive (nuclear) ionizers use ionizing radiation to produce ions for neutralization of static electric charges. The most common radioactive ionizers depend on alpha particle generation from the decay of Polonium-210 (^{210}Po). Performance of radioactive ionizers deteriorates with the decay of the radioactive material. The neutralizers must be registered and installed in accordance with the Nuclear Regulatory Commission regulations and replaced periodically (at least annually) because ionization capability diminishes with radioactive decay. Radioactive ionizers are often used in conjunction with inductive neutralizers to control high charge densities. Although cost and regulatory compliance issues are associated with radioactive ionizers, they are nonincendive, require no wiring, and can reduce static electric charges to the lowest levels.

6.6 Control of Static Charge on Personnel.

The human body is an electrical conductor and can accumulate a static charge if insulated from ground. This charge can be generated by contact and separation of footwear with floor coverings, by induction, or by participation in various manufacturing operations. Where ignitible mixtures exist, a potential for ignition from the charged human body exists, and means to prevent accumulation of static electric charge on the human body might be necessary.

6.6.1 Steps to prevent charge accumulation include use of the following:

(1) Conductive flooring and footwear
(2) Personnel grounding devices
(3) Antistatic or conductive clothing

6.6.2 Conductive Flooring and Footwear.

6.6.2.1 Conductive or antistatic flooring can provide effective dissipation of static electricity from personnel. Materials can be solid or they can be coatings that are selected on the basis of wear characteristics, chemical resistance, and the floor area that needs to be covered. Small areas can be handled with a grounded metal plate. Typical resistance to ground for flooring systems should be less than 10^8 ohms. Accumulation of debris, wax, and other high-resistivity materials will compromise the conductivity of the floor.

6.6.2.2* Electrostatic dissipative (ESD) footwear, used in conjunction with conductive flooring, provides a means to control and dissipate static electric charges from the human body. Resistance to earth through ESD footwear and floor should be between 10^6 ohms and 10^9 ohms. For materials with very low ignition energies, the resistance to earth through footwear and flooring should be less than 10^6 ohms. Resistance can be measured with commercially available footwear conductivity testers.

6.6.2.3 Resistance of footwear can increase with accumulation of debris on footwear, use of orthopedic foot beds, and reduced floor contact area. Conductivity of footwear can be tested on a periodic basis to confirm functionality.

6.6.2.4 Conductive footwear is footwear designed to have a resistance to ground through footwear and floor of less than 10^6 ohms and is typically used where materials of low ignition energy, such as explosives and propellants, are handled. Conductive footwear should not be used where a possibility for electrocution by line voltages exists.

6.6.3* Personnel Grounding Devices.

6.6.3.1 Where ESD footwear will not provide adequate personnel grounding, supplementary devices should be used. Such devices include wrist straps, heel/toe grounders, and conductive overshoes. They should be selected so that accumulation of hazardous static electric charge is prevented, while the risk of electrocution is not increased. In most practical situations, grounding of personnel is achieved by ensuring that the resistance from the skin to ground is approximately 10^8 ohms or less. The need to protect against electrocution via a grounding device imposes a minimum resistance from skin to ground of 10^6 ohms. Based on skin contact and contact with the floor, especially during activities where the entire sole of the footwear is not in contact with the floor (e.g., kneeling), effectiveness can be compromised. Grounding devices should have a minimum resistance of 10^6 ohms for shock protection.

6.6.3.2 The simplest type of commercial device is a grounding bracelet with a built-in resistor, typically giving a resistance to ground of about 10^6 ohms for shock protection. Wrist straps of this type have the greatest utility at ventilation hoods and at other locations where limitation on the operator's mobility can be tolerated. Breakaway wrist tether systems could be necessary where emergency egress is needed. A hood can be equipped with two external coiled grounding cords with removable cuff attachments, the latter being kept by individual users.

6.6.3.3 Ground continuity should be checked periodically to the manufacturer's specified limits using a voltmeter or volt ohmmeter or a commercial tester.

6.6.4 Antistatic or Conductive Clothing.

6.6.4.1 Although silk and most synthetic fibers are excellent insulators, and undergarments made from them exhibit static phenomena, no conclusive evidence exists to indicate that wearing such undergarments constitutes a hazard. However, removal of outer garments is particularly hazardous in work areas, such as in hospital operating rooms, explosives manufacturing facilities, and similar occupancies and where clothing is contaminated by flammable liquids. Outer garments used in these areas should be suitable for the work area and should be antistatic. NFPA 99, *Standard for Health Care Facilities*, provides information on test methods for evaluating the antistatic performance of wearing apparel.

6.6.4.2* Although usually a very small likelihood of ignition by a grounded person due to any type of clothing exists, the charging of personnel (which can occur, for example, when personnel are getting out of a forklift truck) is greatly increased by clothing having high resistivity.

6.6.4.3* In oxygen-enriched atmospheres, such as could be present in liquid oxygen filling plants, vapor from the cooled gas can permeate the employee's clothing, rendering it more combustible. A static electric charge that accumulates on the person and then suddenly discharges can ignite the clothing.

6.6.5 Gloves. Gloves should be antistatic or conductive with the same resistivity as prescribed for footwear. Gloves should be tested in conjunction with the footwear.

6.6.6 Cleaning or Wiping Cloths. Synthetic fabrics used in cleaning or wiping cloths can develop sufficient static electric charge to produce discharges capable of igniting solvent vapors. Flammable liquids and combustible liquids used at temperatures above their flash points, where used with synthetic cleaning or wiping cloths, will increase the risk of fire. Typically, charge generation will increase with the speed and vigor of the wiping action. The material being cleaned or wiped, if nonconductive, can also accumulate an incendive charge.

Cotton or synthetic fabric treated with an antistatic compound should be used if static electric charge generation needs to be controlled, especially if cleaning or wiping with flammable solvents. Conductive solvents are recommended. Test methods for determining the electrostatic-generating properties of fabrics can be found in NFPA 99, *Standard for Health Care Facilities*.

6.7 Maintenance and Testing.

All provisions for personnel static electricity control should be maintained and tested to remain effective. Preventive maintenance procedures and recommendations can be found in NFPA 99, *Standard for Health Care Facilities*, for clothing, footwear, and flooring.

6.8 Discomfort and Injury.

Static shock can result in discomfort and, under some circumstances, injury. While the discharge itself is typically not dangerous to humans, it can cause an involuntary reaction that results in a fall or entanglement with moving machinery. If charge accumulation cannot be avoided and no flammable gases or vapors are present, consideration should be given to the various methods by which contact with metal parts can be eliminated. Such methods include use of nonmetal handrails, insulated doorknobs, and other nonconductive shields.

Chapter 7 Flammable and Combustible Liquids and Their Vapors

7.1 Scope.

This chapter discusses the assessment and control of static electricity hazards involved with the storage, handling, and use of flammable and combustible liquids and their vapors and mists. While focused on flammable and combustible liquids, the principles of this chapter also apply to noncombustible liquids and vapors (e.g., wet steam) where their storage, use, and handling can cause a static electricity ignition hazard. The chapter begins with a discussion of the combustion characteristics of liquids and their vapors and mists, followed by a discussion of charge generation and dissipation in liquids. Emphasis is then given to processes involving the following:

(1) Flow in pipe, hose, and tubing
(2) Storage tanks
(3) Process vessels
(4) Gauging and sampling
(5) Tank cleaning
(6) Portable tanks and containers
(7) Vacuum cleaning

7.2 Combustion Characteristics of Liquids, Vapors, and Mists.

The following combustion properties of liquids need to be understood in order to properly assess the static electricity ignition hazard:

(1) Flash point
(2) Flammable limit and vapor pressure
(3) Ignition energy
(4) Oxidant concentration

7.2.1* Flash Point. Flash point is the minimum temperature at which a liquid gives off sufficient vapor to form an ignitible mixture with air near the surface of the liquid. Flash point is determined using a variety of test procedures and apparatus, the selection of which sometimes depends on other physical characteristics of the liquid. *(See A.7.2.1 for more detailed information.)*

If the flash point of a liquid is at or below typical ambient temperatures, it is likely to evolve an ignitible vapor. The lower the flash point, the higher the vapor pressure and the more likely that a vapor will be present to ignite. Because of the variability in flash point test methods, the published flash point of a particular liquid only approximates the lowest temperature at which ignition is possible for that liquid. Thus, an allowance of 4°C to 9°C below the published flash point should be made when evaluating ignition hazard. The following effects also can generate an ignitible vapor:

(1) Off-gassing of flammable vapors from solids or low-volatility liquids
(2) Processing at pressures below atmospheric pressure
(3) Nonhomogeneity of the vapors above the liquid
(4) Mist, droplets, or foam on the surface of a liquid

7.2.2* Flammable Limits. Vapors or gases in air are flammable only between certain concentrations — the lower flammable limit (LFL) and the upper flammable limit (UFL). The concentrations between these limits constitute the flammable range. Below the LFL, vapors are too lean to burn and above the UFL, they are too rich to burn. Increased pressure (above atmospheric pressure) and increased temperature both widen the flammability range of typical hydrocarbons.

7.2.3 Ignition Energy. The energy needed to ignite a vapor-air mixture varies with the concentration. (See Figure 7.2.3 for a typical relationship between ignition energy and concentration.) For most materials, the lowest ignition energy value occurs at a concentration near the midpoint between those for the LFL and UFL. The lowest value is referred to as the minimum ignition energy (MIE). Some MIEs are given in Appendix B.

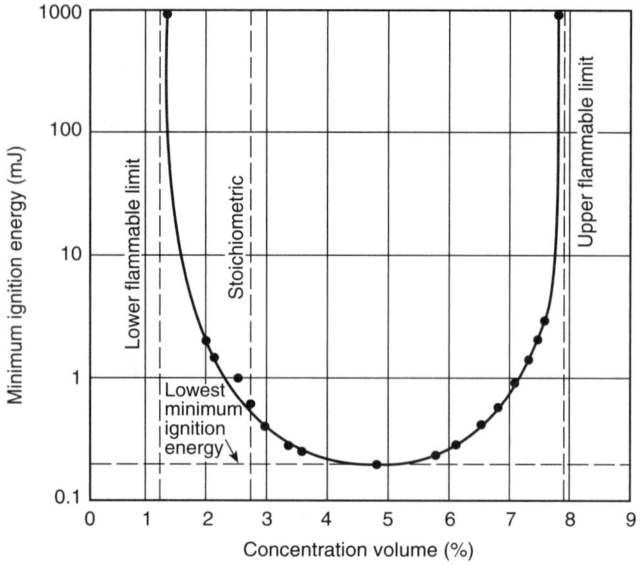

Figure 7.2.3 Minimum Ignition Energy of Benzene as a Function of Concentration. ***(Adapted from Britton, 1992, pp. 56–70.)***

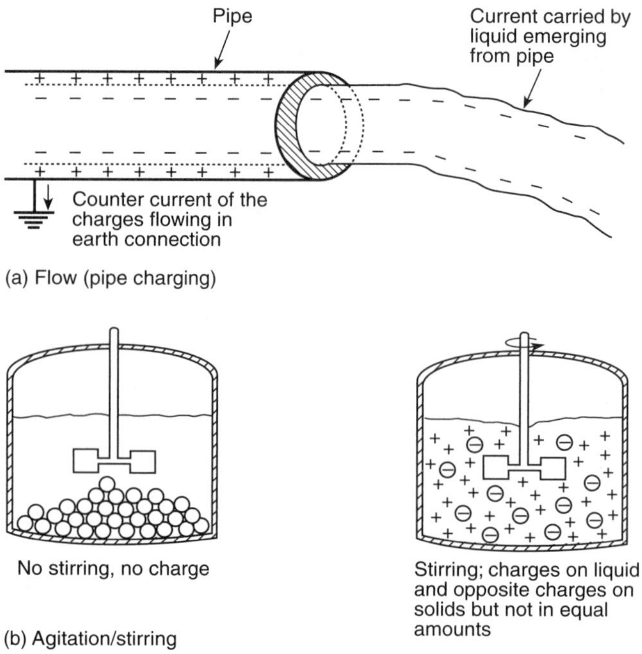

Figure 7.3.1 Examples of Charge Generation in Liquids. ***(Walmsley, 1992, p. 33.)***

7.2.4* Oxidant Concentration. Combustibility is normally determined for atmospheric air, which contains 21 percent oxygen. With an oxygen-enriched atmosphere, the flammable range expands; that is, the LFL decreases and the UFL increases. If the oxygen concentration is sufficiently reduced by inerting, however, an oxygen concentration below which no ignition is possible is reached. This concentration is referred to as the *limiting oxygen concentration* (LOC). By effectively inerting to below the LOC, the hazard of ignition can be eliminated, as explained in NFPA 69, *Standard on Explosion Prevention Systems.* Other oxidants, if present in the mixture, should be addressed similarly. Laboratory testing might be required to evaluate the hazard.

7.3 Generation and Dissipation of Charge in Liquids.

7.3.1* Charge Generation. Charge separation occurs when liquids flow through pipes, hoses, and filters, when splashing occurs during transfer operations, or when liquids are stirred or agitated. The greater the area of the interface between the liquid and surfaces and the higher the flow velocity, the greater the rate of charging. The charges become mixed with the liquid and are carried to receiving vessels where they can accumulate. The charge is often characterized by its bulk charge density and its flow as a streaming current to the vessel. *(See Figure 7.3.1.)*

7.3.2* Charge Relaxation. Static electric charge on a liquid in a grounded conductive container will dissipate at a rate that depends on the conductivity of the liquid. For liquids with conductivity of 1 picosiemens per meter (1 pS/m) or greater, charge relaxation proceeds by exponential, or ohmic, decay, as described for semiconductive materials in 4.2.4. For liquids with conductivity less than 1 pS/m, relaxation occurs more rapidly than would be predicted by the exponential decay model. *(See 4.2.5.)* According to the Bustin relationship *(see A.7.3.2)*, when low viscosity liquids (less than 30×10^{-6} m²/sec) are charged, relaxation proceeds by hyperbolic decay. However, for these same liquids, the exponential decay constant gives a conservative estimate for the relaxation time.

7.3.3* Factors Affecting Liquid Charging. In grounded systems, the conductivity of the liquid phase has the most effect on the accumulation of charge in the liquid or on materials suspended in it. A liquid is considered nonconductive (charge accumulating) if its conductivity is below 50 pS/m, assuming a dielectric constant of 2. *(See A.7.3.3 for a detailed discussion of this subject.)* Appendix B lists values of conductivity for typical liquids. What is important is that charge should decay from the liquid fast enough to avoid ignition hazards. The acceptable conductivity in any particular application can be larger or smaller than this range, depending on flow rate and processing conditions.

Conductive liquids, defined as having conductivities greater than 10^4 pS/m, do not pose a hazard due to static electric charge accumulation in typical hydrocarbon and chemical processing and handling operations. Liquids having conductivities of 50 pS/m to 10^4 pS/m are considered semiconductive by this recommended practice.

The charging characteristics of many industrial liquids, particularly non-polar hydrocarbons, are the result of trace contaminants that are present in the liquid, sometimes in concentrations less than 1 part per million (ppm). Thus, industrial liquids can become more or less conductive by orders of magnitude, depending on the concentration of contaminants that result from process, storage, and handling practices.

Conductive liquids that at first could appear to be safe can present a significant hazard if isolated from ground by an insulating container or if suspended in air. When isolated, essentially all charge on the conductive liquid can be released as an incendive spark. When suspended as a mist, significant static electric fields can lead to incendive brush discharge.

7.3.3.1 In the petroleum industry, for tank loading and distribution operations involving petroleum middle distillates, liquids in the semiconductive category are handled as conductive liquids. Such procedures are possible because regulations prohibit use of nonconductive plastic hoses and tanks and multiphase mixtures and end-of-line polishing filters are not involved.

7.3.3.2 In general chemical operations, semiconductive liquids represent a distinct category in which the tendency to accumulate charge varies greatly with the operation and with liquid conductivity. These operations can involve multiphase mixtures, nonconductive tank linings, and microfilters, all of which promote charge accumulation in equipment.

7.4 Flow in Pipe, Hose, and Tubing.

7.4.1* Metal Piping Systems. All parts of continuous all-metal piping systems should have a resistance to ground that does not exceed 10 ohms. A significantly higher resistance could indicate poor electrical contact, although this will depend on the overall system. For flanged couplings, neither paint on the flange faces nor thin plastic coatings used on nuts and bolts will normally prevent bonding across the coupling after proper torque has been applied. Jumper cables and star washers are not usually needed at flanges. Star washers could even interfere with proper torquing. Electrical continuity of the ground path should be confirmed after assembly and periodically thereafter.

Bonding wires might be needed around flexible, swivel, or sliding joints. Tests and experience have shown that resistance in these joints is normally below 10 ohms, which is low enough to prevent accumulation of static charges. However, it is recommended that the manufacturer's specifications be checked or that these joints be inspected, because a few are fabricated with insulating surfaces. When painted, slip flanges (lap joints) using nonconductive gaskets can cause loss of continuity in the grounding path. This loss of continuity can be remedied by using a conductive gasket, such as a flexible, graphite-filled, spiral-wound gasket or by installing a jumper wire across the joint.

It should be ensured that bonding and grounding do not compromise sections of pipe that are supposed to be isolated. For example, insulating flanges could have been installed to avoid arcs from stray current or from cathodic protection systems, which provide a separate ground path.

Figures 7.4.1(a) and (b) provide guidance in estimating the charge on a nonconductive liquid flowing through a smooth pipe.

7.4.2* Nonconductive Pipe and Lined Pipe. Nonconductive surfaces affect the rates of charge generation and charge dissipation during flow through a pipe. The rate of charge generation is similar in conductive and nonconductive pipes, while the rate of charge loss can be significantly slower in nonconductive pipes. For charged, nonconductive liquids, insulation by the pipe wall can result in charge accumulation of the opposite polarity on the outer surface of the insulating liner or pipe. Charge accumulation can eventually lead to electrical breakdown and pinhole punctures of either the liner or, in the case of nonconductive pipe, the entire wall thickness.

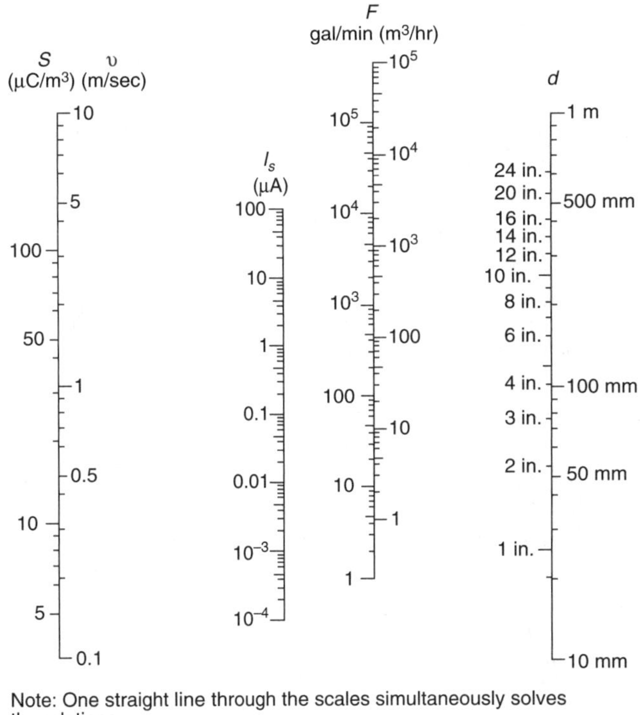

Note: One straight line through the scales simultaneously solves the relations.

$I_s = 2.5 \times 10^{-5}\, \upsilon^2 d^{\ 2}$ $S = 3.18 \times 10^{-5}\, \upsilon$ $F = (\pi/4)\, \upsilon d^{\ 2}$

(To convert from ft/sec to m/sec multiply by 3.28)
(To convert from bbl/hr to m³/hr multiply by 0.159)

Figure 7.4.1(a) Nomograph for Estimating the Charge on a Nonconductive Liquid Flowing Through a Smooth Pipe. ***(Pratt, 1997, p. 112.)***

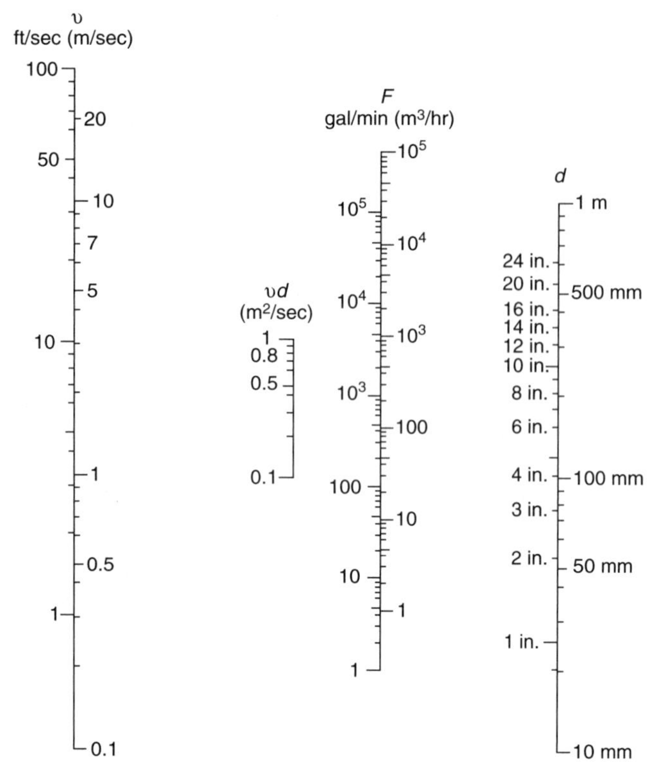

Note: One straight line through the scales solves the flow relation.

$$F = (\pi/4)\, \upsilon d^{2}$$

(To convert from bbl/hr to m^3/hr multiply by 0.159)

Figure 7.4.1(b) Nomograph for Estimating Fluid Flow Parameters. ***(Pratt, 1997, p. 114.)***

7.4.3* Flexible Hoses and Tubing. Flexible hoses and flexible tubing are available in metal, lined metal, nonconductive plastic, reinforced rubber and plastic, and composite-ply types. Where nonconductive hose or tubing must be used because of process conditions, the hazards of static electric charge generation should be thoroughly investigated. As a minimum, all conductive couplings (e.g., end fittings) and components should be bonded and grounded. If hoses are used immediately downstream of filters in nonconductive liquid service, they should be of metal or other conductive material. Semiconductive liners might be necessary to prevent charge accumulation and pinhole damage to the hose. Conductive hoses should be electrically continuous and the continuity should be periodically checked.

7.4.4 Fill Pipes. Fill pipes should be conductive and should be bonded to the filling system. Fill pipes should extend to the bottom of the vessel and can be equipped with either a 45 degree cut tip or a tee to divert flow horizontally near the bottom of the vessel being filled. The design should prevent upward spraying during the initial stage of filling. A "slow start" might be necessary so that the inlet velocity is held to less than 1 m/sec until the outlet of the dip pipe is covered by at least two pipe diameters of liquid.

7.4.5 Filtration.

7.4.5.1 Microfilters. Microfilters typically have pore sizes less than 150 μm. These filters generate very large streaming currents with nonconductive liquids, due to their large contact area. (Conductive liquids typically dissipate their charge to ground through the body of the liquid.) Streaming currents will frequently be greater than those for the pipe flow entering the microfilter by two orders of magnitude and the charge density added to the liquid can exceed 2000 $\mu C/m^3$.

To prevent these charges from entering the receiving vessel, the filter should be placed far enough upstream so that the charge can decay to the same magnitude as it would be in the pipe flow. Common industry practice is to provide 30 seconds of residence time in the pipe or conductive hose downstream of the microfilter, especially if the conductivity of the liquid is not known. For nonconductive liquids that have both very low conductivity (i.e., less than 2 pS/m) and high viscosity (i.e., greater than 30 centistokes) at the lowest intended operating temperature, longer residence times might be appropriate. In these cases, a residence time of up to three times the relaxation time constant of the liquid should be considered.

7.4.5.2 Strainers. Mesh strainers finer than 150 μm should be treated as microfilters. Mesh strainers coarser than 150 μm can also generate significant static electric charge when fouled with accumulated debris. If such coarse strainers are used in services where debris can be expected to accumulate, then these strainers should also be treated as microfilters.

7.4.5.3 Polishing Filters. A polishing filter is sometimes placed at the end of a delivery line to remove debris. This filter might be a bag installed on the end of a hose and directly exposed to the vapor in the tank. Filters used in flammable liquid service should be enclosed in grounded metal housings.

7.4.6 Suspended Material. Immiscible and marginally soluble liquids and slow-dissolving solids can disperse as droplets or as an emulsion. *(For a discussion of examples of such situations in mixing operations, see 7.10.1.)* Where a nonconductive liquid contains a dispersed phase, such as water in oil, the continuous phase determines the charge relaxation behavior. Charge generation is typically greater for such suspensions than that for a single phase.

7.4.7 Miscellaneous Line Restrictions. Piping system components such as orifice plates, valves, elbows, and tees increase turbulence and can increase the rate of charge generation. Brief contact with a plastic component, in particular, can cause significant charge generation. Suspended material such as water *(see 7.4.6)* has also been found to increase this effect.

7.5 Storage Tanks.

7.5.1 General. Liquid flowing into a tank can carry a static electric charge that will accumulate in the tank. This charge can be detected as a potential above the surface of the liquid in the tank. The maximum surface potential attained depends not only on the charge density of the incoming liquid but also on the dimensions of the tank. For commercial tanks of equal volume, the maximum potential will be greater in tanks having smaller cross-sectional areas, because liquid depth increases faster relative to the rate of charge relaxation. Smaller potentials will therefore be generated in, for example, a near-rectangular barge tank than in a cylindrical vertical tank of the same volume.

7.5.2 Conductive Fixed-Roof Storage Tanks. Charge accumulation in the liquid in a tank can lead to static electric discharge between the liquid surface and the tank shell, roof supports, or tank appurtenances. The charge generation rate is influenced by turbulence in the liquid and by the settling of particulate matter, such as water droplets, iron scale, and sediment.

7.5.2.1 Precautions. If the vapor space in the tank is likely to contain an ignitible mixture (e.g., in cases where intermediate vapor pressure products or low vapor pressure products contaminated with high vapor pressure liquids are stored), or where switch loading is practiced, the following protective measures should be taken:

(a) Splash filling and upward spraying should be avoided. The fill pipe should discharge near the bottom of the tank, with minimum agitation of water and sediment on the tank bottom.

(b) If possible, the inlet flow velocity should be limited during the initial stage of tank filling to reduce agitation and turbulence. The flow velocity of the incoming liquid should be no greater than 1 m/sec until the fill pipe is submerged either two pipe diameters or 0.6 m, whichever is less. Since too low a velocity can result in settling out of water at low points in the piping, the inlet flow velocity should be kept as close to 1 m/sec as possible during this period. Otherwise, subsequent re-entrainment of water or other contaminants when the velocity is increased could significantly increase the product's charging tendency.

(c)* For storage tanks greater than 50 m^3 containing liquids that are either nonconductive or whose conductivity is unknown, the inlet flow velocity can be increased to 7 m/sec after the fill pipe is submerged. Where operating experience, such as in the petroleum industry, has shown that the practice is acceptable, the inlet flow velocity can be increased above 7 m/sec, but in no case greater than 10 m/sec. See Figures 7.4.1(a) and (b) for determining flowrate.

(d) If the liquid is nonconductive and contains a dispersed phase, such as entrained water droplets, the inlet flow velocity should be restricted to 1 m/sec throughout the filling operation.

(e) A 30-second minimum residence time should be provided for liquid to flow between upstream microfilter screens and the tank *(see 7.4.5)*.

(f) Tanks should be inspected for ungrounded conductive objects, such as loose gauge floats and sample cans. If these objects are floating on the liquid surface, they can promote sparks.

(g) Lines should not be blown out with air or other gases if the liquid is a Class I liquid or is handled at or above its flash point. Introducing substantial amounts of air or other gas into a tank through the liquid can create a hazard due to charge generation, misting of the liquid, and formation of an ignitible atmosphere.

7.5.2.2 Grounding. Storage tanks for nonconductive liquids should be grounded. Storage tanks on grade-level foundations are considered inherently grounded, regardless of the type of foundation (e.g., concrete, sand, or asphalt). For tanks on elevated foundations or supports, the resistance to ground can be as high as 10^6 ohms and still be considered adequately grounded for purposes of dissipation of static electric charges, but the resistance should be verified. The addition of grounding rods and similar grounding systems will not reduce the hazard associated with static electric charges in the liquid.

7.5.2.3 Spark Promoters. A tank gauging rod, high-level sensor, or other conductive device that projects downward into the vapor space of a tank can provide a location for static electric discharge between itself and the rising liquid. These devices should be bonded securely and directly downward to the bottom of the tank by a conductive cable or rod to eliminate a spark gap or should be installed in a gauging well that is bonded to the tank. They should be inspected periodically to ensure that the bonding system has not become detached. If tank fixtures are nonconductive, the potential for sparking does not exist and no specific measures are needed. Devices that are mounted to the sidewall of the tank (e.g., level switches or temperature probes) and project a short distance into the tank might not pose a static electric discharge hazard. These situations should be evaluated on an individual basis.

7.5.2.4 Tank Mixers. In-tank jet mixing or high-velocity agitator mixing can stir up water and debris and cause splashing at the surface that can generate static electric charges. If an ignitible mixture exists at the surface, ignition is possible.

Surface splashing should be minimized. Gas blanketing or inerting can be employed to eliminate the ignition hazard.

7.5.2.5 Gas Agitation. Gas agitation using air, steam, or other gases is not recommended because it can produce high levels of charge in liquids, mists, or foams. In addition, air agitation can create an ignitible atmosphere in the vapor space of the tank. If gas agitation is unavoidable, the vapor space should be purged prior to mixing and the process should be started slowly to ensure that static electric charge does not accumulate faster than it can dissipate. Note that special precautions need to be taken to prevent agitation with air to dilute any initial inerting. Similarly, while agitation with an inert gas can eventually result in an inert vapor space, the electrostatic charge buildup due to the agitation process can result in a spark and ignition before inerting of the tank vapor space is achieved. A waiting time should be observed prior to any gauging or sampling activities.

7.5.3 Conductive Floating Roof Storage Tanks. Floating roof storage tanks are inherently safe, provided the floating roof is bonded to the tank shell. Bonding is typically achieved by providing shunts between the floating roof or cover and the wall of the tank. The shunts are installed for lightning protection, but they also provide protection from static electric charges that could be generated. If the floating roof is landed on its supports, charge accumulation in the surface of the liquid can occur, and the same precautions as for a fixed roof tank should be followed. If an internal floating roof tank is not adequately ventilated, flammable vapor can accumulate between the floating roof and the fixed roof.

7.5.4 Coated and Lined Tanks. Metal tanks with nonconductive coatings or linings can be treated as a conductive tank if either of the following criteria apply:

(a) The nonconductive coating or lining has a volume resistivity equal to or lower than 10^{10} ohm-m, such as fiberglass-reinforced linings for corrosion prevention, and is no thicker than 2 mm.

(b) The nonconductive lining has a volume resistivity greater than 10^{10} ohm-m, such as polyethylene or rubber linings, but has a breakdown potential less than 4 kV.

Metal tanks with nonconductive coatings or linings that do not meet the criteria of 7.5.4(a) or (b) should be treated as nonconductive tanks. Regardless of the coating or lining thickness or resistivity, the tank should be bonded to the filling system. The coating or lining is not regarded as a barrier to the flow of static electric charges. Its resistivity is of the same order of magnitude as that of the liquid or there could be small bare areas (holidays) in the coating.

A thin coat of paint, a thin plastic liner, or a layer of metal oxide on the inside of piping, vessels, or equipment does not constitute a static electric hazard.

7.5.5 Tanks Constructed of Nonconductive Materials. Tanks constructed of nonconductive materials are not allowed for storing Classes I, II, and IIIA liquids except under special circumstances, as outlined in Section 2.2 of NFPA 30, *Flammable and Combustible Liquids Code. (See 7.10.7 for design and use recommendations.)*

7.6 Loading of Tank Vehicles.

Recommended loading precautions for tank vehicles vary with the characteristics of the liquid being handled and the design of the loading facility. A summary of precautions recommended when a flammable mixture exists in the tank vehicle compartment, based on API RP 2003, *Protection Against Ignitions Arising Out of Static, Lightning, and Stray Currents,* is provided in Table 7.6. These precautions are intended for tank vehicles with conductive (metal) compartments. *(For compartments with nonconductive linings, see 7.10.4. For compartments of nonconductive material, see 7.10.7.)*

7.6.1 Top Filling. Splash filling should be avoided by using a fill pipe that is designed according to the recommendations in 7.4.4.

7.6.2 Bottom Filling. The bottom-filling inlet should be designed with a deflector or a diverter to prevent upward spraying and generation of mist. Using a cap or a tee to direct incoming liquid sideways toward the compartment walls, rather than upward, will achieve this objective.

7.6.3 Switch Loading. The practice of loading a high flash point, low conductivity liquid into a tank that previously contained a low flash point liquid is known as switch loading. This practice can result in the ignition of residual flammable vapor as the tank is filled. The methods of hazard prevention are similar to those for 7.5.2.1(a), (b), (d), (e), (f), and (g). Flow velocities are found in Table 7.6.

7.6.4 Highway Transport. As noted in API RP 2003, *Protection Against Ignitions Arising Out of Static, Lightning, and Stray Currents,* tank vehicles normally create no static electricity hazard during transport, provided they are compartmented or contain baffles. The compartments or baffles minimize sloshing of the liquid in the tank vehicle, which could result in significant charge generation. Clear bore (unbaffled) tank vehicles should not be used for liquids that can generate an ignitible mixture in the vapor space.

7.6.5 Antistatic Additives. Charge accumulation can be reduced by increasing the conductivity of the liquid by adding a conductivity-enhancing agent (antistatic additive). These additives are normally added in parts per million concentrations and should be used in accordance with manu-

Table 7.6 Summary of Precautions for Loading Tank Vehicles

	Liquid Being Loaded			
	Nonconductive			
Recommended Loading Precaution[1]	**Low Vapor Pressure**	**Intermediate Vapor Pressure**	**High Vapor Pressure**[6]	**Conductive**[3,4]
1. BONDING AND GROUNDING. Tank trucks should be bonded to the fill system, and all bonding and grounding should be in place prior to starting operations. Ground indicators, often interlocked with the filling system, are frequently used to ensure bonding is in place. Bonding components, such as clips, and the fill system continuity should be periodically examined and verified. For top loading, the fill pipe should form a continuous conductive path and should be in contact with the bottom of the tank.	Yes[2]	Yes	Yes	Optional
2. INITIAL LOADING. Top loading fill pipes and bottom loading systems should be equipped with spray deflectors and splash filling should be avoided. A slow start (i.e., velocity less than 1 m/sec) should be employed until the inlet into the compartment is covered by a depth equal to two fill line diameters to prevent spraying and to minimize surface turbulence.	Yes	Yes	Yes	Yes
3. MAXIMUM LOADING RATE. The maximum loading rate should be limited so the velocity in the fill pipe or load connection does not exceed 7 m/sec or ($0.5/d$) m/sec (where d = inlet inside diameter in meters), whichever is less.[5] Transition from slow start to normal pumping rate can be achieved automatically using a special loading regulator tip (which shifts rate when submerged to a safe depth). Excessive flow rates should be avoided either procedurally or by system design, which is the recommended method.	Yes[2]	Yes	Optional[3]	Optional
4. CHARGE RELAXATION. A residence time of at least 30 seconds should be provided between any microfilter or strainer and the tank truck inlet.[7] A waiting period of at least 1 minute should be allowed before the loaded tank compartment is gauged or sampled through the dome or hatch. However, sampling and gauging via a sample well (gauge well) can be done at any time.	Yes[2]	Yes	Yes	Optional
5. SPARK PROMOTERS. A tank gauging rod, high-level sensor, or other conductive device that projects downward into the vapor space of a tank can provide a location for static discharge between itself and the rising liquid and should be avoided. These devices should be bonded securely and directly downward to the bottom of the tank by a conductive cable or rod (to eliminate a spark gap), or should be installed in a gauging well that is bonded to the bottom.[8] Periodic inspection should be conducted to ensure that the bonding system does not become detached and that there are no ungrounded components or foreign objects.	Yes	Yes	Yes	Optional

[1] The recommended loading precautions vary with the product being handled. In loading operations where a large variety of products are handled and where it is difficult to control loading procedures, such as at self-service loading racks, following a single, standard procedure that includes all of the precautions is recommended.
[2] Recommended loading precautions need not be applied if only low vapor pressure combustible liquids at ambient temperatures are handled at the loading rack and there is no possibility of switch loading or cross contamination of products. All loading precautions should be followed when handling low vapor pressure products at temperatures near (within 4°C to 9°C) or above their flash point.
[3] When additives are used to increase conductivity, caution should be exercised. *(See 7.6.5.)*
[4] Semiconductive liquids can accumulate charge where charging rates are extremely high or where they are effectively isolated from ground. They could need to be handled as nonconductive. *(See 7.3.3.1 and 7.3.3.2.)*
[5] When the product being handled is a nonconductive, single-component liquid (such as toluene or heptane), the recommended maximum fill rate is ($0.38/d$) m/sec.
[6] If high vapor pressure products are handled at low temperatures (near or slightly below their flash point), all of the recommended loading precautions should be followed.
[7] Very low conductivity and high viscosity products can require additional residence time of up to 100 seconds. *(See 7.4.5.1.)*
[8] If these devices are nonconductive, the potential for sparking does not exist and no specific measures need be taken. Devices that are mounted to the sidewall of the tank (e.g., level switches and temperature probes), that project a short distance into the tank, and that have no downward projection might not pose an electrostatic hazard. These situations should be evaluated on an individual basis.

facturer instructions. When antistatic additives are used as a primary means of minimizing accumulation of static electric charge, the operator should verify the concentration of the additive at critical points in the system.

7.7* Vacuum Trucks.

For control of static electricity, hoses should be conductive or semiconductive. Alternatively, all conductive components should be bonded and the truck should be grounded. In no case should plastic dip pipes or plastic intermediate collection pans or drums be used.

7.8 Railroad Tank Cars.

In general, the precautions for railroad tank cars are similar to those for tank vehicles in Section 7.6. The major exception is the larger volume typical of railroad tank cars (e.g., greater than 87 m^3) compared with that of tank vehicles (e.g., about 50 m^3). This greater volume allows greater maximum filling rates to be used, up to a maximum of $(0.8/d)$ m/sec, where d is the inside diameter of the inlet, in meters. Many tank cars are equipped with nonconductive bearings and nonconductive wear pads located between the car itself and the trucks (wheel assemblies). Consequently, resistance to ground through the rails might not be low enough to prevent accumulation of a static electric charge on the tank car body. Therefore, bonding of the tank car body to the fill system piping is necessary to protect against charge accumulation. In addition, because of the possibility of stray currents, loading lines should be bonded to the rails.

7.9 Marine Vessel and Barge Cargo Tanks.

Marine vessel and barge cargo tanks are beyond the scope of this recommended practice. The recommendations given in the *International Safety Guide for Oil Tankers and Terminals* (ISGOTT) should be followed.

7.10 Process Vessels.

7.10.1 Means of Static Electric Charge Accumulation. Accumulation of static electricity in process vessels occurs by the same methods as described in Section 7.5 for tanks. Where a conductive and a nonconductive liquid are to be blended, the conductive liquid should be added to the vessel first, if possible, so that the conductivity of the mixture is as high as possible throughout the mixing process. Re-entry of recirculation loops should be designed to minimize splashing and surface disruption, for example, by use of subsurface jets that do not break the liquid's surface.

7.10.2* Procedures for Transfer to Tanks. When introducing two or more nonconductive liquids to a blending tank, the less dense liquid should be loaded first to avoid a surface layer comprising the lighter, more highly charged component. Splash recirculation should normally be done only if the vessel is inerted or vapor enriched.

7.10.3 Agitation. Agitators should be covered with sufficient depth of liquid before being operated, to minimize splashing, or should be operated at reduced speed until sufficient depth has been achieved. In cases where hazardous charge accumulation cannot be avoided using the measures discussed in this section, the vessel can be inerted.

7.10.4 Vessels with Nonconductive Linings. The accumulation of static electric charge can result in pinhole damage to equipment such as enamel- or glass-lined reactors. Since static electric discharges often occur at the liquid interface as liquid drains from the wetted wall, a vapor ignition hazard could also exist. In some cases, it is possible to specify static dissipative coatings for the vessel or agitator. It is recommended that conductive vessels and appurtenances be bonded and grounded. In some cases, inerting might be necessary.

7.10.5 Adding Solids. The most frequent cause of static electric ignitions in process vessels is the addition of solids to flammable liquids in the vessels. Even when the vessel is inerted, large additions of solids will introduce air into the vessel while expelling flammable vapor from the vessel. The sudden addition of a large volume of solids can also result in static discharge from a floating pile of charged powder. It is recommended that manual addition of solids through an open port or manway be done only in 25-kg batches. Larger batch additions [e.g., from flexible intermediate bulk containers *(see 9.1.7)*] should be done through an intermediate hopper with a rotary valve or an equivalent arrangement. The hopper can be separately inerted to reduce air entrainment into the mixing vessel while expulsion of vapor into the operating area can be avoided by venting the vessel to a safe location. The addition of solids from nonconductive plastic bags can be hazardous even if the solids are noncombustible (e.g., silica). Bags should be constructed of paper, plies of paper and plastic in which the nonconductive plastic film is covered by paper on both sides, or antistatic plastic. Because grounding clips can be impractical, such bags can be effectively grounded by contact with a grounded conductive vessel or skin contact with a grounded operator. Fiber drums or packages should not have a loose plastic liner that can leave the package and behave like a plastic bag. Metal chimes should be grounded. Personnel in the vicinity of openings of vessels that contain flammable liquids should be grounded and special attention should be paid to housekeeping, because accumulation of nonconductive residues (e.g., resins) on the floor or on items such as grounding clips can impair electrical continuity.

7.10.6 Agitation. When solids are dissolved or dispersed into nonconductive liquids, the rate of charge generation can be large depending on factors such as solids loading, particle size, and agitation rate. Dissipation of the charge is frequently achieved by raising the conductivity of the continuous phase by reformulation with conductive solvents or by the addition of antistatic additives. Ignition hazards can alternatively be controlled by inerting.

7.10.7 Nonconductive Process Vessels. In general, nonconductive process vessels should not be used with flammable liquids. They present external ignition risks if their outer surfaces become charged. If a nonconductive tank is to be used and the possibility exists that the atmosphere around the tank or in the vapor space could be ignitible, the following should be incorporated to ensure the safe dissipation of charge and to prevent discharges:

(a) All conductive components (e.g., a metal rim and hatch cover) should be bonded together and grounded.

(b) Where used to store nonconductive liquids, an enclosing, grounded conductive shield should be provided to prevent external discharges. This shield can be in the form of a wire mesh buried in the tank wall, provided it is grounded. The shield should enclose all external surfaces.

(c) Where used to store nonconductive liquids, the tank should have a metal plate with a surface area not less than $0.05 \text{ cm}^2/\text{m}^3$ of tank volume located at the bottom of the tank and bonded to ground. This plate provides an electrical path between the liquid contents and ground through which the charge can dissipate.

(d) Where used to store conductive liquids, a grounded fill line extending to the bottom of the tank or an internal grounding cable extending from the top to the bottom of the tank and connected to ground should be provided. A grounded fill line that enters at the bottom and does not introduce a spark promoter fulfills this recommendation.

7.11 Gauging and Sampling.

Gauging and sampling operations, including temperature measurement, can introduce spark promoters into a storage tank or compartment. It is recommended that a conductive gauging well for manual sampling and gauging be used. The precautions given in Section 7.11 should be taken where use of a gauging well is not possible, where the material stored is a nonconductor, or where the vapor space of the container could be ignitible. Where these operations are conducted manually, the personnel grounding recommendations of Section 6.6 should be considered.

7.11.1 Materials. Gauging and sampling systems should be either completely conductive or completely nonconductive. For example, conductive sampling and gauging devices should be used with a conductive lowering device, such as a steel tape or cable. Chains are not electrically continuous and should not be used in flammable atmospheres. Conductive sampling and gauging devices, including the sampling container and lowering device, should be properly bonded to the tank or compartment. Such bonding should be accomplished by use of a bonding cable or by maintaining continuous metal-to-metal contact between the lowering device and the tank hatch.

Ideally, if nonconductive hand gauging or sampling devices are used, no waiting period is required after loading or filling. However, it should be noted that these devices might not retain the necessary level of nonconductivity due to environmental factors such as moisture or contamination. Therefore, an appropriate waiting period is also recommended when nonconductive devices are used.

Cord made from synthetic material such as nylon should not be used due to possible charging if it slips rapidly through gloved hands. Although natural cellulosic fiber cord can, in principle, be used, it is found that such cord is frequently composed of a natural synthetic blend, with corresponding charge generating ability.

7.11.2 Gauging. Where possible, gauging should be carried out with automatic gauging systems. These can be used safely in tanks, provided the gauge floats and similar devices are electrically bonded to the tank shell through a conductive lead-in tape or conductive guide wires. Free-floating, unbonded floats can be effective spark promoters and should be avoided. Noncontact gauging devices, such as radar and ultrasonic gauges, are also satisfactory, provided electrical continuity is ensured. Isolated conductive components must be avoided.

7.11.3 Waiting Period. Depending on the size of the compartment and the conductivity of the product being loaded, a sufficient waiting period should be allowed for accumulated charge to dissipate. A 30-minute waiting period is recommended before gauging or sampling storage tanks greater than 40 m^3, unless a gauging well is used. The waiting period before gauging or sampling of smaller vessels can be reduced to 5 minutes for tanks between 20 m^3 and 40 m^3 and to 1 minute for tanks less than 20 m^3. Longer waiting periods might be appropriate for very low conductivity liquids ($\kappa < 2$ pS/m) or nonconductive liquids that contain a second dispersed phase (such as a Class I liquid with more than 0.5 weight percent water). If a gauging well is used, a waiting period is unnecessary.

7.12 Tank Cleaning.

7.12.1 Water Washing. The mist created in a tank by water spraying can be highly charged. This is a particular problem with tanks larger than 100 m^3, due to the size of the mist cloud that can form. Water washing using sprays should only be done in an inerted or nonflammable atmosphere. Although specifically written for marine cargo tanks, the *International Safety Guide for Oil Tankers and Terminals* presents a comprehensive discussion of tank cleaning. Tanks of less than 100 m^3 and with all conductive components grounded have a negligible discharge hazard. Where a possibility of steam entering the tank during the water-washing process exists, the precautions in 7.12.3 should be followed.

7.12.2 Solvent Washing. Mist charge densities created by flammable solvents are similar to those from water washing and similar precautions should be taken regarding grounding of conductive components. Where an ignitible atmosphere or mist cannot be avoided because of the type of solvent or cleaning process used, the tank or vessel being cleaned should be inerted or enriched to reduce the likelihood of ignition during the cleaning process. Where the vessel is not inerted (or enriched) and an ignitible atmosphere is present, the following precautions should be considered when using solvent as a cleaning agent:

(a) The solvent should be conductive. When a solvent blend such as reclaimed solvent is used, the conductivity should be checked periodically.

(b) High flash point materials (at least 9°C above the maximum operating temperature during cleaning) should be used. The flash point should be confirmed on a daily basis.

(c) The cleaning system should be conductive and bonded to the tank. Continuity tests of all bonded equipment should be done periodically.

(d) Ungrounded conductive objects should not be introduced into the tank during the cleaning process or for a sufficient period of time after the cleaning process. This waiting period might have to be several hours, due to generation of mist.

7.12.3 Steam Cleaning. Steam cleaning can create very large charge densities with correspondingly large space charge potentials that increase with the size of the tank. Therefore, the following precautions are recommended:

(1) Tanks larger than 4 m^3 should be inerted before steam cleaning.
(2) All components of the steaming system should be conductive and grounded.
(3) All conductive components of the tank should be bonded and grounded.

7.12.4 Internal Grit Blasting. Where possible, tanks and process vessels should be clean and free of ignitible materials (no more than 10 percent of LFL). Hose used for grit blasting should be grounded and the resistance to ground from any part of the hose assembly, especially the nozzle, should not exceed 10^6 ohms. *(See A.7.4.3.)*

7.13 Portable Tanks, Intermediate Bulk Containers (IBCs), and Containers.

The following practices are recommended to reduce static electricity hazards during filling and emptying of portable tanks, IBCs, and containers.

7.13.1 Metal Portable Tanks and IBCs. Metal portable tanks and IBCs should be bottom-filled, if possible. Where used for nonconductive flammable liquids, filters should be placed at least 30 seconds upstream, as recommended in 7.4.5. The portable tank or IBC should be bonded to the fill system prior to opening and should be closed before being disconnected from the bond. Filling rates should be similar to those normally used for drum filling, about 225 L/min or less, unless the container is inerted. If the fill pipe does not extend close to the bottom and the vessel is not inerted, a slow start velocity of 1 m/sec or less should be used until the fill pipe is submerged to about 150 mm. Portable tanks and IBCs with nonconductive linings present hazards somewhat more severe than with drums, due to the larger capacity and the greater energy that can be stored for equal charge densities.

7.13.2 Nonconductive Portable Tanks and IBCs. Filling a nonconductive portable tank or IBC with combustible liquids at temperatures below their flash points presents no significant static electric ignition hazard. Filling such a vessel with a combustible liquid above or within 9°C of its flash point should be done as if the liquid were flammable. Refilling a vessel that could contain flammable vapors from a previous product should not be allowed. Additionally, the routine handling of nonconductive vessels filled with any type of liquid can generate a charge on the outside surface of the vessel. Nonconductive portable tanks and IBCs should not be used where ignitible ambient vapors are present.

Portable tanks and IBCs constructed of nonconductive materials are prohibited for use with Class I liquids by NFPA 30, *Flammable and Combustible Liquids Code*. Where such containers are used for Class II and Class III liquids, the precautions for filling depend on the size of the container, the container design, and the conductivity of the liquid.

7.13.3 Metal Containers. When being filled, metal containers and associated fill equipment should be bonded together and grounded. Bonding should be done with a clamp having hardened steel points that will penetrate paint, corrosion products, and accumulated material using either screw force or a strong spring. *(See Appendix C for recommendations.)* The clamp should be applied prior to removing the container bungs and at a point on the top chime that is located away from the bung openings. The grounded fill pipe should be cut at approximately 45 degrees and be left relatively sharp to inhibit brush discharges from the liquid surface. The tip of the fill pipe should extend to within 25 mm of the bottom of the drum and remain beneath the liquid surface until the drum is filled. Viscous liquids that flow without splashing can be deflected by a short fill nozzle to flow down the inside wall of the drum. Inerting of the drum is seldom necessary.

When dispensing from a metal container, it should be grounded. Self-closing, metal dispensing valves should be used. When dispensing from an upright drum, the dip pipe, conductive hose, and pump should be bonded to the drum and grounded. *(For funnels and receiving containers, see 7.13.6.)*

7.13.4 Plastic-Lined Metal Containers. The effects of static electricity from thin, internal coatings such as phenolic or epoxy paints can be neglected, provided the lining is not thicker than 2 mm. A container with a thin lining up to 2 mm in thickness can be treated as a metal container. Where the drum has a lining of nonconductive plastic thicker than 2 mm, it should be treated as a nonconductive container, unless it can be shown that the surface resistivity is not greater than 10^{10} ohms per square.

7.13.5 Plastic Containers. The use of plastic containers for Class I liquids is limited by NFPA 30, *Flammable and Combustible Liquids Code*. Where such containers are used for Class II and Class III liquids, the precautions for filling depend on the size of the container, the container design, and the conductivity of the liquid. Since plastic containers cannot be grounded, they should not be used for Class I liquids or handled in flammable atmospheres without expert review of the hazards. For Class II liquids, hazards of static electricity should be addressed as follows:

(1) Where the liquid might exceed its flash point during filling or emptying
(2) Where the container might be stored or handled in an ignitible ambient atmosphere

In 7.13.5(1), options include bottom filling and cooling of the liquid prior to unloading, especially if the container has been in direct sunlight or in a hot storage area. Continuous inerting during unloading can also be considered.

In 7.13.5(2), plastic containers should be stored away from containers of flammable liquids, so that the hazard of static electric discharge from the external surface of the plastic container is avoided.

7.13.6 Hand-Held Containers Not Greater than 20 L Capacity. The fire risk from static electricity increases with the volume of the container and the volatility of liquid handled. Thus, the smallest volume container capable of effectively fulfilling a particular need should normally be selected and should not exceed 20 L. Listed safety cans are recommended, especially those types equipped with a flexible metal dispensing hose so they can be used without a funnel. Because nonconductive containers cannot be grounded, they should be limited to 2 L for Class IA liquids and 5 L for Class IB and Class IC liquids. An exception is gasoline, where approved 20 L plastic cans have been widely used for many years with no reported increase in ignition incidents due to static electricity compared to metal cans. This is in part due to the rapid establishment of rich (above the UFL) gasoline vapor inside the can; these plastic containers should not be used for other flammable liquids without review of the hazards. Unlike gasoline, conductive liquids such as alcohols can become inductively charged by a charged plastic container and give rise to sparks. In addition, the container can contain an ignitible atmosphere.

7.13.6.1 Nonconductive Containers. Subject to the volume limitations described in 7.13.6, it is common to handle flammable liquids in small glass or plastic containers of 0.5 L capacity or less. Where such containers are involved in frequent transfer operations, such as a small-scale solvent blending operation, a grounded metal funnel whose spout extends to the bottom of the container should be used when filling the container. This practice ensures that any charge induced on the liquid by the container, as could happen if the plastic container has been charged by rubbing, is dissipated through the grounded funnel. Plastic or glass funnels should be used only where essential for compatibility reasons.

7.13.6.2 Containers for Sampling. Ignition risk is greatly increased when an ignitible atmosphere is present outside the container; for example, when sampling directly from a tank or transferring a sample near a manway, since this can precipitate a large fire or explosion. A grounded metal sample "thief" or glass bottle in a grounded metal sample cage can be used in such cases. Nonconductive plastic containers should be avoided, except when used in well-ventilated areas, since they are more easily charged than glass. If outdoor sampling is carried out at sample spigots that are located away from tank openings and in freely ventilated areas, and sampled quantities are 1 L or less, the fire risk is, in most cases, insufficient to require any special procedures other than bonding of metal components.

7.13.7* Cleaning. Containers should be bonded and grounded prior to opening for cleaning operations such as steaming. Cleaning equipment should also be bonded or grounded.

7.14* Vacuum Cleaners.

Collecting liquids and solids in an ignitible atmosphere using a vacuum cleaner can create a significant hazard due to ignition from static electric discharge. If it is necessary to use such equipment in a process area, the hazards and the procedures for safe use should be carefully reviewed and clearly communicated to the potential users.

7.15 Clean Gas Flows.

Usually a negligible generation of static electricity occurs in single-phase gas flow. The presence of solids such as pipe scale or suspended liquids such as water or condensate will create charge, which is carried by the gas phase. The impact of the charged stream on ungrounded objects can then create spark hazards. For example, carbon dioxide will form charged solid "snow" when discharged under pressure. This phenomenon can create an ignition hazard in an ignitible atmosphere. For this reason, carbon dioxide from high-pressure cylinders or fire extinguishers should never be used to inert a container or vessel. Gases with very low ignition energies, such as acetylene and hydrogen, that contain suspended material can be ignited by corona discharge when escaping from stacks at high velocity. This phenomenon is associated with electrical breakdown at the periphery of the charged stream being vented. Such discharges can occur even if the equipment is properly grounded.

7.16 Plastic Sheets and Wraps.

Nonconductive plastic sheets and wraps, such as those used to wrap shipping pallets, present hazards similar to those of plastic bags. Such sheets and wraps can generate brush discharges from their surfaces following rubbing or separation of surfaces. Isolated wet patches can also create spark hazards. An additional problem is charging of personnel during handling. It is recommended that plastic sheet and wrap not be brought into areas that can contain ignitible atmospheres. Plastic pallet wrap can be removed outside the area and, if necessary, can be replaced by a suitable tarpaulin or other temporary cover. Antistatic wrap is available. Tear sheets (used outside many clean areas) can generate significant static electric charge when pulled from a dispenser, and precautions are similar to those for plastic sheet. *(Additional information on handling sheet materials is found in Section 9.2.)*

Chapter 8 Powders and Dusts

8.1 General.

Powders include pellets, granules, and dust particles. Pellets have diameters greater than 2 mm, granules have diameters between 0.42 mm and 2 mm, and dusts have diameters of 420 micrometers (microns, μm) or less. It should be noted that aggregates of pellets and granules will often also contain a significant amount of dust. The movement of powders in industrial operations will commonly generate static electric charges. The accumulation of these charges and their subsequent discharge can lead to fires and explosions.

8.2 Combustibility of Dust Clouds.

8.2.1 A combustible dust is defined as any finely divided solid material 420 μm or smaller in diameter (i.e., material that will pass through a U.S. No. 40 standard sieve) that can present a fire or deflagration hazard.

8.2.2 For a static electric discharge to ignite a combustible dust, the following four conditions need to be met:

(1) An effective means of separating charge must be present.
(2) A means of accumulating the separated charges and maintaining a difference of electrical potential must be available.
(3) A discharge of the static electricity of adequate energy must be possible.
(4) The discharge must occur in an ignitible mixture of the dust.

8.2.3 A sufficient amount of dust suspended in air needs to be present in order for an ignition to achieve sustained combustion. This minimum amount is called the *minimum explosible concentration* (MEC). It is the smallest concentration, expressed in mass per unit volume, for a given particle size that will support a deflagration when uniformly suspended in air. *(In this chapter, air is assumed to be the supporting atmosphere unless another oxidizing atmosphere is specified.)*

8.2.4* In order to ignite a dust cloud by a static electric discharge, the discharge needs to have enough energy density, both in space and in time, to effect ignition. However, the term used for discharge ignition is simply that of the energy in the discharge. The minimum ignition energy (MIE) of a dust cloud is the energy in a capacitive discharge at or above which ignition can occur.

8.3 Charging Mechanisms.

8.3.1 Contact static electric charging occurs extensively in the movement of powders, both by surface friction between powders and surfaces and by friction between individual powder particles. The charging characteristics of particles are often determined as much by surface contamination as by their chemical characteristics; thus, the magnitude and polarity of a charge is difficult to predict. Charging can be expected any time a powder comes into contact with another surface, such as in sieving, pouring, scrolling, grinding, micronizing, sliding, and pneumatic conveying. In these operations, the more vigorous the contact, the more charge is generated, as shown in Table 8.3.1. The table shows that a wide range of charge densities is possible in a given operation; the actual values will depend on both the product and the operation.

8.3.2 An upper limit to the amount of charge that can be carried by a powder suspended in a gas exists. This limit is set by the strength of the electric field at the surface of the particle and depends on the surface charge density as well as the particle's size and shape. For well-dispersed particles, the maximum surface charge density is of the order of 10 $\mu C/m^2$. This value can be used to estimate maximum charge-to-mass ratios from particle diameter and density information.

8.4 Charge Retention.

8.4.1 Bulk powder can retain a static electric charge, depending on its bulk resistivity and its bulk dielectric constant. The relaxation time is expressed by the following equation:

$$\tau = \epsilon\epsilon_0\rho$$

Where:

τ = relaxation time constant (seconds)
ϵ = dielectric constant of the bulked powder
ϵ_0 = permittivity of free space (8.85×10^{-12} sec/ohm-m)
ρ = bulk volume resistivity of the powder (ohm-m)

For historical reasons, the ability of a solid to transmit electric charges is characterized by its volume resistivity. For liquids, this ability is characterized by its conductivity.

8.4.2 Powders are divided into the following three groups:

(a) Low-resistivity powders having volume resistivities in bulk of up to 10^8 ohm-m. Examples include metals, coal dust, and carbon black.

Table 8.3.1 Typical Charge Levels on Medium Resistivity Powders Emerging from Various Powder Operations (Before Compaction)

Operation	Mass Charge Density (μC/kg)
Sieving	10^{-3} to 10^{-5}
Pouring	10^{-1} to 10^{-3}
Auger or screw-feed transfer	10^{-2} to 1.0
Grinding	10^{-1} to 1.0
Micronizing	10^{2} to 10^{-1}
Pneumatic conveying	10^{3} to 10^{-1}

Source: British Standard 5958, *Code of Practice for Control of Undesirable Static Electricity,* Part 1, General Considerations.

(b) Medium-resistivity powders having volume resistivities between 10^8 and 10^{10} ohm-m. Examples include many organic powders and agricultural products.

(c) High-resistivity powders having volume resistivities above 10^{10} ohm-m. Examples include organic powders, synthetic polymers, and quartz.

8.4.2.1 Low-resistivity powders can become charged during flow. The charge rapidly dissipates when the powder is conveyed into a grounded container. However, if conveyed into a nonconductive container, the accumulated charge can result in an incendive spark.

8.4.2.2 When a medium-resistivity powder comes to rest in bulk, the charge retained depends on the resistance between the powder and ground. If the powder is placed in a grounded container, charge retention is determined by the bulk volume resistivity of the powder, which includes the interparticle resistance, as governed by the relationship given in 8.4.1. If the powder is placed in a nonconductive container, charge retention is determined by the resistance of the container. The special significance of medium-resistivity powders is that they are relatively safe during handling, because they do not produce bulking brush discharges or sparks.

8.4.2.3 High-resistivity powders do not produce spark discharges in themselves but can produce other types of discharge such as corona, brush, bulking brush, and propagating brush discharges *(see Section 4.3).* High-resistivity powders lose charge at a slow rate, even in properly grounded containers. Many high-resistivity powders are also hydrophobic and in bulk are capable of retaining charge for hours or even days. High-resistivity powders, such as thermoplastic resins, can have bulk resistivities up to about 10^{16} ohm-m.

8.5 Discharges in Powder Operations.

8.5.1 Spark Discharge. Where spark discharges occur from conductors, the energy in the spark can be estimated from the following equations or from the nomograph in Figure 4.3.3.3:

$$W = \frac{1}{2}CV^2 \quad W = \frac{1}{2}QV \quad W = \frac{Q^2}{2C} \quad Q = CV$$

Where:

W = energy (joules)
C = capacitance (farads)
Q = charge (coulombs)
V = potential difference (volts)

It should be noted that these equations apply only to capacitive discharges from conductors and cannot be applied to discharges from insulators. Discharge energies so estimated can be compared with the MIE of the dust to provide an insight into the probability of ignition by capacitive spark discharge *(see 4.3.3)*. Layers of combustible dusts can be ignited by capacitive spark discharge, and this can lead to secondary dust explosions. For a dust layer, there is no correlation with the MIE for dust cloud ignition. Capacitive spark discharges must be avoided by grounding all conductive containers, equipment, products, and personnel.

8.5.2 Corona and Brush Discharge. When handling large amounts of powder having medium or high resistivities, corona and brush discharges are to be expected. No evidence is available, however, that a corona discharge is capable of igniting a dust cloud. Likewise, no evidence is available that a brush discharge can ignite dusts with MIEs greater than 3 mJ, provided no flammable gas or vapor is present in the dust cloud.

8.5.3 Propagating Brush Discharge. Since propagating brush discharges can have energies greater than 1 J, they should be considered capable of igniting both clouds and layers of combustible dusts.

8.5.4 Bulking Brush Discharge. When powders that have resistivities greater than about 10^9 ohm-m are put into grounded conductive containers, they usually dissipate their charges by conduction at a rate that is slower than that of the charge that is accumulated in the loading process. The charge is therefore compacted, and discharges occur from the bulking point (where the particles first contact the heap) to the walls of the container. These discharges are termed *bulking brush discharges*. Experience indicates that these discharges are not capable of igniting dusts having MIEs greater than 10 mJ. However, such discharges have been attributed to explosions of dusts having MIEs less than 10 mJ.

During the compaction process, the energy in the discharge increases as the particle size increases. Therefore, it can be expected that systems most at risk are those involving pellets having an appreciable fraction of fines (dust).

8.6 Pneumatic Transport Systems.

8.6.1 Pneumatic transport of powdered material through pipes or ducts can produce a static electric charge on both the product being transported and the conduit. This static electric charge will remain on the material as it exits the system. Precautions against accumulation of charge should be taken where the material is collected.

8.6.2 Pipes and ducts should be metal and should be grounded. Equipment to which the conduits connect should also be metal and grounded to dissipate the charge impressed upon it by the transport of the material. Where the use of pipe-joining methods or installation of piping components results in an interruption of continuity of the ground path, either a jumper cable should be used to maintain continuity or an independent ground should be provided for the isolated section of the conduit, as shown in Figure 8.6.2.

8.6.3 Use of nonconductive pipe or ductwork is not recommended. The use of short lengths of transparent plastic as flow visualizers is also not recommended, because they have been known to give rise to propagating brush discharges capable of igniting dusts.

8.7* Flexible Hoses.

Hoses made of nonconductive material that incorporate a spiral stiffening wire should be kept in good repair to ensure that the internal wire directly contacts the metal end couplings and that the end couplings make a good connection to ground. Use of hoses with more than one internal spiral is not recommended, because determining if one of the spirals has lost its continuity is not possible.

8.8 Flexible Boots and Socks.

8.8.1 Flexible boots and socks are commonly used for gravity transfer operations. Boots are typically made of plastic or rubber, while socks are typically made of woven fabric. A nonconductive boot could give rise to either brush discharge or propagating brush discharge. Propagating brush discharge cannot happen with a sock, because of the low breakdown strength of the air gaps in the weave. However, there are conditions where socks can produce brush discharges (e.g., where used with flexible intermediate bulk containers.) *(See Section 9.1.)*

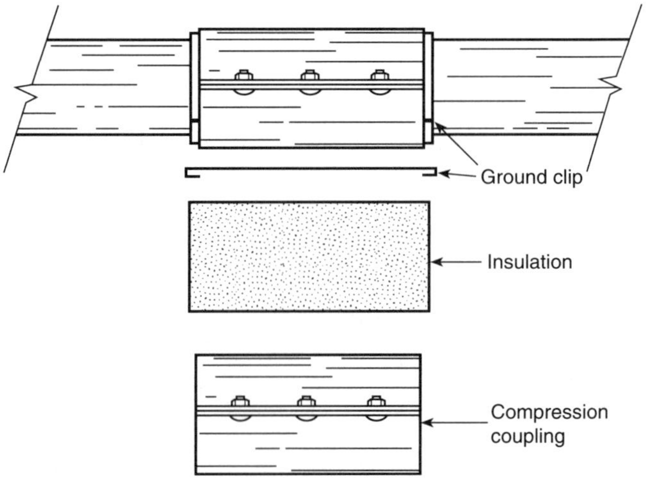

Figure 8.6.2 Compression Fitting for Pneumatic Transport Duct. ***(Pratt, 1997, p. 136.)***

8.8.2 For combustible dusts, the end-to-end resistance of boots and socks should be less than 10^8 ohms and preferably less than 10^6 ohms, measured with a megohmmeter.

8.8.3 Flexible connections should not be depended on for a bond or ground connection between process equipment. Separate bonding or grounding connections should be used.

8.9 Bag Houses.

8.9.1 As dusts are drawn or blown into a bag house, they necessarily carry with them a static electric charge, the magnitude of which depends on the characteristics of the dust and the process, as illustrated in Table 8.3.1. The charge will remain on the dust and will accumulate on the surfaces of the bags. It is therefore important to keep all conductive equipment grounded to prevent the induction of this accumulated charge onto conductive components that could have inadvertently become ungrounded. Such induction is particularly true in the case of cage assemblies.

8.9.2 If cage assemblies are not well grounded, capacitive spark discharge can occur from the ungrounded cages to either the structure of the bag house or to adjacent cage assemblies. Many times the bags have metal braid pigtails attached to their cuffs, the notion being that the pigtail can be simply brought through the cage and bonded to the tube sheet. This method of grounding the cage is not always successful. Furthermore, the reason for the pigtail is often misunderstood. Because the bag is nonconductive, the bag itself is not grounded. It is therefore useless to extend the metal braid down the entire length of the bag. *(See Figure 8.9.2.)*

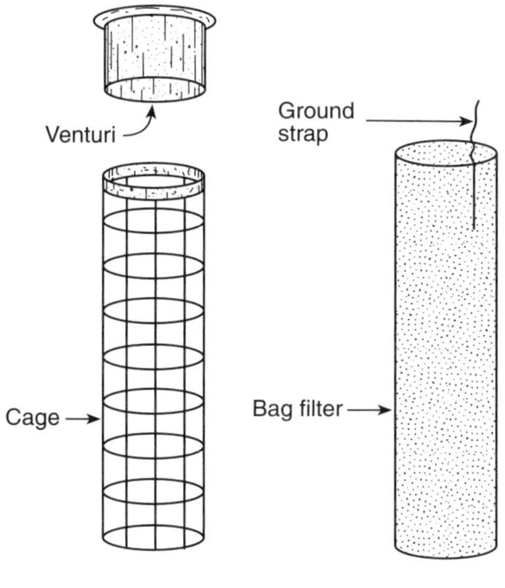

Figure 8.9.2 Arrangement of Cage and Filter Bag. ***(Pratt, 1997, p. 134.)***

8.9.3 Bags and cages should be engineered so that a positive ground connection is always maintained during maintenance, even with inexperienced or inattentive personnel. One way of doing this is by sewing two metal braids into the cuffs of the bags, 180 degrees apart. Each braid is continuous and is sewn up the inside of the cuff, across the top, and down the outside of the cuff. By doing this, the braids will always make a positive contact with the cage, the venturi, and the clamp and will withstand the rigors of the operation. In any case, the resistance between the cage and ground should be less than 10 ohms.

8.9.4 No evidence is available that filter bags made from conductive or antistatic fabric are needed to prevent incendive discharges. On the contrary, such bags could create discharge hazards if sections of the fabric become isolated or if a bag falls into the bottom of the bag house.

8.10* Hybrid Mixtures.

8.10.1 The term *hybrid mixture* applies to any mixture of suspended combustible dust and flammable gas or vapor, where neither the dust itself nor the vapor itself is present in sufficient quantity to support combustion, but the mixture of the two can support combustion. Hybrid mixtures pose particular problems because they combine the problems of the large charge densities of powder-handling operations with the small ignition energies of flammable vapors. The MIE of a hybrid mixture is difficult to assess, but a conservative estimate can be made by assuming that the MIE of the mixture is at or near the MIE of the gas alone. Because hybrid mixtures contain a flammable gas or vapor, they can be ignited by brush discharge.

8.10.2 Powders that contain enough solvent (i.e., greater than 0.2 percent by weight) so that significant concentrations of solvent vapor can accumulate in the operations in which they are handled are termed *solvent-wet powders*. Consideration should be given to applying the recommendations of Chapter 7 to solvent-wet powders, unless the resistivity of the solvent-wet product is less than 10^8 ohm-m.

8.11* Manual Addition of Powders to Flammable Liquids.

8.11.1 The most frequent cause of static electric ignitions in process vessels is the addition of solids to flammable liquids in the vessels. Even when the vessel is inerted, large additions of solids will introduce air into the vessel while expelling flammable vapor from the vessel. The sudden addition of a large volume of solids can also result in static discharge from a floating pile of charged powder. It is recommended that manual addition of solids through an open port or manway be done only in 25-kg batches. Larger batch additions [e.g., from flexible IBCs *(see 9.1.7)*] should be done through an intermediate hopper with a rotary valve or an equivalent arrangement. The hopper can be separately inerted to reduce air entrainment into the mixing vessel while expulsion of vapor into the operating area can be avoided by venting the vessel to a safe location. The addition of solids from nonconductive plastic bags can be hazardous even if the solids are noncombustible (e.g., silica). Bags should be constructed of paper, plies of paper and plastic in which the nonconductive plastic film is covered by paper on both sides, or antistatic plastic. Because grounding clips can be impractical, such bags can be effectively grounded by contact with a grounded conductive vessel or by skin contact with a grounded operator. Fiber drums or packages should not have a loose plastic liner that can leave the package and behave like a plastic bag. Metal chimes should be grounded. Personnel in the vicinity of openings of vessels that contain flammable liquids should be grounded, and special attention should be paid to housekeeping, because accumulation of nonconductive residues (e.g., resins) on the floor or on items such as grounding clips can impair electrical continuity.

8.11.2 Powder should not be emptied from a nonconductive container in the presence of a flammable atmosphere. Direct emptying of powders from nonconductive plastic bags into a vessel that contains a flammable atmosphere should be strictly prohibited.

8.11.3 Where a thorough understanding of the process exists and where the vessel does not contain an ignitible atmosphere, adding the powder to the vessel before adding the liquid might be practical.

8.12 Bulk Storage.

When powders are moved into bulk storage (e.g., silos, rail cars, trucks, IBCs, or flexible IBCs), the powder is compacted by the force of gravity. The compaction process is accompanied by bulking brush discharge, as explained in 8.5.4. In the compaction process, the energy of the discharge increases as the particle size increases. Therefore, the systems most at risk are pellets with an appreciable fraction of fines (dust).

The exact conditions for ignition-capable bulking brush discharge are not well understood. However, the following general factors that are known to increase its probability have been identified by Glor in *Electrostatic Hazards in Power Handling:*

(1) An increase in the resistivity of the powder, greater than 10^{10} ohm-m
(2) An increase in the particle size of the powder, greater than 1 mm
(3) An increase in the charge density of the powder, greater than 1 μC/kg
(4) An increase in filling rate — for granules with a diameter greater than 1 mm to 2 mm, greater than 2×10^3 kg/hr, and for granules with a diameter of about 0.8 mm, greater than 20 to 30 $\times$ 10^3 kg/hr

Chapter 9 Specific Applications

9.1 Intermediate Bulk Containers.

9.1.1 The discussion and precautions for powders and granular solids, as set forth in Chapter 8, also apply to operations that involve handling these materials in intermediate bulk containers (IBCs). Static electric charges are generated in granular materials when filling and emptying IBCs. These charges result from movement and rubbing of the granules against process equipment or against each other. The amount of charge that can be accumulated on a bed of a powdered material in a grounded container depends on the resistivity of the powder, not the resistivity of a block of the material. The higher the resistivity, the lower the apparent conductivity and the longer a charge will be retained. In cases of very large resistivities, charges will relax slowly and can remain on beds of material for appreciable periods. Thus, generation and relaxation occur simultaneously when granular materials are moved about. When the rate of generation exceeds the rate of relaxation, significant charges can accumulate.

9.1.2 When a static electric charge accumulates on bulk containers or associated process equipment, the following four types of discharge can occur:

(1) Spark discharge
(2) Brush discharge
(3) Propagating brush discharge
(4) Bulking brush discharge

(See Section 4.3 for discussions of these four types of discharge.)

9.1.2.1 Spark discharges can take place between two conductors at different potentials and can release energy capable of igniting atmospheres containing flammable gases or vapors or combustible dusts, depending on process conditions. An example of a situation in which such a discharge can occur is a conductor that is isolated from ground and located in a bin of material.

9.1.2.2 Brush discharge is usually not a concern in the normal handling of granular materials. However, brush discharge can be a source of ignition where flammable gases or vapors are present, as in the handling of hybrid mixtures or very rapid discharge of a granular material from a container. Such situations should be avoided if possible.

9.1.2.3 Propagating brush discharges typically contain energies of 1 J or greater, depending on process conditions. These discharges can ignite most flammable atmospheres.

9.1.2.4 Bulking brush discharges contain energies on the order of 10 mJ. In order to minimize the risks associated with bulking brush discharge, powders that have minimum ignition energies of 10 mJ or less should only be loaded into containers of 2 m^3 or less, unless the vessel is inerted.

9.1.3 If a granular material contains only particles larger than 420 μm, then ignitible dust clouds cannot be formed. However, if a granular material consists of fine particles or contains an appreciable fraction of fine particles, then ignitible dust suspensions can be formed and ignition sources cannot be tolerated.

9.1.4 Experience has shown that, where powders have resistivities of less than 10^8 ohm-meters, static electric charges usually relax rapidly enough to prevent their accumulation in the bulk granular material.

9.1.5 Conductive Intermediate Bulk Containers.

9.1.5.1 Conductive IBCs (e.g., those constructed of metal) should be grounded during all operations in areas where an ignitible atmosphere exists. Engineering and administrative controls should be considered to ensure that grounding of conductive containers and associated conductive equipment is accomplished during all operations. For example, metal funnels should be grounded and flexible fill pipes should have any conductive components, including any internal stiffening wires, connected to ground.

9.1.5.2 When a nonconductive material is transferred into a conductive IBC that is grounded, any charge that has accumulated on the material will remain on the material. The process of relaxation is the slow migration of the charges through the material to the opposite charges on the wall. While this relaxation process is occurring, which can last a few seconds or many minutes depending on the conductivity of the material, an electric field still exists at the surface of the material. In the case of IBCs with open tops, this electric field can induce charges on other conductors that might be present, including ungrounded personnel. Thus, induction of surface charges onto other ungrounded conductors can occur even when conductive IBCs are properly grounded. Appropriate precautions should be taken.

9.1.5.3 IBCs made of conductive materials and nonconductive liners should be used only if the liners are essential (e.g., to maintain compatibility between the IBC and the material being handled.) The risk of ignition and the possibility of electric shock from propagating brush discharge depend very much on the thickness and resistivity of the liner, the handling procedure, the electrical properties of the material being handled, and the incendive nature of any combustible material that might be present. In general, propagating brush discharge will not occur, provided the nonconductive liner has a breakdown voltage lower than 4 kV. Each situation should be considered individually.

9.1.6 Nonconductive Intermediate Bulk Containers.

9.1.6.1 The term *nonconductive* applies to any IBC that has a volume resistivity greater than 10^{10} ohm-m or a surface resistivity greater than 10^{11} ohms.

9.1.6.2 When a material is transferred into a nonconductive IBC, the container material will hinder the relaxation to ground of any static electric charge that is present on the material. In this instance, even conductive contents can accumulate charge.

9.1.6.3 Nonconductive IBCs should not be filled or emptied in areas where easily ignitible atmospheres (i.e., minimum ignition energy less than 10 mJ) are present (i.e., gases, flammable vapors, sensitive dusts, and hybrid mixtures.) When powders are to be added to flammable liquids from nonconductive containers, the receiving system should be closed and inerted.

9.1.6.4 Nonconductive IBCs should not be used with solvent wet powders in locations where the ambient temperature is near or above the flash point of the solvent. Powders should be considered to be solvent-wet powders if they contain more than 0.2 percent by weight of solvent.

9.1.6.5 If a nonconductive IBC is moved into a location where flammable gases or vapors are also present, rubbing of the container should be avoided.

9.1.7 Flexible Intermediate Bulk Containers (FIBCs).

9.1.7.1 Description. Flexible intermediate bulk containers (FIBCs) are basically very large fabric bags supported in a frame. They are more convenient than rigid IBCs because they can be fully collapsed after use, taking up little storage space. The fabric is usually polypropylene and the fabric is sewn to form a three-dimensional cube or rectangle with lifting straps. A FIBC can be filled with a powder or granular material and moved about with conventional materials-handling equipment. An advantage of FIBCs is that they can be unloaded very quickly, typically 300 kg to 500 kg in 30 seconds or less. Therefore, rates at which static electric charges are generated can often exceed the rates at which the charges can relax under common conditions of use and accumulation of a static electric charge can be expected. In general, the precautions given in this subsection for IBCs also apply to FIBCs.

9.1.7.2 Charge Generation. Static electric charges can be generated during filling and emptying of FIBCs and can accumulate on both the contents and the fabric of the FIBC. If the accumulated charge is strong enough and is released in the presence of an ignitible atmosphere, ignition can occur.

9.1.7.3 Nonconductive FIBCs. FIBCs constructed of nonconductive materials (e.g., polypropylene fabric with polyester stitching) have no special features incorporated in their design to minimize static electric charging. These nonconductive FIBCs can be used for materials that do not form ignitible atmospheres in normal handling operations. Experience has shown that granular materials that do form dust clouds during handling can also be safely handled in these FIBCs, provided the dust clouds generated cannot be easily ignited by a static electric discharge. In other words, the material should have an MIE for a dust cloud that is greater than 100 mJ. These FIBCs should not be used for granular materials that have a minimum ignition energy less than 100 mJ and should never be used in areas where a flammable gas or vapor is present.

9.1.7.4 Conductive FIBCs.

9.1.7.4.1 FIBCs constructed of conductive fabric can be treated the same as conductive IBCs, as specified in 9.1.5. If conductive FIBCs are used with a liner, the recommendations of 9.1.5.3 also apply. It is essential that these FIBCs be grounded during all operations.

9.1.7.4.2 FIBCs constructed of nonconductive fabric and containing woven, grounded, conductive filaments can be considered to be conductive. One type has conductive filaments that are spaced less than 20 mm apart, each of which is connected at least once to its neighbor, preferably at one end, and are intended to be grounded. Another type has conductive filaments or threads that form an interconnecting grid of not more than 50 mm mesh size and are also intended to be grounded. The recommendations for conductive IBCs given in 9.1.5 also apply to these FIBCs. A grounding tab that is electrically connected to the conductive threads is provided and is intended to be connected to a ground point when the FIBC is filled or emptied.

9.1.7.5 Summary. At the present time, the manufacture and use of FIBCs is in a development stage and many variants and combinations of fabric are being introduced. It is, therefore, not possible to make recommendations for the safe use of all the types being used in areas where an ignitible atmosphere might be present. In particular, at this time, sufficient data are not available for FIBCs constructed of layers of various conductive and nonconductive fabrics.

9.1.8 Container Linings.

9.1.8.1 Both conductive and nonconductive liners are used in containers. Conductive liners (e.g., carbon-filled polyethylene) have been used inside nonconductive containers in order to provide a means for grounding. In instances where product contamination is a concern, nonconductive (e.g., polyethylene) liners have been used in both conductive (e.g., metal) and nonconductive (e.g., polypropylene) containers. And, in instances where very fine powders can leak through the weave of the cloth, nonconductive (e.g., polyethylene) liners have been used.

9.1.8.2 Extreme caution should be exercised when using conductive liners to ensure that they do not become ungrounded. As in the case of conductive containers, conductive liners should be grounded during all operations.

9.1.8.3 In the handling of materials that accumulate static electric charges, the use of liners that have resistivities greater than 10^{11} ohms inside conductive containers could create conditions where propagating brush discharges or capacitor-like discharges can occur. Because the conditions under which these discharges can occur are many and varied, no general recommendations can be given at this time.

9.1.8.4 Where using nonconductive liners in nonconductive containers, the precautions required are the same as for nonconductive containers alone, as given in 9.1.7 and 9.1.8.3.

9.1.8.5 Liners should never be removed from containers (e.g., to shake out a residue) where easily ignitible atmospheres (i.e., MIE less than 30 mJ) are present (e.g., gases, flammable vapors, sensitive dusts, and hybrid mixtures).

9.2 Web and Sheet Processes.

9.2.1 General. In web processes such as printing, coating, spreading, and impregnating, static electricity is a frequent, annoying, and often expensive source of production problems. If flammable solvents are used in the process, static electric charges can constitute an ignition source.

In practice, paper or any other substrate charged with static electricity will attract or repel other objects. This phenomenon can cause difficulty in controlling the sheet or web, which is the continuous substrate that is being printed or coated. It can also cause problems with delivering and handling the printed product due to static attraction between the sheets or folded signatures. Static electric charges can transfer by induction or by contact with various objects (e.g., during handling of the paper or substrate by personnel). This static electric charge can accumulate on a person who is not adequately grounded.

9.2.2 Substrates.

9.2.2.1 Paper. The characteristics of the surface of the paper have a great deal to do with the amount of static electric charge that is generated during processing. Generally, printing on paper causes fewer problems than printing on plastic substrates and other synthetic materials. Static electric charge accumulates on paper during the handling process. On presses and in other handling operations, static electric charge can be generated by belts driving the paper rolls, sliding of the web over idler rollers and angle bars, motion of the web through a nip, and motion of brushes and delivery belts in the folder. In some operations, static electric charge is deliberately deposited on the web to improve certain operations such as material deposition and sheet transfer. In gravure printing, for example, electrostatic assist is used to improve the transfer of ink. On high-speed offset and high-speed gravure presses, ribbon tacking is used to control the ribbons and signatures in the folder.

9.2.2.2 Plastics. Most plastic films are characterized by extremely high surface and bulk resistivities. This resistivity allows static electric charge to accumulate on the web after contact with machine parts, such as rollers and belts, with little dissipation occurring.

9.2.2.3 Fabrics and Nonwovens. Fabrics are usually made of blends of natural fibers (usually hygroscopic and capable of relaxing a charge) and synthetic fibers (usually highly resistive and capable of holding a charge). The less the proportion of natural fibers, the greater the incidence of static electric problems in subsequent operations. Fabrics are thin, like paper and plastic films, and accumulate static electricity in a similar manner. Nonwovens often have a loft that gives them a three-dimensional structure. They are almost exclusively synthetic, so they tend to generate and hold substantial charges in the forming process. These charges can be more difficult to remove due to the depth of the loft. In a subsequent coating or saturating process, large amounts of charge can again be generated due to relative motion of the fibers and, again, it can be difficult to remove if the loft returns. The solvent-wet batt contains a relatively large volume of flammable vapor and electrostatic discharge can cause ignition.

9.2.3 Inks and Coatings.

9.2.3.1 Inks used in letter presses and offset presses are typically Class IIIB liquids that have flash points above 200°F and present little fire or explosion hazard. However, inks used in silk screen, rotogravure, and flexograph printing are usually Class IB and Class IC liquids with flash points less than 100°F. Fires can occur in these inks due to the use of solvents whose vapors can be ignited by static electric discharge, as well as by other ignition sources.

9.2.3.2 The solutions and suspensions that are used to coat and saturate webs are diverse. While they are still wet, water-based coatings are generally conductive enough to dissipate any charge that is generated in the process, even though there might be minor concentrations of solvent present that can create an ignitible vapor layer on the web. When dry, however, these coatings are not always capable of dissipating the charge, but vapors are seldom left at this point.

9.2.3.3 Flammable solvent-based inks and coatings should be considered nonconductive and, therefore, incapable of dissipating a charge. Conductivity enhancers in the ink or coating cannot be relied on to assist dissipation of charge at high processing speeds. Measurement of coating solution conductivity can provide additional data to determine static generation and dissipation characteristics.

9.2.3.4 Black inks used in gravure printing are generally nonconductive. When accumulations of black ink, particularly those used on uncoated papers, are washed or cleaned off the rubber impression rollers, the resin can be washed out of the ink buildup, leaving a residue of conductive carbon (i.e., the pigment). If this conductive residue is not thoroughly wiped off the rollers, sparking and arcing from the roller to the cylinder or other grounded press parts can occur.

9.2.4 Processes.

9.2.4.1 Printing Presses. All other factors being equal, printing presses that operate at higher speeds generate more static electricity. A rotogravure press, for example, can generate static electric where the rubber roll presses the substrate against an engraved roll, which is wetted with the ink. Charge can be transferred from the engraved roll to the substrate. In a multicolored press, there is a similar arrangement for each color. The generation of charge is a function of the pressure between the rolls and the angle to the roll. The electrostatic assist (ESA) process, when used, deposits large amounts of charge onto the substrate. Note that ESA equipment must be suitable for Class I, Division 1 locations.

9.2.4.2 Coating. Coating of web materials is done using a wide variety of equipment. Some of these processes can generate hazardous amounts of static electric charge due to their design while others cause little effect. The operating conditions that cause high rates of charge generation include high forces between rollers and the web, such as in gravure coating. This high rate of generation is aggravated by maintaining a tension difference across the coating roller, which causes slippage. The result is a sizable charge at the point where a large amount of ignitible vapor and liquid surface area is present. The rubber backup roller can accumulate enough electrostatic charge to present an ignition hazard. An electrostatic charge neutralizer might be required on the roller.

9.2.4.3 Saturating. Saturating is the process of immersion of a web in a liquid so that the liquid fills the pores in the web. The excess liquid is then squeezed or wiped from both sides of the web. Electrostatic charging during saturating operations is not usually a problem for most webs. When the web is a nonwoven with substantial loft and the liquid is flammable and of low conductivity, a static electric hazard can be created.

9.2.4.4 Calendaring. Calendaring is a process by which a substrate is squeezed at high pressure between rollers that are generally smooth. This process is used to create a dense product with a smooth surface, such as magazine cover stock. It is also used to mill and form webs from materials such as rubber and plastics. The intimate contact caused by the high pressures and the working of the materials between the nipped rollers creates charge on the web. Charging can be high enough to form corona discharge at the exit of the nip.

Because flammable solvents are not usually present, the effect of static electric charge is to cause operator shock and web-handling problems. Static neutralizers can effectively remove the charge.

9.2.4.5 Web Handling and Converting. The path of the web through processing machinery often is guided over many rollers. Movement of the web over these rollers produces static electric charge due to friction. A freely turning idler roller imparts little charge to the web. As the speed of the process increases above 200 ft/min, air is drawn between the web and the roller, reducing the intimacy of contact and, thus, the rate of charge generation. If the roller does not turn freely, however, the web slips on the roller surface and can generate a large static electric charge. Periodic inspections and maintenance should be performed to ensure that the rollers are always free-turning.

Converting large rolls of web goods to finished product involves operations such as slitting, sheeting, folding, and packaging. These operations involve many contacting surfaces, and materials used to construct machinery for them are chosen for durability and resistance to wear more so than for static electric characteristics. The machinery is designed to occupy a minimum of floor area for economy of plant space. The machines are dense with rollers, belts, and framework that variously generate charge and suppress static electric charge and provide minimal space to install static electricity neutralizers. The result is a static electric problem that can be hard to measure and hard to control. Problems include the following:

(1) Product sticking or flying
(2) Machine jams
(3) Misaligned product stacks
(4) Bad product rolls
(5) Bad packages

With photosensitive materials, the charge can cause product damage, which is only revealed when the customer complains.

9.2.4.6 Ribbon Tacking. On high-speed offset and gravure presses, high-voltage tacking is used to improve the delivery of signatures to the folder. These high-voltage devices should be suitable for Class II, Division 2 locations if subject to accumulations of settled paper dust.

9.2.5 Control of Static Electricity in Web Processes.

9.2.5.1 Higher operating speeds result in the potential for increased volumes of flammable vapors in the immediate vicinity of the area where flammable liquids are applied to the substrate. Coating technology such as the enclosed ink gravure fountain has minimized the volume of flammable vapors in the application area. Solvent vapors can be diluted below the LFL by means of mechanical ventilation. The performance of the ventilation system is optimized by capturing vapors as close to their source as possible. The ventilation system should be interlocked with the equipment to ensure safe operation of the process. Vapors will always be within the flammable range in the areas close to the application process and the substrate.

9.2.5.2 A common method of removing static electric charge from processing machinery is by grounding the machine. All conductive parts of the machine should be electrically bonded together for this method to be effective. It should be noted that the charge on the material being processed is not removed by this grounding and bonding.

9.2.5.3 The ionization method of charge neutralization is the most effective means for controlling static electric charge on the webs in presses and coaters. Neutralizers, also called static elimination bars or inductive ionizers, are commonly used close to the substrate, but they should not touch the substrate. The reduction of charge at any one point in the operation does not prevent generation of charge in later steps of the process. Static neutralizers might be necessary at a number of locations *(see Figure 9.2.5.3)*.

High-speed operations might require a second static neutralizer at a single location, but providing more than two usually gains no additional benefit. Neutralizers should extend across the full width of the web. For fast-running

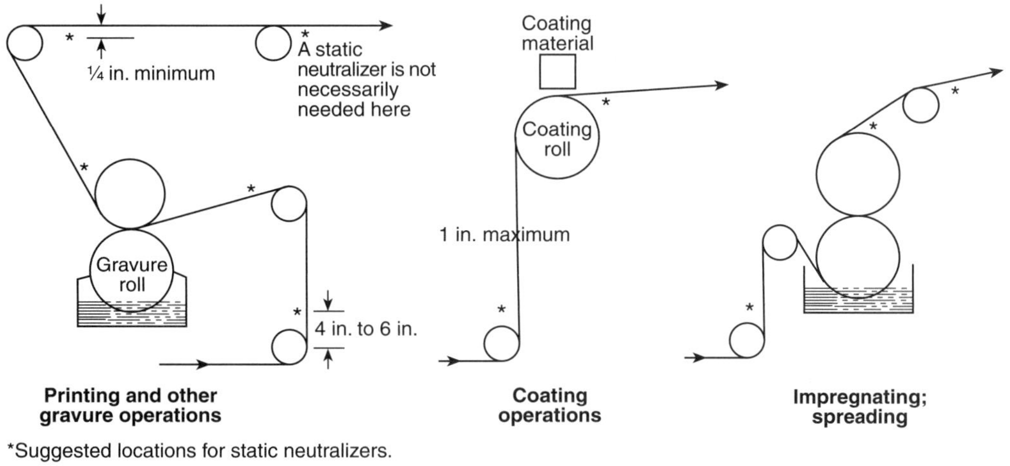

Figure 9.2.5.3 Typical Locations for Static Control.

presses, inductive neutralizers, such as tinsel wrapped around a bar or grounded needle points, spaced 12 mm to 25 mm apart and placed at the inlet and delivery side of each impression roller, have been found to be effective.

As with all neutralizers, positioning is important and effectiveness of individual installations should be confirmed by field measurement of residual charge or voltage. An area of web as small as 0.01 m^2 or an area of 115 mm in diameter might be capable of releasing an incendive discharge. The neutralizer should be installed as far as possible from grounded metal machine parts or areas where the web is supported by rollers. The recommended location of the initial inductive ionizer is 100 mm to 175 mm from the roller tangent (i.e., web exit) point and 6 mm to 25 mm from the web. Thus, the first neutralizer after the roller might have to be placed so close to the roller that it is partially suppressed from ionizing but can minimize the area of high charge. An additional ionizer or two downstream from the roller might also be necessary to reduce the charge to the desired extent. Final location near critical rollers should be determined by adjustment and measurement to minimize web charge, as indicated by apparent voltage.

9.2.5.4 Tinsel or needle points remain functional only as long as they are clean and sharp. Accumulation of contaminants (e.g., ink, coating solution, or paper dust) and corrosion products should be controlled by an effective maintenance program. Visual inspection and performance verification of the neutralizers should be done at intervals that will depend on how process conditions affect performance of the neutralizers.

9.2.5.5 Humidification has been used in the past to decrease static electric charge on materials in the environment, such as in garments and floor coverings. However, with modern high-speed operations, nonconductors, such as paper and plastic films, cannot adsorb enough moisture from the atmosphere during the brief time that they are exposed to humid air to increase their surface conductivity. Plastics generally do not become sufficiently conductive, even in humid atmospheres, due to lack of ionized molecules in an acquired surface layer of moisture. Therefore, humidification is not a recommended method of control of static electricity in such processes.

9.3 Spray Application.

9.3.1 Processes that involve spray application of liquids or powders (e.g., paints, coatings, lubricants, and adhesives) can cause accumulation of static electric charges on the spray apparatus and on the surfaces of the objects being sprayed and other objects in the spray area. If the material sprayed is ignitible, then static electric discharge can result in ignition.

9.3.2* Personnel involved in operating or servicing electrostatic spray application equipment should be trained in the operating procedures recommended by the manufacturer of the equipment and in the control of hazards associated with the materials being sprayed and their residues.

9.4 Belts and Conveyors.

9.4.1 General. Flat or "V"-shaped rubber or leather belts used for transmission of power or belts used for the transportation of solid materials can generate static electric charges that warrant corrective measures if a possibility exists that ignitible concentrations of flammable gases or vapors or combustible dusts or fibers might be present. The amount of charge generated will increase as any of the following increase:

(1) Belt speed
(2) Belt tension
(3) Width of the contact area

9.4.2 Flat Belts.

9.4.2.1* Synthetic, rubber, or leather flat belts are usually good insulators because they are dry when operated and because they operate at elevated temperatures due to friction. Generation of static electric charge occurs where the belt leaves the pulley and can occur with either conductive or nonconductive pulleys.

9.4.2.2* Accumulation of static electric charge on belts can be prevented by using belts made of conductive materials or by applying a conductive dressing to the belt. If a dressing is used, it should be reapplied frequently or the conductivity of the belt will diminish. In either case, belts should be kept free of accumulations. It should be noted that both conductive and nonconductive belts can generate and deposit charge on the material or objects being conveyed.

9.4.2.3 An electrostatic neutralizer placed so that the points are close to the inside of the belt and a few inches away from the point where the belt leaves the pulley will also be effective in draining away most of the charge. *(See 9.2.5.3.)*

9.4.3* "V" Belts. V belts are not as susceptible to hazardous accumulations of static electric charge as are flat belts. Under certain conditions of temperature and humidity, however, a V belt can generate a significant charge. Where ignitible mixtures of gases, vapors, dusts, or fibers are present, the preferred method to limit ignition by static electric discharge is to use a direct drive instead of a belt. If a V belt is necessary for other reasons, it should be protected in accordance with 9.4.2.2.

9.4.4 Conveyor Belts.

9.4.4.1 Belts used for the transport of solid materials usually move at speeds that are slow enough to prevent an accumulation of static electric charges. However, if the material being transported is very dry or if the belt operates in a heated environment and at high speeds, significant charges can be generated.

9.4.4.2 Material that is spilled from the end of a conveyor belt into a hopper or chute can carry a significant charge. In these cases, the belt support and terminal pulleys should be electrically grounded or bonded to the hopper or chute. A passive or active neutralizer installed close to the end of the conveyor can also help reduce the charge. Conductive or antistatic belts cannot be expected to remove static from the conveyed product.

9.4.5 Pulleys and Shafts.

9.4.5.1 Metal pulleys can accumulate a charge equal and opposite to that carried by the belt that runs over them, and this charge typically will pass to the supporting shaft and then through bearings to the equipment and to ground. Where nonconductive components isolate metal parts, separately bonding or grounding those parts could be necessary.

9.4.5.2 Lubricated bearings are still sufficiently conductive to allow dissipation of a static electric charge from the shaft. However, the conductivity across bearings that operate at very high speed is not necessarily sufficient to prevent accumulation of charge when the rate of generation is also high. For this reason, shafts that rotate at high speed should be checked for accumulation of static electric charge and should be bonded or grounded by means of a sliding metal contact to the housing, if necessary.

9.4.5.3 The effective resistance across a bearing while in operation can be measured with a common ohmmeter. One of the probes should be placed on the grounded machine frame and the other probe should be allowed to rest against the rolling member. A value of approximately 10^4 ohms is to be expected. If a value greater than 10^5 ohms is found, an auxiliary grounding brush or shoe might be needed to prevent changes with time that could exceed 10^6 ohms. The grounding brush needs periodic checking and maintenance. Critical bearings without a brush should be measured periodically.

9.4.6 Maintenance of Belts and Conveyors. Belts and conveyors should be inspected frequently for slipping or jamming to lessen the chance of generation of static electricity. Drive systems should be designed to stall without slipping when operating in hazardous environments. Lubricant does not prevent the removal of static electric charges. Therefore, all bearings should be properly lubricated. However, the flow of static electricity across the film of lubricant sometimes results in pitting of bearing surfaces. A conductive grounded brush running on the shaft or pulley will prevent pitting of the bearings.

9.5 Explosives.

Most explosives and materials used as solid propellants contain enough oxidizer to sustain an explosive reaction without any outside contribution. These materials typically are sensitive to static electric discharge and can be extremely hazardous to handle if suitable precautions are not taken. In addition to the recommendations contained in this recommended practice, the following documents should be consulted for more specific information:

(1) NFPA 495, *Explosive Materials Code*
(2) NFPA 498, *Standard for Safe Havens and Interchange Lots for Vehicles Transporting Explosives*
(3) NFPA 1124, *Code for the Manufacture, Transportation, and Storage of Fireworks and Pyrotechnic Articles*

(4) NFPA 1125, *Code for the Manufacture of Model Rocket and High Power Rocket Motors*
(5) U.S. Department of Defense Standard 6055.9, *Ammunition and Explosive Safety Standards*
(6) U.S. Department of Defense Standard 4145.26M, *Contractors' Safety Manual for Ammunition and Explosives*
(7) IME Safety Library Publication No. 3, *Suggested Code of Regulations for the Manufacture, Transportation, Storage, Sale, Possession, and Use of Explosive Materials*
(8) IME Safety Library Publication No. 17, *Safety in the Transportation, Storage, Handling, and Use of Explosive Materials*

9.6 Cathode Ray Tube Video Display Terminals.

9.6.1 A static electric charge is commonly present on the face of cathode ray tube (CRT) video display terminals, particularly on color monitors and color television screens. This charge is the direct result of the CRT's high-energy electron beam "writing" the image on the inside surface of the screen. The charge accumulates on the nonconductive surface of the screen and can reach energies capable of igniting a flammable atmosphere, if discharge occurs. Such an atmosphere can be created by wiping the screen of an operating or recently operated CRT with a cloth or tissue that is wet with commercial cleaners that typically contain a flammable liquid such as isopropyl alcohol or by using spray-on aerosol cleaners that use a flammable gas propellant. The static electric charge can be removed from the screen of the CRT by accessories or by proper procedures *(see 9.6.3)*. Other video display terminals, such as liquid crystal displays, gas plasma displays, and vacuum fluorescent displays, do not present similar static electric effects. This does not mean, however, that these displays are intrinsically safe for use in hazardous locations.

9.6.2 In an industrial environment that is classified as hazardous in accordance with Article 500 of NFPA 70, *National Electrical Code®*, only engineering methods are acceptable for controlling the hazard. Due to the high voltages present, the CRT should be enclosed in a purged or pressurized enclosure, as described in NFPA 496, *Standard for Purged and Pressurized Enclosures for Electrical Equipment*. To protect against a static electric discharge from the screen's surface, the screen cannot be exposed to the surrounding environment but should be located behind a window in the enclosure.

9.6.3 In nonhazardous locations, a commercial static electric dissipating screen that overlays the CRT screen can be used to drain the static electric charge by means of a connection to ground. The ground connection from the overlay should be secure to prevent shock or an ignition-capable spark. One safe procedure for reducing the charge on the CRT screen is to wipe the screen with a water-wet cloth or tissue immediately before using any solvent-based cleaners. This action will drain off the excess charge through the operator's body. Ideally, nonflammable or low-volatility cleaning agents should be used.

Chapter 10 Referenced Publications

10.1

The following documents or portions thereof are referenced within this recommended practice and should be considered as part of its recommendations. The edition indicated for each referenced document is the current edition as of the date of the NFPA issuance of this recommended practice. Some of these documents might also be referenced in this recommended practice for specific informational purposes and, therefore, are also listed in Appendix E.

10.1.1 NFPA Publications.

National Fire Protection Association, 1 Batterymarch Park, P.O. Box 9101, Quincy, MA 02269-9101.

NFPA 30, *Flammable and Combustible Liquids Code*, 2000 edition.

NFPA 33, *Standard for Spray Application Using Flammable or Combustible Materials*, 2000 edition.

NFPA 53, *Recommended Practice on Materials, Equipment, and Systems Used in Oxygen-Enriched Atmospheres*, 1999 edition.

NFPA 69, *Standard on Explosion Prevention Systems*, 1997 edition.

NFPA 70, *National Electrical Code®*, 1999 edition.

NFPA 99, *Standard for Health Care Facilities*, 1999 edition.

NFPA 326, *Standard for the Safeguarding of Tanks and Containers for Entry, Cleaning, or Repair*, 1999 edition.

NFPA 495, *Explosives Materials Code*, 1996 edition.

NFPA 496, *Standard for Purged and Pressurized Enclosures for Electrical Equipment*, 1998 edition.

NFPA 498, *Standard for Safe Havens and Interchange Lots for Vehicles Transporting Explosives*, 1996 edition.

NFPA 1124, *Code for the Manufacture, Transportation, and Storage of Fireworks and Pyrotechnic Articles*, 1998 edition.

NFPA 1125, *Code for the Manufacture of Model Rocket and High Power Rocket Motors*, 1995 edition.

10.1.2 Other Publications.

10.1.2.1 AIChE Publication.

American Institute of Chemical Engineers, 3 Park Avenue, New York, NY 10016-5901.

Britton, L. G., "Using Material Data in Static Hazard Assessment," *Plant/Operations Progress*, April, 1992.

10.1.2.2 API Publication.

American Petroleum Institute, 1220 L Street, NW, Washington, DC 20005.

RP 2003, *Protection Against Ignitions Arising Out of Static, Lightning, and Stray Currents*, 6th edition, 1998.

10.1.2.3 ASTM Publication.

American Society for Testing and Materials, 100 Barr Harbor Drive, West Conshohocken, PA 19428-2959.

ASTM D 257, *Standard Test Methods for DC Resistance or Conductance of Insulating Materials*, 1999.

10.1.2.4 CENELEC Publication.

CENELEC, Rue de Stassart, 35, B - 1050 Brussels, Belgium.

EN 61241-2-2, *Electrical Apparatus for Use in the Presence of Combustible Dust — Part 2: Test Methods; Section 2: Method for Determining the Electrical Resistivity of Dust in Layers*, International Electrotechnical Commission, Brussels, 1993.

10.1.2.5 IME Publications.

Institute of Makers of Explosives, 1120 Nineteenth Street, NW, Suite 310, Washington, DC 20036-3605.

Safety Library Publication No. 3, *Suggested Code of Regulations for the Manufacture, Transportation, Storage, Sale, Possession, and Use of Explosive Materials.*

Safety Library Publication No. 17, *Safety in the Transportation, Storage, Handling, and Use of Explosive Materials.*

10.1.2.6 JIS Publication.

Japan Industrial Standards.JIS B 9915, *Measuring Methods for Dust Resistivity (with Parallel Electrodes)*, Japan Industrial Standards, Tokyo, 1989.

10.1.2.7 NPCA Publication.

National Paint and Coatings Association, 1500 Rhode Island Avenue, NW, Washington, DC 20005.

Generation and Control of Static Electricity, 3rd edition, 1988.

10.1.2.8 U.S. Department of Defense Publications.

U.S. Government Printing Office, Washington, DC 20402.

Standard 4145.26M, *Contractors' Safety Manual for Ammunition and Explosives.*

Standard 6055.9, *Ammunition and Explosive Safety Standards.*

10.1.3 Additional Publications.

10.1.3.1

BS 5958, *Code of Practice for Control of Undesirable Static Electricity*, Part 1, General Considerations, British Standards Institution, London, 1991.

10.1.3.2

Glor, M., *Electrostatic Hazards in Powder Handling*, Research Studies Press, Ltd., Letchworth, Hertfordshire, England, 1988.

10.1.3.3

International Safety Guide for Oil Tankers and Terminals, Fourth Edition, Witherby and Co., Ltd., London, 1996.

10.1.3.4

Pratt, T. H., *Electrostatic Ignitions of Fires and Explosions*, Burgoyne, Inc., Marietta, GA, 1997.

10.1.3.5

Walmsley, H. L., "Avoidance of Electrostatic Hazards in the Petroleum Industry," *Journal of Electrostatics*, Vol. 27, No. 1 and No. 2, Elsevier, New York, 1992.

Appendix A Explanatory Material

Appendix A is not a part of the recommendations of this NFPA document but is included for informational purposes only. This appendix contains explanatory material, numbered to correspond with the applicable text paragraphs.

A.1.1.2 See NFPA 70, *National Electrical Code*®, for additional information on grounding to prevent shock hazards.

A.1.1.3 For information on the hazards of static electricity in hospital operating rooms and similar areas, see NFPA 99, *Standard for Health Care Facilities.*

A.1.1.4 For information on the hazards of lightning, see NFPA 780, *Standard for the Installation of Lightning Protection Systems.*

A.1.1.5 For information on the hazards of stray electrical currents and induced radio frequency currents, see API RP 2003, *Protection Against Ignitions Arising Out of Static, Lightning, and Stray Currents.*

A.1.1.6 For information on the hazards of automotive and marine craft fueling, see NFPA 30A, *Code for Motor Fuel Dispensing Facilities and Repair Garages*, and NFPA 302, *Fire Protection Standard for Pleasure and Commercial Motor Craft*. For information on aircraft refueling, see NFPA 407, *Standard for Aircraft Fuel Servicing.*

A.1.1.7 For information on the hazard of static electricity in clean rooms, see NFPA 318, *Standard for the Protection of Cleanrooms.*

A.3.1.5 Capacitance. The property of a system of conductors and nonconductors that permits the storage of electrically separated charges when potential differences exist between the conductors. For a given potential difference, the higher the capacitance, the greater the amount of charge that can be stored. Quantitatively, it is the ratio of the charge on one of the conductors of a capacitor (there being an equal and opposite charge on the other conductor) to the potential difference between the conductors. The unit of capacitance is the farad. Because the farad is so large a quantity, capacitance is usually reported in microfarads (μF) or picofarads (pF).

1 farad = 10^6 microfarads = 10^{12} picofarads. *(See Table A.3.1.5 for examples.)*

Capacitance is the constant of proportionality between the charge and potential difference for a system of conductive bodies.

A.4.3.3.4 MIEs can be determined for pure materials and their mixtures. The actual ignition energy could be higher than the MIE by an order of magnitude or more if the mixture varies significantly from the most easily ignited concentration. For hazard evaluation, the MIE should be considered as the worst case.

Table A.3.1.5 Examples of Capacitance of Various Items

Item	Capacitance (pF)
Tank car	1000
Automobile	500
Person	100 to 300
Oil/solvent drum	10 to 100
Metal scoop	10 to 20
Needle electrode	1
Dust particle	10^{-7}

A.4.3.8 See Britton, *Avoiding Static Ignition Hazards in Chemical Operations,* for additional information.

A.6.6.2.2 See ANSI Z41, *American National Standard for Personal Protection — Protective Footwear.*

A.6.6.3 See Britton, *Avoiding Static Ignition Hazards in Chemical Operations*, for additional information.

A.6.6.4.2 See Britton, *Avoiding Static Ignition Hazards in Chemical Operations*, for additional information.

A.6.6.4.3 See NFPA 53, *Recommended Practice on Materials, Equipment, and Systems Used in Oxygen-Enriched Atmospheres.*

A.7.2.1 Class I flammable liquids [i.e., those that have flash points of less than 100°F (37.8°C)] can form ignitible vapor-air mixtures under most ambient conditions. Class II and Class III combustible liquids, which have flash points of 100°F (37.8°C) or greater, typically require some degree of preheating before they evolve enough vapor to form an ignitible mixture. Certain liquids of low fire hazard, such as solvent formulations consisting of mostly water, might be classed as combustible liquids, yet they can generate ignitible vapor-air mixtures in closed containers at less than 100°F (37.8°C). Similarly, certain liquids that do not have a flash point could be capable of generating an ignitible vapor-air mixture as a result of degassing or slow decomposition, especially where the vapor space is small compared with the liquid volume.

Errors in Flash Point Testing. The reported flash point of a liquid might not represent the minimum temperature at which a pool of liquid will form an ignitible atmosphere. Typical closed-cup test methods involve downward flame propagation, which is more difficult than upward propagation, and the region where the test flame is introduced is normally fuel-lean relative to the liquid surface. Also, the volume of the test apparatus is too small to allow flame propagation of certain flammable vapors such as halogenated hydrocarbons. Limitations of flash point test methods are discussed in ASTM E 502, *Standard Test Method for Selection and Use of ASTM Standards for the Determination of Flash Point of Chemicals by Closed Cup Methods.* In most cases, closed-cup flash points are lower than open-cup values.

Safety Margin for Use of Flash Point. The temperature of interest in determining the hazard is the temperature at the exposed liquid surface, not that of the bulk liquid, because vapor is in equilibrium with the liquid at the surface. However, the surface temperature is difficult to determine in many instances. While the surface temperature should be considered to the extent possible, most hazard evaluations are, by necessity and practicality, based on bulk temperature.

Therefore, a safety factor should be applied where the hazard is assessed using the flash point. For pure liquids in containers, the vapor should be considered potentially ignitible if the liquid temperature is within 4°C of the reported flash point. For mixtures whose composition is less certain, such as hydrocarbon mixtures, the safety factor should be at least 9°C. Where combinations of adverse effects are identified, the safety factors might have to be increased accordingly.

Effect of Bulk Liquid Temperature. The surface temperature of a quiescent liquid in a tank can significantly exceed the temperature of the bulk liquid, due to heat transfer from the unwet upper walls of the tank, which in some cases could be heated by sunlight to as much as 140°F. Because vapor-liquid equilibrium is established at the vapor-liquid interface, this higher surface temperature can result in a vapor concentration that is elevated compared to the concentration based on the bulk liquid temperature. This elevated concentration means that vapor in the tank could be ignitible even if the bulk liquid temperature is less than the reported flash point, which can be a significant hazard during sampling. Vapor vented from large storage tanks could be at a concentration that is only 30 percent to 50 percent of theoretical saturation, based on bulk liquid temperature. This vapor also could be a significant hazard, if tank vapor is assumed to be above the UFL.

Effect of Ambient Pressure. The vapor pressure above a liquid depends only on the temperature at the surface and the time necessary to reach equilibrium. The fraction of the total pressure exerted by a vapor determines the composition of the vapor-air mixture. Thus, when the total pressure is reduced, as could be the case at high altitudes, the vapor concentration in air increases. Because flash points are reported at a pressure of 1 atm (760 mm Hg or 101.3 kPa), an ambient pressure less than this value will lower the actual effective flash point. The flash point correction given in ASTM E 502 is expressed as follows:

$$FP_{corr}\ (°C) = C + 0.25(101.3 - A)$$

$$FP_{corr}\ (°C) = C + 0.03(760 - B)$$

$$FP_{corr}\ (°F) = F + 0.06(760 - B)$$

Where:

FP_{corr} = corrected flash point
F = observed flash point (°F)
C = observed flash point (°C)
A = ambient barometric pressure (kPa)
B = ambient barometric pressure (mm Hg)

Effect of Low Concentration of Volatiles. Small concentrations of volatile components in a liquid mixture can accumulate in the vapor space of a container over time, which can reduce the flash point to a temperature below the reported value. This effect can result from off-gassing, chemical reaction, or some other mechanism. An example is bitumen. Similarly, if a tank truck is not cleaned after delivering gasoline and a higher flash point liquid, such as kerosene or diesel fuel, is transferred to it, the residual gasoline will create an ignitible atmosphere both in the truck's tank and possibly in the receiving tank as well. Solids that contain more than 0.2 wt percent flammable solvent need to be evaluated for their potential to form ignitible vapor in containers.

Liquid Mist. If a liquid is dispersed in air in the form of a mist, it could be possible to propagate a flame through the mist, even at temperatures below the liquid's flash point. In such cases, the mist droplets behave like dust particles. The flash point of the liquid is irrelevant when determining combustibility of mists. Even at very low liquid temperatures, frozen liquid droplets can burn in this manner. Ease of ignition and rate of combustion both increase as the droplet size of the mist decreases. Depending on the volatility of the liquid, droplets with a diameter less than 20 to 40 μm typically vaporize and ignite ahead of a flame front, and their overall combustion behavior is similar to that of a vapor. Because mists are usually produced by some form of shear process and these same processes also generate static electricity, it is good practice to avoid splashing and other procedures that generate mist inside equipment.

A.7.2.2 Operating a process at less than the LFL is often safer than operating above the UFL, particularly for tanks and other large vessels. Even if liquid in a tank rapidly generates sufficient vapor for operation above the UFL, flammable mixtures can still be present at tank openings, such as sampling ports, and the flammable range could be traversed inside the tank during start-up or some other operating condition. Often, the atmosphere in the vessel can be inerted, as described in NFPA 69, *Standard on Explosion Prevention Systems.* This technique reduces the oxygen concentration below that required to sustain combustion. Inerting might not be effective near tank openings, especially in cases where additions of solids can entrain air. Also, for storage tanks, the inert gas supply should be capable of compensating for changes in temperature or in-breathing of air during tank emptying.

Vapor Pressure. The vapor pressure can be used together with a measured LFL to estimate the flash point. Usually, the calculated flash point is less than the measured value due to limitations in the flash point test technique. Conversely, only an approximate estimate of the LFL can be made from the flash point. The vapor pressure can be used to replace the "concentration" axis in Figure 7.2.3 with the corresponding temperatures required to generate the concentrations shown in the figure. This method allows one to determine the equilibrium liquid temperature at which vapor ignition is most probable, corresponding to generation of the vapor-air mixture having the lowest ignition energy. For many liquids, this point is approximately halfway between the LFLs and UFLs. For example, benzene generates its lowest MIE vapor-air mixture at about 7°C (4.8 percent benzene vapor in air) and toluene at about 2.6°C (4.1 percent toluene vapor in air). Therefore, for operations conducted at room temperatures, toluene is more prone to ignition from a low energy static electric discharge than is benzene. In closed containers at equilibrium, benzene becomes too rich to burn (the concentration of vapor exceeds its UFL of 8 percent) at temperatures above about 16°C. Conversely, at about 7°C, benzene is more easily ignited than toluene, because the latter will generate a vapor composition not far above its LFL. Some lowest MIE compositions are given in Appendix B.

High Vapor Pressure Liquids. High vapor pressure liquids are defined in API RP 2003, *Protection Against Ignitions Arising Out of Static, Lightning, and Stray Currents,* as having a Reid vapor pressure greater than 4.5 psia (31 kPa absolute). At normal handling temperatures, rapid evaporation of these liquids minimizes the duration of a flammable atmosphere above the liquid during loading and the UFL is soon exceeded. However, if there is no initial heel in the tank and the tank is not inerted, the flammable range will be traversed prior to attaining vapor equilibrium. The duration of the ignitible atmosphere is minimal for liquefied gases such as propylene, but could be considerable for certain petroleum distillate fuels. Inerting might be considered when loading high vapor pressure nonconductive liquids to tanks containing air with no liquid heel.

Intermediate Vapor Pressure Liquids. Intermediate vapor pressure liquids are defined in API RP 2003 as having a Reid vapor pressure less than 4.5 psia (31 kPa absolute) and a closed-cup flash point below 100°F (37.8°C). They are most likely to generate ignitible mixtures in vessels at ordinary temperatures. Although graphical methods have been proposed to estimate whether liquids are likely to generate ignitible atmospheres at various temperatures, based on their Reid vapor pressures, such graphs were originally derived for petroleum fuel mixtures and do not always apply to other flammable liquids.

Low Vapor Pressure Liquids. Low vapor pressure liquids are Class II and Class III combustible liquids [i.e., those with closed-cup flash points above 100°F (37.8°C)] and will generate ignitible atmospheres only if handled at elevated temperature, suspended as a mist, or subject to slow vapor evolution. However, static electricity generated during handling could ignite vapors present from previous operations.

A.7.2.4 Preventing an ignitible atmosphere can be accomplished using any of the methods described in NFPA 69, *Standard on Explosion Prevention Systems.* Of these methods, the most common is to add a suitable inert gas, such as nitrogen, so that the resulting concentration of oxygen is not sufficient to support a flame. A safety factor is usually applied. For most flammable gases and vapors, inerting typically requires reducing the oxygen concentration to about 5 percent by volume.

A.7.3.1 The system of two layers having opposite net charge is referred to as an *electrical double layer.* For conductive liquids such as water, the diffuse layer is only a few molecules thick. But for nonconductive liquids such as light petroleum distillates, the layer could be many millimeters thick. Ionic species present in liquids undergo charge separation at interfaces in a manner that results in one sign of charge being more strongly bound at the contacted surface than the other. This results in a bound layer of liquid close to the contact surface. Farther away from the contact surface is a "diffuse layer" that has a charge of opposite polarity. Any process that shears the liquid, such as pipe flow, moves the diffuse layer downstream with the bulk of the liquid, while the bound layer charge relaxes to the wall, provided the wall is grounded. This process, in effect, allows the diffuse layer to result in charge accumulation in the liquid. When small droplets having dimensions smaller than the thickness of the double layer are formed, the formation of the droplet can pinch off a net charge. This can result in charged sprays and mists for both conductive and nonconductive liquids. The larger the area of the interface, the greater is the rate of charging. Examples of such processes are fine filtration, agitation of two-phase systems such as water and oil, and suspension of powder in liquid.

Streaming Current. The charges that are carried in the bulk of the flowing liquid create a current referred to as a *streaming current.* Although the charge is separated at the wall, flow mixes the charge into the bulk of the liquid and a charge density measured in coulombs per cubic meter can be achieved. Streaming current, in coulombs per second or amperes, is equal to the volume flow rate, in cubic meters per second, multiplied by the liquid charge density, in coulombs per cubic meter.

A.7.3.2 Charge relaxation is characterized by a time constant, which is the time required for a charge to dissipate to e^{-1} (approximately 37 percent) of its original value, assuming that charge relaxation follows exponential decay. This time constant is determined from the following equation:

$$\tau = \frac{\epsilon\epsilon_0}{\kappa}$$

Where:

τ = time constant
ϵ = dielectric constant for the liquid
ϵ_0 = 8.854×10^{-12}, permittivity of free space (farads per meter)
κ = liquid conductivity (picosiemens per meter)

Overall, the time constant provides some indication of a liquid's potential for accumulating a static electric charge.

Exponential, or "ohmic," decay has been experimentally confirmed for hydrocarbon liquids having conductivities of 1 pS/m or greater and is described by the following equation:

$$Q_t = Q_0 e^{-tK/\epsilon\epsilon_0}$$

Where:

Q_t = charge density (coulombs per cubic meter)
Q_0 = initial charge density (coulombs per cubic meter)
e = 2.718, base of natural logarithms
t = time (seconds)
κ = liquid conductivity (picosiemens per meter)
ϵ = liquid dielectric constant
ϵ_0 = 8.854×10^{-12}, permittivity of free space (farads per meter)

According to W. M. Bustin (Bustin, W. M., et al, 1964), the rate at which charge is lost depends on the conductivity of the liquid. The lower the conductivity, the slower the relaxation. Liquids with conductivity of less than 1 pS/m relax differently when they are highly charged. The usual relationship described by Ohm's law does not apply. Instead, for nonviscous liquids (i.e., less than 30×10^{-6} m²/sec), relaxation precedes hyperbolic decay. The Bustin theory of charge relaxation has been experimentally confirmed for low-conductivity hydrocarbon liquids, both in small-scale laboratory experiments and in full-scale tests and is described by the following equation:

$$Q_t = \frac{Q_0}{(1 + \mu Q_0 t/\epsilon\epsilon_0)}$$

Where:

Q_t = charge density (coulombs per cubic meter)
Q_0 = initial charge density (coulombs per cubic meter)
μ = ion mobility, about 1×10^{-8} m²/V-sec for charged distillate oil (square meters per volt-second)
t = time (seconds)
$\epsilon\epsilon_0$ = electrical permittivity (farads per meter)

The Bustin theory of charge relaxation depends only on the initial charge density, Q_0, and ion mobility, μ. The conductivity of the uncharged liquid is not a factor. In addition, Bustin charge decay theory is not very sensitive to initial charge density when the initial charge density is greater than about 100 microcoulombs per cubic meter.

A.7.3.3 The mechanism of charge generation is highly complex. For flow of liquid in pipes, the charging current depends on the liquid's electrical conductivity and dielectric constant and its viscosity and flow characteristics, which involve factors such as flow velocity, pipe diameter, and surface roughness. For equal flow characteristics, electrical conductivity is the dominant factor. This is most pronounced for low conductivity liquids, due to trace contaminants. Trace contaminants have negligible effect on the liquid's dielectric constant and viscosity, but have a dominant effect on conductivity. Conductive liquids are much less affected by trace contaminants. In many systems, such as long pipes, the charge density reaches a steady state at which the rate of charge generation is balanced by the rate of charge relaxation back to ground.

Classification of Liquids Based on Conductivity. The conductivity of most flammable and combustible liquids varies from about 10^{-2} pS/m to 10^{10} pS/m (i.e., by 12 orders of magnitude). Dielectric constants usually range from 2 to 40 — the higher values being generally exhibited by polar liquids, which also exhibit higher conductivity. Because relaxation behavior is primarily governed by conductivity, conductivity can be used to classify liquids relative to their potential for charge accumulation as nonconductive, semiconductive, and conductive. Because conductivity is so sensitive to purity and temperature, class demarcations can be given only to within an order of magnitude. It should be kept in mind that conductivity under actual conditions could be less than what is measured in the laboratory. *(See Appendix B for conductivity values and relaxation times for some typical liquids.)*

Nonconductive Liquids. Liquids that have relaxation time constants greater than 0.36 seconds (equivalent to a conductivity of less than 50 pS/m for typical hydrocarbons having dielectric constants of about 2) are considered nonconductive. Examples include purified toluene and most low-sulfur diesel oils. They are highly susceptible to variation due to trace contamination. Corona and brush discharges, rather than spark discharges, are observed from charged nonconductive liquids. Because only partial discharge is possible, induction charging from highly charged plastic containers is not a significant hazard. Nonconductive liquids are most prone to accumulate charge in grounded metallic containers. For the purposes of this recommended practice, the criterion of 50 pS/m is not iron-clad; the dielectric constant also plays a role. For example, the dielectric constant of ethyl ether is 4.6 versus 2.3 for benzene. Therefore, the relaxation time constant for ethyl ether at a conductivity of 100 pS/m is approximately the same as that for benzene at a conductivity of 50 pS/m. It is the relaxation time constant, not the conductivity alone, that determines the rate of loss of charge.

Semiconductive Liquids. Liquids that have relaxation time constants ranging from 0.36 sec down to 0.002 sec (equivalent to a conductivity range between 50 and 10^4 pS/m for typical hydrocarbons having dielectric constants of about 2) are considered conductive. Examples include crude oil and butyl acetate. They tend not to accumulate charge, except where charging rates are extremely high or where they are effectively isolated from ground, such as when flowing through a rubber hose or end-of-line "polishing" filters. Spark discharges are possible from the more conductive of these liquids.

Conductive Liquids. Liquids that have relaxation time

constants less than 0.002 sec (equivalent to a conductivity greater than 10^4 pS/m for typical hydrocarbons having dielectric constants of about 2) are considered highly conductive. These liquids tend not to accumulate charge except where handling conditions isolate them from ground. Such conditions include complete isolation in the form of a droplet suspended in air, partial isolation by suspension in another liquid, and containment in a plastic or other highly resistive container. Conductive liquids are most prone to induction charging by plastic containers and are sufficiently conductive to lose much of the induced charge in the form of a spark.

Changes in Conductivity Caused by Solidification. Liquids can undergo a sudden and dramatic decrease in conductivity at their freezing points, which in some cases can cause unexpected static electricity hazards. For example, the conductivity of biphenyl decreases by 4 orders of magnitude between the liquid phase (above 69°C) and the solid phase. A static electric ignition was reported when biphenyl at 120°C was loaded into a tank containing a thick layer of solid biphenyl from a previous operation.

Normally, hot biphenyl is conductive enough to rapidly dissipate charge when loaded into a grounded metal tank. But due to the presence of the thick, insulating layer of solid biphenyl, charge was able to accumulate and a brush discharge occurred from the liquid surface to the fill pipe.

A.7.4.1 See Britton, *Avoiding Static Ignition Hazards in Chemical Operations*, for additional information.

Various theoretical and empirical models have been derived expressing either charge density or charging current in terms of flow characteristics, such as pipe diameter and flow velocity. Liquid dielectric and physical properties appear in more complex models. For turbulent flow of a nonconductive liquid through a pipe under conditions where the residence time is long compared with the relaxation time, the charging current, I_s, can be expressed by the following:

$$I_s = N\,(v^x)(d^y)$$

Where:

I_s = charging current
N = constant (characterizing flow conditions, see text)
x = approximately 2
y = approximately 2
v = flow velocity (m/sec)
d = diameter of conduit (m)

Various values for the constants can be found in the literature. For I_s, in amperes, the constant N has been reported to range from 3.75×10^{-6} C-sec/m^4 to 25×10^{-6} C-sec/m^4. The low value corresponds to turbulent flow through a long, smooth pipe, while the high value corresponds to turbulent flow through spiral-wound composite hose. An order of magnitude value for N is 1×10^{-5} C-sec/m^4. While more recent studies suggest that y is equal to 1, it has been most commonly reported that both x and y are approximately equal to 2, so that the charging current is roughly proportional to the square of ($v \times d$). An important outcome of the studies is that ($v \times d$) can be used as a means of characterizing the charging current in pipe flow and as a basis for setting flow limits when filling tanks. *(See Sections 7.5 and 7.6.)*

A.7.4.2 All-plastic nonconductive pipe is not recommended for handling nonconductive or semiconductive liquids, except where it can be shown that the advantages outweigh any risks associated with external static electric discharge or leaks from pinholes or where tests have demonstrated that the phenomena will not occur. Grounded, plastic-lined metal pipe does not pose either of these risks directly, but tolerance for liner pinholes should be considered. For example, if the liquid is corrosive to metal piping, gradual loss of metal because of pinholes could lead to unacceptable product contamination and eventual loss of containment. Conversely, minor pinhole damage might be acceptable, if the liner is intended only to minimize product discoloration caused by rust and scale.

Where nonconductive and partly conductive liquids need to be transferred through plastic piping systems, mitigating strategies include the following:

(1) Reducing the rate of charging by decreasing flow velocity
(2) Eliminating or relocating microfilters further upstream
(3) Reducing wall resistivity, possibly to less than 10^8 ohm-m
(4) Increasing the breakdown strength of the pipe wall by increasing the thickness or changing the material of construction
(5) Incorporating an external grounded conductive layer on the piping

Combinations of these strategies can be considered. For example, in many cases, the presence of an external conductive layer on a plastic pipe will not by itself eliminate puncturing of the internal plastic wall, and, if the layer does not provide containment, it will not prevent external leakage.

A.7.4.3 For all-metal conductive hoses, the resistance to ground from any point should normally be 10 ohms or less, except where insulating flanges are used to avoid sparks from stray current. For conductive hoses that contain a continuous bonding element, such as wire or braid, the resistance to ground from any metal connector should normally be 1000 ohms per meter or less, with the same exception being applicable. Resistance to ground through semiconductive hoses whose current-limiting design eliminates a low-resistance bonding element and resistance to ground through insulating flanges should be between 10^3 ohms/m and 10^5 ohms/m. In either case, the total resistance to ground from a metal hose connector should not exceed 10^6 ohms.

While a resistance to ground of less than 10^6 ohms will prevent accumulation of static electric charge in most cases, if periodic testing reveals a significant increase in the as-installed resistance, this could be the result of corrosion or other damage that could lead to sudden loss of continuity. The hose or insulating flange, or both, should be inspected to determine the need for replacement. Where conductive hoses have double spirals, one for bonding and the other for mechanical strength, continuity between the end connectors only confirms the continuity of one spiral. A fire was reported during draining of toluene from a tank vehicle through such a hose. It was found that the inner spiral was not only broken but was not designed to be bonded to the end connectors. For handling nonconductive liquids, one option is to use a hose with a semiconductive or conductive liner, so that a broken inner spiral cannot become isolated from ground and form a spark gap. Ideally, the inner spiral should be separately bonded to the end connectors.

It is especially important to ensure continuity with end connectors (or nozzles) where a hose is used in an ignitible atmosphere. In general, it is safer to use a properly designed fixed fill system, such as a dip pipe arrangement, for filling tank vehicles, rather than to use a hose.

Utility Hoses. Where used in flammable atmospheres, such as inside tanks, utility hoses should be conductive or semiconductive. In particular, all metal connectors and nozzles should be grounded. Ungrounded hose connectors on nonconductive hose can become charged by a variety of means, such as by inserting a nitrogen hose into a tank containing charged liquid or mist, by rubbing, or by steam impact. While clean and dry gases do not generate charge, a nonconductive hose will become highly charged by the flow of steam.

A.7.5.2.1(c) See API 2003, *Protection Against Ignitions Arising Out of Static, Lightning, and Stray Currents*, for further information.

A.7.7 See API Publication 2219, *Safe Operation of Vacuum Trucks in the Petroleum Service*, for general recommendations.

A.7.10.2 See NFPA 69, *Standard on Explosion Prevention Systems*, for additional information.

A.7.13.7 See NFPA 326, *Standard for the Safeguarding of Tanks and Containers for Entry, Cleaning, or Repair.*

A.7.14 If used for flammable liquid spills, which could involve a second phase such as spill control granules or debris, wet-dry vacuum cleaners pose a number of problems, including the following:

(1) Generation of static electricity
(2) Electrical classification of powered equipment
(3) Chemical compatibility
(4) Industrial hygiene (relative to the exhaust from the vacuum cleaner)

Commercial machines for Class I, Group D, and Class II, Groups E, F, and G atmospheres are typically air-operated via a venturi, so they contain no electrical power. Air supply and liquid recovery hoses should be conductive and constructed of semiconductive fabric. Filters are also semiconductive or conductive. The design is such that all parts are continuously bonded and grounded. Normally ground continuity at prescribed checkpoints is established before each use. Floats or similar mechanisms are employed to shut off suction once the recovery tank has reached capacity level. Additional precautions might be needed to avoid overflow via siphoning (if the recovery hose is completely submerged in liquid) or when defoaming agents are not used. For flammable liquid spills in particular, measures should be taken, including training and personnel grounding, to ensure personnel are not a source of ignition.

A.8.2.4 The MIE of a dust cloud is determined using a sample that is representative of the dust in a process. The equipment and procedures used over the years have been many and varied. Therefore, MIE data might not be comparable from one data set to the next. Furthermore, the conditions under which laboratory data are acquired can be different than that of the process being examined. For these reasons, comparisons of MIE data are sometimes qualitative rather than quantitative. Nevertheless, comparisons can be quite useful.

The MIE of a dust decreases with decreasing particle size and with increasing temperature. The MIE could increase with increasing moisture content of the dust. The MIE of a dust varies little with the humidity of the supporting atmosphere, excluding problems with hygroscopic dusts. The factors affecting MIE should be considered in a hazards analysis of a process.

A.8.7 Some flexible hoses can be cut to length and put into service by simply slipping them over a pipe with a hose clamp. It is important that the spiral wires be in good metal-to-metal contact with the pipes in order to maintain a proper ground of the spiral. This contact can be done by stripping the spiral and bending it under the hose next to the pipe and under the clamp. In cases where hoses with metal spirals are connected to plastic pipes, the spirals should be independently grounded.

The reason for discouraging the use of hose with more than one spiral is that, if one of the spiral wires is broken in such a way that it is disconnected from ground, it can become a source of spark gap ignition.

A.8.10 See Britton, *Avoiding Static Ignition Hazards in Chemical Operations*, for additional information.

A.8.11 See Britton, *Avoiding Static Ignition Hazards in Chemical Operations*, for additional information.

A.9.3.2 See NFPA 33, *Standard for Spray Application Using Flammable or Combustible Materials*, for further information.

A.9.4.2.1 Investigations into the static electric hazard in grain elevators have shown that danger exists when the voltage reaches 30,000 volts on the belt. These studies also show that low relative humidity is an important factor in that it allows voltages to increase very rapidly at temperatures below freezing.

A.9.4.2.2 The surface resistivity of a grain conveyor belt affects its ability to accumulate a charge. Tests have shown that belts with a resistance of 10^6 ohms to 10^8 ohms are conductive enough to prevent significant accumulation of charge. According to the draft standard *Safety of Machinery — Electrotechnical Aspects*, published by the European Committee for Electrotechnical Standardization (CENELEC), a belt is considered to be sufficiently conductive if the surface resistance is less than 3×10^8 ohms.

A.9.4.3 According to the draft standard *Safety of Machinery — Electrotechnical Aspects*, published by the European Committee for Electrotechnical Standardization (CENELEC), V belts and similar transmission belts are considered to be sufficiently conductive if the following criterion is met:

$$R \times B < 10^5 \text{ ohm-m}$$

Where:

R = resistance measured at the inner side of the mounted transmission belt between an electrode halfway between the two pulleys and ground

B = width of a flat belt or twice the depth of the side face of a V belt

For belts constructed of different materials, the belt is considered to be sufficiently conductive if the resistance across the belt does not exceed 10^9 ohms, measured at 23°C and 50 percent relative humidity.

Appendix B Physical Characteristics of Materials

This appendix is not a part of the recommendations of this NFPA document but is included for informational purposes only.

B.1 Combustibility Parameters of Gases and Vapors.

Table B.1 lists typical gases and vapors and the lowest value of their minimum ignition energies in millijoules; the stoichiometric composition, expressed as percent by volume in air (or other oxidant); and the flammable limits, also expressed as percent by volume in air (or other oxidant). (The data are taken from Britton, L. G., "Using Material Data in Static Hazard Assessment," *Plant/Operations Progress*, American Institute of Chemical Engineers, New York, NY, Vol. 11, No. 2, April, 1992, pp. 56–70.)

B.2 Static Electric Characteristics of Liquids.

Table B.2 lists typical flammable and combustible liquids and their conductivities, dielectric constants, and relaxation time constants. (The data are taken from Britton, L. G., "Using Material Data in Static Hazard Assessment," *Plant/Operations Progress*, American Institute of Chemical Engineers, New York, NY, Vol. 11, No. 2, April, 1992, pp. 56–70.)

Appendix C Recommended Means for Providing Bonding and Grounding

This appendix is not a part of the recommendations of this NFPA document but is included for informational purposes only.

C.1 Diagrams.

Figures C.1(a) through C.1(k) are reprinted from *Generation and Control of Static Electricity* by permission of the National Paint and Coatings Association, 1500 Rhode Island Avenue, NW, Washington, DC 20005-5597. Refer to this publication for additional diagrams.

Appendix D Glossary of Terms

This appendix is not a part of the recommendations of this NFPA document but is included for informational purposes only.

D.1 General.

This glossary contains terms and definitions that are not included in Chapter 3. The terms are presented here to assist the user.

D.1.1 Air Ionizer. A device for producing ions in air. Ions from an air ionizer can be attracted to stationary charges on nonconductive items (or items insulated from ground) to eliminate the charge imbalance. Other air ionizers are used to inject ions of a single polarity into an enclosure.

Table B.1 Combustibility Parameters of Gases and Vapors

Gas or Vapor (in air at standard temperature and pressure, unless otherwise noted)	Lowest Minimum Ignition Energy (mJ)	Stoichiometric Mixture (% by volume)	Flammable Limits (% by volume)
acetaldehyde	0.37	7.73	4.0–57.0
acetone	1.15 @ 4.5%	4.97	2.6–12.8
acetylene	0.017 @ 8.5%	7.72	2.5–100
acetylene in oxygen	0.0002 @ 40%	—	—
acrolein	0.13	5.64	2.8–31
acrylonitrile	0.16 @ 9.0%	5.29	3.0–17.0
allyl chloride	0.77	—	2.9–11.1
ammonia	680	21.8	15–28
benzene	0.2 @ 4.7%	2.72	1.3–8.0
1,3–butadiene	0.13 @ 5.2%	3.67	2.0–12
butane	0.25 @ 4.7%	3.12	1.6–8.4
n–butyl chloride	1.24	3.37	1.8–10.1
carbon disulfide	0.009 @ 7.8%	6.53	1.0–50.0
cyclohexane	0.22 @ 3.8%	2.27	1.3–7.8
cyclopentadiene	0.67	—	—
cyclopentane	0.54	2.71	1.5 – nd
cyclopropane	0.17 @ 6.3%	4.44	2.4–10.4
dichlorosilane	0.015	17.36	4.7–96
diethyl ether	0.19 @ 5.1%	3.37	1.85–36.5
diethyl ether in oxygen	0.0012	—	2.0–82
diisobutylene	0.96	—	1.1–6.0
diisopropyl ether	1.14	—	1.4-7.9
dimethoxymethane	0.42	—	2.2–13.8
2,2–dimethylbutane	0.25 @ 3.4%	2.16	1.2–7.0
dimethyl ether	0.29	—	3.4–27.0
2,2–dimethyl propane	1.57	—	1.4–7.5
dimethyl sulfide	0.48	—	2.2–19.7
di–t–butyl peroxide	0.41	—	—
ethane	0.24 @ 6.5%	5.64	3.0–12.5
ethane in oxygen	0.0019	—	3.0–66
ethyl acetate	0.46 @ 5.2%	4.02	2.0–11.5
ethylamine	2.4	5.28	3.5–14.0
ethylene	0.07	—	2.7–36.0
ethylene in oxygen	0.0009	—	3.0–80
ethyleneimine	0.48	—	3.6–46
ethylene oxide	0.065 @ 10.8%	7.72	3.0–100
furan	0.22	4.44	2.3–14.3
heptane	0.24 @ 3.4%	1.87	1.05–6.7
hexane	0.24 @ 3.8%	2.16	1.1–7.5
hydrogen	0.016 @ 28%	29.5	4.0–75
hydrogen in oxygen	0.0012	—	4.0–94
hydrogen sulfide	0.068	—	4.0–44
isooctane	1.35	—	0.95–6.0
isopentane	0.21 @ 3.8%	—	1.4–7.6
isopropyl alcohol	0.65	4.44	2.0–12.7
isopropyl chloride	1.08	—	2.8–10.7
isopropylamine	2.0	—	—
isopropyl mercaptan	0.53	—	—
methane	0.21 @ 8.5%	9.47	5.0–15.0
methane in oxygen	0.0027	—	5.1–61
methanol	0.14 @ 14.7%	12.24	6.0–36.0
methylacetylene	0.11 @ 6.5%	—	1.7–nd

Table B.1 Continued

Gas or Vapor (in air at standard temperature and pressure, unless otherwise noted)	Lowest Minimum Ignition Energy (mJ)	Stoichiometric Mixture (% by volume)	Flammable Limits (% by volume)
methylene chloride	>1000	—	14–22
methyl butane	<0.25	—	1.4–7.6
methyl cyclohexane	0.27 @ 3.5%	—	1.2–6.7
methyl ethyl ketone	0.53 @ 5.3%	3.66	2.0–12.0
methyl formate	0.4	—	4.5–23
n–pentane	0.28 @ 3.3%	2.55	1.5–7.8
2–pentane	0.18 @ 4.4%	—	—
propane	0.25 @ 5.2%	4.02	2.1–9.5
propane in oxygen	0.0021	—	—
propionaldehyde	0.32	—	2.6–17
n–propyl chloride	1.08	—	2.6–11.1
propylene	0.28	—	2.0–11.0
propylene oxide	0.13 @ 7.5%	—	2.3–36.0
tetrahydrofuran	0.54	—	2.0–11.8
tetrahydropyran	0.22 @ 4.7%	—	—
thiophene	0.39	—	—
toluene	0.24 @ 4.1%	2.27	1.27–7.0
trichlorosilane	0.017	—	7.0–83
triethylamine	0.75	2.10	—
2,2,3–trimethyl butane	1.0	—	—
vinyl acetate	0.7	4.45	2.6–13.4
vinyl acetylene	0.082	—	1.7–100
xylene	0.2	1.96	1.0–7.0

nd = not determined.

Table B.2 Static Electric Characteristics of Liquids

Liquid	Conductivity (pS/m)	Dielectric Constant	Relaxation Time Constant (sec)
Conductive Liquids: Conductivity >10^4 pS/m			
acetaldehyde (15°C)	1.7×10^{8}	21.1	1.1×10^{-6}
acetamide	8.8×107	59	5.9×10^{-6}
acetic acid (0°C)	5×10^{5}	6.15	1.1×10^{-4}
acetic acid (25°C)	1.12×10^{6}	6.15	4.9×10^{-5}
acetic anhydride (25°C)	4.8×10^{7}	n/a	n/a
acetone (25°C)	6×10^{6}	20.7	3×10^{-5}
acetonitrile (20°C)	7×10^{8}	37.5	5×10^{-7}
acetophenone (25°C)	3.1×10^{5}	17.39	5.0×10^{-4}
acetyl bromide (25°C)	2.4×10^{8}	n/a	n/a
acetyl chloride (25°C)	4×10^{7}	n/a	n/a
acrolein	1.55×10^{7}	n/a	n/a
acrylonitrile	7×10^{5}	38	4.8×10^{-4}
allyl alcohol (25°C)	7×10^{8}	n/a	n/a
aminoethyl-ethanolamine	$> 1 \times 10^{6}$	n/a	n/a
n-aminoethyl piperazine	2.4×10^{5}	n/a	n/a
ammonia (−79°C)	1.3×10^{7}	n/a	n/a
iso-amyl alcohol	1.4×10^{5}	14.7	9.3×10^{-4}

(continues)

Table B.2 Continued

Liquid	Conductivity (pS/m)	Dielectric Constant	Relaxation Time Constant (sec)
aniline (25°C)	2.4×10^{6}	6.89	2.5×10^{-5}
anthracene (25°C)	3×10^{4}	n/a	n/a
arsenic tribromide (25°C)	1.5×10^{8}	n/a	n/a
arsenic trichloride (25°C)	1.2×10^{8}	n/a	n/a
benzaldehyde (25°C)	1.5×10^{7}	n/a	n/a
benzoic acid (125°C)	3×10^{5}	n/a	n/a
benzonitrile (25°C)	5×10^{6}	25.2	4.5×10^{-5}
benzyl alcohol (25°C)	1.8×10^{8}	n/a	n/a
benzylamine (25°C)	$< 1.7 \times 10^{6}$	n/a	n/a
benzyl benzoate (25°C)	$< 1 \times 10^{5}$	n/a	n/a
benzyl cyanide	$< 5 \times 10^{6}$	18.7	$> 3.3 \times 10^{5}$
biphenyl (liquid above 120°C)	$> 1 \times 10^{4}$	n/a	n/a
bromoform (25°C)	$< 2 \times 10^{6}$	4.39	$> 1.9 \times 10^{-5}$
iso-butyl alcohol	9.12×10^{5}	17.51	1.7×10^{-4}
sec-butyl alcohol	$< 1 \times 10^{7}$	16.56	$> 1.5 \times 10^{-5}$
t-butyl alcohol	2.66×10^{6}	12.47	4.2×10^{-5}
iso-butyl chloride	1×10^{4}	6.49	5.7×10^{-3}
sec-butyl chloride	1×10^{4}	7.09	6.3×10^{-3}
capronitrile (25°C)	3.7×10^{8}	n/a	n/a
m-chloroaniline (25°C)	5×10^{6}	n/a	n/a
chlorohydrin (25°C)	5×10^{7}	n/a	n/a
m-cresol	1.397×10^{6}	11.8	7.5×10^{-5}
o-cresol	1.27×10^{5}	11.5	8.0×10^{-4}
p-cresol	1.378×10^{6}	9.91	6.4×10^{-5}
cyanogen	$< 7 \times 10^{5}$	n/a	n/a
cyclohexanone	5×10^{5}	n/a	n/a
cymene (25°C)	$< 2 \times 10^{6}$	n/a	n/a
dibutyl-o-phthalate	1.8×10^{5}	6.436	3.2×10^{-4}
dichloroacetic acid (25°C)	7×10^{6}	n/a	n/a
cis-dichloroethylene	8.5×10^{5}	9.20	9.6×10^{-5}
dichlorohydrin (25°C)	1.2×10^{9}	n/a	n/a
diethylamine (-33.5°C)	2.2×10^{5}	n/a	n/a
diethyl carbonate (25°C)	1.7×10^{6}	2.82	1.5×10^{-5}
diethylene glycol	5.86×10^{7}	31.69	4.8×10^{-6}
diethylenetriamine	$> 1 \times 10^{6}$	n/a	n/a
diethyl oxalate (25°C)	7.6×10^{7}	n/a	n/a
diethyl sulfate (25°C)	2.6×10^{7}	n/a	n/a
dimethyl acetamide	1.1×10^{7}	n/a	n/a
dimethyl formamide	6×10^{6}	36.71	5.4×10^{-5}
dimethyl sulfate (0°C)	1.6×10^{7}	n/a	n/a
dimethyl sulfoxide	2×10^{5}	46.68	2.1×10^{-3}
diphenyl oxide	$< 1.7 \times 10^{6}$	4.22	$> 2.2 \times 10^{-5}$
epichlorohydrin (25°C)	3.4×10^{6}	22.6	5.9×10^{-5}
ethanolamine	1.1×10^{9}	37.72	3.0×10^{-7}
ethylacetate (25°C)	4.6×10^{4}	6.02	1.2×10^{-3}
ethyl acetoacetate (25°C)	4×10^{6}	15.7	3.5×10^{-5}
ethyl acrylate	3.35×10^{5}	n/a	n/a
ethyl alcohol (25°C)	1.35×10^{5}	24.55	1.6×10^{-3}
ethylamine (0°C)	4×10^{7}	n/a	n/a
ethyl benzoate (25°C)	$< 1 \times 10^{5}$	6.02	$> 5.3 \times 10^{-4}$
ethyl bromide (25°C)	$< 2 \times 10^{6}$	9.39	$> 4.2 \times 10^{-5}$
ethyl chloride	$< 3 \times 10^{5}$	9.45	$> 2.8 \times 10^{-4}$
ethyl cyanoacetate	6.9×10^{7}	26.7	3.4×10^{-6}

Table B.2 Continued

Liquid	Conductivity (pS/m)	Dielectric Constant	Relaxation Time Constant (sec)
ethylene carbonate	$< 1 \times 10^7$	89.6	$> 7.9 \times 10^{-5}$
ethylenediamine	9×10^6	12.9	1.3×10^{-5}
ethylene dibromide (25°C)	$< 2 \times 10^4$	4.78	$> 2.1 \times 10^{-3}$
ethylene glycol	1.16×10^8	37.7	2.9×10^{-6}
ethylene glycol monobutyl ether	4.32×10^7	9.30	1.9×10^{-6}
ethylene glycol monoethyl ether	9.3×10^6	29.6	2.8×10^{-5}
ethylene glycol monomethyl ether	1.09×10^8	16.93	1.4×10^{-6}
ethyleneimene	8×10^8	18.3	2.0×10^{-7}
ethylene oxide	4×10^6	12.7	2.8×10^{-5}
ethyl formate	1.45×10^5	7.16	4.4×10^{-4}
ethylidene chloride	2.0×10^5	10.0	4.4×10^{-4}
ethyl isothiocyanate (25°C)	1.26×10^7	n/a	n/a
ethyl lactate	1.0×10^8	13.1	1.2×10^{-6}
ethyl nitrate (25°C)	5.3×10^7	n/a	n/a
ethyl oxalate	7.12×10^7	n/a	n/a
ethyl propionate	8.33×10^{10}	5.65	6×10^{-10}
ethyl thiocyanate (25°C)	1.2×10^8	n/a	n/a
eugenol (25°C)	$< 1.7 \times 10^6$	n/a	n/a
formamide (25°C)	4×10^8	111.0	2×10^{-6}
formic acid (25°C)	6.4×10^9	58.5	8.1×10^{-8}
furfural (25°C)	1.5×10^8	n/a	n/a
glycerol (25°C)	6.4×10^6	42.5	5.9×10^{-5}
guaiacol (25°C)	2.8×10^7	n/a	n/a
hydrogen bromide (-80°C)	8×10^5	n/a	n/a
hydrogen chloride (-96°C)	1×10^6	n/a	n/a
hydrogen cyanide (0°C)	3.3×10^8	n/a	n/a
hydrogen iodide (at boiling point)	2×10^7	n/a	n/a
iodine (110°C)	1.3×10^4	n/a	n/a
mercury (0°C)	1.063×10^{18}	n/a	n/a
methoxy triglycol	$> 1 \times 10^6$	n/a	n/a
methyl acetamide	2×10^7	191.3	8.5×10^{-5}
methyl acetate (25°C)	3.4×10^8	6.68	1.7×10^{-7}
methyl alcohol (18°C)	4.4×10^7	32.70	6.6×10^{-6}
methyl cyanoacetate	4.49×10^7	29.30	5.8×10^{-6}
methyl ethyl ketone (25°C)	1×10^7	18.51	1.6×10^{-5}
methyl formamide	8×10^7	182.4	2.0×10^{-5}
methyl formate	1.92×10^8	8.5	3.9×10^{-7}
methyl iodide (25°C)	$< 2 \times 10^6$	n/a	n/a
methyl isobutyl ketone	$< 5.2 \times 10^6$	13.11	$> 2.2 \times 10^{-5}$
methyl nitrate (25°C)	4.5×10^8	n/a	n/a
n-methyl-2-pyrolidone	2×10^6	32.0	1.4×10^{-4}
methyl thiocyanate (25°C)	1.5×10^8	n/a	n/a
naphthalene (82°C)	4×10^4	n/a	n/a
nitrobenzene (0°C)	5×10^5	34.82	6.2×10^{-4}
nitroethane	5×10^7	28.06	5.0×10^{-6}
nitromethane (18°C)	6×10^7	35.87	5.3×10^{-6}
1-nitropropane	3.3×10^7	23.24	6.2×10^{-6}
2-nitropropane	5×10^7	25.52	4.5×10^{-6}
nitrotoluene (25°C) (ortho or meta)	$< 2 \times 10^7$	n/a	n/a
octyl alcohol	1.39×10^7	10.34	6.9×10^{-6}
phenetole (25°C)	$< 1.7 \times 10^6$	n/a	n/a
phenol	1×10^6	9.78	8.7×10^{-5}
phenyl isothiocyante (25°C)	1.4×10^8	n/a	n/a

(continues)

Table B.2 Continued

Liquid	Conductivity (pS/m)	Dielectric Constant	Relaxation Time Constant (sec)
phosgene (25°C)	7×10^5	n/a	n/a
phosphorus (25°C)	4×10^8	n/a	n/a
phosphorus oxychloride (25°C)	2.2×10^8	n/a	n/a
pinene (23°C)	$< 2 \times 10^4$	n/a	n/a
piperidine (25°C)	$< 2 \times 10^7$	n/a	n/a
propionaldehyde (25°C)	8.5×10^7	18.5	1.9×10^{-6}
propionic acid (25°C)	$< 1 \times 10^5$	3.44	$> 3.0 \times 10^{-4}$
propionitrile	8.51×10^6	27.2	2.8×10^{-5}
propyl acetate (i- or n-)	$1 - 6 \times 10^4$	6.002	n/a
n-propyl alcohol (25°C)	2×10^6	20.33	9×10^{-5}
iso-propyl alcohol (25°C)	3.5×10^8	19.92	5×10^{-7}
propyl formate	5.5×10^9	7.72	1.2×10^{-8}
pyridine (18°C)	5.3×10^6	12.4	2.1×10^{-5}
quinoline (25°C)	2.2×10^6	9.0	3.6×10^{-5}
salicylaldehyde (25°C)	1.6×10^7	13.9	7.5×10^{-6}
succinonitrile	5.64×10^{10}	56.5	8.9×10^{-9}
sulfolane	$< 2 \times 10^6$	43.3	$> 1.9 \times 10^{-4}$
sulfonyl chloride (25°C)	2×10^8	n/a	n/a
sulfuric acid (25°C)	1×10^{12}	n/a	n/a
tetraethylene-pentamine	$> 1 \times 10^6$	n/a	n/a
tetramethylurea	$< 6 \times 10^6$	23.06	$> 3.4 \times 10^{-5}$
m-toluidine	5.5×10^4	9.91	1.6×10^{-3}
o-toluidine	3.79×10^7	6.34	1.5×10^{-6}
p-toluidine (100°C)	6.2×10^6	4.98	7.1×10^{-6}
trichloroacetic acid (25°C)	3×10^5	n/a	n/a
1,1,1-trichloroethane	7.3×10^5	7.53	9.1×10^{-5}
triethylene glycol	8.4×10^6	23.69	2.5×10^{-5}
triethylenetetramine	$> 1 \times 10^6$	n/a	n/a
trimethylamine (−35°C)	2.2×10^4	n/a	n/a
vinyl acetate	2.6×10^4	n/a	n/a
water (extremely pure)	4.3×10^6	80.4	1.7×10^{-4}
water (air distilled)	$\sim 1 \times 10^9$	80.4	7.1×10^{-4}
Semiconductive Liquids: Conductivity from 50 pS/m to 10^4 pS/m			
amyl acetate	2160	4.75	1.9×10^{-2}
armeen	470	n/a	n/a
biphenyl (liquid at 69°C − 120°C)	2500-10,000	n/a	n/a
bromobenzene	1200	5.40	4×10^{-2}
1-bromonaphthalene	3660	4.83	1.1×10^{-2}
butyl acetate (i- or n-)	4300	n/a	n/a
butyl acrylate	3580	n/a	n/a
chlorobenzene	7000	5.621	7.1×10^{-3}
chloroform	< 10,000	4.806	$> 4.3 \times 10^{-3}$
dibutyl sebacate	1700	4.54	2.4×10^{-2}
dichlorobenzene	3000	9.93	2.9×10^{-3}
ethylene dichloride	4000	10.36	2.2×10^{-2}
2-ethylhexyl acrylate	610	n/a	n/a
gasoline (leaded)	> 50	2.3	< 0.41
hydrogen sulfide (at boiling point)	1000	n/a	n/a
methylene chloride	4300	8.93	1.8×10^{-2}
pentachloroethane	100	3.83	0.3
sulfur (130°C)	5000	n/a	n/a

Table B.2 Continued

Liquid	Conductivity (pS/m)	Dielectric Constant	Relaxation Time Constant (sec)
1,2,4-trichlorobenzene	200	4.08	0.18
trichloroethylene	800	3.42	3.7×10^{-2}
vinyltrimethoxysilane (<2% methanol)	5900	n/a	n/a
Nonconductive Liquids: Conductivity <50 pS/m			
anisole	10	4.33	3.8
benzene (pure)	5×10^{-3}	2.3	~100 (dissipation)
biphenyl (solid <69°C)	0.17	n/a	n/a
bromine (17.2°C)	13	n/a	n/a
butyl stearate	21	3.111	1.3
caprylic acid	< 37	2.45	> 0.58
carbon disulfide (1°C)	7.8×10^{-4}	2.6	~100 (dissipation)
carbon tetrachloride	4×10^{-4}	2.238	~100 (dissipation)
chlorine (−70°C)	< 0.01	n/a	n/a
cyclohexane	< 2	2.0	> 8.8
decalin	6	2.18	3.2
dichlorosilane	n/a	n/a	n/a
diesel oil (purified)	~0.1	~2	~100 (dissipation)
diethyl ether	30	4.6	1.4
1,4-dioxane	0.1	2.2	~100 (dissipation)
ethyl benzene	30	2.3	0.68
gasoline (straight run)	~0.1	~2	~100 (dissipation)
gasoline (unleaded)	< 50 (varies)		
heptane (pure)	3×10^{-2}	2.0	~100 (dissipation)
hexane (pure)	1×10^{-5}	1.90	~100 (dissipation)
hexamethyldisilazane	29	n/a	n/a
isovaleric acid	40	2.64	> 0.58
jet fuel	0.01 - 50	2.2	0.39-100
kerosene	1 - 50	2.2	0.39-19
pentachlorodiphenyl	0.8	5.06	~100
silicon tetrachloride	n/a	n/a	n/a
stearic acid (80°C)	< 40	n/a	n/a
styrene monomer	10	2.43	2.2
sulfur (115°C)	100	n/a	n/a
toluene	< 1	2.38	21
trichlorosilane	n/a	n/a	n/a
turpentine	22	n/a	n/a
xylene	0.1	2.38	~100

D.1.2 Antistatic Additives. Additives used to change the electrostatic properties of solid and liquid materials. Extrinsic and intrinsic antistatic additives can be distinguished according to the method of addition. Based on the permanence of their effect, antistatic treatments can be short- or long-term.

D.1.3 Brush Discharge. A higher energy form of corona discharge characterized by low-frequency bursts or by streamers. Brush discharges can form between charged nonconductive surfaces and grounded conductors acting as electrodes. For positive electrodes, pre-onset or breakdown streamers are observed and the maximum effective energy is a few millijoules. For negative electrodes, the maximum effective energy is a few tenths of a millijoule. Brush discharges can ignite flammable gas and hybrid mixtures but not dust in air.

D.1.4 Bulking Brush Discharge. A partial surface discharge created during bulking of powder in containers, appearing as a luminous, branched channel flashing radially from the wall toward the center of the container. Its maximum effective energy is 10 mJ to 25 mJ. It can ignite flammable gas, hybrid mixtures, and some fine dusts in air.

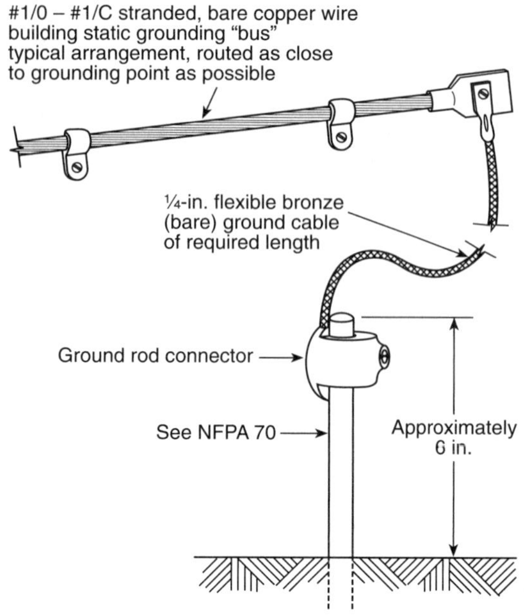

Figure C.1(a) Ground Bus Connection to Ground Rod.

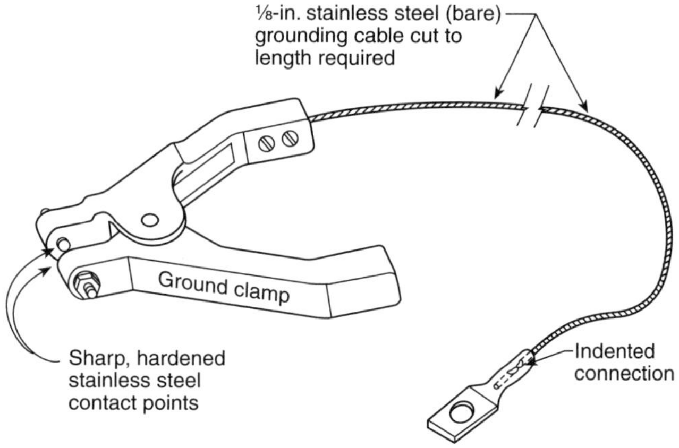

Figure C.1(b) Small Ground Clamp.

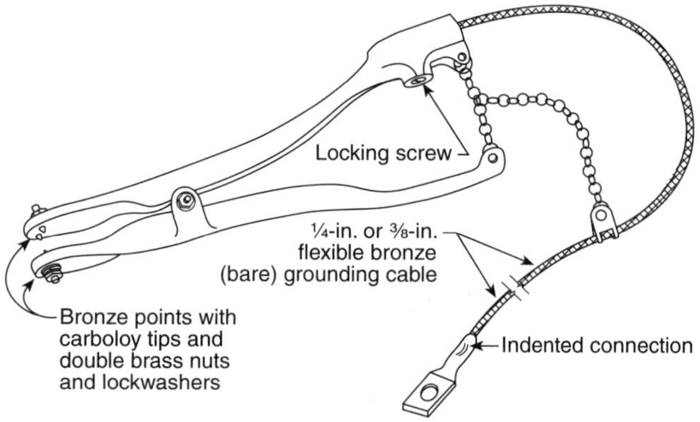

Figure C.1(c) Large Ground Clamp.

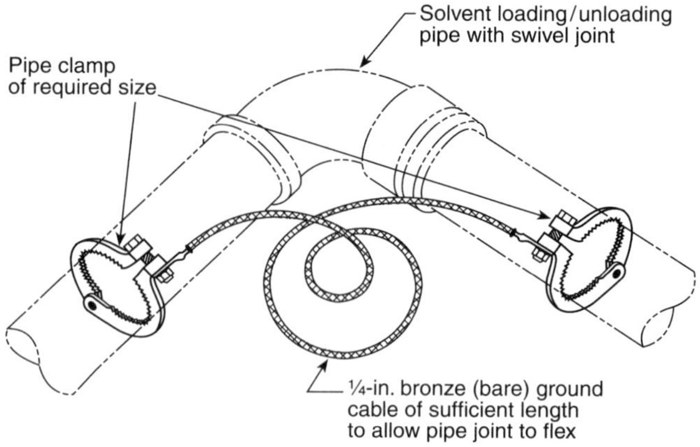

Figure C.1(d) Pipe Grounding Jumper.

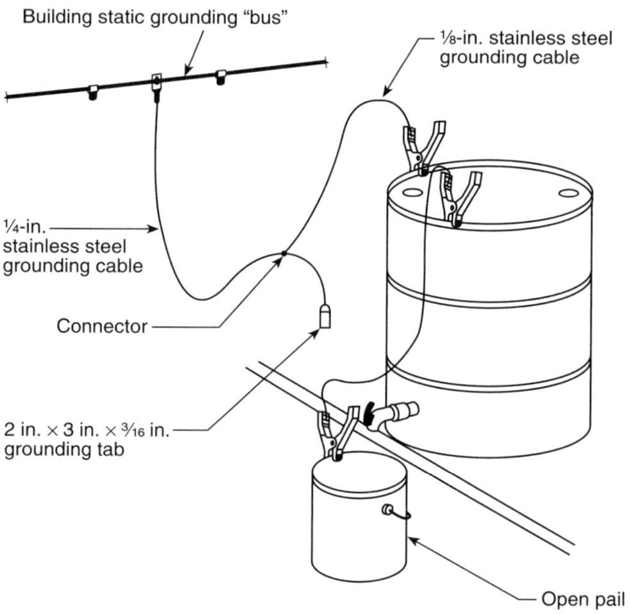

Figure C.1(e) Typical Grounding System for Small Volume Solvent Dispensing via Drum Tap.

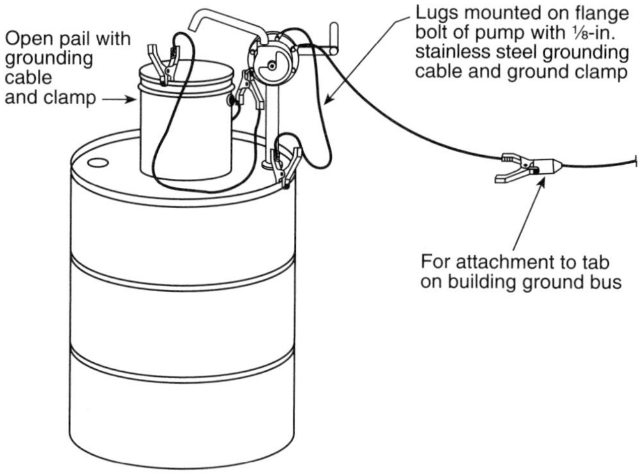

Figure C.1(f) Typical Grounding System for Small Volume Solvent Dispensing via Drum Pump.

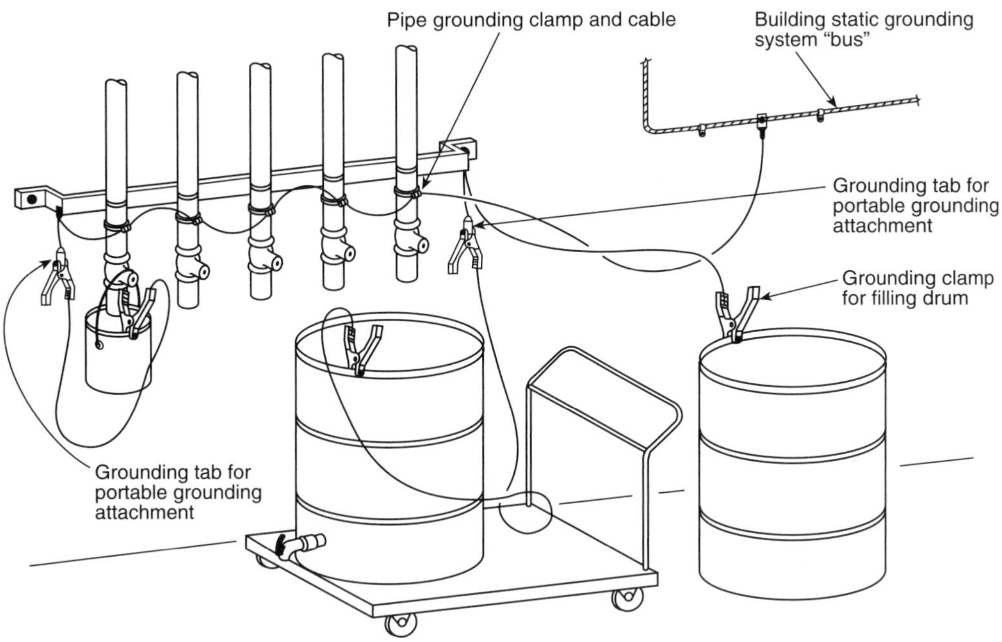

Figure C.1(g) Typical Grounding System for Small Volume Solvent Handling at Dispensing Station.

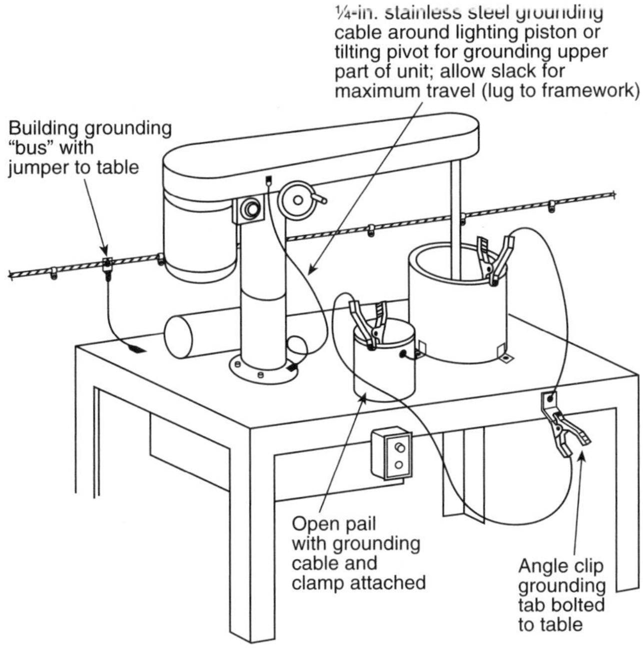

Figure C.1(h) Typical Grounding System for Small Equipment.

D.1.5 Charge. A collection or imbalance of electrons or of positive or negative ions that can accumulate on both conductors and insulators and that has both magnitude and polarity. Movement of charge constitutes an electric current. Excess or deficiency of electrons is expressed in coulombs. An electron carries a charge of -1.6×10^{-19} coulombs.

D.1.6 Charge Decay Time. The time for static electric charge to be reduced to a given percent of the charge's original level.

D.1.7 Charge Density. The charge per unit area on a surface or the charge per unit volume in space. Surface charge density is measured in coulombs per square meter. Volume charge density, also called space charge density or space charge, is measured in coulombs per cubic meter.

D.1.8 Charge Relaxation. The process by which separated charges recombine or excess charge is lost from a system.

D.1.9 Charging Current (I_c). The rate of flow of charge into a given system per unit of time, expressed in amperes.

D.1.10 Charging, Field. Charging of particles in an electric field as the result of ions from a corona or other source; also known as ion-bombardment charging. The maximum charge that a particle can gain by field charging is proportional to the cross-sectional area of the particle and the strength of the electric field. Field charging is a dominant particle-charging mechanism for particles larger than a few tenths of a micron.

D.1.11 Charging, Induction. The act of charging an object by bringing it near another charged object, then touching the first object to ground; also known as induction. Charge polarization is induced on a grounded object in the vicinity of a charged surface due to the electric field existing between

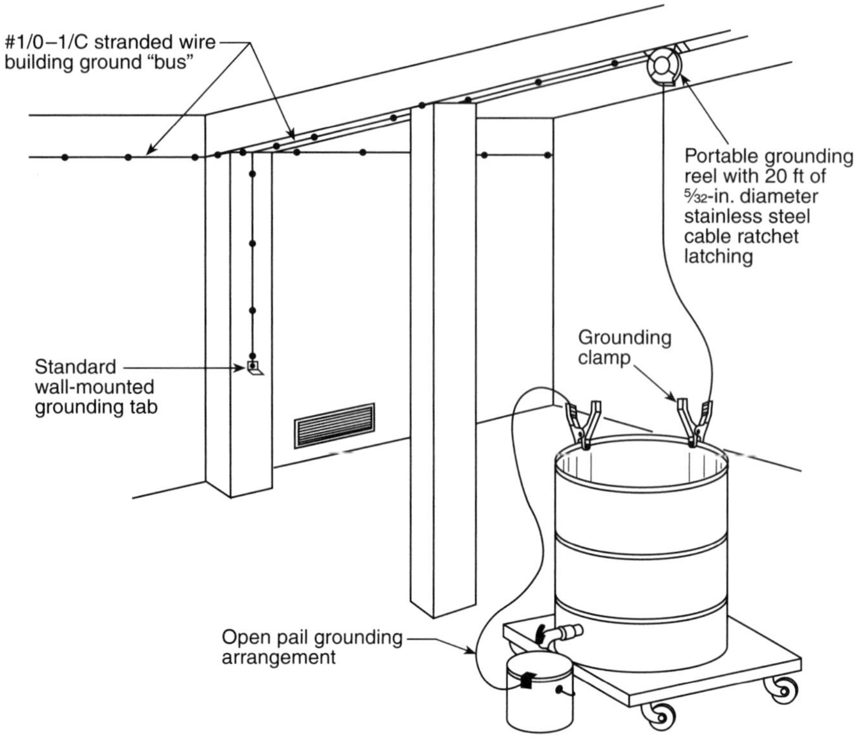

Figure C.1(i) Typical Grounding System for Small Volume for Portable Tank and Drum Transfer Area.

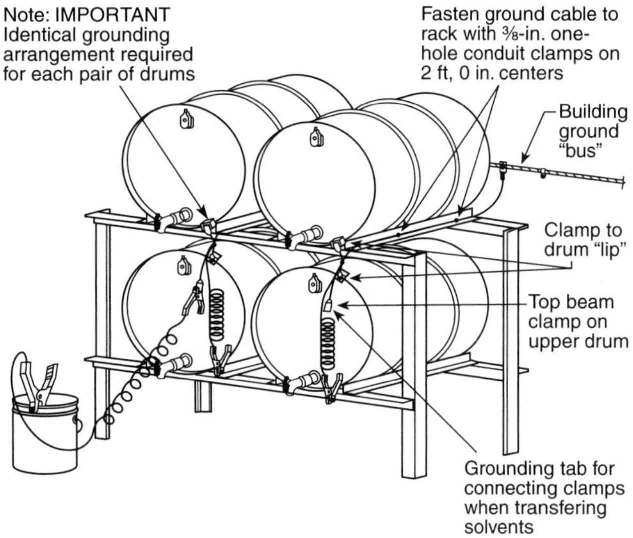

Figure C.1(j) Typical Grounding System for Drum Rack.

the object and the surface. If the ground connection is removed from the object during this period, the induced charge remains on the object. Induction charging will occur when a person walks from a conductive floor covering onto an insulating floor in the presence of an electric field.

D.1.12 Charging, Triboelectric. Static electric charging that results from contact or friction between two dissimilar materials; also known as frictional charging and contact-separation charging. Triboelectric charging is the most familiar, yet least understood, charge generation mechanism. Triboelectric charging results from contact or friction between two unlike materials. The amount of charge generated in this way is dependent primarily on the nature of contact, the intrinsic electrical and static electric properties of the materials involved, and the prevailing conditions of humidity and temperature. An indication of the tendency of materials to accept static electric charge in this way is given by the triboelectric series. Recent studies indicate that the amount of charge transferred depends not only on the composition of the materials but also the capacitance of the junction. Examples of triboelectric charging are as follows:

(1) Pneumatic transport of powders along pipes
(2) High-speed webs of synthetic materials moving over rollers
(3) Charging of the human body as a person walks across a carpet
(4) Extrusion of plastics or the ejection of plastic parts from a mold

Table D.1.12 illustrates typical electrostatic voltages observed as a result of triboelectric charging at two levels of relative humidity (RH).

D.1.13 Conductive Floor. Flooring that has an average resistance between 2.5×10^3 ohms and 1×10^6 ohms when

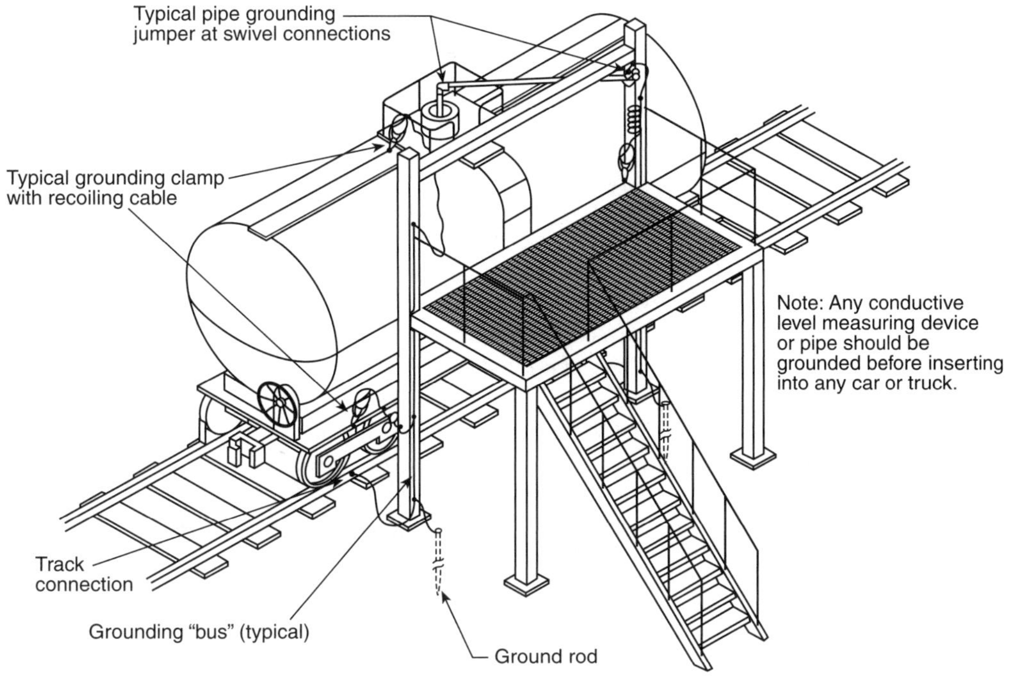

Figure C.1(k) Typical Grounding System for Tank Car or Tank Truck Loading/Unloading Station.

measured using specified electrodes placed a specified distance apart. *(See NFPA 99, Standard for Health Care Facilities.)*

D.1.14 Conductive Hose. Hose that has an electrical resistance of less than 10^3 ohms per meter, when measured between the end connectors.

D.1.15 Conductivity (κ)(l/resistivity).

(a) An intrinsic property of a solid or liquid that governs the way electrical charges move across its surface or through its bulk.

Table D.1.12 Electrostatic Voltages (kV) Resulting from Triboelectric Charging at Two Levels of Relative Humidity

Situation	RH 10–20%	RH 65–90%
Walking across carpet	35	1.5
Walking over vinyl floor	12	0.25
Working at bench	6	0.1
Vinyl envelopes for work instructions	7	0.6
Poly bag picked up from bench	20	1.2
Work chair padded with polyurethane foam	18	1.5

This property can be dramatically affected by the temperature and especially by the presence of moisture or antistats. Conductors such as metals and aqueous solutions have a high conductivity (low resistivity) and will lose charge quickly when grounded. Insulators have a high resistivity and lose charge very slowly, even when grounded. Conductors isolated from ground can store charge and can rise to fairly high potentials in many industrial situations, resulting in hazardous spark discharges.

The unit of conductivity is the ohm/m [note basis in reciprocal of resistivity with units of ohm-m; ohm/m is the same as siemens per meter (S/m)]. Some authors express volume resistivity data in terms of ohm-m (100 ohm-cm = 1 ohm-m).

Conductivity is not generally assigned to a gas. In gases, free electrons or ions can be formed or injected by external means. In the presence of an electric field, the ions migrate to electrodes or charged surfaces and constitute a current. Conductivity is rarely an intrinsic property of gas.

(b) Conductivity (electrical conductivity or specific conductance) is the ratio of the electric current density to the electric field in a material.

D.1.16 Cone Discharge. See definition D.1.4, *Bulking Brush Discharge.*

D.1.17 Corona Discharge. An electrical discharge in the microampere range that results from a localized electrical breakdown of gases by charges on surfaces such as sharp edges, needle points, and wires. The charges can arise on conductors at high voltage or on grounded conductors that are situated near a charged surface. Corona is accompanied by a faint luminosity.

D.1.18 Coulomb. A quantity of electrons equal to 6.24×10^{18} electrons; also, the quantity of electricity on the positive plate of a capacitor of 1 farad capacitance when the potential difference across the plates is 1 volt.

D.1.19 Current (*I*). A measure of the rate of transport of electric charge past a specified point or across a specified surface. The symbol I is generally used for constant currents and i for time-variable currents. The unit of current is the ampere. One ampere equals 1 coulomb per second.

D.1.20 Density, Bulk. The mass per unit volume of a dust pile or a dust deposit.

D.1.21 Dielectric Breakdown. A voltage-dependent failure mechanism that occurs when a potential difference is applied across a dielectric region that exceeds the region's inherent breakdown characteristics.

D.1.22 Dielectric Constant. The ratio of the permittivity of a material to the permittivity of a vacuum that indicates a material's ability, relative to a vacuum, to store electrical energy or charge, when the material is placed in an electric field. Typical dielectric constants and dielectric strengths are shown in Table D.1.22.

A dielectric is not necessarily an insulator. For example, water, which has a high dielectric constant, is not a very good insulator. The measure of a good dielectric is its polarizability rather than its conductivity.

D.1.23 Dielectric Strength. The maximum electrical potential gradient (electric field) that a material can withstand without rupture, usually specified in volts per millimeter of thickness; also known as electric strength or breakdown strength. See Table B.2 for typical data. The maximum dielectric field strength in atmospheric air is around 3×10^6 V/m. This figure implies the existence of maximum value of surface charge density. Oscillating electric fields produce a significantly greater stress on materials than do time-independent fields. For this reason, insulation systems based on DC ratings will fail quickly when used with AC power units.

Table D.1.22 Dielectric Properties of Selected Materials

Material	Dielectric Constant	Dielectric Strength (V/m)
Bakelite	4.9	2.4×10^7
Cellulose acetate	3.8	1.0×10^7
Mica	5.4	1.0×10^8
Plexiglas, Lucite	3.4	4.0×10^7
Polystyrene	2.5	2.4×10^7
Porcelain	7	6.0×10^6
Titanium dioxide	90	6.0×10^6
Barium titanate	1200	5.0×10^6

D.1.24 Dissipative. Used to describe materials having a surface resistivity between 10^5 ohms per square and 10^{11} ohms per square.

D.1.25 Double Layer. A phenomenon usually associated with a solid/liquid interface where ions of one charge type are fixed to the surface of the solid and an equal number of mobile ions of opposite charge are distributed through the neighboring region of the liquid. In such a system, the movement of liquid causes a displacement of the mobile ions with respect to the fixed charges on the solid surface.

D.1.26 Electrometer. A device used to measure static electric charge with high-input impedance, typically greater than 10^{13} ohms, which draws negligible current from the measured object.

D.1.27 Electrostatic Field. The electric force per unit charge that is produced by a distribution of charge; also called the electric field intensity or potential gradient. An electric field can be most easily thought of as lines of influence originating from unit positive charges and terminating at unit negative charges or at infinity. It is therefore closely related to charges on surfaces and in gas streams.

D.1.28 Electrostatic Field Meter. A device that estimates the electric field from a charged object by sensing the charge and the charge polarity on the surface of a conductor or insulator.

D.1.29 Faraday Cage. An electrically continuous, conductive enclosure that provides a shield from static electricity (region of no electrostatic field). The cage or shield is usually, although it need not be, grounded.

D.1.30 Field Suppression. The observed reduction in electric field away from an object when it is brought near a grounded object. The effect results from a repositioning of the electric field lines toward the grounded surface.

D.1.31 Ignition Energy. The energy required to effect ignition of a specified fuel mixture under specified conditions of test.

D.1.32 Incendive. The ability to cause ignition.

D.1.33 Induction. The process by which charges are moved to new locations on a conductor by the action of an electric field or the movement of a conductor into the influence of an existing electric field.

D.1.34 Induction Bar. A passive corona device that is connected to ground and that produces corona from the grounded electrode when highly charged objects are brought close to it. Ions (corona) will be produced by the electrodes if the charge on an object and its placement is sufficiently close that corona onset is achieved.

D.1.35 Ionization. A process by which a neutral atom or molecule loses or gains electrons, thereby acquiring a net charge and becoming an ion.

D.1.36 Joule. A unit of work and energy equal to 1 watt-second.

D.1.37 Limiting Oxidant Concentration (LOC). The lowest molar (or volume) gas-phase concentration of oxidant at which a specified fuel can propagate a flame.

D.1.38 Lower Flammable Limit (LFL). The lowest molar (or volume) concentration of a combustible substance in an oxidizing medium that will propagate a flame.

D.1.39 Megohmmeter. A meter used to measure high resistance values and that typically operates at higher test voltages than standard ohmmeters, usually in the range of 100 volts to 1000 volts.

D.1.40 Minimum Explosible Concentration (MEC). The lowest concentration of a combustible dust in air, expressed in grams per cubic meter, that will propagate a flame.

D.1.41 Minimum Ignition Energy (MIE). The energy, expressed in joules, stored in a capacitor that, upon discharge, is just sufficient to effect ignition of the most ignitible mixture of a given fuel mixture under specified test conditions.

D.1.42 Ohm. A unit of electric resistance equal to the resistance through which a current of 1 ampere will flow when there is a potential difference of 1 volt across the current.

D.1.43 Ohms per Square. A unit of measure used to describe surface resistivity that reflects the resistance value between two electrodes that form two sides of a square and is independent of the size of the square. The resulting value indicates how easily electrons can flow across a surface. It is normally used as a resistivity measurement of a thin conductive layer or material over a relatively insulative base material.

D.1.44 Propagating Brush Discharge. An energetic discharge caused by electrical breakdown of the dielectric layer in a capacitor. The capacitor is typically formed by charged plastic coating on a metal substrate, although plastic pipe and tote bins can also form the required charged double layer. The effective energy can exceed 1000 mJ, causing both shock to personnel and ignition hazards for a wide variety of materials, including dusts in air.

D.1.45 Relaxation Time Constant (τ). The time, in seconds, for charge to decay by Ohm's law to e^{-1} (36.7 percent) of its initial value. For a capacitor, the relaxation time constant is the product of resistance (ohms) and capacitance (farads).

D.1.46 Resistance (*R*). The opposition that a device or material offers to the flow of direct current, equal to the voltage drop across the element divided by the current through the element; also known as electrical resistance.

D.1.47 Resistivity, Surface. The electric resistance of the surface of an insulator, in ohms per square, as measured between the opposite sides of a square on the surface and whose value in ohms is independent of the size of the square or the thickness of the surface film.

D.1.48 Resistivity, Volume. The resistance of a sample of material, expressed in ohm-meters, having unit length and unit cross-sectional area.

D.1.49 Spark. A short-duration electric discharge due to a sudden breakdown of air or some other insulating material separating two conductors at different electric potentials, accompanied by a momentary flash of light; also known as electric spark, spark discharge, or sparkover.

D.1.50 Static Neutralizer, Alternating Current. A static neutralizer that uses an alternating current corona to generate positive and negative ions. Ions from this equipment are drawn from this ionizer to neutralize charged (usually insulating) surfaces. The frequency for AC static neutralizers is usually the line frequency (50/60 Hz).

D.1.51 Static Neutralizer, Electrically Powered. A static neutralizer that uses one or more electrified needles rigidly held in a housing and a high-potential supply that powers the neutralizer. Ion generation occurs in the air space surrounding the highly charged needle points.

D.1.52 Surface-Charge Density. The charge per unit area of solid surface expressed in coulombs per square meter. In air, the maximum surface charge density before air breakdown occurs is 2.65×10^{-5} C/m^2. Only a fraction of this value is achieved in most practical situations.

D.1.53 Surface Streamer. A surface-to-wall discharge observed on charged liquids during tanker filling, appearing as a flash up to 30 cm long and accompanied by a crackling sound. The effective energy is unknown but is possibly as high as approximately 10 mJ. Surface streamers are also observed on tribocharged solid surfaces, such as between aircraft windshields and metal airframe, where puncturing does not occur.

D.1.54 Tribocharging. Charge separation caused by the rubbing of surfaces, creating triboelectricity; also see definition D.1.12, *Charging, Triboelectric*.

D.1.55 Triboelectric Series. When different materials are rubbed together, charge is transferred. Many researchers working in the field of triboelectricity have set up triboelectric series to show which combination of materials would receive which charge. There is some agreement on the locations of some materials, but most of the series are very dissimilar, even when the same materials are used.

D.1.56 Upper Flammable Limit (UFL). The highest molar (or volume) concentration of a combustible substance in an oxidizing medium that will propagate a flame.

Appendix E Referenced Publications

E.1

The following documents or portions thereof are referenced within this recommended practice for informational purposes only and are thus not considered part of its recommendations. The edition indicated here for each reference is the current edition as of the date of the NFPA issuance of this recommended practice.

E.1.1 NFPA Publications.

National Fire Protection Association, 1 Batterymarch Park, P.O. Box 9101, Quincy, MA 02269-9101.

NFPA 30A, *Code for Motor Fuel Dispensing Facilities and Repair Garages*, 2000 edition.

NFPA 69, *Standard on Explosion Prevention Systems*, 1997 edition.

NFPA 70, *National Electrical Code®*, 1999 edition.

NFPA 99, *Standard for Health Care Facilities*, 1999 edition.

NFPA 302, *Fire Protection Standard for Pleasure and Commercial Motor Craft*, 1998 edition.

NFPA 318, *Standard for the Protection of Cleanrooms*, 2000 edition.

NFPA 407, *Standard for Aircraft Fuel Servicing*, 1996 edition.

NFPA 780, *Standard for the Installation of Lightning Protection Systems*, 1997 edition.

E.1.2 Other Publications.

E.1.2.1 AIChE Publications.

American Institute of Chemical Engineers, 3 Park Avenue, New York, NY 10016-5901.

Britton, L. G., "Using Material Data in Static Hazard Assessment," *Plant/Operations Progress*, April, 1992.

Britton, L. G. *Avoiding Static Ignition Hazards in Chemical Operations*, AIChE, 1999.

E.1.2.2 ANSI Publication.

American National Standards Institute, Inc., 11 West 42nd Street, 13th floor, New York, NY 10036.

ANSI Z41, *American National Standard for Personal Protection — Protective Footwear*, 1991.

E.1.2.3 API Publications.

American Petroleum Institute, 1220 L Street, NW, Washington, DC 20005.

Bustin, W. M., et al, *New Theory for Static Relaxation from High Resistivity Fuel*, API Refining Division Proceedings, Vol. 44, No. 3, 1964.

RP 2003, *Protection Against Ignitions Arising Out of Static, Lightning, and Stray Currents*, sixth edition, September, 1998.

API 2219, *Safe Operation of Vacuum Trucks in the Petroleum Service*, 2nd edition, 1999.

E.1.2.4 ASTM Publication.

American Society for Testing and Materials, 100 Barr Harbor Drive, West Conshohocken, PA 19428-2959.

ASTM E 502, *Standard Test Method for Selection and Use of ASTM Standards for the Determination of Flash Point of Chemicals by Closed Cup Methods*, 1984 (reapproved 1994).

E.1.2.5 CENELEC Publication.

European Committee for Electrotechnical Standardization (CENELEC), Rue de Stassart, 35, B-1050 Brussels, Belgium.

CENELEC Draft Standard *Safety of Machinery — Electrotechnical Aspects*, 1996.

E.1.2.6 NPCA Publication.

National Paint and Coatings Association, 1500 Rhode Island Avenue, NW, Washington, DC 20005-5597.

Generation and Control of Static Electricity, 1998.

NFPA 99B, *Standard for Hypobaric Facilities,* 2005

1.1 Scope.

1.1.1* This standard shall apply to all hypobaric facilities in which humans will be occupants or are intended to be occupants of the hypobaric chamber.

A.1.1.1 There is currently a widespread interest in high-altitude flight and space exploration. For this purpose, high-altitude chambers and space simulators have been developed and put to use. Equipment, experimental animals, and humans have been exposed to various artificial atmospheres under varying pressures ranging from 760 mm Hg (101.3 kPa) atmospheric pressure to close to 0 mm Hg (0 kPa).

In some chambers, the atmosphere might be enriched with oxygen or contain 100 percent oxygen. The increased combustibility of materials in those oxygen-enriched atmospheres has resulted in several fires in such chambers, with loss of life. See NFPA 53, *Recommended Practice on Materials, Equipment, and Systems Used in Oxygen-Enriched Atmospheres,* for descriptions of some of these accidents.

There is continual need for human diligence and expertise in the establishment, operation, and maintenance of hypobaric facilities.

The partial pressure of oxygen present in the atmosphere of a hypobaric facility is one of the determining factors of the amount of available oxygen. This pressure will rise if the percentage of oxygen increases proportionately more than the fall in total pressure. Even more important than partial pressure of oxygen from the standpoint of fire hazards compared with normal air, however, is the decrease in percentage of nitrogen available. The absence of the inerting effect of this nitrogen will generally lower the ignition energy and markedly elevate the burning rate of combustible and flammable substances. *(See B.1.2.2.1 and B.1.2.2.2.)*

It is the responsibility of the chief administrator or commanding officer of the facility possessing a hypobaric chamber to adopt and enforce appropriate regulations for hypobaric facilities. In formulating and administering the program, full use should be made of technical personnel highly qualified in hypobaric facility operations and safety.

It is essential that hypobaric chamber personnel having responsibility for the hypobaric facility establish and enforce appropriate programs to fulfill the provisions of this standard.

Potential hazards can be controlled only when continually recognized and understood by all pertinent personnel. The Technical Committee on Hyperbaric and Hypobaric Facilities realizes that such facilities are not normally used to treat patients. Nevertheless, human beings are being exposed; hence the need for preparation of this standard.

This standard was prepared with the intent of offering standards for the design, maintenance, and operation of such facilities.

This standard covers the recognition of, and protection against, hazards of an electrical, explosion, and implosion nature, as well as fire hazards.

Medical complications of hypobaric procedures are discussed primarily to acquaint rescue personnel with these problems.

1.1.2 This standard shall not apply to hypobaric facilities used for animal experimentation if the size of the hypobaric chamber does not allow for human occupancy.

1.2 Purpose.

1.2.4* Hypobaric chambers shall be classified according to the following criteria:

(1) Class D – Human rated, air atmosphere not oxygen enriched
(2) Class E – Human rated, oxygen-enriched atmosphere (partial pressure of oxygen is above 0.235 ATA)

A.1.2.4 Chapter 20, "Hyperbaric Facilities," in NFPA 99, *Standard for Health Care Facilities*, classifies hyperbaric chambers as A, B, or C. To avoid confusion, hypobaric facilities are classified as D and E.

Chambers designed for animal experimentation equipped for access of personnel to care for the animals are classified as Class D and E for the purpose of this chapter depending upon atmosphere. Animal chambers of a size that cannot be entered by humans are not included in this standard.

Both Class D and E chambers are human-rated; however, chambers used for high altitude training involving oxygen breathing are classified as Class D for the purpose of this standard.

Chapter 4 Construction and Equipment

4.3 Illumination.

4.7* Electrical Systems.

A.4.7 It is the intention of Chapter 4 that no electrical equipment be installed or used within the chamber that is not intrinsically safe or designed and tested for use under hypobaric conditions. Control devices, wherever possible, should be installed outside of the chamber.

4.7.2 Electrical Wiring and Equipment.

4.7.2.5 All electrical wiring installed in a Class E hypobaric chamber shall comply with the requirements of NFPA 70, *National Electrical Code*, Articles 500 and 501, Class I, Division 1.

4.7.2.6 Wiring installed Class E hypobaric chambers shall be approved for use in Class I, Group C atmospheres at the maximum proposed vacuum and oxygen concentration.

4.7.2.8 All boxes, fittings, and joints used in Class E hypobaric chambers shall be explosionproof.

4.7.2.9 Fixed electrical equipment within a Class E hypobaric chamber enclosure shall comply with the requirements of NFPA 70, *National Electrical Code*, Articles 500 and 501, Class I, Division 1.

4.7.2.10 Equipment installed within a Class E hypobaric chamber shall be approved for use in Class I, Group C atmospheres at the maximum vacuum and oxygen concentration attainable.

4.7.2.11.3 For equipment used inside Class E hypobaric chambers, each circuit shall have its own individual overcurrent protection in accordance with Section 240-11 of NFPA 70, *National Electrical Code.*

NFPA 120, *Standard for Fire Prevention and Control in Coal Mines,* 2003

1.1 Scope.

1.1.1* This standard shall cover minimum requirements for reducing loss of life and property from fire and explosion in the following:

(1) Underground bituminous coal mines
(2) Coal preparation plants designed to prepare coal for shipment
(3) Surface building and facilities associated with coal mining and preparation
(4) Surface coal and lignite mines

A.1.1.1 In the development of this document, the data in NIOSH Information Circular 9470, "Analysis of Mine Fires for All Underground and Surface Coal Mining Categories: 1990–1999," were examined. Table A.1.1.1 shows the number of fires for underground coal mines, surface fires at underground coal mines, at surface coal mines, and at coal preparation plants, as well as the number of fire injuries and coal production for the time period from 1990 to 1999.

Analysis of the data shows a general decrease in the number of fires over the 10-year period, particularly from 1996 to 1999, while coal production increased slightly. The largest number of fires over the 10-year period, as well as for each 2-year time period, occurred at surface coal mines. There were 164 injuries due to fire during the 10-year period, with the number decreasing significantly over the last 4 years. There were two fatalities in 1991.

1.1.2 This standard shall not apply to the following:

(1) Flammable and combustible liquids produced in underground coal mines
(2) Other equipment and processes, such as coal pulverizers, used to condition coal for firing in boilers at power-generating plants or gasification plants or for utilization in certain special processes

Chapter 6 Coal Processing

6.2.1 Electrical Classification of Hazard.

6.2.1.1 Plant areas of open construction where coal dust or any combustible gases liberated from the coal are dispersed to the open atmosphere shall be classified nonhazardous.

Table A.1.1.1 Number of Coal Mine and Preparation Plant Fires, Injuries Due to Fire, and Coal Production from 1990 to 1999

	Number of Fires*					
Years	**Underground Coal Mines**	**Surface at Underground Coal Mines**	**Surface Coal Mines**	**Coal Preparation Plants**	**Number of Fire Injuries***	**Coal Production† (10^6 short tons)**
1990–1991	25	17	67	23	59	2,004
1992–1993	18	14	37	22	29	1,928
1994–1995	23	16	47	18	39	2,059
1996–1997	6	7	40	8	19	2,155
1998–1999	15	11	24	20	18	2,218
1990–1999 Total	87	65	215	91	164	10,364

*Derived from MSHA "Fire Accident Abstract" and "Fire Accident Report" publications.
†Derived from MSHA "Injury Experience in Coal Mining" publications.

6.2.1.2 Plant areas isolated from the coal process, such as control rooms, electrical equipment rooms, or substations, that are provided with ventilation to prevent the accumulation of combustible gases or coal dust shall be classified nonhazardous.

6.2.1.3 Enclosed areas of processing plants where coal is wet to prevent particles from becoming airborne or where dry coal dust does not accumulate shall be classified nonhazardous.

6.2.1.4.1* Electrical equipment approved as "permissible" by the Mine Safety and Health Administration (MSHA) shall be acceptable in locations classified Class I, Division 1.

A.6.2.1.4.1 Electrical equipment classified as "permissible" is certified as meeting the requirements of 30 CFR Part 18, Chapter 1.

6.2.1.5 Areas of a processing plant normally designated as Class I shall be permitted to be considered nonhazardous, provided the following conditions are met:

(1) Ventilation to prevent an accumulation of an explosive or ignitable mixture of gases
(2) Failsafe continuous methane monitoring designed to sound an alarm when the methane–air mixture reaches 20 percent (1 percent methane by volume) of the lower explosive level (LEL)
(3) An interlock to stop the process equipment automatically when the methane–air mixture reaches 40 percent (2 percent methane by volume) of the LEL
(4) An electrical system arranged so that when methane concentrations reach 40 percent of the LEL, all electrical circuits including control circuit conductors are de-energized
(5) Any equipment that is needed to restore the plant to a methane–air mixture of less than 20 percent (1 percent methane by volume) of the LEL, such as lighting, ventilation, or sump pumps, installed in accordance with Class I, Division 1 requirements

6.2.1.6* Enclosed areas in which coal dust is not in suspension in explosive or ignitable quantities or in which coal dust might be present in explosive or ignitable quantities or might be in suspension in the air due to a malfunction shall be designated as Class II, Division 2 in accordance with Article 500 of NFPA 70, *National Electrical Code.*

A.6.2.1.6 Approved, intrinsically safe electrical equipment can be permitted to be used in any areas classified as "hazardous."

6.2.1.9 Electrical Equipment Rooms. Positive pressure shall be maintained in electrical equipment rooms, such as switchgear rooms, motor control centers, and cable-spreading rooms, to prevent the entry of fugitive dust.

6.2.1.9.1 Thermographic scanning shall be performed on switchgear and motor starters on an annual basis.

Chapter 7 Storage and Use of Compressed Gases and Flammable and Combustible Liquids

7.1.5.6 All electrical equipment within an enclosed storage room containing flammable gases shall be Class I, Division 2, Group A, B, C, or D.

7.5 Flammable Liquids Stored and Used Underground.

7.5.1 General.

7.5.1.1* Electrical equipment in flammable liquid storage areas shall be classified as one of the following:

(1) Class I, Division 1 as specified in NFPA 70, *National Electrical Code*
(2) "Permissible" electrical equipment

A.7.5.1.1 Electrical equipment classified as "permissible" is certified as meeting the requirements of 30 CFR Part 18, Chapter 1.

Chapter 9 Coal Conveyance and Storage

9.1* Conveyors — General.

A.9.1 Unless the conveyor is very long, burning coal on a moving belt is not likely to ignite the belt. Also, if the belt should ignite, the burning of the belt is likely to be extinguished after the burning coal has been discharged and the belt continues to run. No reports of running conveyor belts in and around preparation plants that have caught fire and burned have been located. Every reported case of belts catching fire and burning has occurred after the belts have been stopped.

Some preparation plants use the froth flotation process to separate impurities from fine coal. The agents typically used in froth flotation are Class II combustible liquids. The coal recovered from the froth cells is coated minimally with these agents. It has been found that frothed coal carried on conveyor belts will coat the belting with the agents, causing the coated belting to ignite easily and the flame spread to become significantly more rapid than usual. It is recommended that belts that carry frothed coal be protected with automatic sprinklers. While the froth flotation process operates as a water slurry and presents no risk of fire, the reagents normally used are No. 2 fuel oil and methyl isobutyl carbinol (MIBC), which are Class II combustible liquids.

Fortunately, evidence of heating is easy to detect. During the early stage of heating, the odor is unmistakable. When heating is more advanced, smoke and steam also might be apparent. If the hot coal is in an exposed storage pile, the hot material can be dug out and wetted. If the hot material has to be loaded onto a conveyor belt, the loading areas should be hosed down, and water should be applied to the hot material before or as it is loaded onto the belt.

Tunnels under silos or storage piles should be ventilated adequately and should be protected with a system of automatic sprinklers. Separate hose for fire fighting should be provided. The main tunnel should have exit routes at opposite ends of the tunnel.

9.1.1* Belt conveyors shall meet the following minimum requirements:

(8) Electrical equipment shall be classified as Class I, Division 2, Group F in all areas where required by NFPA 70, *National Electrical Code.*

A.9.1.1 The U.S. Mine Safety and Health Administration standards found in 30 CFR for fire-retardant conveyor belt materials should be used as a guide. Fire-retardant belt materials will burn and, therefore, might require additional fire protection.

Stockpile conveyors, reclaim conveyors, and conveyors going to loadout silos or bins should be fire-resistant belt.

Rip detection also can be considered for long runs or critical belt systems. Critical factors to be considered should be the impact on production if the belt is lost, the cost of the belt itself, availability of spare belting, length of time to repair the belt, and alternatives to bypass the belt if it is lost.

NFPA 122, *Standard for Fire Prevention and Control in Metal/ Nonmetal Mining and Metal Mineral Processing Facilities,* 2004

1.1* Scope.

A.1.1 Because of the uniqueness and often remoteness of metal and nonmetal mines and ore processing facilities, provisions in this standard could differ from commonly accepted fire protection standards and guides devised for other types of occupancies. The provisions of this document are considered necessary to provide a reasonable level of protection from loss of life and property from fire and explosions. They reflect situations and the state of the art at the time the standard was issued.

As of 2001, there were 12,479 metal/nonmetal mining and processing operation in the United States. In the most recent 12-year period, approximately 515 fires of all types were reported.

Fires and explosions in mines and mineral processing plants have caused major loss of property, production equipment, buildings, and business interruption. In the five-year period from 1994 to 1998, mines and quarries of all types averaged $12.3 million a year in direct damage in fires reported to U.S. local fire departments. In the same period, nonmetallic mineral processing and product manufacturing facilities averaged $16.1 million a year in direct damage in fires reported to U.S. local fire departments. *(For more information, see the NFPA Fire Protection Handbook, 2003 edition, Chapter 29.)*

Fires adversely affect all areas of mining and mineral processing operations, including underground and surface self-propelled and mobile mining equipment, underground fuel storage areas, surface ore concentrating and processing buildings and equipment, and support facilities associated with these activities.

Fire and related hazards in metal ore processing facilities include but are not limited to conveyor belts; rubber lined equipment; combustible and flammable reagents; gaseous, liquid, or solid fuels; mineral extraction solvents and carriers; dielectric, thermal, and lubricating oils; hydraulic fluids; grouped plastic electric cables; and combustible construction. Significant fire and explosions have occurred in concentrator mills due to these hazards.

Ignition sources for these hazards are present and cannot always be controlled. The most common ignition source in this industry is uncontrolled hot work.

Control and awareness of combustible loading, including "hidden" combustibles like rubber or plastic lined equipment, is important to understanding these hazards. Automatic fire suppression systems coupled with effective emergency response have been effective in limiting fire damage in processing facilities.

Most fires involving mobile or self-propelled mining equipment — whether underground or surface — occur on or near engine exhaust systems, high-speed drive lines, malfunctioning high-pressure–high-temperature hydraulic systems, or faulty electrical components. Total elimination of fire hazards on equipment is impossible since sources of ignition and fuel for fires are inherent in the basic equipment design. The fire problem is further complicated by the collection of environmental debris. Therefore, efforts to reduce fire losses on mobile equipment must be aimed at fire prevention and fire suppression.

To improve fire protection and prevention on mining equipment, some manufacturers of mining equipment have placed emphasis on the reduction of the fire potential of specific items in the original design of their equipment. Such items include turbochargers, exhaust manifolds and exhaust pipe shielding and insulation, location of combustible and flammable liquid reservoirs, and hydraulic and fuel line routing.

Most mining equipment is required to have at least one hand-portable extinguisher mounted in a readily accessible location. Extinguishers are most effective where used by trained operators. However, considering the size and configuration of machines found at a mine, fires can be difficult or impossible to fight with a hand-held extinguisher. For this reason, fire suppression systems have been developed to aid in suppressing those fires that are hard to access and thereby to reduce "off-road" equipment fire losses.

The key to operator protection is early detection of fires to provide a warning to the operator, fuel shutoff to minimize fuel for the fire, and fire suppression during its earliest stages. Specialized systems to perform these functions can be required to protect the operator and the machines. To be totally effective, however, system operation must be fully understood by owners and operators, and provisions must be made for periodic inspection and maintenance.

Fire suppression systems, including hand-portable extinguishers, offer the mining industry a cost-effective tool by which personnel and investments in mining equipment can be protected.

It could be necessary for those charged with purchasing, testing, approving, and maintaining fire protection equipment for the mining industry to consult an experienced fire protection specialist.

1.1.1 This standard covers minimum requirements for safeguarding life and property against fire and related hazards associated with metal and nonmetal underground and surface mining and metal mineral processing plants.

1.1.2 As applies to underground mining, this standard shall cover only the following:

(1) Diesel-powered equipment
(2) Storage and handling of flammable and combustible liquids

1.1.3 As applies to underground mining, this standard shall not cover flammable and combustible liquids produced in underground mines, such as shale oil mines.

1.1.4 As applies to surface mining, this standard shall cover only the following:

(1) Mobile equipment in use without its own motive power train and normally moved by self-propelled equipment
(2) Self-propelled equipment that contains a motive power train as an integral part of the unit and is not rail-mounted

1.1.5 This standard shall not cover buildings or employee housing and support facilities for a mining operation, or preparation or use of explosives.

1.1.6* As applies to metal mineral processing, this standard shall cover fire and related hazards associated with surface metal mineral processing plants including but not limited to conveying, crushing, fine milling, beneficiation, flotation, drying, filtering, ore and concentrate storage, and support facilities for the mineral processing activity.

A.1.1.6 A typical metal mineral processing plant — also called a concentrating or dressing mill — is physically separated from the mining operation, although it can be connected by conveyor systems. Typical metals produced using concentrator plants are gold, silver, platinum, nickel, zinc, lead, molybdenum, and copper. Essentially any metal can be concentrated in this manner. Some concentrating mills are located on floating dredges, such as those used in titanium mining.

The general purpose of the processing plant is to receive crushed ore, further reduce it in size by additional crushing, milling, and screening, and separate waste materials (gangue) from desirable metal mineral values. Most metal mineral mills are similar in that they have large semi-autogenous, ball, or roll mills for fine grinding the ore into a pulp or slurry. Once ground, the slurry is processed by flotation or beneficiation using reagents. After flotation, the concentrate — which can be in the 20 percent to 30 percent metals value range — is filtered or thermally dried and stored. Some metals, like molybdenum, feature combustible thermal oils in the drying process. Concentrate is sent to metallurgical refineries to recover the final pure product. The refinery might be adjacent to the mill but is usually separate.

By-products produced in a typical metal concentrator mill include tailings, which consist of waste gangue and entrained water and process chemicals. This waste is sent to a tailings disposal facility.

1.1.7* As applies to surface metal mineral processing plants, this standard shall not cover the following:

(1) Solvent extraction plants
(2) Pressure-leaching processes
(3) Alumina refineries
(4) Nonmetal mineral processing plants
(5) Metal smelters including roasting, sintering, and calcining
(6) Metal refineries such as electrowinning or electro-refining processes
(7) Gas, liquid, or solid waste handling or storage systems

A.1.1.7 There are number of processes associated with concentration or refining of metal ores that are not applicable to this standard but deserve mention due to their hazards and integration with the concentration process. These include solvent extraction–electrowinning (SX–EW); pressure leaching processes (using high-pressure autoclave reactors); alumina refineries; metal smelters, including flash furnaces; roasting, sintering, calcining, and electro-refining processes; and gas, liquid, or solid waste handling systems. There are also nonmetal mineral processing plants such as those used for recovering phosphates, nitrates, potash, and soda ash. All of these processes are chemical in nature, and all have serious fire and explosion hazards.

Of particular mention and importance from a potential fire hazard standpoint are solvent extraction (SX) plants.

An SX plant is a separations process using combustible solvents like kerosene or alcohol for separating valuable metal minerals. An SX process facility often features thousands of gallons of solvent in plastic tanks using plastic piping and can be located outdoors or inside buildings. SX plants are common at copper mines where the oxide content of the ore body allows acid leaching in heaps. They are also common for uranium, nickel, and cobalt.

While kerosene is usually a Class II combustible liquid and in a cold state is relatively difficult to ignite, once ignited it burns similar to other lower flash point hydrocarbons. At high elevations, the flash point can render the material a Class I flammable liquid. In very hot climates, the material can be above its flash point and the potential for heating is increased when the solution is carried in black plastic piping subject to solar heating.

Protection of SX plants needs to consider response time of fire fighters and types of fire fighting appliances and suppression agents. Because of the large quantities of combustible liquids and use of plastic piping and process systems — which can fail prematurely due to fire impingement and rapidly release additional combustible liquids — a well-developed and large area fire could occur in minutes, and responding fire fighters could be faced with protecting exposures rather than suppression of the incipient event. For this reason the use of fast-acting automatic detection and suppression systems is advised. Foam-water systems have proven effective in suppressing combustible liquid fires. If used, consideration needs to be given to providing protection over and under mixer-settlers and tanks, in tunnels with plastic piping, under pipe racks, over pumps, and inside mixer-settlers.

Provision of drainage, confinement, control of static electricity by bonding and grounding, and selection of stout piping systems, such as stainless steel or structural fiberglass reinforced plastic instead of less robust polyethylene, is advised.

A mineral SX plant should not be confused with an agricultural SX plant that uses low flash point flammable solvents, like hexane, for recovering oils from soybeans, canola, and corn, and that has a higher hazard. NFPA 36, *Standard for Solvent Extraction Plants*, applies to protection of agricultural solvent extraction plants but does not apply to protection of mineral solvent extraction plants. There currently are no NFPA standards on mineral SX plants.

1.1.8 Nothing in this standard is intended to prohibit the use of new methods or devices, provided sufficient technical data are submitted to the authority having jurisdiction to demonstrate that the new method or device is equivalent in quality, effectiveness, durability, and safety to that specified by this standard.

Chapter 3 Definitions

3.3 General Definitions.

3.3.4.1 ***Large Combustible Liquid Storage Area.*** An area used for storage of Class II and Class III combustible liquids where the aggregate quantity present is greater than 3785 L (1000 gal). Handling of liquids incidental to transfer can take place within a storage area.

Chapter 9 Flammable Liquid Storage in Underground Mines

9.1.1* Electrical equipment in large flammable liquid storage areas shall be Class I, Division 1, as specified in NFPA 70, *National Electrical Code*, or shall be classified as "permissible" electrical equipment.

A.9.1.1 Electrical equipment classified as "permissible" is certified as meeting the requirements of 18 CFR, Chapter 1.

9.4 Large Flammable Liquid Storage Areas.

9.4.2.1* Electrical equipment within 15.2 m (50 ft) from the storage area shall be Class I, Division 1, as specified in NFPA 70, *National Electrical Code*, or shall be classified as "permissible" electrical equipment.

A.9.4.2.1 Electrical equipment classified as "permissible" is certified as meeting the requirements of 18 CFR, Chapter 1.

NFPA 303, *Fire Protection Standard for Marinas and Boatyards,* 2000

1.1 Scope.

This standard applies to the construction and operation of marinas, boatyards, yacht clubs, boat condominiums, docking facilities associated with residential condominiums, multiple-docking facilities at multiple-family residences, and all associated piers, docks, and floats. This standard is not intended to apply to a private, non-commercial docking facility constructed or occupied for the use of the owners or residents of the associated single-family dwelling.

1.1.1 This standard also applies to support facilities and structures used for construction, repair, storage, hauling and launching, or fueling of vessels if fire on a pier would pose an immediate threat to these facilities, or if a fire at a referenced facility would pose an immediate threat to a docking facility.

1.1.2 This standard applies to marinas and facilities:

(1) Servicing small recreational and commercial craft, yachts, and other craft of not more than 300 gross tons
(2) Not covered by NFPA 307, *Standard for the Construction and Fire Protection for Marine Terminals, Piers, and Wharves*; or NFPA 30A, *Code for Motor Fuel Dispensing Facilities and Repair Garages*

1.1.3 No requirement in this standard shall be construed as reducing applicable building, fire, and electrical codes.

Chapter 3 Electrical Wiring and Equipment

3.9 Hazardous Locations.

The entire electrical system installed in a hazardous (classified) location shall comply with the requirements given in Article 500, NFPA 70, *National Electrical Code*, and where required by the conditions, to the requirements of this standard related to damp and wet locations.

NFPA 318, *Standard for the Protection of Semiconductor Fabrication Facilities,* 2002

1.1 Scope.

This standard applies to semiconductor fabrication facilities and comparable research and development areas in which hazardous chemicals are used, stored, and handled and containing what is herein defined as a cleanroom or clean zone, or both.

Chapter 6 Construction

6.5* Electrical Classification.

A.6.5 The hand delivery and pouring of combustible and flammable chemicals have been reduced to a minimum in large state-of-the-art factories. Storage, located in storage rooms, is remote from the cleanroom. The majority of chemicals are dispensed automatically by way of bulk delivery systems. The hazards associated with spills in the cleanroom are minimal, considering the amount of air being recirculated.

6.5.1 General. The fabrication area or cleanroom shall be considered unclassified electrically with respect to Article 500 of NFPA 70, *National Electrical Code®*, where all of the following requirements are met:

(1) Chemical storage and handling meet the requirements of Chapter 7 of this standard.
(2) Ventilation and exhaust systems meet the requirements of Chapter 5 of this standard.
(3) The average air change is not less than 0.176 L/sec/m^2 (4 ft^3/min · ft^2) of floor area and the number of air changes at any location is not less than 0.132 L/sec/m^2 (3 ft^3/min · ft^2) of floor area. The use of recirculated air shall be allowed.

NFPA 407, *Standard for Aircraft Fuel Servicing,* 2001

1.1 Scope.

This standard applies to the fuel servicing of all types of aircraft using liquid petroleum fuel. It does not apply to any of the following:

(1) In-flight fueling
(2) Fuel servicing of flying boats or amphibious aircraft on water
(3) Draining or filling of aircraft fuel tanks incidental to aircraft fuel system maintenance operations or manufacturing

Chapter 4 Design

4.3.7 Vehicle or Cart Lighting and Electrical Equipment.

4.3.7.4* Motors, alternators, generators, and associated control equipment located outside of the engine compartment or vehicle cab shall be of a type listed for use in accordance with NFPA 70, *National Electrical Code®*, Class I, Division 1, Group D locations.

A.4.3.7.4 *Electrical Equipment in Aircraft Fuel Servicing Vehicle or Cart Engine Compartments.* Equipment contained in the engine compartment or vehicle cab and located 457 mm (18 in.) or more above ground can be permitted to be of the general purpose type.

4.3.7.5 Electrical equipment and wiring located within a closed compartment shall be of a type listed for use in accordance with NFPA 70, *National Electrical Code*, Class I, Division 1, Group D locations.

4.3.7.6 Lamps and switching devices, other than those covered in 4.3.7.4 and 4.3.7.5, shall be of the enclosed, gasketed, weatherproof type. Other electrical components shall be of a type listed for use in accordance with NFPA 70, *National Electrical Code*, Class I, Division 2, Group D locations.

4.3.7.7 Electrical service wiring between a tractor and trailer shall be designed for heavy-duty service. The connector shall be of the positive-engaging type. The trailer receptacle shall be mounted securely.

NFPA 409, *Standard on Aircraft Hangars,* 2004

1.1 Scope.

This standard contains the minimum requirements for the proper construction of aircraft hangars and protection of aircraft hangars from fire.

Chapter 5 Construction of Group I and Group II Aircraft Hangars

5.9 Landing Gear Pits, Ducts, and Tunnels.

5.9.1* Landing gear pits, ducts, and tunnels located below floor level shall be designed on the premise that flammable liquids and vapor will be present at all times. Materials and equipment shall be impervious to liquids and shall be fire resistant or noncombustible.

A.5.9.1 Landing gear pits, ducts, and tunnels located beneath the hangar floor should be avoided if possible because of the danger of accumulation of flammable liquids or vapors; where their use is essential, the protection measures specified in Section 5.9 should be followed. For floor drainage, see 5.11.2.

5.9.2 Electrical equipment for all landing gear pits, ducts, and tunnels located below hangar floor level shall be suitable for use in Class I, Division 1, Group D hazardous locations in compliance with Article 501 of NFPA 70.

NFPA 423, *Standard for Construction and Protection of Aircraft Engine Test Facilities,* 2004

1.1 Scope.

1.1.1 This standard establishes the minimum fire safety practices regarding location, construction, services, utilities, fire protection, operation, and maintenance of aircraft engine test facilities.

1.1.3 This standard does not apply to engines and engine accessories or to engine test facilities where fuels other than hydrocarbon fuels are used.

Chapter 5 Service and Utilities

5.3 Electrical Requirements.

5.3.1 Any pits, depressions, or other below-floor-level locations of engine test cells, fuel-handling areas, and hydraulic rooms shall be classified as Class I, Division 1 hazardous locations as defined in Article 500 of NFPA 70, and such classification shall extend up to floor level.

5.3.2 The engine test cell, including intake and exhaust plenums, fuel-handling areas, and hydraulic rooms, shall be classified as a Class I, Division 2 hazardous location as defined in Article 500 of NFPA 70, and such classification shall extend to a level 0.46 m (18 in.) above the floor.

5.3.3* All wiring and equipment that are installed or operated within any of the hazardous locations specified in Section 5.3 shall comply with applicable provisions of Article 501 of NFPA 70.

A.5.3.3 It is common practice to locate limit switches for elevating work platforms below the floor level. An accidental shorting or grounding of these circuits should not allow the elevator to move or overrun, which could result in damage to engine fuel lines and in ensuing fire.

5.3.5 All wiring in the exhaust plenum that is not located within the hazardous location as specified in 5.3.2 shall be installed in rigid conduit.

NFPA 484, *Standard for Combustible Metals,* 2002

1.1 Scope.

1.1.1* This standard shall apply to the production, processing, finishing, handling, storage, and use of all metals and alloys that are in a form that is capable of combustion or explosion.

A.1.1.1 Under the proper conditions, most metals in the elemental form will react with oxygen to form an oxide. These reactions are exothermic. The conditions of the exposure are affected by the temperature of the metal, whether it is in large pieces or in the form of small particles, the ratio of its surface area to its total weight, the extent or presence of an oxide coating, the temperature of the surrounding atmosphere, the oxygen content of the atmosphere, moisture content of the atmosphere, and the presence of flammable vapors.

Chapter 4 Aluminum

4.1 Aluminum Powder Production Plants.

4.1.6 Electrical Power and Control.

4.1.6.2* Powder-manufacturing areas shall be classified, where applicable, in accordance with Article 500 of NFPA 70, *National Electrical Code*.

A.4.1.6.2 For additional information on classification of dusty locations, see NFPA 499, *Recommended Practice for the Classification of Combustible Dusts and of Hazardous (Classified) Locations for Electrical Installations in Chemical Process Areas.*

4.1.6.2.1 Offices and similar areas within the aluminum powder–manufacturing building that are segregated and reasonably free from dust shall not be classified.

4.1.6.2.2 Control equipment meeting the requirements of NFPA 496, *Standard for Purged and Pressurized Enclosures for Electrical Equipment*, shall be permitted.

4.2 Aluminum Powder Handling and Use.

4.2.4 Machinery and Operations.

4.2.4.2 Electrical Equipment.

4.2.4.2.4* Wet solvent milling areas or other areas where combustible or flammable liquids are present shall be classified where applicable, in accordance with Article 500 of NFPA 70, *National Electrical Code*, with the exception of control equipment meeting the requirements of NFPA 496, *Standard for Purged and Pressurized Enclosures for Electrical Equipment.*

A.4.2.4.2.4 For additional information on classification of areas containing solvent vapors, see NFPA 497, *Recommended Practice for the Classification of Flammable Liquids, Gases, or Vapors and of Hazardous (Classified) Locations for Electrical Installations in Chemical Process Areas.*

Chapter 6 Magnesium

6.1 Location and Construction of Magnesium Powder Production Plants.

6.1.6 Electrical Power.

6.1.6.1 All electrical equipment and wiring shall be installed in accordance with NFPA 70, *National Electrical Code.*

6.1.6.2* All parts of manufacturing buildings shall be classified.

A.6.1.6.2 See NFPA 499, *Recommended Practice for the Classification of Combustible Dusts and of Hazardous (Classified) Locations for Electrical Installations in Chemical Process Areas.*

6.3 Machining, Finishing, and Fabrication of Magnesium.

6.3.4 Electrical Equipment.

6.3.4.1* Dust-producing machines, including areas containing dust-collection equipment, shall be classified in accordance with Article 500 of NFPA 70, *National Electrical Code.*

A.6.3.4.1 See NFPA 499, *Recommended Practice for the Classification of Combustible Dusts and of Hazardous (Classified) Locations for Electrical Installations in Chemical Process Areas*, for guidance on classified areas for Class II materials.

Chapter 7 Tantalum

7.1 Construction of Production Plants.

7.1.3 Electrical Power.

7.1.3.1 All electrical equipment and wiring shall be installed in accordance with NFPA 70, *National Electrical Code,* and

NFPA 496, *Standard for Purged and Pressurized Enclosures for Electrical Equipment.*

7.1.3.2* In local areas of a plant where a hazardous quantity of dust accumulates or is present in suspension in the air, the area shall be classified, and all electrical equipment and installations in those local areas shall comply with Article 500 of NFPA 70, *National Electrical Code.*

A.7.1.3.2 Refer to NFPA 499, *Recommended Practice for the Classification of Combustible Dusts and of Hazardous (Classified) Locations for Electrical Installations in Chemical Process Areas.*

Chapter 8 Titanium

8.6 Titanium Powder Production and Use.

8.6.2 Titanium Powder Handling.

8.6.2.4 Electrical Installations. All titanium powder production, drying, and packing areas shall be evaluated for fire and explosion hazards associated with the operation and shall be provided with approved electrical equipment for the hazardous location present, which shall be installed in accordance with the requirements of NFPA 70, *National Electrical Code.*

Chapter 9 Zirconium

9.6* Zirconium Powder Production and Use.

A.9.6 Not all methods of producing metal powder are applicable to zirconium. Reduction of zirconium hydride and some forms of milling are generally used to produce the limited amounts of powder now needed commercially. To reduce oxidation and possible ignition hazards, milling can be performed under water or in an inert atmosphere of helium or argon. Some powders are given a very light copper coating during the manufacturing process. See Figure A.4.3.4.4(a), Figure A.4.3.4.4(b), Figure A.4.3.4.4(c), and Figure A.8.4.4.1.2(b).

Like many other metal powders, zirconium is capable of forming explosive mixtures in air. The ignition temperatures of dust clouds, under laboratory test conditions, range from 330°C to 590°C (626°F to 1094°F). The minimum explosive concentration is 45.1 g/m^3 (0.045 oz/ft^3). The maximum pressure produced in explosions in a closed bomb at a concentration of 500 g/m^3 (0.5 oz/ft^3) ranged from 317 kPa to 558 kPa (46 psi to 81 psi). The average rate of pressure rise in these tests ranged from 1724 kPa/s to 29,670 kPa/s (250 psi/s to 4300 psi/s). The maximum rate of pressure rise ranged from 3792 kPa/s to over 69,000 kPa/s (550 psi/s to over 10,000 psi/s). The minimum energy of electrical condenser discharge sparks necessary for ignition of a dust cloud was 10 mJ. For a dust layer, the minimum value was 8 μJ. Some samples of zirconium powder were ignited by electric sparks in pure carbon dioxide, as well as in air. In some cases, zirconium at elevated temperatures was found to react in nitrogen as well as in carbon dioxide. Zirconium powder is considered a flammable solid.

9.6.2 Zirconium Powder Handling.

9.6.2.4 Electrical Installations.

9.6.2.4.1 All zirconium powder production, drying, and packing areas shall be evaluated for fire and explosion hazards associated with the operation and shall be provided with approved electrical equipment for the hazardous location.

NFPA 495, *Explosive Materials Code,* 2001

1.1 Scope.

1.1.1 This code shall apply to the manufacture, transportation, storage, sale, and use of explosive materials.

1.1.2 This code shall not apply to the transportation of explosive materials where under the jurisdiction of the U.S. Department of Transportation (DOT). It shall apply, however, to state and municipal supervision of compliance with "Hazardous Materials Regulations," U.S. Department of Transportation, Title 49, *Code of Federal Regulations*, Parts 100–199.

1.1.3 This code shall not apply to the transportation and use of military explosives by federal or state military agencies, nor shall it apply to the transportation and use of explosive materials by federal, state, or municipal agencies while engaged in normal or emergency performance of duties.

1.1.4 This code shall not apply to the manufacture of explosive materials under the jurisdiction of the U.S. Department of Defense. This code also shall not apply to the distribution of explosive materials to or storage of explosive materials by military agencies of the United States, nor shall it apply to arsenals, navy yards, depots, or other establishments owned by or operated by or on behalf of the United States.

1.1.5 This code shall not apply to pyrotechnics such as flares, fuses, and railway torpedoes. It also shall not apply to fireworks and pyrotechnic special effects as defined in NFPA 1123, *Code for Fireworks Display*; NFPA 1124, *Code for the Manufacture, Transportation, and Storage of Fireworks and Pyrotechnic Articles*; and NFPA 1126, *Standard for the Use of Pyrotechnics before a Proximate Audience.*

1.1.6 This code shall not apply to model and high power rocketry as defined in NFPA 1122, *Code for Model Rocketry*; NFPA 1125, *Code for the Manufacture of Model Rocket and High Power Rocket Motors*; and NFPA 1127, *Code for High Power Rocketry.*

1.1.7 This code shall not apply to the use of explosive materials in medicines and medicinal agents in the forms prescribed by the United States Pharmacopeia or the National Formulary.

Chapter 5 Blasting Agents

5.1 Scope.

5.1.1 Unless otherwise specified in this chapter, blasting agents shall be transported, stored, and used in the same manner as other explosive materials.

5.1.2 Water gels, slurries, and emulsion explosives shall not be subject to the requirements of this chapter. *(See Chapter 6.)*

5.2 Fixed Location Mixing.

5.2.6 All electrical switches, controls, motors, and lights located in the mixing room shall comply with NFPA 70, *National Electrical Code*®, Article 502.

Exception: This requirement shall not apply to electrical wiring and equipment located outside the mixing building.

Chapter 8 Aboveground Storage of Explosive Materials

8.5 Magazine Construction — Basic Requirements.

8.5.6 Electric lighting, electric safety flashlights, or electric safety lanterns shall be permitted to be used within a magazine. The installation of electric lighting shall meet the following requirements:

(7) Magazines containing explosive materials that could release flammable vapors shall have wiring and fixtures that meet the requirements of NFPA 70, *National Electrical Code*, Article 501.

NFPA 496, *Standard for Purged and Pressurized Enclosures for Electrical Equipment,* 2003

Editor's Note: *This document is reprinted here in its entirety, including its annexes.*

Chapter 1 Administration

1.1 Scope.

1.1.1 This standard applies to purging and pressurizing for the following:

(1) Electrical equipment located in areas classified as hazardous by Article 500 or Article 505 of NFPA 70
(2) Electrical equipment containing sources of flammable vapors or gases and located in either classified or unclassified areas
(3) Control rooms or buildings located in areas classified as hazardous by Article 500 or Article 505 of NFPA 70
(4) Analyzer rooms containing sources of flammable vapors or gases and located in areas classified as hazardous by Article 500 or Article 505 of NFPA 70

1.1.2* This standard does not apply to electrical equipment located in:

(1) Areas classified as Class I, Zone 0
(2) Areas classified as Class III
(3) Areas where flammable liquids may be splashed or spilled on the electrical equipment

1.2 Purpose.

This standard provides information on the methods for purging and pressurizing enclosures to prevent ignition of a flammable atmosphere. Such an atmosphere may be introduced into the enclosure by a surrounding external atmosphere or by an internal source. By these means, electrical equipment that is not otherwise acceptable for a flammable atmosphere may be utilized in accordance with Article 500 or Article 505 of NFPA 70.

1.3 Application.

1.3.1 Chapters 4, 5, and 6 of this standard apply to electrical instrument and process control equipment, motors, motor controllers, electrical switchgear, and similar equipment that is installed in Class I or Class II locations and that does not contain an internal source of flammable vapor, gas, or liquid.

1.3.2 Chapter 7 of this standard applies to control rooms that are located in Class I or Class II locations and that do not contain an internal source of flammable vapor, gas, or liquid.

1.3.3* Chapter 8 of this standard applies to electrical instrument and process control equipment and similar enclosed equipment, such as a gas chromatograph or a gas analyzer, that does contain an internal source of flammable vapor, gas, or liquid.

1.3.4 Chapter 9 of this standard applies to analyzer rooms and buildings.

1.4 Equivalency.

Nothing in this standard is intended to prevent the use of systems, methods, or devices of equivalent or superior quality, strength, fire resistance, effectiveness, durability, and safety over those prescribed by this standard.

1.4.1 Technical documentation shall be submitted to the authority having jurisdiction to demonstrate equivalency.

1.4.2 The system, method, or device shall be approved for the intended purpose by the authority having jurisdiction.

1.5 Units of Measurement.

1.5.1 SI Units. Metric units of measurement in this standard are in accordance with the modernized metric system known as the International System of Units (SI).

1.5.2 Compliance. Compliance with the numbers shown in either the SI system or the inch-pound system shall constitute compliance with this standard.

1.6 Mandatory Rules, Permissive Rules, and Explanatory Material.

1.6.1 Mandatory Rules. Mandatory rules of this standard are those that identify actions that are specifically required or prohibited and are characterized by the use of the terms *shall* or *shall not*.

1.6.2 Permissive Rules. Permissive rules of this standard are those that identify actions that are allowed but not required, are normally used to describe options or alternative methods, and are characterized by the use of the terms *shall be permitted* or *shall not be required*.

1.6.3 Explanatory Material. Explanatory material is located in Annex A. The information contained in Annex A is explanatory only and is not enforceable as part of this standard.

Chapter 2 Referenced Publications

2.1* General.

The documents or portions thereof listed in this chapter are referenced within this standard and shall be considered part of the requirements of this document.

2.2 NFPA Publication.

National Fire Protection Association, 1 Batterymarch Park, P.O. Box 9101, Quincy, MA 02269-9101.

NFPA 70, *National Electrical Code®*, 2002 edition.

2.3 Other Publications.

2.3.1 ISA Publication.

ISA-Instrumentation, Systems, and Automation Society, 67 Alexander Drive, P.O. Box 12277, Research Triangle Park, NC 27709.

ANSI/ISA 12.13.01-2000, *Performance Requirements for Combustible Gas Detectors*.

Chapter 3 Definitions

3.1 General.

The definitions contained in this chapter shall apply to the terms used in this standard. Where terms are not included, common usage of the terms shall apply.

3.2 NFPA Official Definitions.

3.2.1* Approved. Acceptable to the authority having jurisdiction.

3.2.2* Authority Having Jurisdiction (AHJ). An organization, office, or individual responsible for enforcing the requirements of a code or standard, or for approving equipment, materials, an installation, or a procedure.

3.3 General Definitions.

3.3.1* Alarm. A piece of equipment that generates a visual or audible signal that is intended to attract attention.

3.3.2 Analyzer Room or Building. A specific room or building containing analyzers, one or more of which is piped to the process.

3.3.3 Enclosure Volume. The volume of the empty enclosure without internal equipment. The enclosure volume for motors, generators, and other rotating electric machinery is the volume within the enclosure minus the volume of the internal components, e.g., rotors, stators, and field coils.

3.3.4 Ignition-Capable Equipment. Equipment that, under normal operation, produces sparks, hot surfaces, or a flame that can ignite a specific flammable atmosphere.

3.3.5* Ignition Temperature. The autoignition temperature of a flammable gas or vapor or the lower of either the layer ignition temperature or cloud ignition temperature of a combustible dust.

3.3.6 Indicator. A piece of equipment that shows flows or pressure and is monitored periodically, consistent with the requirement of the application.

3.3.7 Power Equipment. Equipment that utilizes power greater than 2500 VA or switches loads greater than 2500 VA.

3.3.8 Pressurization. The process of supplying an enclosure with a protective gas with or without continuous flow at sufficient pressure to prevent the entrance of a flammable gas or vapor, a combustible dust, or an ignitable fiber.

3.3.8.1 ***Type X Pressurizing.*** Reduces the classification within the protected enclosure from Division 1 or Zone 1 to unclassified.

3.3.8.2 ***Type Y Pressurizing.*** Reduces the classification within the protected enclosure from Division 1 to Division 2 or Zone 1 to Zone 2.

3.3.8.3 ***Type Z Pressurizing.*** Reduces the classification within the protected enclosure from Division 2 or Zone 2 to unclassified.

3.3.9* Pressurizing System. A grouping of components used to pressurize and monitor a protected enclosure.

3.3.10 Protected Enclosure. An enclosure pressurized by a protective gas.

3.3.11 Protected Equipment. The electrical equipment internal to the protected enclosure.

3.3.12 Protective Gas. The gas used to maintain pressurization or to dilute a flammable gas or vapor.

3.3.13 Protective Gas Supply. The compressor, blower, or compressed gas container that provides the protective gas at a positive pressure. The supply includes inlet (suction) pipes or ducts, pressure regulators, outlet pipes or ducts, and any supply valves not adjacent to the pressurized enclosure.

3.3.14 Purging. The process of supplying an enclosure with a protective gas at a sufficient flow and positive pressure to reduce the concentration of any flammable gas or vapor initially present to an acceptable level.

3.3.15* Specific Particle Density. The density of individual dust particles, as opposed to the bulk density of the material.

3.3.16 Ventilated Equipment. Equipment, such as motors, that requires airflow for heat dissipation as well as pressurization to prevent entrance of flammable gases, vapors, or dusts.

3.4 *NEC*® Extracted Definitions.

3.4.1 Class I, Division 1. A Class I, Division 1 location is a location: (1) In which ignitable concentrations of flammable gases or vapors can exist under normal operating conditions, or (2) In which ignitable concentrations of such gases or vapors may exist frequently because of repair or maintenance operations or because of leakage, or (3) In which breakdown or faulty operation of equipment or processes might release ignitable concentrations of flammable gases or vapors and might also cause simultaneous failure of electrical equipment in such a way as to directly cause the electrical equipment to become a source of ignition. [**70:**500.5(B)(1)]

3.4.2 Class I, Division 2. A Class I, Division 2 location is a location: (1) In which volatile flammable liquids or flammable gases are handled, processed, or used, but in which the liquids, vapors, or gases will normally be confined within closed containers or closed systems from which they can escape only in case of accidental rupture or breakdown of such containers or systems or in case of abnormal operation of equipment, or (2) In which ignitable concentrations of gases or vapors are normally prevented by positive mechanical ventilation, and which might become hazardous through failure or abnormal operation of the ventilating equipment, or (3) That is adjacent to a Class I, Division 1 location, and to which ignitable concentrations of gases or vapors might occasionally be communicated unless such communication is prevented by adequate positive-pressure ventilation from a source of clean air and effective safeguards against ventilation failure are provided. [**70:**500.5(B)(2)]

3.4.3 Class II, Division 1. A Class II, Division 1 location is a location: (1) In which combustible dust is in the air under normal operating conditions in quantities sufficient to produce explosive or ignitable mixtures, or (2) Where mechanical failure or abnormal operation of machinery or equipment might cause such explosive or ignitable mixtures to be produced, and might also provide a source of ignition through simultaneous failure of electric equipment, through operation of protection devices, or from other causes, or (3) In which combustible dusts of an electrically conductive nature may be present in hazardous quantities. [**70:**500.5(C)(1)]

3.4.4 Class II, Division 2. A Class II, Division 2 location is a location: (1) Where combustible dust is not normally in the air in quantities sufficient to produce explosive or ignitable mixtures, and dust accumulations are normally insufficient to interfere with the normal operation of electrical equipment or other apparatus, but combustible dust may be in suspension in the air as a result of infrequent malfunctioning of handling or processing equipment and (2) Where combustible dust accumulations on, in, or in the vicinity of the electrical equipment may be sufficient to interfere with the safe dissipation of heat from electrical equipment or may be ignitable by abnormal operation or failure of electrical equipment. [**70:**500.5(C)(2)]

3.4.5 Class I, Zone 0. A Class I, Zone 0 location is a location in which: (1) Ignitable concentrations of flammable gases or vapors are present continuously, or (2) Ignitable concentrations of flammable gases or vapors are present for long periods of time. [**70:**505.5(B)(1)]

3.4.6 Class I, Zone 1. A Class I, Zone 1 location is a location: (1) In which ignitable concentrations of flammable gases or vapors are likely to exist under normal operating conditions; or (2) In which ignitable concentrations of flammable gases or vapors may exist frequently because of repair or maintenance operations or because of leakage; or (3) In which equipment is operated or processes are carried on, of such a nature that equipment breakdown or faulty operations could result in the release of ignitable concentrations of flammable gases or vapors and also cause simultaneous failure of electrical equipment in a mode to cause the electrical equipment to become a source of ignition; or (4) That is adjacent to a Class I, Zone 0 location from which ignitable concentrations of vapors could be communicated, unless communication is prevented by adequate positive pressure ventilation from a source of clean air and effective safeguards against ventilation failure are provided. [**70:**505.5(B)(2)]

3.4.7 Class I, Zone 2. A Class I, Zone 2 location is a location: (1) In which ignitable concentrations of flammable gases or vapors are not likely to occur in normal operation and, if they do occur, will exist only for a short period; or (2) In which volatile flammable liquids, flammable gases, or flammable vapors are handled, processed, or used but in which the liquids, gases, or vapors normally are confined within closed containers of closed systems from which they can escape, only as a result of accidental rupture or breakdown of the containers or system, or as a result of the abnormal operation of the equipment with which the liquids or gases are handled, processed, or used; or (3) In which ignitable concentrations of flammable gases or vapors normally are prevented by positive mechanical ventilation but which may become hazardous as a result of failure or abnormal operation of the ventilation equipment; or (4) That is adjacent to a Class I, Zone 1 location, from which ignitable concentrations of flammable gases or vapors could be communicated, unless such communication is prevented by adequate positive-pressure ventilation from a source of clean air and effective safeguards against ventilation failure are provided. [**70:**505.5(B)(3)]

Chapter 4 General Requirements for Pressurized Enclosures

4.1 Applicability.

This chapter contains the general requirements for pressurized enclosures containing electrical equipment.

4.2 Enclosure.

4.2.1 The protected enclosure, including windows, shall be constructed of material that is not likely to be damaged under the conditions to which it may be subjected.

4.2.1.1 Precautions shall be taken to protect the enclosure from excessive pressure of the protective gas supply.

4.2.1.2 Excess-pressure-relieving devices, where required to protect the enclosure in the case of a control failure, shall be designed to prevent the discharge of ignition-capable particles to a Division 1 location.

4.2.2* Normal discharge of the protective gas from a designated enclosure outlet shall be to an unclassified location, unless the discharge meets the conditions specified in 4.2.2.1 or 4.2.2.2.

4.2.2.1 The discharge shall be permitted to be to a Division 2 or Zone 2 location if the equipment does not create ignition-capable particles during normal operation.

4.2.2.2 The discharge shall be permitted to be to a Division 1 or Division 2 location or to a Zone 1 or Zone 2 location if the outlet is designed to prevent the discharge of ignition-capable particles during normal operation.

4.2.3* In Division 1 and Zone 1 locations, where the conduit or raceway entry into a pressurized enclosure is not pressurized as part of the approved protection system, an explosionproof conduit seal shall be installed as close as practicable to, but not more than 450 mm (18 in.) from, the pressurized enclosure.

4.2.4 In Division 2 and Zone 2 locations, an explosionproof conduit seal shall not be required at the pressurized enclosure.

4.3 Pressurizing System.

4.3.1* The protected enclosure shall be constantly maintained at a positive pressure of at least 25 Pa (0.1 in. of water) above the surrounding atmosphere during operation of the protected equipment.

4.3.2 Where the protective gas supply is used to supply other than Type X pressurized equipment, an alarm shall be provided to indicate failure of the protective gas supply to maintain the required pressure.

4.3.3 All pressurizing system components that may be energized in the absence of the protective gas shall be approved for the classified location in which they are installed.

4.3.4 Installation, operating, and maintenance instructions shall be provided for the pressurizing system.

4.4 Protective Gas System.

4.4.1* The protective gas shall be essentially free of contaminants or foreign matter and shall contain no more than trace amounts of flammable vapor or gas.

4.4.1.1 All protective gas supplies shall be designed to minimize chances for contamination.

4.4.1.2* Air of normal instrument quality, nitrogen, or other nonflammable gas shall be permitted as a protective gas.

4.4.2 Piping for the protective gas shall be protected against mechanical damage.

4.4.3 Where compressed air is used, the compressor intake shall be located in an unclassified location.

4.4.4* Where the compressor intake line passes through a classified location, it shall be constructed of noncombustible material, designed to prevent leakage of flammable gases, vapors, or dusts into the protective gas, and protected against mechanical damage and corrosion.

4.4.5 The electrical power for the protective gas supply (blower, compressor, etc.) shall be supplied either from a separate power source or from the protected enclosure power supply before any service disconnects to the protected enclosure.

4.4.6 Where "double pressurization" is used (e.g., a Division 1 enclosed area pressurized to a Division 2 classification that contains ignition-capable equipment also protected by pressurization), the protective gas supplies shall be independent.

4.5* Determination of Temperature Marking.

4.5.1* The temperature class (T Code) marked on the enclosure shall represent (under normal conditions) the highest of the following:

(1) The hottest enclosure external surface temperature
(2)* The hottest internal component surface temperature
(3) The temperature of the protective gas leaving the enclosure

Exception: The surface temperature of the internal components shall be permitted to exceed the marked temperature class (T Code) in accordance with any one of the following conditions:

(1) The enclosure is marked as required in 4.11.4 with the time period sufficient to permit the component to cool to the marked temperature class (T Code).

(2) The small component has been shown to be incapable of igniting a test gas associated with a lower temperature class (T Code) or will not ignite the flammable vapor, gas, or dust involved.

(3) The component is separately housed so that the surface temperature of the housing is below the marked temperature class (T Code), and the housing complies with (a) and (b).

(a) The housing shall be pressurized or sealed.

(b) Where the housing can be readily opened, then the housing shall be marked as required in 4.11.4.

4.5.2 It shall be permitted to mark the actual temperature in degrees Celsius in place of the temperature class (T Code).

4.5.3 Temperature class (T Codes) shall be as shown in Table 4.5.3 for equipment marked for Class I, Divisions 1 or 2, or Class II, Divisions 1 or 2.

4.5.4 Temperature class (T Codes) shall be as shown in Table 4.5.4 for equipment marked for Class I, Zones 1 or 2.

4.6* Ventilated Equipment.

The flow of protective gas shall keep the equipment adequately cooled.

Table 4.5.3 Temperature Class (T Codes) for Class I, Divisions 1 or 2, or for Class II, Divisions 1 or 2 Locations

Maximum Temperature		Temperature Class (T Code)
°C	°F	
450	842	T1
300	572	T2
280	536	T2A
260	500	T2B
230	446	T2C
215	419	T2D
200	392	T3
180	356	T3A
165	329	T3B
160	320	T3C
135	275	T4
120	248	T4A
100	212	T5
85	185	T6

[**70:**Table 500.8(B)]

Table 4.5.4 Temperature Class (T Codes) for Class I, Zone 1 and 2 Applications

Temperature Class (T Code)	Maximum Surface Temperature (°C)
T1	≤450
T2	≤300
T3	≤200
T4	≤135
T5	≤100
T6	≤85

[**70:**Table 505.9(D)(1)]

4.7* Power Equipment.

Enclosures containing power equipment shall be of substantially noncombustible construction and shall be reasonably tight. Gaskets shall be permitted.

4.8* Type Z Pressurizing.

4.8.1 Detection shall be provided to indicate failure to maintain positive pressure within a protected enclosure.

4.8.1.1* Failure to maintain the positive pressure within a protected enclosure shall be communicated by an alarm or an indicator.

4.8.1.2 It shall not be required to de-energize the protected equipment upon detection of the failure to maintain positive pressure within a protected enclosure.

4.8.2 Any protected enclosure that can be isolated from the protective gas supply shall be equipped with an alarm.

Exception: The protected enclosure shall be permitted to be equipped with an indicator where the isolation is done with a valve(s) that complies with the following:

(1) The valve is immediately adjacent to the protected enclosure.
(2) The valve(s) is intended for use only during servicing of the protected enclosure.
(3) The valve(s) is marked as required in 4.11.5.

4.8.3 Where an alarm is used:

(1) Thc alarm shall be located at a constantly attended location.
(2) The alarm actuator shall take its signal from the protected enclosure and shall not be installed between the enclosure and the protective gas supply.
(3) The alarm actuator shall be mechanical, pneumatic, or electrical.
(4) Electrical alarms and electrical alarm actuators shall be approved for the location in which they are installed.
(5) No valves shall be permitted between the alarm actuator and the enclosure.
(6) The alarm shall be permitted to satisfy the requirement in 4.3.2 to provide an alarm on the protected gas supply.

4.8.4 Where an indicator is used:

(1) The indicator shall be located for convenient viewing.
(2) The indicator shall not be installed between the enclosure and the protective gas supply.
(3) The indicator shall indicate either pressure or flow.
(4) No valves shall be permitted between the indicator and the enclosure.
(5) The protective gas supply shall have an alarm that is located at a constantly attended location to fulfill the requirement in 4.3.2.

4.9* Type Y Pressurizing.

4.9.1 All of the requirements in Section 4.8 shall apply.

4.9.2 Equipment within a protected enclosure shall be approved for Division 2 or Zone 2 locations.

4.9.3 Ventilated equipment that would develop temperatures higher than the marked temperature class (T Code) upon failure of the ventilation shall be automatically de-energized when the flow of protective gas stops.

4.10* Type X Pressurizing.

4.10.1* A cutoff switch shall be incorporated to de-energize power automatically from all circuits within the protected enclosure not approved for Division 1 or Zone 1 upon failure of the protective gas supply to maintain positive pressure.

Exception: Power to the circuits shall be permitted to be continued for a short period if immediate loss of power would result in a more hazardous condition, and if both audible and visual alarms are provided at a constantly attended location.

4.10.1.1 The cutoff switch provided to de-energize power upon failure of the protective gas supply to maintain positive pressure shall be either flow actuated or pressure actuated.

4.10.1.2 The cutoff switch shall be approved for use in the location in which it is installed.

4.10.1.3 No valves shall be permitted between the cutoff switch and the protected enclosure.

4.10.1.4 The cutoff switch shall take its signal from the protected enclosure and shall not be installed between the enclosure and the protective gas supply.

4.10.2* Equipment, such as motors or transformers, that may be overloaded shall be provided with devices to detect any increase in temperature of the equipment beyond its design limits and shall de-energize the equipment automatically.

Exception: Power to the circuits shall be permitted to be continued for a short period if immediate loss of power would result in a more hazardous condition, and if both audible and visual alarms are provided at a constantly attended location.

4.10.3 For ventilated equipment, the flow of protective gas shall provide sufficient cooling even during overload conditions, or the equipment subject to overloading shall be provided with devices to detect any increase in temperature beyond its design limits and to de-energize that equipment automatically.

Exception: Power to the circuits shall be permitted to be continued for a short period if immediate loss of power would result in a more hazardous condition, and if both audible and visual alarms are provided at a constantly attended location.

4.11 Markings.

4.11.1 A permanent marking shall be on the protected enclosure in a prominent location so that it is visible before the protected enclosure can be opened.

4.11.2 The marking required by 4.11.1 shall include the information specified as follows:

(1) The following statement, or an equivalent statement:

WARNING — PRESSURIZED ENCLOSURE — This enclosure must not be opened unless the area atmosphere is known to be below the ignitable concentration of combustible materials or unless all devices within have been de-energized.

(2) The external area classification for the protected enclosure
(3) The pressurization type (e.g., Type X, Type Y, or Type Z)
(4) The temperature class (T Code) or the operating temperature in degrees Celsius as determined in Section 4.5

Exception No. 1: The temperature class (T Code) or operating temperature marking shall not be required where the highest temperature does not exceed 100°C.

Exception No. 2: For equipment marked for use in a specific gas or dust atmosphere, the temperature class (T Code) or operating temperature marking shall not be required where the highest temperature does not exceed 80 percent of the ignition temperature (in degrees Celsius) of the flammable vapor, gas, or dust involved. If the dust involved is an organic dust that may dehydrate or carbonize, the higher temperature shall not be permitted to exceed the lower of either 80 percent of the layer/cloud ignition temperature, or 165°C.

4.11.3 The additional markings specified in Sections 5.3 and 6.3 may also be included in the permanent marking described in 4.11.2.

4.11.4 Where 4.5.1, Exception (1) or (2) is used, the following or equivalent statement shall appear as permanent marking:

WARNING — HIGH TEMPERATURE INTERNAL PARTS — This enclosure must not be opened unless the area atmosphere is known to be below the ignitable concentration of combustible materials or unless all equipment within has been de-energized for ________ minutes.

4.11.5 Where 4.8.2, Exception is used, the following or equivalent statement shall appear in a permanent marking:

WARNING — PROTECTIVE GAS SUPPLY VALVE — This valve must be kept open unless the area atmosphere is known to be below the ignitable concentration of combustible materials, or unless all equipment within the protected enclosure is de-energized.

Chapter 5 Pressurized Enclosures for Class I

5.1 Applicability.

This chapter applies to enclosures containing electrical equipment that are located in Class I locations, and in conjunction with the requirements of Chapter 8, to enclosures located in unclassified locations that contain an internal source of flammable gas or vapor.

5.2 General Requirements.

5.2.1 The requirements of Chapter 4 shall be met.

5.2.2 Where the enclosure has been opened or if the protective gas supply has failed to maintain the required positive pressure, the enclosure shall be purged.

5.2.3 Airflow through the enclosure during purging shall be designed to avoid air pockets.

5.2.4 Once the enclosure has been purged of flammable concentrations, only positive pressure shall be required to be maintained within the enclosure.

5.2.5 A specific flow rate shall not be required for the positive pressure required by 5.2.4.

5.2.6* Compartments within the main enclosure or adjacent enclosures connected to the main enclosure shall be considered separately, and protection shall be provided by one of the following methods:

(1) The internal compartment shall be vented to the main enclosure by nonrestricted top and bottom vents that are common to the main enclosure. Each vent shall provide not less than 6.5 cm^2 (1.0 $in.^2$) of vent area for each 6560 cm^3 (400 $in.^3$), with a minimum vent size of 6.3 mm (¼ in.) diameter.
(2) The internal compartment or adjacent enclosure shall be purged in series or shall be purged separately.
(3)* The equipment in the internal compartment or adjacent enclosure shall be protected by other means (e.g., explosionproof, intrinsic safety, hermetic sealing, nonincendive, encapsulation, and so forth).

5.2.6.1 Components with a free internal volume less than 20 cm^3 (1.22 $in.^3$) shall not be required to be considered as internal compartments requiring protection, provided the total volume of all such components is not a significant portion of the protected enclosure volume.

5.2.6.2 It shall not be required to include components considered to be environmentally sealed such as transistors, microcircuits, capacitors, and so forth, in the percent of volume analysis.

5.3 Markings.

A permanent marking containing the start-up conditions shall be on the protected enclosure in a prominent location.

Exception: Start-up conditions shall be permitted to alternately be marked on an adjacent pressurizing system if referenced on the protected enclosure.

5.3.1 The marking shall contain the following, or an equivalent, statement:

WARNING — Power must not be restored after enclosure has been opened until enclosure has been purged for ________ minutes at a flow rate of ____________.

5.3.2 The minimum pressure shall be permitted to be used in place of the flow rate where the pressure is a positive indication of the correct flow.

5.4* Additional Requirements for Type Y or Type Z Pressurizing.

The protected equipment shall be energized only under the conditions specified in 5.4.1 or 5.4.2.

5.4.1 Protected equipment shall not be energized until at least four enclosure volumes of the protective gas (ten volumes for motors, generators, and other rotating electric machinery) have passed through the enclosure while maintaining an internal pressure of at least 25 Pa (0.1 in. of water).

5.4.2 Protected equipment shall be permitted to be energized immediately where a pressure of at least 25 Pa (0.1 in. of water) exists, and the atmosphere within the enclosure is known to be below the ignitable concentration of the combustible material.

5.5 Additional Requirements for Type X Pressurizing.

5.5.1 A means shall be used to prevent energizing of electrical equipment within the protected enclosure until at least four enclosure volumes of the protective gas (ten volumes for motors, generators, and other rotating electric machinery) have passed through the enclosure while maintaining an internal pressure of at least 25 Pa (0.1 in. of water).

5.5.2* If the enclosure can be readily opened without the use of a key or tools, an interlock shall be provided to immediately de-energize all circuits within the enclosure that are not approved for the location when the enclosure is opened.

5.5.2.1 The interlock, even though located within the enclosure, shall be approved for external area classification.

5.5.2.2 Protected enclosures that contain hot parts requiring a cool-down period shall be designed to require the use of a key or tool for opening.

Chapter 6 Pressurized Enclosures for Class II

6.1 Applicability.

This chapter applies to enclosures containing electrical equipment that are located in Class II locations.

6.2 General Requirements.

6.2.1 The requirements in Chapter 4 for each type of pressurizing shall be complied with, except as modified in this chapter.

6.2.2* Where combustible dust has accumulated within the protected enclosure, the protected enclosure shall be opened and the dust removed before pressurizing.

6.2.3 Adjacent enclosures connected to the main enclosure shall be permitted to be collectively pressurized to prevent the entrance of dust if there is communication to maintain the specified pressure at all points.

6.2.4* The protected enclosure shall be constantly maintained at a pressure above the surrounding atmosphere, depending on the specific particle density during operation of the protected equipment, and shall not be less than that specified in Table 6.2.4.

Table 6.2.4 Minimum Enclosure Pressure Versus Dust Density

Specific Particle Density			Minimum Enclosure Pressure	
kg/m³	lb/ft³	Specific Gravity	Pa	in. of water
<2083	<130	<2.083	25	0.1
>2083	>130	>2.083	125	0.5

6.2.5* Where the ignition temperature of the dust is not known, maximum surface temperatures shall not exceed those stated in Table 6.2.5.

6.3 Markings.

Start-up conditions shall be permanently marked in a prominent location on the protected enclosure.

6.4 Marking Information.

The marking required by Section 6.3 shall contain the following, or an equivalent, statement:

WARNING — Power must not be restored after the enclosure has been opened until combustible dusts have been removed and the enclosure repressurized.

6.5* Additional Requirements for Type X Pressurizing.

An alarm, provided at a constantly attended location, shall be permitted to be used in place of the cutoff switch specified in 4.10.1, if the enclosure is tightly sealed to prevent the entrance of dust.

6.6 Additional Requirements for Ventilated Equipment.

The discharge of protective gas shall not create a combustible atmosphere by disturbing layers of dusts.

Chapter 7 Pressurized Control Rooms

7.1* Applicability.

This chapter applies to buildings or portions of buildings commonly referred to as control rooms.

Table 6.2.5 Class II Temperatures

Class II Group	Equipment Not Subject to Overloading		Equipment (Such as Motors or Power Transformers) that May Be Overloaded			
			Normal Operation		Abnormal Operation	
	°C	°F	°C	°F	°C	°F
E	200	392	200	392	200	392
F	200	392	150	302	200	392
G	165	329	120	248	165	329

[**70:**Table 500.8(C)(2)]

7.2 Protective Gas.

7.2.1 The protective gas shall be air.

7.2.2* The air shall be essentially free of contaminants or foreign matter and shall contain no more than trace amounts of flammable vapor or gas.

7.2.3* The source of air shall be determined from the nature of the process and the physical layout but shall not be from a classified location.

7.2.4 Any ducts shall be constructed of noncombustible materials.

7.2.5 The fan suction line shall be free of leaks and shall be given protection from mechanical damage and corrosion to prevent ignitable concentrations of flammable gases, vapors, or dusts from being drawn into the control room.

7.3 Considerations Relating to Positive Pressure Ventilation.

7.3.1 The following factors shall be considered in designing a control room suitable for safe operation in a hazardous (classified) location:

(1) The number of people to be housed
(2) The type of equipment to be housed
(3) The location of the control room relative to the direction of the prevailing wind and to the location of process units (e.g., relief valves, vent stacks, and emergency relief systems)

7.3.2* If the control room is in a classified location, it shall be designed to minimize the entry of flammable vapors, gases, liquids, or dusts.

7.4 Requirements for Positive Pressure Air Systems.

7.4.1* The positive pressure air system shall meet the requirements of the following:

(1) Maintain a pressure of at least 25 Pa (0.1 in. of water) in the control room with all openings closed.
(2) Provide a minimum outward velocity of 0.305 m/sec (60 ft/min) through all openings capable of being opened. The velocity shall be measured with all these openings simultaneously open, and a drop in pressure below the 25 Pa (0.1 in. of water) specified in 7.4.1(1) shall be permitted while meeting this requirement.

Exception No. 1: Doorways or other openings that are used solely for infrequent movement of equipment in or out of pressurized control rooms or analyzer rooms shall be permitted to remain closed where all of the following conditions are met:

(1) The control room is under management control.
(2) These doors are marked to restrict use.
(3) These doors are not used for egress.
(4) These doors are secured in the closed position.

Exception No. 2: Gland or bulkhead plates or other similar covers that cannot be removed without the use of a key or tool shall be permitted to remain closed.

7.4.2 The positive pressure air system shall be permitted to include heating, ventilation, and air conditioning equipment, as well as any auxiliary equipment necessary to comply with 7.4.1.

7.4.3 Where there is not a separate air supply source to an air-consuming device (such as a compressor or laboratory hood) in the control room, air shall be supplied to accommodate its needs as well as the needs of the positive pressure air system.

7.4.4 The positive pressure air system shall be designed to provide the required pressure and flow rate for all areas of the control room.

7.4.5 For Type X pressurizing, a cutoff switch shall be incorporated to de-energize power automatically from all circuits within the control room, not approved for the external area classification, upon failure of the positive pressure air system.

Exception: Power to the circuits shall be permitted to be continued for a short period if immediate loss of power would result in a more hazardous condition.

7.4.6 For Type Y and Type Z pressurizing, power to the control room shall not be required to be de-energized upon failure of the positive pressure air system.

7.4.7* Failure of the positive pressure air system shall be detected at the discharge end of the fan and shall activate an alarm at a constantly attended location.

7.4.8* Provisions shall be made to energize the control room safely after interruption of the positive pressure air system. Such provisions shall include:

(1) Checking the atmosphere in the control room with a flammable vapor detector *(see ANSI/ISA 12.13.01-2000)* to determine that the atmosphere contains less than the ignitable concentration of gases or vapors.
(2) Removing accumulations of combustible dust.

7.4.9 The switch, electrical disconnect, and motor for the air system fan shall be approved for the external area classification.

7.4.10 The electrical power for the positive pressure air system shall be taken off the main power line ahead of any service disconnects to the control room or shall be supplied from a separate power source.

Chapter 8 Pressurized Enclosures Having an Internal Source of Flammable Gas or Vapor

8.1 Applicability.

This chapter applies to instruments such as gas chromatographs, gas analyzers, and other enclosures that contain an internal source of flammable gas or vapor.

8.2* General Requirements.

8.2.1 The requirements of Chapters 4, 5, and 6 shall apply, except as modified in this chapter.

8.2.2* For the purpose of this chapter, every protected enclosure shall be considered to have a "normal" per 8.2.2.1 and an "abnormal" per 8.2.2.2 condition, and the electrical equipment in the enclosure is assumed to be operating correctly in both conditions.

8.2.2.1 "Normal" shall mean the anticipated release of flammable gas or vapor within the enclosure when the system that supplies the flammable gas or vapor is operating properly. The magnitude of this anticipated release is one of the following:

(1) None — There is no release of flammable gas or vapor, or the release of flammable gas or vapor is documented to reflect that it is of such a low level that without ventilation and/or purge the concentration is not capable of reaching 25 percent of the lower flammable limit.
(2) Limited — There is a release of flammable gas or vapor, but the release is limited to an amount that can be diluted by the pressurizing system to a concentration less than 25 percent of the lower flammable limit.

8.2.2.2 "Abnormal" shall mean the anticipated release of flammable gas or vapor within the enclosure when the system that supplies the flammable gas or vapor is either leaking or is otherwise operating abnormally. The magnitude of this anticipated release is one of the following:

(1) Limited — The release of flammable gas or vapor is limited to an amount that can be diluted by the pressurizing system to a concentration less than 25 percent of the lower flammable limit.
(2) Unlimited — The release of flammable gas or vapor is of such magnitude that it cannot be diluted by the pressurizing system to a concentration less than 25 percent of the lower flammable limit.

8.2.2.3 Precautions shall be taken if the abnormal condition release may be great enough to adversely affect an external area classification.

8.2.3* Pressurizing requirements shall be established according to Table 8.2.3.

8.2.4 Protected enclosures containing an open flame shall comply with 8.2.4.1 and 8.2.4.2.

8.2.4.1 Protected enclosures containing an open flame shall be considered to have equipment suitable for unclassified locations for the purposes of determining the pressurizing requirement according to Table 8.2.3.

8.2.4.2 Open flames within a protected enclosure shall be automatically extinguished upon failure of the pressurization system, regardless of the type of pressurizing.

8.3 Specific Requirements.

8.3.1 Where a release of flammable gas or vapor within an enclosure can occur either in normal operation or under abnormal conditions, protection shall be provided by one of the following:

(1) Diluting with air to maintain the concentration of flammable gas, vapor, or mixture to less than 25 percent of its lower flammable limit, based on the lowest value of the lower flammable limit of any individual flammable gas or vapor entering the enclosure
(2) Diluting or pressurizing with inert gas to reduce the oxygen content in the enclosure to a level of not more than 5 percent by volume or to 50 percent of the minimum concentration of oxygen required to form a flammable mixture, whichever is lower.

8.3.2 Where the protected enclosure is located in a Class I or Class II area, the pressurizing system shall also prevent entrance of the external atmosphere by providing a minimum internal pressure of 25 Pa (0.1 in. of water).

8.3.3 The locations and sizes of gas or vapor outlets in the protected enclosure shall be designed to allow effective removal of both the flammable gas or vapor and the protective gas.

8.3.4 Where an inert protective gas is used, the outlets shall be permitted to be closed after purging to prevent undue loss

Table 8.2.3 Pressurizing Requirements for Enclosures Subject to Internal Release

(1) External Area Classification	(2) Classified or Unclassified Location that Internal Equipment Is Suitable for	(3) Pressurizing Requirements for Limited Release Under Abnormal Conditions		(4) Additional Requirements for Unlimited Release Under Abnormal Conditions
		No Release Under Normal Conditions	Limited Release Under Normal Conditions	
Class I, Division 1 (Class I, Zone 1)	Class I, Division 1 (Class I, Zone 1)	None	None	None
	Class I, Division 2 (Class I, Zone 2)	Y	Y	None
	Unclassified	X	X	Inert[1]
Class I, Division 2 (Class I, Zone 2)	Class I, Division 1 (Class I, Zone 1)	None	None	None
	Class I, Division 2 (Class I, Zone 2)	None	Z	None
	Unclassified	Z	X	Inert[1]
Class II	Class I, Division 1 (Class I, Zone 1)	None	None	None
	Class I, Division 2 (Class I, Zone 2)	None	Z	None
	Unclassified	Z	X	Inert[1]
None	Class I, Division 1 (Class I, Zone 1)	None	None	None[2]
	Class I, Division 2 (Class I, Zone 2)	None	Z	None[2]
	Unclassified	Z	X	Inert[1]

Note: To determine the pressurizing requirements according to Table 8.2.3:
(1) Find the external area classification in column (1).
(2) Find the internal equipment type in column (2).
(3) Determine the pressurizing requirement for limited release under abnormal conditions by using the appropriate normal condition in column (3).
(4) Determine any additional requirements from column (4) if the abnormal condition is unlimited release.
[1] See A.8.2.3.
[2] See 8.2.2.3.

of inert protective gas, provided that this does not constitute a further danger such as inadequate flow of protective gas or excessive pressure buildup.

8.3.5 In applications where flammable mixtures shall be permitted to be piped into the enclosure through the flammable gas or vapor system, precautions shall be taken to prevent propagation of an explosion back to the process equipment.

8.3.6 The flow rate of protective gas shall be sufficient to maintain the requirements of 8.3.1 and to ensure adequate mixing, so that the release of a flammable gas or vapor is limited.

8.3.7 To achieve proper pressurization with air, caution shall be required to ensure that the air pressure used within the enclosure does not exceed the pressure of the flammable gas or vapor system supplying the enclosure, as air could enter the process, causing possible problems such as explosive concentrations of the flammable gas or vapor, corrosion, or oxidation.

8.3.8 Precautions shall be taken to protect the enclosure from excessive pressure of the protective gas supply.

Chapter 9 Pressurized Analyzer Rooms Containing a Source of Flammable Gas, Vapor, or Liquid

9.1 Applicability.

This chapter applies to analyzer rooms and buildings containing electrical equipment having process streams of flammable liquid, vapor, or compressed flammable gas piped into the equipment.

9.2 General.

9.2.1 For the purpose of this chapter, every pressurized analyzer room containing a source of flammable gas, vapor, or liquid shall be considered to have one of the following types of anticipated releases:

(1) None — There is no release of flammable gas or vapor, or the release of flammable gas or vapor is documented to reflect that it is of such a low level that, without ventilation and/or purge, the concentration is not capable of reaching 25 percent of the lower flammable limit.
(2) Limited — There is a release of flammable gas or vapor, but the release is limited to an amount that can be diluted by the pressurizing system to a concentration less than 25 percent of the lower flammable limit.
(3) Unlimited — There is a release of flammable gas or vapor, and the release is of such a magnitude that it cannot be diluted by the pressurizing system to a concentration less than 25 percent of the lower flammable limit.

9.2.2 The magnitude of the anticipated release within the analyzer room shall be "none" or "limited" based on the largest single failure.

9.2.3 Where the analyzer room is in a hazardous (classified) location, it shall be designed to prevent the entry of flammable gases and vapors, flammable liquids, and combustible dusts.

9.2.4 The requirements of Chapter 7 for control rooms shall apply except as modified in this chapter.

9.2.5* Analyzer rooms shall be separated from control rooms by distance or by a wall impermeable to vapors.

9.2.6 Flow of air through the room shall ensure adequate air distribution.

9.2.7* Flammable vapors shall be removed as close to their source as is practical.

9.2.8* The following shall apply where personnel can enter an analyzer room that is purged with inert gases:

(1) Administrative controls combined with training and safe entry procedures shall be established.
(2) Warning signs advising of the hazard of inert gas shall be posted.
(3)* Inert gas shall not be used for purging an entire analyzer room where personnel shall be permitted to enter.

9.3 Specific Requirements.

9.3.1 Flow-Limiting Devices.

9.3.1.1 To prevent an unlimited release in the analyzer room, process streams shall have orifices or other flow-limiting devices on the inlets and on the outlet, if the outlet can constitute a source of uncontrolled leakage from the process.

9.3.1.2 Orifices or other flow-limiting devices shall be located outside and close to the wall of the building or room.

9.3.2* Where flammable vapor, gas, or liquid is discharged from an enclosure (e.g., analyzer enclosure), it shall not create a hazard within the analyzer house or to the surroundings.

9.3.3 Sample conditioning equipment (such as equipment used for heating, cooling, or drying) shall be suitable for the area electrical classification.

9.3.4 Process piping within the analyzer room shall be minimized.

9.3.5 Means for emergency isolation of the process from the analyzers shall be provided outside the analyzer building.

9.3.6 False ceilings and floors shall not be used in analyzer rooms.

9.3.7* Ventilation fans shall be constructed to minimize the possibility of sparking.

9.3.8 In the event of pressurizing failure the following shall be required:

(1) An audible and visual alarm shall be activated at a constantly attended location.
(2)* Electrical power to ignition-capable equipment within the analyzer room shall be automatically shut down.
(3) Open flames shall be automatically extinguished.
(4) Power shall not be restored until the analyzer room is below ignitable concentration of the combustible material.

Exception: Automatic shutdown shall not be required under any of the following conditions:

(1) If the anticipated release is "none," the analyzer room is unclassified, and the area outside the analyzer room is unclassified.
(2) If the anticipated release is "none," the analyzer room

is unclassified, and the area outside the analyzer room is classified Class I, Division 2 or Class I, Zone 2.

(3) If the anticipated release is "limited," and the analyzer room is classified as Class I, Division 2 or Class I, Zone 2.

(4) If the analyzer room is classified as Class I, Division 1 or Class I, Zone 1.

9.3.9 Where gas or vapor mixtures within the flammable range are piped to the analyzer room, precautions shall be taken to prevent propagation of an explosion back to the process equipment.

Annex A Explanatory Material

Annex A is not a part of the requirements of this NFPA document but is included for informational purposes only. This annex contains explanatory material, numbered to correspond with the applicable text paragraphs.

A.1.1.2 Electrical equipment should be located in an area having as low a degree of hazard classification as is practical. Where there is probability of flammable liquid exposure, additional means should be taken to avoid ingress.

A.1.3.3 The flammable gas or vapor is piped internally to the enclosure so that process parameters can be measured. The source of release could be fittings or vents. It is not intended that fumes or vapors from components within the electrical equipment be considered (e.g., from decomposing insulation).

A.2.1 Editions of the referenced documents are the editions used in preparation of this document. It is important that the authority having jurisdiction (AHJ) be aware that later editions might exist. Compliance with later editions should be considered when the requirements of the most current edition of the referenced documents have changed.

A.3.2.1 Approved. The National Fire Protection Association does not approve, inspect, or certify any installations, procedures, equipment, or materials; nor does it approve or evaluate testing laboratories. In determining the acceptability of installations, procedures, equipment, or materials, the authority having jurisdiction may base acceptance on compliance with NFPA or other appropriate standards. In the absence of such standards, said authority may require evidence of proper installation, procedure, or use. The authority having jurisdiction may also refer to the listings or labeling practices of an organization that is concerned with product evaluations and is thus in a position to determine compliance with appropriate standards for the current production of listed items.

A.3.2.2 Authority Having Jurisdiction (AHJ). The phrase "authority having jurisdiction," or its acronym AHJ, is used in NFPA documents in a broad manner, since jurisdictions and approval agencies vary, as do their responsibilities. Where public safety is primary, the authority having jurisdiction may be a federal, state, local, or other regional department or individual such as a fire chief; fire marshal; chief of a fire prevention bureau, labor department, or health department; building official; electrical inspector; or others having statutory authority. For insurance purposes, an insurance inspection department, rating bureau, or other insurance company representative may be the authority having jurisdiction. In many circumstances, the property owner or his or her designated agent assumes the role of the authority having jurisdiction; at government installations, the commanding officer or departmental official may be the authority having jurisdiction.

A.3.3.1 Alarm. An alarm is intended to alert the user that the pressurizing system should be immediately repaired or that the electrical equipment protected by the failed pressurizing system should be removed from service.

A.3.3.5 Ignition Temperature. Normally, the minimum ignition temperature of a layer of a specific dust is lower than the minimum ignition temperature of a cloud of that dust. Since this is not universally true, the lower of the two minimum ignition temperatures is listed in NFPA 499.

A.3.3.9 Pressurizing System. The pressurizing system may include components such as the alarm actuator, indicator, cutoff switch, or components of the protective gas supply. The components may be mounted in a separate enclosure/panel or be included within the protected enclosure.

A.3.3.15 Specific Particle Density. Specific particle density (sometimes referred to as the true density) is the mass per unit volume or, more commonly, weight per unit volume and is expressed as kilograms per cubic meter (pounds per cubic foot). It refers only to the material making up the particle. The term *bulk density* is obtained by placing granular or powdered material in a specified volume and calculating the density. Bulk density includes the void space between the particles created because of the irregular particle shape. As an example, the specific particle density of sulfur is about 2083 kg/m^3 (130 lb/ft^3) while the bulk density of pulverized sulfur dust is about 801 kg/m^3 (50 lb/ft^3).

A.4.2.2 During brief periods of purging, the area around the vent may contain a concentration of flammables that requires caution.

A.4.2.3 Pressurized raceways do not need to be sealed if they have been properly designed as part of pressurized systems with the required alarms or indicators. The exception is not meant to allow the user to install the equipment and ignore proper installation of classified location wiring. The exception allows the same raceway to be used for electrical wiring and the protective gas. The design must consider the restriction of protective gas flow when conductors are installed in the raceway.

A.4.3.1 The reason for requiring that a positive pressure be maintained is to prevent flammable vapors or gases from being forced into the enclosure by external air currents.

A.4.4.1 Air filtration may be desirable.

A.4.4.1.2 Ordinary plant compressed air is usually not suitable for purge or pressurizing systems, due to contaminants that may cause equipment to malfunction.

A.4.4.4 The compressor suction line should not pass through any area having a hazardous atmosphere, unless it is not practical to do otherwise.

A.4.5 The T Code is based on the ambient temperature surrounding the pressurized equipment not exceeding 40°C (104°F). The maximum ambient temperature rating of the equipment must not be exceeded.

A.4.5.1 Because a high-temperature source of ignition is not immediately removed by de-energizing the equipment, additional precautions are necessary for hot components. If the external temperature of the enclosure is greater than the autoignition temperature (in degrees Celsius) of the gas or vapor, it is obvious that purging will not prevent an explosion. Thus, it is essential that excess surface temperature be prevented, unless it has been specifically shown to be safe by a qualified testing laboratory. Dust that is carbonized or excessively dry is highly susceptible to spontaneous ignition. Sources of internal temperatures above the autoignition temperature (in degrees Celsius) of the gas or vapor involved, such as vacuum tube filaments, are hermetically sealed to prevent them from contacting the atmosphere that may become hazardous. However, it is essential that the surface of the glass envelope does not exceed the 80 percent limit, unless shown by test to be safe.

A.4.5.1(2) The ignition temperature of gases and vapors that is listed in reference documents such as NFPA 497 is determined under conditions where a significant volume of gas is at the same temperature. The condition specified in 4.5.1(2), Exception (2) indicates that when ignition is attempted with a small component, convection effects and partial oxidation at the surface of the component decrease the rate of heat transfer to the gas. Therefore, the component must be at a temperature much higher than the quoted ignition temperature to ignite the flammable mixture. Typical transistors, resistors, and similar small components must have a surface temperature of 220°C to 300°C (428°F to 572°F) to ignite diethyl ether whose ignition temperature is 160°C (320°F). Similar values have been measured in ignition tests of carbon disulfide whose ignition temperature is 100°C (212°F).

A.4.6 Airflow required for cooling may be more than that required for purging.

A.4.7 Enclosures containing power equipment are more likely to produce ignition-capable particles. It is necessary to have an enclosure through which these particles cannot burn or escape from openings other than the vents. Other techniques may be used to ensure that ignition-capable particles are contained in the enclosure. Nonmetallic enclosure flammability ratings of 94 V-0 or 94 5V are considered as substantially noncombustible. *(See ANSI/UL94-1996 for description of flammability ratings.)*

A.4.8 Type Z pressurizing reduces the classification within an enclosure from Division 2 to unclassified. With Type Z pressurizing, a hazard is created only if the pressurizing system fails simultaneously with the area outside of the enclosure becoming ignitable due to the release of normally contained flammable liquids or gases, or combustible dust. For this reason, it is not considered essential to remove power from the equipment upon failure of the pressurizing system.

Alarm and indicator configurations for Types Y and Z pressurizing are shown in Figure A.4.8.

A.4.8.1.1 An alarm is preferred, but an indicator is acceptable if the protected enclosure is much less likely to fail than the protective gas supply. Excessive leakage from the protected enclosure is only likely during servicing, at which time the indicator will assist the maintenance personnel in determining when the enclosure is adequately sealed to maintain pressure.

A.4.9 Type Y pressurizing reduces the classification within an enclosure from Division 1 to Division 2. Equipment and devices within the enclosure must be suitable for Division 2. This requires that the enclosure does not contain an ignition source. Thus, a hazard is created within the enclosure only upon simultaneous failure of the pressurizing system and of the equipment within the enclosure. For this reason, it is not considered essential to remove power from the equipment upon the failure of the pressurizing system.

A.4.10 Type X pressurizing reduces the classification within an enclosure from Division 1 to unclassified. Because the probability of a hazardous atmosphere external to the enclo-

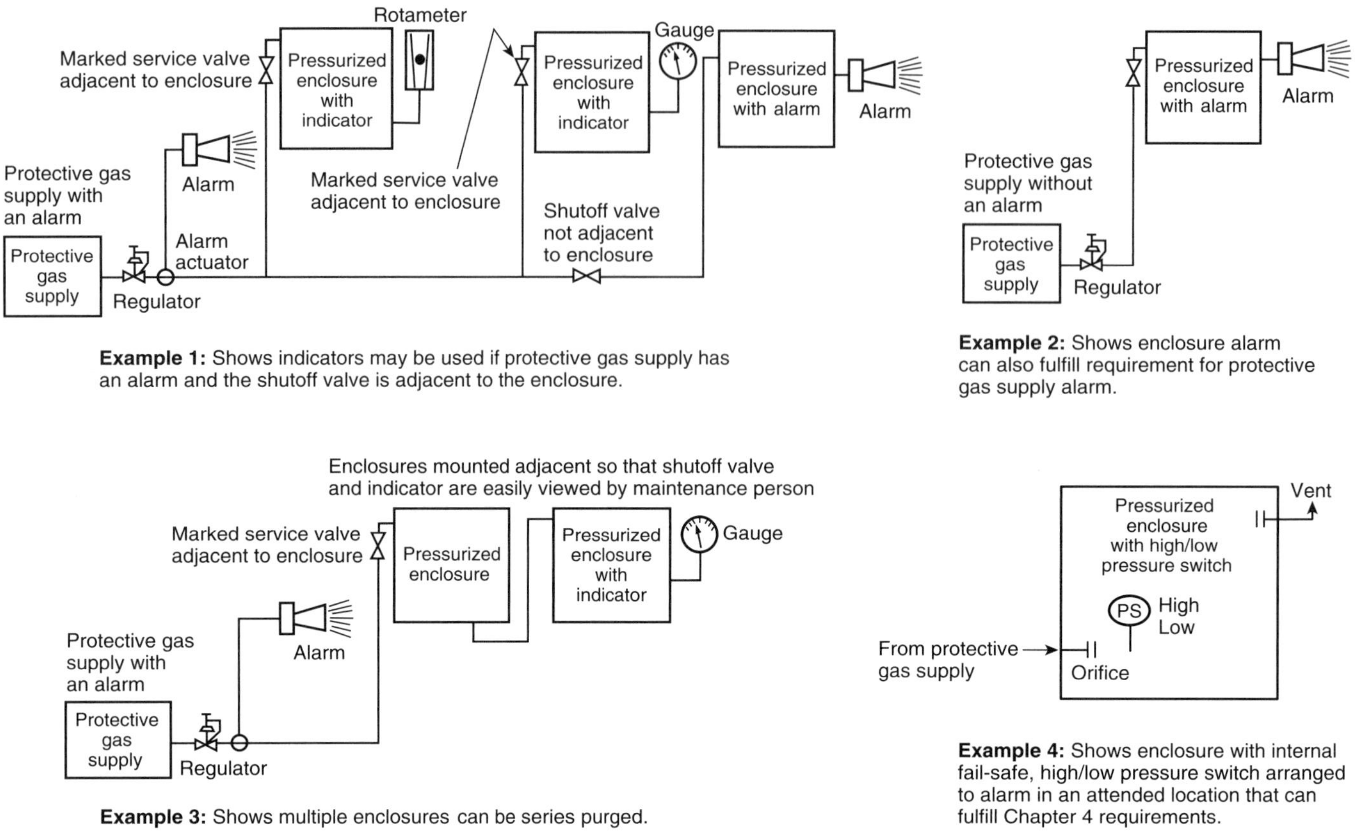

FIGURE A.4.8 Typical Alarm and Indicator Configurations for Types Y and Z Pressurizing. Purge outlet devices that could be provided are not shown for clarity.

sure is high and the enclosure normally contains a source of ignition, it is essential that any interruption of the pressurizing results in de-energizing of the equipment. Also, it is essential that the enclosure be tight enough to prevent escape of molten metal particles or sparks.

A.4.10.1 Power to the circuits may be continued for a short period where the Division 1 location only has a flammable concentration on an intermittent basis, and where entrance of the external atmosphere would be slow because the protected enclosure is tightly sealed. Where flammable concentrations occur frequently or enclosure failure may be catastrophic, Type Y pressurizing should be used if it is necessary to continue operating the process to prevent a more hazardous condition.

A.4.10.2 Overload conditions need only be a concern where the motor load or the transformer load is not determined by the product but by external variable loading in the actual application.

A.5.2.6 In order for any internal or adjacent enclosure to be automatically purged as the main enclosure is purged, adequate vents must be provided to permit air circulation between the two enclosures. The area required to provide adequate venting will depend on the volume of the internal or adjacent enclosure. It is considered that meeting this requirement will prevent the formation of unpurged pockets of gas or vapor within the enclosure. This does not imply that internal or adjacent enclosures not meeting these requirements are prohibited, but that such enclosures must be provided with their own purge systems.

A.5.2.6(3) Cathode ray tubes (CRTs) are hermetically sealed components.

A.5.4 Any time the enclosure has been opened or the pressurizing system has failed, the possibility exists that an ignitable mixture may have accumulated in the enclosure. For enclosures that are effectively subdivided by internal parts, a greater purge volume may be necessary.

A.5.5.2 It is essential that any door access that can be opened by untrained personnel be protected with interlock switches. Consistent with the practice that has been established with explosionproof enclosures, it is considered that the commonly displayed warning nameplate is adequate protection for an enclosure that requires the use of a tool to be opened.

A.6.2.2 Cleaning should be done using a method that will not create a dust cloud (e.g., vacuuming or brushing). Use of compressed air should be avoided. Protected enclosures should normally be kept closed whether the equipment is in operation or not.

A.6.2.4 The density of 2083 kg/m^3 (130 lb/ft^3) is slightly greater than that of sulfur dust, which was one of the dusts used in performing the tests on which the values in Table 6.2.4 are based. The pressures in the table are based on the assumption that the maximum crack width exposed to falling dust is 0.4 mm (1⁄64 in.). The ability of a dust to enter an opening due to the force of gravity against an outward velocity of gas is directly proportional to its specific particle density.

A.6.2.5 Equipment installed in Class II locations should be able to function at full rating without developing surface temperatures high enough to cause excessive dehydration or gradual carbonization of any organic dust deposits that may occur.

A.6.5 A hazard is created within an enclosure only after the pressure has failed and enough dust to be ignitable penetrates into the enclosure. This takes an appreciable length of time with any normally tight enclosure. Because of this, it is not always considered essential to remove the power from the equipment automatically upon failure of the pressurizing. It is necessary only to provide an adequate warning so that operations will not continue indefinitely without pressurizing. It is essential that the enclosure be tight enough to prevent escape of sparks or burning material. Examples of enclosures that are tightly sealed to prevent the entrance of dust are Type 3, Type 3S, Type 4, Type 4X, Type 6, Type 6P, Type 12, or Type 13 enclosures.

A.7.1 Control rooms commonly house one or more of the following facilities:

(1) Process control instruments and panels
(2) Data processing equipment
(3) Communications equipment
(4) Lighting, power equipment, and related equipment
(5) Emergency power equipment
(6) Lunch, restroom, and locker facilities
(7) Offices and maintenance facilities
(8) Heating and ventilating equipment

A.7.2.2 Air filtration may be desirable.

A.7.2.3 Ordinarily, air can be taken from an area to one side of a process area where there is a minimum chance of flammable gases or vapors or combustible dusts being found. The elevation of the fan suction depends on the density of the gases, vapors, or dust under handling temperatures and adverse atmospheric conditions. For a control room in the center of a process area, ducting may be necessary.

A.7.3.2 To prevent entry of flammable vapors, gases, or dusts, positive pressure ventilation using a source of clean air may be used, and the equipment in the control room need not be housed in special enclosures. To prevent entry of flammable liquids, differences in elevation or use of dikes, and so forth, may be required.

A.7.4.1 A minimum number of doors should be provided so that positive pressures can be maintained, but, at the same time, the number of doors should be adequate for safe exit.

A.7.4.7 Suitable devices for detecting loss of air pressure include velocity pressure switches, static pressure switches, and plenum chambers with orifices. Electrical interlocks on the fan motor are not adequate, since belt slippage, loose impellers, or backward rotation of the fan would not be detected.

A.7.4.8 An enforced purge wherein an interlock timer requires proof of purging for a set period of time prior to energizing the control room should be considered.

A.8.2 The consequences of a release of flammable gas or vapor into an enclosure are substantially more serious than a similar release to the open atmosphere. Through the use of a pressurizing system, these consequences may be reduced, and electrical equipment not otherwise acceptable for a flammable atmosphere may be utilized. The effect of a temporary leak in the open is a transient rise in concentration of flammable gas or vapor in the atmosphere. A leak inside an enclosure, in the absence of purging, remains within the enclosure, and if undetected will slowly raise the concentration inside the enclosure until its atmosphere becomes ignitable. This increase in concentration is likely to be slowed only slightly by breathing and diffusion.

A.8.2.2 Because of the confining property of electrical equipment enclosures, it is necessary to view "normal" and "abnormal" conditions in terms of a longer time span than is necessary in considering releases in the open. "Normal" must include consideration of the probable operation of the apparatus after some years of service and includes degradation of the system components over time. For no release within an enclosure under normal conditions, there must be a minimum risk (i.e., very low probability) that flammable material will escape from its containment system during the time the apparatus is in service and within the range of service conditions to which it is likely to be subjected. Therefore, materials and types of construction that degrade in service or with age and that are not likely to be maintained or replaced cannot be considered to permit a "normal condition — no release" as defined in 8.2.2.

Although specific rules that will apply to all designs cannot be written, in general, a design will be considered to have no normal release if the flammable gas or vapor is enclosed in metallic pipes, tubes, vessels, or elements such as bourdons, bellows, or spirals; in systems that contain no moving seals; and if prototype systems do not leak when tested at 1.5 times their rated pressure, except in cases where another safety factor is applicable. Joints made with pipe threads, welding, metallic compression fittings, or other equally reliable methods would usually be considered to have no normal release. Windows, elastomeric seals, and nonmetallic flexible tubing would in most cases not meet the requirement for "normal condition — no release" unless it can be demonstrated that time and environment will not degrade them below the leakage level expected of the operating pipe threads and compression seals. Systems that cannot meet a stringent interpretation of these guidelines should be considered "normal condition — limited release." Seals, rotating or sliding seals, flanged joints, and flexible nonmetallic tubing can be assumed to leak minutely after a period of service. Attention must be given to the possibility that expected degradation of components may result in release of flammable gas or vapor at a rate faster than that which the dilution system can handle. Such situations are not common but, when encountered, they should not be classified as "normal condition — limited release." The prime criterion for "normal condition — limited release" is that the dilution capability of the protective system must not be exceeded. In enclosures having open flames in normal operation, it is assumed that flame extinguishment is a normal occurrence and should be classified as a normal release unless loss of flame automatically stops the flow of flammable gas or vapor.

A limited abnormal release is one that, by design, is maintained at a level within the dilution capability of the protective system. The limiting element may be a restriction in the flow line. In the case of designs using elastomeric seals, the limiting flow may be considered to be the flow that would exist were the seal not in place.

A.8.2.3 Electrical equipment permitted in unclassified locations may contain arcing or sparking contacts or may have hot surfaces. If there is no normal release within the equipment enclosure, a single failure of the system containing the flammable gas may provide an ignitable atmosphere. The ignition source is always present, by virtue of the electrical equipment. Purging is, therefore, required, and Type Z pressurizing will provide adequate protection. If, however, there is a limited release under normal conditions and there is limited release under abnormal conditions, then Type X pressurizing is required. In this case, purging with air is satisfactory. Type X pressurizing requires that the electrical power to the protected enclosure be disconnected upon failure of the pressurizing system. Disconnection is required because, under the conditions described, an ignitable atmosphere may be generated in the presence of arcing or sparking equipment or hot surfaces. Electrical equipment suitable for Class I, Division 2 or Class I, Zone 2 locations may present a source of ignition only upon failure or other abnormal conditions. If there is no normal release within the equipment enclosure, no purging is required because there is not normally a source of ignition present, even if the system containing the flammable gas fails. If there is limited normal release and limited abnormal release, then Type Z pressurizing, with air, provides adequate protection. If, however, the abnormal release is unlimited (i.e., beyond the dilution from a single failure of the containment system), then air is not permitted as the protective gas for such enclosures when the electrical equipment is only suitable for unclassified locations. Inert gas must be used so that an ignitable atmosphere is prevented from developing (unless, of course, the pressurizing system itself fails).

For an air-purged enclosure containing equipment suitable for Class I, Division 2 or Class I, Zone 2 locations, although an unlimited abnormal release results in a ignitable flammable atmosphere, the electrical equipment is assumed to be operating normally and therefore does not present a source of ignition. However, if a failure of the containment system is not obvious, inert gas purging should be used because of the danger that a flammable atmosphere may exist for a prolonged period of time, during which the electrical equipment may also fail and provide the source of ignition. Whether electrical equipment is located in a Class I, Division 2, a Class I, Zone 2, or an unclassified location does not affect the need for a pressurizing system. For Class I, Division 2 or Class I, Zone 2 locations, the pressurizing system serves two purposes: (1) to prevent the external atmosphere from entering the enclosure, and (2) to dilute any flammable gas released within the enclosure. In an unclassified location, the purging serves only to dilute any flammable gas released within the enclosure.

A.9.2.5 Flammable gases, vapors, or liquids for analysis should not be piped into control rooms because of the danger of ignition.

A.9.2.7 Flammable hydrocarbon vapors are usually heavier than air and should be removed at floor level. Lighter-than-air gases such as hydrogen and methane should be removed at the ceiling level.

A.9.2.8 Leakage of inert gases used for purging or pressurizing of enclosures in an analyzer room can deplete the room's oxygen.

A.9.2.8(3) See NFPA 69.

A.9.3.2 Flammable gases or vapors should be discharged at a safe point outside the analyzer room in an upward or horizontal direction to aid in dispersion. The vent discharge should be located at least 1.5 m (5 ft) away from building openings and at least 3.7 m (12 ft) above grade level. The vent design and location should further consider possible trapping of vapors by eaves or other obstructions. *(See NFPA 30.)*

A.9.3.7 The fan motor and associated control equipment should be located external to the ductwork or should be suitable for the location.

A.9.3.8(2) It is assumed that the flow of flammable vapors or liquids will continue in case of failure of the ventilation system, and that the atmosphere in the analyzer room will reach the flammable range. In these situations, power must be removed to avoid ignition.

Annex B Informational References

B.1 Referenced Publications.

The following documents or portions thereof are referenced within this standard for informational purposes only and are thus not part of the requirements of this document unless also listed in Chapter 2.

B.1.1 NFPA Publications. National Fire Protection Association, 1 Batterymarch Park, P.O. Box 9101, Quincy, MA 02269-9101.

NFPA 30, *Flammable and Combustible Liquids Code*, 2003 edition.

NFPA 69, *Standard on Explosion Prevention Systems*, 2002 edition.

NFPA 497, *Recommended Practice for the Classification of Flammable Liquids, Gases, or Vapors and of Hazardous (Classified) Locations for Electrical Installations in Chemical Process Areas*, 1997 edition.

NFPA 499, *Recommended Practice for the Classification of Combustible Dusts and of Hazardous (Classified) Locations for Electrical Installations in Chemical Process Areas*, 1997 edition.

B.1.2 Other Publications.

B.1.2.1 UL Publication. Underwriters Laboratories, 333 Pfingsten Road, Northbrook, IL 60062.

ANSI/UL94-1996, *Safety Test for Flammability of Plastic Materials for Parts in Devices and Appliances*.

B.2 Informational References.

The following documents or portions thereof are listed here as informational resources only. They are not a part of the requirements of this document.

B.2.1 ASTM Publications. American Society for Testing and Materials, 100 Barr Harbor Drive, West Conshohocken, PA 19428-2959.

ASTM E 659-1994, *Standard Test Method for Determining the Autoignition Temperature of Liquid Chemicals*.

ASTM D 2155-66-1976, *Method of Test for Autogenous Ignition Temperatures of Petroleum Products*.

B.2.2 Other Publications.

Dorsett, H. G., et al., 1960. *Laboratory Equipment and Test Procedures for Evaluating Explosibility of Dusts*, RI 5624. Pittsburgh, PA: U.S. Bureau of Mines.

Electrical Safety Practices, Monograph 112. 1969. McCarron, R., "Report of an Investigation of the Effect of Internal Arcing Versus External Spot Temperatures of Metal Instrument Cases," Instrument Society of America (Instrumentation, Systems and Automation Society).

ISA RP12.4-1996, *Pressurized Enclosures,* Instrumentation, Systems and Automation Society, Research Triangle Park, NC 27709.

Magison, Ernest. 1998. *Electrical Instruments in Hazardous Locations*, 4th Edition, Instrumentation, Systems and Automation Society, Research Triangle Park, NC.

McMillan, Alan. 1998. *Electrical Installations in Hazardous Areas*, Butterworth-Heinemann 225 Wildwood Ave, Woburn, MA 01801-2041.

Perry, R. H., and D. Green. 1984. *Chemical Engineer's Handbook*, 7th Edition. McGraw-Hill, New York, NY.

Schram, Peter J. and Earley, Mark W. 1997, *Electrical Installations in Hazardous Locations*, 2nd Edition, NFPA, 1 Batterymarch Park, P.O. Box 9101, Quincy, MA 02269-9101.

Zenz, F. A., and D. F. Othmer. 1960. *Fluidization and Fluid Particle Systems*. Reinhold.

B.3 References for Extracts.

The following documents are listed here to provide reference information, including title and edition, for extracts given throughout this standard as indicated by a reference in brackets [ ] following a section or paragraph. These documents are not a part of the requirements of this document unless also listed in Chapter 2 for other reasons.

NFPA 70, *National Electrical Code®*, 2002 edition.

Formal Interpretation

NFPA 496

Purged and Pressurized Enclosures for Electrical Equipment

2003 Edition

Reference: 7.4.1
F.I.

Paragraph 7.4.1 requires a positive pressure purge system to:

(1) Maintain a pressure of at least 25 Pa (0.1 in. of water) in the control room with all openings closed.
(2) Provide a minimum outward velocity of 0.305 m/sec (60 ft/min) through all openings capable of being opened. The velocity shall be measured with all these openings simultaneously open and a drop in pressure below the 25 Pa (0.1 in. of water) specified in 7.4.1(1) shall be permitted while meeting this requirement.

Exception No. 1: Doorways or other openings that are used solely for infrequent movement of equipment in or out of pressurized control rooms or analyzer rooms shall be permitted to remain closed where all of the following conditions are met:

(1) The control room is under management control.

(2) These doors are marked to restrict use.

(3) These doors are not used for egress.

(4) These doors are secured in the closed position.

Question: Is it the intent of 7.4.1 to require that the purge system function as "(1)" and "(2)" simultaneously?

Answer: No.

Issue Edition: 1974
Reference: 3-3.1
Date: December 1977

Copyright © 2003 All Rights Reserved
NATIONAL FIRE PROTECTION ASSOCIATION

NFPA 497, *Recommended Practice for the Classification of Flammable Liquids, Gases, or Vapors and of Hazardous (Classified) Locations for Electrical Installations in Chemical Process Areas,* 2004

Editor's Note: *This document is reprinted here in its entirety, including its annexes.*

Chapter 1 Administration

1.1 Scope.

1.1.1 This recommended practice applies to those locations where flammable gases or vapors, flammable liquids, or combustible liquids are processed or handled; and where their release into the atmosphere could result in their ignition by electrical systems or equipment.

1.1.2 This recommended practice provides information on specific flammable gases and vapors, flammable liquids, and combustible liquids, whose relevant combustion properties have been sufficiently identified to allow their classification into the groups established by NFPA 70, *National Electrical Code*® (*NEC*®), for proper selection of electrical equipment in hazardous (classified) locations. The tables of selected combustible materials contained in this document are not intended to be all-inclusive.

1.1.3 This recommended practice applies to chemical process areas. As used in this document, a chemical process area could be a large, integrated chemical process plant or it could be a part of such a plant. It could be a part of a manufacturing facility where flammable gases or vapors, flammable liquids, or combustible liquids are produced or used in chemical reactions, or are handled or used in certain unit operations such as mixing, filtration, coating, spraying, and distillation.

1.1.4 This recommended practice does not apply to situations that could involve catastrophic failure of or catastrophic discharge from process vessels, pipelines, tanks, or systems.

1.1.5 This recommended practice does not apply to oxygen-enriched atmospheres or pyrophoric materials.

1.2 Purpose.

The purpose of this recommended practice is to provide the user with a basic understanding of the parameters that determine the degree and the extent of the hazardous (classified) location. This recommended practice also provides the user with examples of the applications of these parameters.

1.2.1 Information is provided on specific flammable gases and vapors, flammable liquids, and combustible liquids, whose relevant properties determine their classification into groups. This will assist in the selection of special electrical equipment for hazardous (classified) locations where such electrical equipment is required.

1.2.2 This recommended practice is intended as a guide and should be applied with sound engineering judgment. Where all factors are properly evaluated, a consistent area classification scheme can be developed.

1.3 Relationship to NFPA Codes and Standards.

This recommended practice is not intended to supersede or conflict with the following NFPA codes and standards:

(1) NFPA 30, *Flammable and Combustible Liquids Code*, 2003 edition.
(2) NFPA 33, *Standard for Spray Application Using Flammable or Combustible Materials*, 2003 edition.
(3) NFPA 34, *Standard for Dipping and Coating Processes Using Flammable or Combustible Liquids*, 2003 edition.
(4) NFPA 35, *Standard for the Manufacture of Organic Coatings*, 1999 edition.
(5) NFPA 36, *Standard for Solvent Extraction Plants*, 2004 edition.
(6) NFPA 45, *Standard on Fire Protection for Laboratories Using Chemicals*, 2000 edition.
(7) NFPA 50A, *Standard for Gaseous Hydrogen Systems at Consumer Sites*, 1999 edition.
(8) NFPA 50B, *Standard for Liquefied Hydrogen Systems at Consumer Sites*, 1999 edition.
(9) NFPA 58, *Liquefied Petroleum Gas Code*, 2004 edition.
(10) NFPA 59A, *Standard for the Production, Storage, and Handling of Liquefied Natural Gas (LNG)*, 2001 edition.

Chapter 2 Referenced Publications

2.1 General.

The documents or portions thereof listed in this chapter are referenced within this recommended practice and should be considered part of the recommendations of this document.

2.2 NFPA Publications.

National Fire Protection Association, 1 Batterymarch Park, Quincy, MA 02169-7471.

NFPA 30, *Flammable and Combustible Liquids Code,* 2003 edition.

NFPA 33, *Standard for Spray Application Using Flammable or Combustible Materials,* 2003 edition.

NFPA 34, *Standard for Dipping and Coating Processes Using Flammable or Combustible Liquids,* 2003 edition.

NFPA 35, *Standard for the Manufacture of Organic Coatings,* 1999 edition.

NFPA 36, *Standard for Solvent Extraction Plants,* 2004 edition.

NFPA 45, *Standard on Fire Protection for Laboratories Using Chemicals,* 2000 edition.

NFPA 50A, *Standard for Gaseous Hydrogen Systems at Consumer Sites,* 1999 edition.

NFPA 50B, *Standard for Liquefied Hydrogen Systems at Consumer Sites,* 1999 edition.

NFPA 58, *Liquefied Petroleum Gas Code,* 2004 edition.

NFPA 59A, *Standard for the Production, Storage, and Handling of Liquefied Natural Gas (LNG),* 2001 edition.

NFPA 70, *National Electrical Code®*, 2002 edition.

2.3 Other Publications.

2.3.1 API Publications.

American Petroleum Institute, 1220 L Street, N.W., Washington, DC 20005-4070.

ANSI/API RP 500, *Recommended Practice for Classification of Locations for Electrical Installations at Petroleum Facilities Classified as Class I, Division 1 and Division 2,* 1998. (Reaffirmed, November 2002.)

ANSI/API RP 505, *Recommended Practice for Classification of Locations for Electrical Installations at Petroleum Facilities Classified as Class I, Zone 0, Zone 1, and Zone 2,* 1998.

2.3.2 ASHRAE Publication.

American Society of Heating, Refrigeration and Air-Conditioning Engineers, Inc., 1791 Tullie Circle NE, Atlanta, GA 30329.

ASHRAE 15, *Safety Code for Mechanical Refrigeration,* 1994.

2.3.3 ASTM Publication.

American Society for Testing and Materials, 100 Barr Harbor Drive, West Conshohocken, PA 19428-2959.

ASTM D 323, *Standard Method of Test for Vapor Pressure of Petroleum Products (Reid Method)*, 1999.

2.3.4 CGA Publication.

Compressed Gas Association, 4221 Walney Road, 5th floor, Chantilly, VA 20151-2923.

ANSI/CGA G2.1, *American National Standard Safety Requirements for the Storage and Handling of Anhydrous Ammonia*, 1999.

2.3.5 IEC Publication.

International Electrotechnical Commission, 3, rue de Varembé, P.O. Box 131, CH-1211 Geneva 20, Switzerland.

IEC/TR3 60079-20, *Electrical apparatus for explosive gas atmospheres — Part 20: Data for flammable gases and vapours, relating to the use of electrical apparatus*, 1996.

Chapter 3 Definitions

3.1 General.

The definitions contained in this chapter apply to the terms used in this recommended practice. Where terms are not included, common usage of the terms applies.

3.2 NFPA Official Definitions.

3.2.1 Recommended Practice. A document that is similar in content and structure to a code or standard but that contains only nonmandatory provisions using the word "should" to indicate recommendations in the body of the text.

3.2.2 Should. Indicates a recommendation or that which is advised but not required.

3.3 General Definitions.

3.3.1 Adequate Ventilation. A ventilation rate that affords either 6 air changes per hour, or 1 cfm per square foot of floor area, or other similar criteria that prevent the accumulation of significant quantities of vapor–air concentrations from exceeding 25 percent of the lower flammable limit.

3.3.2* Autoignition Temperature (AIT). The minimum temperature required to initiate or cause self-sustained combustion of a solid, liquid, or gas independently of the heating or heated element.

3.3.3 CAS. Chemical Abstract Service.

3.3.4 Combustible Liquid. Any liquid that has a closed-cup flash point at or above 100°F (37.8°C), as determined by the test procedures and apparatus set forth in NFPA 30. Combustible liquids are classified as Class II or Class III combustible liquids. [**30:**3.3]

3.3.4.1 Class II Liquid. Any liquid that has a flash point at or above 100°F (37.8°C) and below 140°F (60°C). [**30:**3.3]

3.3.4.2 Class IIIA Liquid. Any liquid that has a flash point at or above 140°F (60°C), but below 200°F (93°C). [**30:**3.3]

3.3.4.3 Class IIIB Liquid. Any liquid that has a flash point at or above 200°F (93°C). [**30:**3.3]

3.3.5 Combustible Material. A generic term used to describe a flammable gas, flammable liquid produced vapor, or combustible liquid produced vapor mixed with air that may burn or explode.

3.3.5.1* Combustible Material (Class I, Division). Class I, Division combustible materials are divided into Groups A, B, C, and D.

3.3.5.1.1 Group A. Acetylene.

3.3.5.1.2 Group B. Flammable gas, flammable liquid produced vapor, or combustible liquid produced vapor mixed with air that may burn or explode, having either a maximum experimental safe gap (MESG) value less than or equal to 0.45 mm or a minimum igniting current ratio (MIC ratio) less than or equal to 0.40. Note: A typical Class I, Group B material is hydrogen.

3.3.5.1.3 Group C. Flammable gas, flammable liquid produced vapor, or combustible liquid produced vapor mixed with air that may burn or explode, having either a maximum experimental safe gap (MESG) value greater than 0.45 mm and less than or equal to 0.75 mm, or a minimum igniting current ratio (MIC ratio) greater than 0.40 and less than or equal to 0.80. Note: A typical Class I, Group C material is ethylene.

3.3.5.1.4 Group D. Flammable gas, flammable liquid produced vapor, or combustible liquid produced vapor mixed with air that may burn or explode, having either a maximum experimental safe gap (MESG) value greater than 0.75 mm or a minimum igniting current ratio (MIC ratio) greater than 0.80. Note: A typical Class I, Group D material is propane.

3.3.5.2* Combustible Material (Class I, Zone). Class I, Zone combustible materials are divided into Groups IIC, IIB, and IIA.

3.3.5.2.1 Group IIC. Atmospheres containing acetylene, hydrogen, or flammable gas, flammable liquid produced vapor, or combustible liquid produced vapor mixed with air that may burn or explode, having either a maximum experimental safe gap (MESG) value less than or equal to 0.50 mm or minimum igniting current ratio (MIC ratio) less than or equal to 0.45.

3.3.5.2.2 Group IIB. Atmospheres containing acetaldehyde, ethylene, or flammable gas, flammable liquid produced vapor, or combustible liquid produced vapor mixed with air that may burn or explode, having either maximum experimental safe gap (MESG) values greater than 0.50 mm and less than or equal to 0.90 mm or minimum igniting current ratio (MIC ratio) greater than 0.45 and less than or equal to 0.80.

3.3.5.2.3 Group IIA. Atmospheres containing acetone, ammonia, ethyl alcohol, gasoline, methane, propane, or flammable gas, flammable liquid produced vapor, or combustible liquid produced vapor mixed with air that may burn or explode, having either a maximum experimental safe gap (MESG) value greater than 0.90 mm or minimum igniting current ratio (MIC ratio) greater than 0.80.

3.3.6 Flammable Liquid. Any liquid that has a closed-cup flash point below 100°F (37.8°C), as determined by the test procedures and apparatus set forth in NFPA 30. Flammable liquids are classified as Class I liquids, Class IA liquids, Class IB liquids, and Class IC liquids. [**30:**3.3]

3.3.6.1 Class I Liquid. Any liquid that has a closed-cup flash point below 100°F (37.8°C) and a Reid vapor pressure not exceeding 40 psia (2068.6 mm Hg) at 100°F (37.8°C), as determined by ASTM D 323, *Standard Method of Test for Vapor Pressure of Petroleum Products (Reid Method).* [**30:**3.3]

3.3.6.2 Class IA Liquids. Those liquids that have flash points below 73°F (22.8°C) and boiling points below 100°F (37.8°C). [**30:**3.3]

3.3.6.3 Class IB Liquids. Those liquids that have flash points below 73°F (22.8°C) and boiling points at or above 100°F (37.8°C). [**30:**3.3]

3.3.6.4 Class IC Liquids. Those liquids that have flash points at or above 73°F (22.8°C), but below 100°F (37.8°C). [**30:**3.3]

3.3.7 Flash Point. The minimum temperature at which a liquid gives off vapor in sufficient concentration to form an ignitible mixture with air near the surface of the liquid, as specified by test.

3.3.8 Ignitible Mixture. A combustible material that is within its flammable range.

3.3.9 Maximum Experimental Safe Gap (MESG). The maximum clearance between two parallel metal surfaces that has been found, under specified test conditions, to prevent an explosion in a test chamber from being propagated to a secondary chamber containing the same gas or vapor at the same concentration.

3.3.10* Minimum Igniting Current (MIC) Ratio. The ratio of the minimum current required from an inductive spark discharge to ignite the most easily ignitible mixture of a gas or vapor, divided by the minimum current required from an inductive spark discharge to ignite methane under the same test conditions.

3.3.11 Minimum Ignition Energy (MIE). The minimum energy required from a capacitive spark discharge to ignite the most easily ignitible mixture of a gas or vapor.

Chapter 4 Classification of Combustible Materials

4.1 *National Electrical Code*® Criteria.

4.1.1 Articles 500 and 505 of the *NEC* classify a location in which a combustible material is or may be present in the atmosphere in sufficient concentrations to produce an ignitible mixture.

4.1.2* In a Class I hazardous (classified) location, the combustible material present is a flammable gas or vapor.

4.1.3 Class I is further subdivided into either Class I, Division 1 or Class I, Division 2; or Class I, Zone 0, Zone 1, or Zone 2 as detailed in 4.1.3.1 through 4.1.3.5.

4.1.3.1 Class I, Division 1. A Class I, Division 1 location is a location

(1) In which ignitible concentrations of flammable gases or vapors can exist under normal operating conditions, or
(2) In which ignitible concentrations of such gases or vapors may exist frequently because of repair or maintenance operations or because of leakage, or
(3) In which breakdown or faulty operation of equipment or processes might release ignitible concentrations of flammable gases or vapors and might also cause simultaneous failure of electrical equipment in such a way as to directly cause the electrical equipment to become a source of ignition. [**70:**500.5(B)(1)]

4.1.3.2 Class I, Division 2. A Class I, Division 2 location is a location

(1) In which volatile flammable liquids or flammable gases are handled, processed, or used, but in which the liquids, vapors, or gases will normally be confined within closed containers or closed systems from which they can escape only in case of accidental rupture or breakdown of such containers or systems or in case of abnormal operation of equipment, or
(2) In which ignitible concentrations of gases or vapors are normally prevented by positive mechanical ventilation,

and which might become hazardous through failure or abnormal operation of the ventilating equipment, or

(3) That is adjacent to a Class I, Division 1 location, and to which ignitible concentrations of gases or vapors might occasionally be communicated unless such communication is prevented by adequate positive-pressure ventilation from a source of clean air and effective safeguards against ventilation failure are provided. [**70:**500.5(B)(2)]

4.1.3.3 Class I, Zone 0. A Class I, Zone 0 location is a location in which

(1) Ignitible concentrations of flammable gases or vapors are present continuously, or
(2) Ignitible concentrations of flammable gases or vapors are present for long periods of time. [**70:**505.5(B)(1)]

4.1.3.4 Class I, Zone 1. A Class I, Zone 1 location is a location

(1) In which ignitible concentrations of flammable gases or vapors are likely to exist under normal operating conditions; or
(2) In which ignitible concentrations of flammable gases or vapors may exist frequently because of repair or maintenance operations or because of leakage; or
(3) In which equipment is operated or processes are carried on, of such a nature that equipment breakdown or faulty operations could result in the release of ignitible concentrations of flammable gases or vapors and also cause simultaneous failure of electrical equipment in a mode to cause the electrical equipment to become a source of ignition; or
(4) That is adjacent to a Class I, Zone 0 location from which ignitible concentrations of vapors could be communicated, unless communication is prevented by adequate positive pressure ventilation from a source of clean air and effective safeguards against ventilation failure are provided. [**70:**505.5(B)(2)]

4.1.3.5 Class I, Zone 2. A Class I, Zone 2 location is a location

(1) In which ignitible concentrations of flammable gases or vapors are not likely to occur in normal operation and, if they do occur, will exist only for a short period; or
(2) In which volatile flammable liquids, flammable gases, or flammable vapors are handled, processed, or used but in which the liquids, gases, or vapors normally are confined within closed containers of closed systems from which they can escape only as a result of accidental rupture or breakdown of the containers or system, or as a result of the abnormal operation of the equipment with which the liquids or gases are handled, processed, or used; or
(3) In which ignitible concentrations of flammable gases or vapors normally are prevented by positive mechanical ventilation but which may become hazardous as a result of failure or abnormal operation of the ventilation equipment; or
(4) That is adjacent to a Class I, Zone 1 location, from which ignitible concentrations of flammable gases or vapors could be communicated, unless such communication is prevented by adequate positive-pressure ventilation from a source of clean air and effective safeguards against ventilation failure are provided. [**70:** 505.5(B)(3)]

4.1.4 The intent of Articles 500 and 505 of the *NEC* is to prevent combustible material from being ignited by electrical equipment and wiring systems.

4.1.5 Electrical installations within hazardous (classified) locations can use various protection techniques. No single protection technique is best in all respects for all types of equipment used in a chemical plant.

4.1.5.1 Explosionproof enclosures, pressurized equipment, and intrinsically safe circuits are applicable to both Division 1 and Division 2 locations.

4.1.5.2 Nonincendive equipment is permitted in Division 2 locations.

4.1.5.3 Nonsparking electrical equipment and other less restrictive equipment, as specified in the *NEC*, are permitted in Division 2 locations.

4.1.5.4 Factors such as corrosion, weather, maintenance, equipment standardization and interchangeability, and possible process changes or expansion frequently dictate the use of special enclosures or installations for electrical systems. However, such factors are outside the scope of this recommended practice, which is concerned entirely with the proper application of electrical equipment to avoid ignition of combustible materials.

4.1.6 For the purpose of this recommended practice, areas not classified either as Class I, Division 1 or Class I, Division 2, or as Class I, Zone 0, Zone 1, or Zone 2, are "unclassified" areas.

4.2 Behavior of Class I (Combustible Material) Gases, Vapors, and Liquids.

4.2.1 Lighter-than-Air (Vapor Density Less than 1.0) Gases. These gases tend to dissipate rapidly in the atmosphere. They will not affect as great an area as heavier-than-air gases or vapors. Except in enclosed spaces, such gases seldom accumulate to form an ignitible mixture near grade level, where most electrical installations are located. A lighter-than-air gas that has been cooled sufficiently could behave as a heavier-than-air gas until it absorbs heat from the surrounding atmosphere.

4.2.2 Heavier-than-Air (Vapor Density Greater than 1.0) Gases. These gases tend to fall to grade level when released. The gas could remain for a significant period of time, unless dispersed by natural or forced ventilation. A heavier-than-air gas that has been heated sufficiently to decrease its density could behave as a lighter-than-air gas until cooled by the surrounding atmosphere.

4.2.3 Applicable to All Densities. As the gas diffuses into the surrounding air, the density of the mixture approaches that of air.

4.2.4 Compressed Liquefied Gases. These gases are stored above their normal boiling point but are kept in the liquid state by pressure. When released, the liquid immediately expands and vaporizes, creating large volumes of cold gas. The cold gas behaves like a heavier-than-air gas.

4.2.5 Cryogenic Flammable Liquids and Other Cold Liquefied Combustible Materials. Cryogenic liquids are generally handled below −150°F (−101°C). These behave like flammable liquids when they are spilled. Small liquid spills will immediately vaporize, but larger spills may remain in the liquid state for an extended time. As the liquid absorbs heat, it vaporizes and could form an ignitible mixture. Some liquefied combustible materials (not cryogenic) are stored at low temperatures and at pressures close to atmospheric pressure; these include anhydrous ammonia, propane, ethane, ethylene, and propylene. These materials will behave as described in 4.2.1 or 4.2.2.

4.2.6 Flammable Liquids. When released in appreciable quantity, a Class I liquid will begin to evaporate at a rate that depends on its volatility: the lower the flash point, the greater the volatility; hence, the faster the evaporation. The vapors of Class I liquids form ignitible mixtures with air at ambient temperatures more or less readily. Even when evolved rapidly, the vapors tend to disperse rapidly, becoming diluted to a concentration below the lower flammable limit (LFL). Until this dispersion takes place, however, these vapors will behave like heavier-than-air gases. Class I liquids normally will produce ignitible mixtures that will travel some finite distance from the point of origin; thus, they will normally require area classification for proper electrical system design.

4.2.7 Combustible Liquids. A combustible liquid will form an ignitible mixture only when heated above its flash point.

4.2.7.1 With Class II liquids, the degree of hazard is lower because the vapor release rate is low at the normal handling and storage temperatures. In general, these liquids will not form ignitible mixtures with air at ambient temperatures unless heated above their flash points. Also, the vapors will not travel as far because they tend to condense as they are cooled by ambient air. Class II liquids should be considered capable of producing an ignitible mixture near the point of release when handled, processed, or stored under conditions where the liquid could exceed its flash point.

4.2.7.2 Class IIIA liquids do not form ignitible mixtures with air at ambient temperatures unless heated above their flash points. Furthermore, the vapors cool rapidly in air and condense. Hence, the extent of the area requiring electrical classification will be very small or nonexistent.

4.2.7.3 Class IIIB liquids seldom evolve enough vapors to form ignitible mixtures even when heated, and they are seldom ignited by properly installed and maintained general purpose electrical equipment. A Class IIIB liquid will cool below its flash point very quickly when released. Therefore, area classification is seldom needed and Class IIIB liquids are not included in Table 4.4.2.

4.3 Conditions Necessary for Ignition.

In a Class I area, the following three conditions must be satisfied for the combustible material to be ignited by the electrical installation:

(1) A combustible material must be present.
(2) It must be mixed with air in the proportions required to produce an ignitible mixture.
(3) There must be a release of sufficient energy to ignite the mixture.

4.4 Classification of Class I Combustible Materials.

4.4.1 Combustible materials are classified into four Class I, Division Groups: A, B, C, and D; or three Class I, Zone Groups: IIC, IIB, and IIA, depending on their properties.

4.4.2* An alphabetical listing of selected combustible materials, with their group classification and relevant physical properties, is provided in Table 4.4.2.

4.4.3 Table 4.4.3 provides a cross-reference of selected chemicals sorted by their Chemical Abstract Service (CAS) numbers.

4.4.4 Annex C lists references that deal with the testing of various characteristics of combustible materials.

Table 4.4.2 Selected Chemicals

Chemical	CAS No.	Class I Division Group	Type[a]	Flash Point (°C)	AIT (°C)	%LFL	%UFL	Vapor Density (Air = 1)	Vapor Pressure[b] (mm Hg)	Class I Zone Group[c]	MIE (mJ)	MIC Ratio	MESG (mm)
Acetaldehyde	75-07-0	C[d]	I	-38	175	4.0	60.0	1.5	874.9	IIA	0.37	0.98	0.92
Acetic Acid	64-19-7	D[d]	II	43	464	4.0	19.9	2.1	15.6	IIA	2.67	1.76	
Acetic Acid-tert-Butyl Ester	540-88-5	D	II			1.7	9.8	4.0	40.6				
Acetic Anhydride	108-24-7	D	II	54	316	2.7	10.3	3.5	4.9				
Acetone	67-64-1	D[d]	I		465	2.5	12.8	2.0	230.7	IIA	1.15	1.00	1.02
Acetone Cyanohydrin	75-86-5	D	IIIA	74	688	2.2	12.0	2.9	0.3				
Acetonitrile	75-05-8	D	I	6	524	3.0	16.0	1.4	91.1	IIA			1.50
Acetylene	74-86-2	A[d]	GAS		305	2.5	99.9	0.9	36600	IIC	0.017	0.28	0.25
Acrolein (Inhibited)	107-02-8	B(C)[d]	I		235	2.8	31.0	1.9	274.1	IIB	0.13		
Acrylic Acid	79-10-7	D	II	54	438	2.4	8.0	2.5	4.3				
Acrylonitrile	107-13-1	D[d]	I	-26	481	3.0	17.0	1.8	108.5	IIB	0.16	0.78	0.87
Adiponitrile	111-69-3	D	IIIA	93	550			1.0	0.002				
Allyl Alcohol	107-18-6	C[d]	I	22	378	2.5	18.0	2.0	25.4				0.84
Allyl Chloride	107-05-1	D	I	-32	485	2.9	11.1	2.6	366			1.33	1.17
Allyl Glycidyl Ether	106-92-3	B(C)[e]	II		57			3.9					
Alpha-Methyl Styrene	98-83-9	D	II		574	0.8	11.0	4.1	2.7				
n-Amyl Acetate	628-63-7	D	I	25	360	1.1	7.5	4.5	4.2				1.02
sec-Amyl Acetate	626-38-0	D	I	23		1.1	7.5	4.5		IIA			
Ammonia	7664-41-7	D[d,f]	I		498	15.0	28.0	0.6	7498.0	IIA	680	6.85	3.17
Aniline	62-53-3	D	IIIA	70	615	1.3	11.0	3.2	0.7	IIA			
Benzene	71-43-2	D[d]	I	-11	498	1.2	7.8	2.8	94.8	IIA	0.20	1.00	0.99
Benzyl Chloride	98-87-3	D	IIIA		585	1.1		4.4	0.5				
Bromopropyne	106-96-7	D	I	10	324	3.0							
n-Butane	3583-47-9	D[d,g]	GAS		288	1.9	8.5	2.0			0.25	0.94	1.07
1,3-Butadiene	106-99-0	B(D)[d,e]	GAS	-76	420	2	12	1.9		IIB	0.13	0.76	0.79
1-Butanol	71-36-3	D[d]	I	36	343	1.4	11.2	2.6	7.0	IIA			0.91
2-Butanol	78-92-2	D[d]	I	36	405	1.7	9.8	2.6		IIA			
Butylamine	109-73-9	D	GAS	-12	312	1.7	9.8	2.5	92.9			1.13	
Butylene	25167-67-3	D	I		385	1.6	10.0	1.9	2214.6				
n-Butyraldehyde	123-72-8	C[d]	I	-12	218	1.9	12.5	2.5	112.2				0.92
n-Butyl Acetate	123-86-4	D[d]	I	22	421	1.7	7.6	4.0	11.5	IIA		1.08	1.04
sec-Butyl Acetate	105-46-4	D	II	-8		1.7	9.8	4.0	22.2				
tert-Butyl Acetate	540-88-5	D	II			1.7	9.8	4.0	40.6				
n-Butyl Acrylate (Inhibited)	141-32-2	D	II	49	293	1.7	9.9	4.4	5.5				
n-Butyl Glycidyl Ether	2426-08-6	B(C)[e]	II										
n-Butyl Formal	110-62-3	C	IIIA						34.3				
Butyl Mercaptan	109-79-5	C	I	2				3.1	46.4				
Butyl-2-Propenoate	141-32-2	D	II	49		1.7	9.9	4.4	5.5				
para tert-Butyl Toluene	98-51-1	D	IIIA										
n-Butyric Acid	107-92-6	D[d]	IIIA	72	443	2.0	10.0	3.0	0.8				
Carbon Disulfide	75-15-0	[d,h]	I	-30	90	1.3	50.0	2.6	358.8	IIC	0.009	0.39	0.20
Carbon Monoxide	630-08-0	C[d]	GAS		609	12.5	74.0	0.97		IIA			0.84
Chloroacetaldehyde	107-20-0	C	IIIA	88					63.1				
Chlorobenzene	108-90-7	D	I	29	593	1.3	9.6	3.9	11.9				
1-Chloro-1-Nitropropane	2425-66-3	C	IIIA										
Chloroprene	126-99-8	D	GAS	-20		4.0	20.0	3.0					
Cresol	1319-77-3	D	IIIA	81	559	1.1		3.7					
Crotonaldehyde	4170-30-3	C[d]	I	13	232	2.1	15.5	2.4	33.1	IIB			0.81
Cumene	98-82-8	D	I	36	424	0.9	6.5	4.1	4.6	IIA			
Cyclohexane	110-82-7	D	I	-17	245	1.3	8.0	2.9	98.8	IIA	0.22	1.0	0.94

Table 4.4.2 Continued

Chemical	CAS No.	Class I Division Group	Type[a]	Flash Point (°C)	AIT (°C)	%LFL	%UFL	Vapor Density (Air = 1)	Vapor Pressure[b] (mm Hg)	Class I Zone Group[c]	MIE (mJ)	MIC Ratio	MESG (mm)
Cyclohexanol	108-93-0	D	IIIA	68	300			3.5	0.7	IIA			
Cyclohexanone	108-94-1	D	II	44	245	1.1	9.4	3.4	4.3	IIA			0.98
Cyclohexene	110-83-8	D	I	-6	244	1.2		2.8	89.4			0.97	
Cyclopropane	75-19-4	D[d]	I		503	2.4	10.4	1.5	5430	IIB	0.17	0.84	0.91
p-Cymene	99-87-6	D	II	47	436	0.7	5.6	4.6	1.5	IIA			
Decene	872-05-9	D	II		235			4.8	1.7				
n-Decaldehyde	112-31-2	C	IIIA						0.09				
n-Decanol	112-30-1	D	IIIA	82	288			5.3	0.008				
Decyl Alcohol	112-30-1	D	IIIA	82	288			5.3	0.008				
Diacetone Alcohol	123-42-2	D	IIIA	64	603	1.8	6.9	4.0	1.4				
Di-Isobutylene	25167-70-8	D[d]	I	2	391	0.8	4.8	3.8			0.96		
Di-Isobutyl Ketone	108-83-8	D	II	60	396	0.8	7.1	4.9	1.7				
o-Dichlorobenzene	955-50-1	D	IIIA	66	647	2.2	9.2	5.1		IIA			
1,4-Dichloro-2,3 Epoxybutane	3583-47-9	D[d]	I			1.9	8.5	2.0			0.25	0.98	1.07
1,1-Dichloroethane	1300-21-6	D	I		438	6.2	16.0	3.4	227				1.82
1,2-Dichloroethylene	156-59-2	D	I	97	460	5.6	12.8	3.4	204	IIA			
1,1-Dichloro-1-Nitroethane	594-72-9	C	IIIA	76				5.0					
1,3-Dichloropropene	10061-02-6	D	I	35		5.3	14.5	3.8					
Dicyclopentadiene	77-73-6	C	I	32	503				2.8				0.91
Diethylamine	109-87-9	C[d]	I	-28	312	1.8	10.1	2.5		IIA			1.15
Diethylaminoethanol	100-37-8	C	IIIA	60	320			4.0	1.6	IIA			
Diethyl Benzene	25340-17-4	D	II	57	395			4.6					
Diethyl Ether (Ethyl Ether)	60-29-7	C[d]	I	-45	160	1.7	48	2.6	538	IIB	0.19	0.88	0.83
Diethylene Glycol Monobutyl Ether	112-34-5	C	IIIA	78	228	0.9	24.6	5.6	0.02				
Diethylene Glycol Monomethyl Ether	111-77-3	C	IIIA	93	241				0.2				
n-n-Dimethyl Aniline	121-69-7	C	IIIA	63	371	1.0		4.2	0.7				
Dimethyl Formamide	68-12-2	D	II	58	455	2.2	15.2	2.5	4.1				1.08
Dimethyl Sulfate	77-78-1	D	IIIA	83	188			4.4	0.7				
Dimethylamine	124-40-3	C	GAS		400	2.8	14.4	1.6		IIA			
2,2-Dimethylbutane	75-83-2	D[g]	I	-48	405				319.3				
2,3-Dimethylbutane	78-29-8	D[g]	I		396								
3,3-Dimethylheptane	1071-26-7	D[g]	I		325				10.8				
2,3-Dimethylhexane	31394-54-4	D[g]	I		438								
2,3-Dimethylpentane	107-83-5	D[g]	I		335				211.7				
Di-N-Propylamine	142-84-7	C	I	17	299				27.1				
1,4-Dioxane	123-91-1	C[d]	I	12	180	2.0	22.0	3.0	38.2	IIB	0.19		0.70
Dipentene	138-86-3	D	II	45	237	0.7	6.1	4.7					1.18
Dipropylene Glycol Methyl Ether	34590-94-8	C	IIIA	85		1.1	3.0	5.1	0.5				
Diisopropylamine	108-18-9	C	GAS	-6	316	1.1	7.1	3.5					
Dodecene	6842-15-5	D	IIIA	100	255								
Epichlorohydrin	3132-64-7	C[d]	I	33	411	3.8	21.0	3.2	13.0				
Ethane	74-84-0	D[d]	GAS	-29	472	3.0	12.5	1.0		IIA	0.24	0.82	0.91
Ethanol	64-17-5	D[d]	I	13	363	3.3	19.0	1.6	59.5	IIA		0.88	0.89

(continues)

Table 4.4.2 Continued

Chemical	CAS No.	Class I Division Group	Type[a]	Flash Point (°C)	AIT (°C)	%LFL	%UFL	Vapor Density (Air = 1)	Vapor Pressure[b] (mm Hg)	Class I Zone Group[c]	MIE (mJ)	MIC Ratio	MESG (mm)
Ethylamine	75-04-7	D[d]	I	-18	385	3.5	14.0	1.6	1048		2.4		
Ethylene	74-85-1	C[d]	GAS	0	450	2.7	36.0	1.0		IIB	0.070	0.53	0.65
Ethylenediamine	107-15-3	D[d]	I	33	385	2.5	12.0	2.1	12.5				
Ethylenimine	151-56-4	C[d]	I	-11	320	3.3	54.8	1.5	211		0.48		
Ethylene Chlorohydrin	107-07-3	D	IIIA	59	425	4.9	15.9	2.8	7.2				
Ethylene Dichloride	107-06-2	D[d]	I	13	413	6.2	16.0	3.4	79.7				
Ethylene Glycol Monoethyl Ether Acetate	111-15-9	C	II	47	379	1.7		4.7	2.3			0.53	0.97
Ethylene Glycol Monobutyl Ether Acetate	112-07-2	C	IIIA		340	0.9	8.5		0.9				
Ethylene Glycol Monobutyl Ether	111-76-2	C	IIIA		238	1.1	12.7	4.1	1.0				
Ethylene Glycol Monoethyl Ether	110-80-5	C	II		235	1.7	15.6	3.0	5.4				0.84
Ethylene Glycol Monomethyl Ether	109-86-4	D	II		285	1.8	14.0	2.6	9.2				0.85
Ethylene Oxide	75-21-8	B(C)[d,e]	I	-20	429	3.0	99.9	1.5	1314	IIB	0.065	0.47	0.59
2-Ethylhexaldehyde	123-05-7	C	II	52	191	0.8	7.2	4.4	1.9				
2-Ethylhexanol	104-76-7	D	IIIA	81		0.9	9.7	4.5	0.2				
2-Ethylhexyl Acrylate	103-09-3	D	IIIA	88	252				0.3				
Ethyl Acetate	141-78-6	D[d]	I	-4	427	2.0	11.5	3.0	93.2		0.46		0.99
Ethyl Acrylate (Inhibited)	140-88-5	D[d]	I	9	372	1.4	14.0	3.5	37.5	IIA			0.86
Ethyl Alcohol	64-17-5	D[d]	I	13	363	3.3	19.0	1.6	59.5				0.89
Ethyl Sec-Amyl Ketone	541-85-5	D	II	59									
Ethyl Benzene	100-41-4	D	I	21	432	0.8	6.7	3.7	9.6				
Ethyl Butanol	97-95-0	D	II	57		1.2	7.7	3.5	1.5				
Ethyl Butyl Ketone	106-35-4	D	II	46				4.0	3.6				
Ethyl Chloride	75-00-3	D	GAS	-50	519	3.8	15.4	2.2					
Ethyl Formate	109-94-4	D	GAS	-20	455	2.8	16.0	2.6		IIA			0.94
Ethyl Mercaptan	75-08-1	C[d]	I	-18	300	2.8	18.0	2.1	527.4			0.90	0.90
n-Ethyl Morpholine	100-74-3	C	I	32				4.0					
2-Ethyl-3-Propyl Acrolein	645-62-5	C	IIIA	68				4.4					
Ethyl Silicate	78-10-4	D	II					7.2					
Formaldehyde (Gas)	50-00-0	B	GAS	60	429	7.0	73.0	1.0					0.57
Formic Acid	64-18-6	D	II	50	434	18.0	57.0	1.6	42.7				1.86
Fuel Oil 1	8008-20-6	D	II	72	210	0.7	5.0						
Furfural	98-01-1	C	IIIA	60	316	2.1	19.3	3.3	2.3				0.94
Furfuryl Alcohol	98-00-0	C	IIIA	75	490	1.8	16.3	3.4	0.6				
Gasoline	8006-61-9	D[d]	I	-46	280	1.4	7.6	3.0					
n-Heptane	142-82-5	D[d]	I	-4	204	1.0	6.7	3.5	45.5	IIA	0.24	0.88	0.91
n-Heptene	81624-04-6	D[g]	I	-1	204			3.4					0.97
n-Hexane	110-54-3	D[d,g]	I	-23	225	1.1	7.5	3.0	152	IIA	0.24	0.88	0.93
Hexanol	111-27-3	D	IIIA	63				3.5	0.8	IIA			0.98
2-Hexanone	591-78-6	D	I	35	424	1.2	8.0	3.5	10.6				
Hexene	592-41-6	D	I	-26	245	1.2	6.9		186				
sec-Hexyl Acetate	108-84-9	D	II	45				5.0					
Hydrazine	302-01-2	C	II	38	23		98.0	1.1	14.4				

Table 4.4.2 Continued

Chemical	CAS No.	Class I Division Group	Type[a]	Flash Point (°C)	AIT (°C)	%LFL	%UFL	Vapor Density (Air = 1)	Vapor Pressure[b] (mm Hg)	Class I Zone Group[c]	MIE (mJ)	MIC Ratio	MESG (mm)
Hydrogen	1333-74-0	B[d]	GAS		520	4.0	75.0	0.1		IIC	0.019	0.25	0.28
Hydrogen Cyanide	74-90-8	C[d]	GAS	-18	538	5.6	40.0	0.9		IIB			0.80
Hydrogen Selenide	7783-07-5	C	I						7793				
Hydrogen Sulfide	7783-06-4	C[d]	GAS		260	4.0	44.0	1.2			0.068		0.90
Isoamyl Acetate	123-92-2	D	I	25	360	1.0	7.5	4.5	6.1				
Isoamyl Alcohol	123-51-3	D	II	43	350	1.2	9.0	3.0	3.2				1.02
Isobutane	75-28-5	D[g]	GAS		460	1.8	8.4	2.0					
Isobutyl Acetate	110-19-0	D[d]	I	18	421	2.4	10.5	4.0	17.8				
Isobutyl Acrylate	106-63-8	D	I		427			4.4	7.1				
Isobutyl Alcohol	78-83-1	D[d]	I	-40	416	1.2	10.9	2.5	10.5			0.92	0.98
Isobutyraldehyde	78-84-2	C	GAS	-40	196	1.6	10.6	2.5					
Isodecaldehyde	112-31-2	C	IIIA					5.4	0.09				
Isohexane	107-83-5	D[g]			264				211.7			1.00	
Isopentane	78-78-4	D[g]			420				688.6				
Isooctyl Aldehyde	123-05-7	C	II		197				1.9				
Isophorone	78-59-1	D		84	460	0.8	3.8	4.8	0.4				
Isoprene	78-79-5	D[d]	I	-54	220	1.5	8.9	2.4	550.6				
Isopropyl Acetate	108-21-4	D	I		460	1.8	8.0	3.5	60.4				
Isopropyl Ether	108-20-3	D[d]	I	-28	443	1.4	7.9	3.5	148.7		1.14		0.94
Isopropyl Glycidyl Ether	4016-14-2	C	I										
Isopropylamine	75-31-0	D	GAS	-26	402	2.3	10.4	2.0			2.0		
Kerosene	8008-20-6	D	II	72	210	0.7	5.0			IIA			
Liquefied Petroleum Gas	68476-85-7	D	I		405								
Mesityl Oxide	141-97-9	D[d]	I	31	344	1.4	7.2	3.4	47.6				
Methane	74-82-8	D[d]	GAS	-223	630	5.0	15.0	0.6		IIA	0.28	1.00	1.12
Methanol	67-56-1	D[d]	I	12	385	6.0	36.0	1.1	126.3	IIA	0.14	0.82	0.92
Methyl Acetate	79-20-9	D	GAS	-10	454	3.1	16.0	2.6		IIB		1.08	0.99
Methyl Acrylate	96-33-3	D	GAS	-3	468	2.8	25.0	3.0				0.98	0.85
Methyl Alcohol	67-56-1	D[d]	I		385	6.0	36.0	1.1	126.3				0.91
Methyl Amyl Alcohol	108-11-2	D	II	41		1.0	5.5	3.5	5.3				1.01
Methyl Chloride	74-87-3	D	GAS	-46	632	8.1	17.4	1.7					1.00
Methyl Ether	115-10-6	C[d]	GAS	-41	350	3.4	27.0	1.6				0.85	0.84
Methyl Ethyl Ketone	78-93-3	D[d]	I	-6	404	1.4	11.4	2.5	92.4		0.53	0.92	0.84
Methyl Formal	534-15-6	C[d]	I	1	238			3.1					
Methyl Formate	107-31-3	D	GAS	-19	449	4.5	23.0	2.1					0.94
2-Methylhexane	31394-54-4	D[g]	I		280								
Methyl Isobutyl Ketone	141-79-7	D[d]	I	31	440	1.2	8.0	3.5	11				
Methyl Isocyanate	624-83-9	D	GAS	-15	534	5.3	26.0	2.0					
Methyl Mercaptan	74-93-1	C	GAS	-18		3.9	21.8	1.7					
Methyl Methacrylate	80-62-6	D	I	10	422	1.7	8.2	3.6	37.2	IIA			0.95
Methyl N-Amyl Ketone	110-43-0	D	II	49	393	1.1	7.9	3.9	3.8				
Methyl Tertiary Butyl Ether	1634-04-4	D	I	-80	435	1.6	8.4	0.2	250.1				
2-Methyloctane	3221-61-2				220				6.3				
2-Methylpropane	75-28-5	D[g]	I		460				2639				
Methyl-1-Propanol	78-83-1	D[d]	I	-40	416	1.2	10.9	2.5	10.1				0.98
Methyl-2-Propanol	75-65-0	D[d]	I	10	360	2.4	8.0	2.6	42.2				
2-Methyl-5-Ethyl Pyridine	104-90-5	D		74		1.1	6.6	4.2					

(continues)

Table 4.4.2 Continued

Chemical	CAS No.	Class I Division Group	Type[a]	Flash Point (°C)	AIT (°C)	%LFL	%UFL	Vapor Density (Air = 1)	Vapor Pressure[b] (mm Hg)	Class I Zone Group[c]	MIE (mJ)	MIC Ratio	MESG (mm)
Methylacetylene	74-99-7	C[d]	I			1.7		1.4	4306		0.11		
Methylacetylene-Propadiene	27846-30-6	C	I										0.74
Methylal	109-87-5	C	I	-18	237	1.6	17.6	2.6	398				
Methylamine	74-89-5	D	GAS		430	4.9	20.7	1.0		IIA			1.10
2-Methylbutane	78-78-4	D[g]		-56	420	1.4	8.3	2.6	688.6				
Methylcyclohexane	208-87-2	D	I	-4	250	1.2	6.7	3.4			0.27		
Methylcyclohexanol	25630-42-3	D		68	296			3.9					
2-Methycyclohexanone	583-60-8	D	II					3.9					
2-Methylheptane		D[g]			420								
3-Methylhexane	589-34-4	D[g]			280				61.5				
3-Methylpentane	94-14-0	D[g]			278								
2-Methylpropane	75-28-5	D[g]	I		460				2639				
2-Methyl-1-Propanol	78-83-1	D[d]	I	-40	223	1.2	10.9	2.5	10.5				
2-Methyl-2-Propanol	75-65-0	D[d]	I		478	2.4	8.0	2.6	42.2				
2-Methyloctane	2216-32-2	D[g]			220								
3-Methyloctane	2216-33-3	D[g]			220				6.3				
4-Methyloctane	2216-34-4	D[g]			225				6.8				
Monoethanolamine	141-43-5	D		85	410			2.1	0.4	IIA			
Monoisopropanolamine	78-96-6	D		77	374			2.6	1.1				
Monomethyl Aniline	100-61-8	C			482				0.5				
Monomethyl Hydrazine	60-34-4	C	I	23	194	2.5	92.0	1.6					
Morpholine	110-91-8	C[d]	II	35	310	1.4	11.2	3.0	10.1				0.95
Naphtha (Coal Tar)	8030-30-6	D	II	42	277					IIA			
Naphtha (Petroleum)	8030-30-6	D[d,i]	I	42	288	1.1	5.9	2.5		IIA			
Neopentane	463-82-1	D[g]		-65	450	1.4	8.3	2.6	1286				
Nitrobenzene	98-95-3	D		88	482	1.8		4.3	0.3				0.94
Nitroethane	79-24-3	C	I	28	414	3.4		2.6	20.7	IIA			0.87
Nitromethane	75-52-5	C	I	35	418	7.3		2.1	36.1	IIA		0.92	1.17
1-Nitropropane	108-03-2	C	I	34	421	2.2		3.1	10.1				0.84
2-Nitropropane	79-46-9	C[d]	I	28	428	2.6	11.0	3.1	17.1				
n-Nonane	111-84-2	D[g]	I	31	205	0.8	2.9	4.4	4.4	IIA			
Nonene	27214-95-8	D	I			0.8		4.4					
Nonyl Alcohol	143-08-8	D				0.8	6.1	5.0	0.02	IIA			
n-Octane	111-65-9	D[d,g]	I	13	206	1.0	6.5	3.9	14.0	IIA			0.94
Octene	25377-83-7	D	I	8	230	0.9		3.9					
n-Octyl Alcohol	111-87-5	D						4.5	0.08	IIA			1.05
n-Pentane	109-66-0	D[d,g]	I	-40	243	1.5	7.8	2.5	513		0.28	0.97	0.93
1-Pentanol	71-41-0	D[d]	I	33	300	1.2	10.0	3.0	2.5	IIA			
2-Pentanone	107-87-9	D	I	7	452	1.5	8.2	3.0	35.6				0.99
1-Pentene	109-67-1	D	I	-18	275	1.5	8.7	2.4	639.7				
2-Pentene	109-68-2	D	I	-18				2.4					
2-Pentyl Acetate	626-38-0	D	I	23		1.1	7.5	4.5					
Phenylhydrazine	100-63-0	D		89				3.7	0.03				
Process Gas > 30% H_2	1333-74-0	B[j]	GAS		520	4.0	75.0	0.1			0.019	0.45	
Propane	74-98-6	D[d]	GAS	-104	450	2.1	9.5	1.6		IIA	0.25	0.82	0.97
1-Propanol	71-23-8	D[d]	I	15	413	2.2	13.7	2.1	20.7	IIA			0.89
2-Propanol	67-63-0	D[d]	I	12	399	2.0	12.7	2.1	45.4		0.65		1.00
Propiolactone	57-57-8	D				2.9		2.5	2.2				
Propionaldehyde	123-38-6	C	I	-9	207	2.6	17.0	2.0	318.5				

Table 4.4.2 Continued

Chemical	CAS No.	Class I Division Group	Type[a]	Flash Point (°C)	AIT (°C)	%LFL	%UFL	Vapor Density (Air = 1)	Vapor Pressure[b] (mm Hg)	Class I Zone Group[c]	MIE (mJ)	MIC Ratio	MESG (mm)
Propionic Acid	79-09-4	D	II	54	466	2.9	12.1	2.5	3.7				
Propionic Anhydride	123-62-6	D		74	285	1.3	9.5	4.5	1.4				
n-Propyl Acetate	109-60-4	D	I	14	450	1.7	8.0	3.5	33.4				1.05
n-Propyl Ether	111-43-3	C[d]	I	21	215	1.3	7.0	3.5	62.3				
Propyl Nitrate	627-13-4	B[d]	I	20	175	2.0	100.0						
Propylene	115-07-1	D[d]	GAS	-108	455	2.0	11.1	1.5			0.28		0.91
Propylene Dichloride	78-87-5	D	I	16	557	3.4	14.5	3.9	51.7				1.32
Propylene Oxide	75-56-9	B(C)[d,e]	I	-37	449	2.3	36.0	2.0	534.4		0.13		0.70
Pyridine	110-86-1	D[d]	I	20	482	1.8	12.4	2.7	20.8	IIA			
Styrene	100-42-5	D[d]	I	31	490	0.9	6.8	3.6	6.1	IIA		1.21	
Tetrahydrofuran	109-99-9	C[d]	I	-14	321	2.0	11.8	2.5	161.6	IIB	0.54		0.87
Tetrahydronaphthalene	119-64-2	D	IIIA		385	0.8	5.0	4.6	0.4				
Tetramethyl Lead	75-74-1	C	II	38				9.2					
Toluene	108-88-3	D[d]	I	4	480	1.1	7.1	3.1	28.53	IIA	0.24		
n-Tridecene	2437-56-1	D	IIIA			0.6		6.4	593.4				
Triethylamine	121-44-8	C[d]	I	-9	249	1.2	8.0	3.5	68.5	IIA	0.75		
Triethylbenzene	25340-18-5	D		83			56.0	5.6					
2,2,3-Trimethylbutane		D[g]			442								
2,2,4-Trimethylbutane		D[g]			407								
2,2,3-Trimethylpentane		D[g]			396								
2,2,4-Trimethylpentane		D[g]			415								
2,3,3-Trimethylpentane		D[g]			425								
Tripropylamine	102-69-2	D	II	41				4.9	1.5				1.13
Turpentine	8006-64-2	D	I	35	253	0.8			4.8				
n-Undecene	28761-27-5	D	IIIA			0.7		5.5					
Unsymmetrical Dimethyl Hydrazine	57-14-7	C[d]	I	-15	249	2.0	95.0	1.9					0.85
Valeraldehyde	110-62-3	C	I	280	222			3.0	34.3				
Vinyl Acetate	108-05-4	D[d]	I	-6	402	2.6	13.4	3.0	113.4	IIA	0.70		0.94
Vinyl Chloride	75-01-4	D[d]	GAS	-78	472	3.6	33.0	2.2					0.96
Vinyl Toluene	25013-15-4	D		52	494	0.8	11.0	4.1					
Vinylidene Chloride	75-35-4	D	I		570	6.5	15.5	3.4	599.4				3.91
Xylene	1330-20-7	D[d]	I	25	464	0.9	7.0	3.7		IIA	0.2		
Xylidine	121-69-7	C	IIIA	63	371	1.0		4.2	0.7				

Notes:

[a]Type is used to designate if the material is a gas, flammable liquid, or combustible liquid. *(See 4.2.6 and 4.2.7.)*

[b]Vapor pressure reflected in units of mm Hg at 25°C (77°F) unless stated otherwise.

[c]Class I, Zone Groups are based on 1996 IEC TR3 60079-20, *Electrical apparatus for explosive gas atmospheres — Part 20: Data for flammable gases and vapors, relating to the use of electrical apparatus*, which contains additional data on MESG and group classifications.

[d]Material has been classified by test.

[e]Where all conduit runs into explosionproof equipment are provided with explosionproof seals installed within 450 mm (18 in.) of the enclosure, equipment for the group classification shown in parentheses is permitted.

[f]For classification of areas involving ammonia, see ASHRAE 15, *Safety Code for Mechanical Refrigeration*, and ANSI/CGA G2.1, *Safety Requirements for the Storage and Handling of Anhydrous Ammonia*.

[g]Commercial grades of aliphatic hydrocarbon solvents are mixtures of several isomers of the same chemical formula (or molecular weight). The autoignition temperatures of the individual isomers are significantly different. The electrical equipment should be suitable for the AIT of the solvent mixture. *(See A.4.4.2.)*

[h]Certain chemicals have characteristics that require safeguards beyond those required for any of the above groups. Carbon disulfide is one of these chemicals because of its low autoignition temperature and the small joint clearance necessary to arrest its flame propagation.

[i]Petroleum naphtha is a saturated hydrocarbon mixture whose boiling range is 20°C to 135°C (68°F to 275°F). It is also known as benzine, ligroin, petroleum ether, and naphtha.

[j]Fuel and process gas mixtures found by test not to present hazards similar to those of hydrogen may be grouped based on the test results.

Table 4.4.3 Cross-Reference of Chemical CAS Number to Chemical Name

CAS No.	Chemical Name	CAS No.	Chemical Name
50-00-0	Formaldehyde (Gas)	75-86-5	Acetone Cyanohydrin
57-14-7	Unsymmetrical Dimethyl Hydrazine	77-78-1	Dimethyl Sulfate
57-57-8	Propiolactone	78-10-4	Ethyl Silicate
60-29-7	Diethyl Ether (Ethyl Ether)	78-59-1	Isophorone
60-34-4	Monomethyl Hydrazine	78-78-4	Isopentane
62-53-3	Aniline	78-78-4	Methylbutane
64-17-5	Ethanol	78-79-5	Isoprene
64-17-5	Ethyl Alcohol	78-83-1	Isobutyl Alcohol
64-18-6	Formic Acid	78-83-1	Methyl-1-Propanol
64-19-7	Acetic Acid	78-84-2	Isobutyraldehyde
67-56-1	Methanol	78-87-5	Propylene Dichloride
67-56-1	Methyl Alcohol	78-93-3	Methyl Ethyl Ketone
67-63-0	2-Propanol	78-96-6	Monoisopropanolamine
67-64-1	Acetone	79-09-4	Propionic Acid
68-12-2	Dimethyl Formamide	79-10-7	Acrylic Acid
71-23-8	1-Propanol	79-20-9	Methyl Acetate
71-36-3	1-Butanol	79-24-3	Nitroethane
71-36-5	2-Butanol	79-46-9	2-Nitropropane
71-41-0	1-Pentanol	80-62-6	Methyl Methacrylate
71-43-2	Benzene	96-14-0	3-Methylpentane
74-82-8	Methane	96-33-3	Methyl Acrylate
74-84-0	Ethane	97-95-0	Ethyl Butanol
74-85-1	Ethylene	98-00-0	Furfuryl Alcohol
74-86-2	Acetylene	98-01-1	Furfural
74-87-3	Methyl Chloride	98-51-1	tert-Butyl Toluene
74-89-5	Methylamine	98-82-8	Cumene
74-90-8	Hydrogen Cyanide	98-83-9	Alpha-Methyl Styrene
74-93-1	Methyl Mercaptan	98-87-3	Benzyl Chloride
74-98-6	Propane	98-95-3	Nitrobenzene
74-99-7	Methylacetylene	99-87-6	p-Cymene
75-00-3	Ethyl Chloride	100-41-4	Ethyl Benzene
75-01-4	Vinyl Chloride	100-42-5	Styrene
75-04-7	Ethylamine	100-61-8	Monomethyl Aniline
75-05-8	Acetonitrile	100-63-0	Phenylhydrazine
75-07-0	Acetaldehyde	100-74-3	n-Ethyl Morpholine
75-08-1	Ethyl Mercaptan	102-69-2	Tripropylamine
75-15-0	Carbon Disulfide	103-09-3	Ethyl Hexyl Acrylate
75-19-4	Cyclopropane	104-76-7	Ethylhexanol
75-21-8	Ethylene Oxide	104-90-5	2-Methyl-5-Ethyl Pyridine
75-28-5	Isobutane	105-46-4	sec-Butyl Acetate
75-28-5	2-Methylpropane	106-35-4	Ethyl Butyl Ketone
75-28-5	3-Methylpropane	106-63-8	Isobutyl Acrylate
75-31-0	Isopropylamine	106-88-7	Butylene Oxide
75-35-4	Vinylidene Chloride	106-92-3	Allyl Glycidyl Ether
75-52-5	Nitromethane	106-96-7	Bromopropyne
75-56-9	Propylene Oxide	106-99-0	1,3-Butadiene
75-65-0	2-Methyl-2-Propanol	107-02-8	Acrolein (Inhibited)
75-74-1	Tetramethyl Lead	107-05-1	Allyl Chloride
75-83-2	Dimethylbutane	107-06-2	Ethylene Dichloride
75-83-2	Neohexane	107-07-3	Ethylene Chlorohydrin

Table 4.4.3 Continued

CAS No.	Chemical Name	CAS No.	Chemical Name
107-13-1	Acrylonitrile	112-07-2	Ethylene Glycol Monobutyl Ether Acetate
107-15-3	Ethylenediamine	112-30-1	n-Decanol
107-18-6	Allyl Alcohol	112-31-2	Isodecaldehyde
107-20-0	Chloroacetaldehyde	112-31-2	n-Decaldehyde
107-31-3	Methyl Formate	115-07-1	Propylene
107-83-5	Dimethylpentane	115-10-6	Methyl Ether
107-83-5	Isohexane	119-64-2	Tetrahydronaphthalene
107-83-5	2-Methylpentane	121-44-8	Triethylamine
107-87-9	2-Pentanone	123-05-7	Ethylhexaldehyde
107-92-6	n-Butyric Acid	123-05-7	Isooctyl Aldehyde
108-03-2	1-Nitropropane	123-38-6	Propionaldehyde
108-05-4	Vinyl Acetate	123-51-3	Isoamyl Alcohol
108-11-2	Methyl Amyl Alcohol	123-62-6	Propionic Anhydride
108-18-9	Diisopropylamine	123-72-8	n-Butyraldehyde
108-20-3	Isopropyl Ether	123-86-4	n-Butyl Acetate
108-21-4	Isopropyl Acetate	123-91-1	1,4-Dioxane
108-24-7	Acetic Anhydride	123-92-2	Isoamyl Acetate
108-84-9	sec-Hexyl Acetate	124-40-3	Dimethylamine
108-88-3	Toluene	126-99-8	Chloroprene
108-90-7	Chlorobenzene	138-86-3	Dipentene
108-93-0	Cyclohexanol	140-88-5	Ethyl Acrylate (Inhibited)
108-94-1	Cyclohexanone	141-32-2	n-Butyl Acrylate (Inhibited)
109-60-4	n-Propyl Acetate	141-43-5	Monoethanolamine
109-66-0	n-Pentane	141-78-6	Ethyl Acetate
109-67-1	1-Pentene	141-79-7	Methyl Isobutyl Ketone
109-68-2	2-Pentene	141-97-9	Mesityl Oxide
109-73-9	Butylamine	142-82-5	n-Heptane
109-79-5	Butyl Mercaptan	143-08-8	Nonyl Alcohol
109-86-4	Ethylene Glycol Monomethyl Ether	151-56-4	Ethylenimine
109-87-5	Methylal	208-87-2	Methylcyclohexane
109-94-4	Ethyl Formate	302-01-2	Hydrazine
109-99-9	Tetrahydrofuran	463-82-1	Dimethylpropane
110-19-0	Isobutyl Acetate	463-82-1	Neopentane
110-43-0	Methyl n-Amyl Ketone	534-15-6	Methyl Formal
110-54-3	n-Hexane	540-88-5	tert-Butyl Acetate
110-62-3	n-Butyl Formal	541-85-5	Ethyl Sec-Amyl Ketone
110-62-3	Valeraldehyde	589-34-4	3-Methylhexane
110-80-5	Ethylene Glycol Monoethyl Ether	591-78-6	Hexanone
110-82-7	Cyclohexane	592-41-6	Hexene
110-83-8	Cyclohexene	624-83-9	Methyl Isocyanate
110-86-1	Pyridine	626-38-0	sec-Amyl Acetate
110-91-8	Morpholine	627-13-4	Propyl Nitrate
111-15-9	Ethylene Glycol Monoethyl Ether Acetate	628-63-7	n-Amyl Acetate
111-27-3	Hexanol	630-08-0	Carbon Monoxide
111-43-3	n-Propyl Ether	645-62-5	Ethyl-3-Propyl Acrolein
111-65-9	n-Octane	1068-19-5	Methylheptane
111-69-3	Adiponitrile	1071-26-7	Dimethylheptane
111-76-2	Ethylene Glycol Monobutyl Ether	1319-77-3	Cresol
111-84-2	n-Nonane	1330-20-7	Xylene
111-87-5	n-Octyl Alcohol	1333-74-0	Hydrogen

(continues)

Table 4.4.3 Continued

CAS No.	Chemical Name
1333-74-0	Process Gas > 30% H_2
1634-04-4	Methyl Tertiary Butyl Ether
2216-32-2	2-Methyloctane
2216-33-3	3-Methyloctane
2216-34-4	4-Methyloctane
2425-66-3	1-Chloro-1-Nitropropane
2426-08-6	n-Butyl Glycidyl Ether
2437-56-1	Tridecene
3132-64-7	Epichlorohydrin
3221-61-2	2-Methyloctane
3583 47 9	Butane
4016-14-2	Isopropyl Glycidyl Ether
4170-30-3	Crotonaldehyde
6842-15-5	Dodecene
7664-41-7	Ammonia
7783-06-4	Hydrogen Sulfide
7783-07-5	Hydrogen Selenide
8006-61-9	Gasoline
8006-64-2	Turpentine
8008-20-6	Fuel Oil 1
8008-20-6	Kerosene
8030-30-6	Naphtha (Coal Tar)
8030-30-6	Naphtha (Petroleum)
25013-15-4	Vinyl Toluene
25167-67-3	Butylene
25340-18-5	Triethylbenzene
25377-83-7	Octene
25630-42-3	Methylcyclohexanol
26952-21-6	Isooctyl Alcohol
27214-95-8	Nonene
27846-30-6	Methylacetylene-Propadiene
28761-27-5	Undecene
31394-54-4	Dimethylhexane
31394-54-4	2-Methylhexane
34590-94-8	Dipropylene Glycol Methyl Ether
68476-85-7	Liquefied Petroleum Gas
81624-04-6	Heptene

Chapter 5 Classification of Class I (Combustible Material) Areas

5.1 General.

The decision to classify an area as hazardous is based on the possibility that an ignitible mixture may occur. Having decided that an area should be classified, the next step is to determine which classification methodology should be utilized: the U.S. traditional *NEC* Articles 500 and 501, Class, Division, Group; or the *NEC* Article 505, Class, Zone, Group.

5.1.1 Refer to Sections 5.2 and 5.4 for use with the U.S. traditional *NEC* Article 500 Class, Division criteria to determine the degree of hazard: Is the area Division 1 or Division 2?

5.1.2 Refer to Sections 5.3 and 5.4 for using *NEC* Article 505 Class, Zone criteria to determine the degree of hazard: Is the area Zone 0, Zone 1, or Zone 2?

5.2 Class, Division, Classified Locations.

5.2.1 Division 1 Classified Locations.

5.2.1.1 A condition for Division 1 is whether the location is likely to have an ignitible mixture present under normal conditions. For instance, the presence of a combustible material in the immediate vicinity of an open dip tank is normal and requires a Division 1 classification.

5.2.1.2 *Normal* does not necessarily mean the situation that prevails when everything is working properly. For instance, there could be cases in which frequent maintenance and repair are necessary. These are viewed as normal and, if quantities of a flammable liquid or a combustible material are released as a result of the maintenance, the location is Division 1.

5.2.1.3 However, if repairs are not usually required between turnarounds, the need for repair work is considered abnormal. In any event, the classification of the location, as related to equipment maintenance, is influenced by the maintenance procedures and frequency of maintenance.

5.2.2 Division 2 Classified Locations. The criterion for a Division 2 location is whether the location is likely to have ignitible mixtures present only under abnormal conditions. The term *abnormal* is used here in a limited sense and does not include a major catastrophe.

5.2.2.1 As an example, consider a vessel containing liquid hydrocarbons (the source) that releases combustible material only under abnormal conditions. In this case, there is no Division 1 location because the vessel is normally tight. To release vapor, the vessel would have to leak, and that would not be normal. Thus, the vessel is surrounded by a Division 2 location.

5.2.2.2 Chemical process equipment does not often fail. Furthermore, the electrical installation requirements of the *NEC* for Division 2 locations is such that an ignition-capable spark or hot surface will occur only in the event of abnormal operation or failure of electrical equipment. Otherwise, sparks and hot surfaces are not present or are contained in enclosures. On a realistic basis, the possibility of process equipment and electrical equipment failing simultaneously is remote.

5.2.2.3 The Division 2 classification is also applicable to conditions not involving equipment failure. For example, consider an area classified as Division 1 because of the normal presence of an ignitible mixture. Obviously, one side of the Division 1 boundary cannot be normally hazardous and the opposite side never hazardous. When there is no wall, a surrounding transition Division 2 location separates a Division 1 location from an unclassified location.

5.2.2.4 In cases in which an unpierced barrier, such as a blank wall, completely prevents the spread of the combustible material, the area classification does not extend beyond the barrier.

5.3 Class I, Zone Classified Locations.

5.3.1 Zone 0 Classified Locations. A condition for Zone 0 is whether the location has an ignitible mixture present continuously or for long periods of time.

5.3.1.1 Zone 0 classified locations include the following:

(1) Inside vented tanks or vessels containing volatile flammable liquids
(2) Inside inadequately vented spraying or coating enclosures where volatile flammable solvents are used
(3) Between the inner and outer roof sections of a floating roof tank containing volatile flammable liquids
(4) Inside open vessels, tanks, and pits containing volatile flammable liquids
(5) The interior of an exhaust duct that is used to vent ignitible concentrations of gases or vapors
(6) Inside inadequately ventilated enclosures containing normally venting instruments utilizing or analyzing flammable fluids and venting to the inside of the enclosures

5.3.1.2 It is not good practice to install electrical equipment in Zone 0 locations except when the equipment is essential to the process or when other locations are not feasible.

5.3.2 Zone 1 Classified Locations.

5.3.2.1 The criteria for a Zone 1 location include the following:

(1) Is the location likely to have ignitible mixtures present under normal conditions?
(2) Is the location likely to have ignitible mixtures exist frequently because of repair or maintenance operations or because of leakage?
(3) Does the location have conditions in which equipment is operated or processes are carried on, where equipment breakdown or faulty operations could result in the release of ignitible concentrations of flammable gases or vapors, and also could cause simultaneous failure of electrical equipment in a mode to cause the electrical equipment to become a source of ignition?
(4) Is the location adjacent to a Class I, Zone 0 location from which ignitible concentrations of vapors could be communicated, unless communication is prevented by adequate positive-pressure ventilation from a source of clean air and effective safeguards against ventilation failure are provided?

5.3.2.2 Zone 1 classified locations include the following:

(1) Locations where volatile flammable liquids or liquefied flammable gases are transferred from one container to another, in areas in the vicinity of spraying and painting operations where flammable solvents are used
(2) Adequately ventilated drying rooms or compartments for the evaporation of flammable solvents
(3) Adequately ventilated locations containing fat and oil extraction equipment using volatile flammable solvents
(4) Portions of cleaning and dyeing plants where flammable liquids are used
(5) Adequately ventilated gas generator rooms and other portions of gas manufacturing plants where flammable gas may escape
(6) Inadequately ventilated pump rooms for flammable gas or for flammable liquids
(7) The interiors of refrigerators and freezers in which volatile flammable materials are stored in open, lightly stoppered, or easily ruptured containers
(8) Other locations where ignitible concentrations of flammable vapors or gases are likely to occur in the course of normal operation, but not classified Zone 0

5.3.3 Zone 2 Locations.

5.3.3.1 The criteria for a Zone 2 location include the following:

(1) Ignitible mixtures are not likely to occur in normal operation, and, if they do occur, will exist only for a short period.
(2) Ignitible mixtures are handled, processed, or used in the area, but liquids, gases, or vapors normally are confined within closed containers or closed systems from which they can escape only as a result of accidental rupture or breakdown of the containers or system, or as the result of the abnormal operation of the equipment with which the liquids or gases are handled, processed, or used.

(3) Ignitible mixtures normally are prevented by positive mechanical ventilation, but may become hazardous as the result of failure or abnormal operation of the ventilation equipment.
(4) The location is adjacent to a Class I, Zone 1 location, from which ignitible concentrations of flammable gases or vapors could be communicated, unless such communication is prevented by adequate positive-pressure ventilation from a source of clean air, and effective safeguards against ventilation failure are provided.

5.3.3.2 The Zone 2 classification usually includes locations where flammable liquids or flammable gases or vapors are used, but which would become hazardous only in case of an accident or of some unusual operating condition.

5.4 Unclassified Locations.

5.4.1 Experience has shown that the release of ignitible mixtures from some operations and apparatus is so infrequent that area classification is not necessary. For example, it is not usually necessary to classify the following locations where combustible materials are processed, stored, or handled:

(1) Locations that have adequate ventilation, where combustible materials are contained within suitable, well-maintained, closed piping systems
(2) Locations that lack adequate ventilation, but where piping systems are without valves, fittings, flanges, and similar accessories that may be prone to leaks
(3) Locations where combustible materials are stored in suitable containers

5.4.2 Locations considered to have adequate ventilation include the following:

(1) An outside location
(2) A building, room, or space that is substantially open and free of obstruction to the natural passage of air, either vertically or horizontally. (Such areas could be roofed over with no walls, could be roofed over and closed on one side, or could be provided with suitably designed windbreaks.)
(3) An enclosed or partly enclosed space provided with ventilation equivalent to natural ventilation. (The ventilation system must have adequate safeguards against failure.)

5.4.3 Open flames and hot surfaces associated with the operation of certain equipment, such as boilers and fired heaters, provide inherent thermal ignition sources. Electrical classification is not appropriate in the immediate vicinity of these facilities. However, it is prudent to avoid installing electrical equipment that could be a primary ignition source for potential leak sources in pumps, valves, and so forth, or in waste product and fuel feed lines.

5.4.4 Experience indicates that Class IIIB liquids seldom evolve enough vapors to form ignitible mixtures even when heated, and are seldom ignited by properly installed and maintained general-purpose electrical equipment.

5.4.5 Experience has shown that some halogenated liquid hydrocarbons, such as trichloroethylene; 1,1,1-trichloroethane; methylene chloride; and 1,1-dichloro-1-fluoroethane (HCFC-141b), which do not have flash points, but do have a flammable range, are for practical purposes nonflammable and do not require special electrical equipment for hazardous (classified) locations.

5.5 Extent of Classified Locations.

5.5.1 The extent of a Division 1 or Division 2 location or a Zone 0, Zone 1, or Zone 2 location requires careful consideration of the following factors:

(1) The combustible material
(2) The vapor density of the material
(3) The temperature of the material
(4) The process or storage pressure
(5) The size of release
(6) The ventilation

5.5.2* The first step is to identify the materials being handled and their vapor densities. Hydrocarbon vapors and gases are generally heavier than air, whereas hydrogen and methane are lighter than air. The following guidelines apply:

(1) In the absence of walls, enclosures, or other barriers, and in the absence of air currents or similar disturbing forces, the combustible material will disperse. Heavier-than-air vapors will travel primarily downward and outward; lighter-than-air vapors will travel upward and outward. If the source of the vapors is a single point, the horizontal area covered by the vapor will be a circle.
(2) For heavier-than-air vapors released at or near grade level, ignitible mixtures are most likely to be found below grade level; next most likely at grade level; with decreasing likelihood of presence as height above grade increases. For lighter-than-air gases, the opposite is true: there is little or no hazard at and below grade but greater hazard above grade.
(3) In cases where the source of the combustible material is above grade or below grade or in cases where the combustible material is released under pressure, the limits of the classified area are altered substantially. Also, a very mild breeze could extend these limits. However,

a stronger breeze could accelerate dispersion of the combustible material so that the extent of the classified area is greatly reduced. Thus, dimensional limits recommended for either Class I, Division 1 or Division 2; or Class I, Zone 0, Zone 1, or Zone 2 classified areas must be based on experience rather than relying solely on the theoretical diffusion of vapors.

5.5.3 The size of a building and its design could influence considerably the classification of the enclosed volume. In the case of a small, inadequately ventilated room, it could be appropriate to classify the entire room as Class I, Division 1 or Class I, Zone 1.

5.5.4 When classifying buildings, careful evaluation of prior experience with the same or similar installations should be made. It is not enough to identify only a potential source of the combustible material within the building and proceed immediately to defining the extent of either the Class I, Division 1 or Division 2; or Class I, Zone 1 or Zone 2 classified areas. Where experience indicates that a particular design concept is sound, a more hazardous classification for similar installations may not be justified. Furthermore, it is conceivable that an area be reclassified from either Class I, Division 1 to Class I Division 2, or from Class I, Division 2 to unclassified, or from Class I, Zone 1 to Class I, Zone 2, or from Class I, Zone 2 to unclassified, based on experience.

5.5.5 Correctly evaluated, an installation will be found to be a multiplicity of Class I, Division 1 areas of very limited extent. The same will be true for Class I, Zone 1 areas. Probably the most numerous of offenders are packing glands. A packing gland leaking a quart per minute (0.95 L/min), or 360 gallons per day, would certainly not be commonplace. Yet, if a quart bottle were emptied each minute outdoors, the zone made hazardous would be difficult to locate with a combustible gas detector.

5.5.6 The volume of combustible material released is of extreme importance in determining the extent of a hazardous (classified) location, and it is this consideration that necessitates the greatest application of sound engineering judgment. However, one cannot lose sight of the purpose of this judgment; the area is classified solely for the installation of electrical equipment.

5.6 Discussion of Diagrams and Recommendations.

5.6.1 This chapter contains a series of diagrams that illustrate how typical sources of combustible material should be classified and the recommended extent of the various classifications. Some of the diagrams are for single-point sources; others apply to multiple sources in an enclosed space or in an operating area. The basis for the diagrams is explained in Section 5.7.

5.6.2 The intended use of the diagrams is to aid in developing electrical classification maps of operating units, process plants, and buildings. Most of the maps will be plan views. Elevations or sectional views could be required where different classifications apply at different levels.

5.6.3 An operating unit could have many interconnected sources of combustible material, including pumps, compressors, vessels, tanks, and heat exchangers. These in turn present sources of leaks such as flanged and screwed connections, fittings, valves, meters, and so forth. Thus, considerable judgment will be required to establish the boundaries of Division 1 and Division 2 or Zone 0, Zone 1, and Zone 2 locations.

5.6.4 In some cases, individual classification of a multitude of point sources within an operating unit is neither feasible nor economical. In such cases, the entire unit could be classified as a single-source entity. However, this should be considered only after a thorough evaluation of the extent and interaction of the various sources, both within the unit and adjacent to it.

5.6.5 In developing these diagrams, vapor density is generally assumed to be greater than that of air. Lighter-than-air gases, such as hydrogen and methane, will quite readily disperse, and the diagrams for lighter-than-air gases should be used. However, if such gases are being evolved from the cryogenic state [such as liquefied hydrogen or liquefied natural gas (LNG)], caution must be exercised, because for some finite period of time, these gases will be heavier than air due to their low temperature when first released.

5.7 Basis for Recommendations.

5.7.1 The practices of the petroleum refining industry are published by the American Petroleum Institute, in ANSI/API RP 500, *Recommended Practice for Classification of Locations for Electrical Installations at Petroleum Facilities Classified as Class I, Division 1 and Division 2,* and ANSI/API RP 505, *Recommended Practice for Classification of Locations for Electrical Installations at Petroleum Facilities Classified as Class I, Zone 0, Zone 1, and Zone 2.* These practices are based on an analysis of the practices of a large segment of the industry, experimental data, and careful weighing of pertinent factors. Petroleum facility operations are characterized by the handling, processing, and storage of large quantities of materials, often at elevated temperatures. The recommended limits of classified locations for petroleum facility installations could therefore be stricter than are warranted for more traditional chemical processing facilities that handle smaller quantities.

5.7.2 Various codes, standards, and recommended practices of the National Fire Protection Association include recommendations for classifying hazardous (classified) locations. These recommendations are based on many years of experience. NFPA 30, *Flammable and Combustible Liquids Code*, and NFPA 58, *Liquefied Petroleum Gas Code*, are two of these documents.

5.7.3 Continuous process plants and large batch chemical plants could be almost as large as refineries and should therefore follow the practices of the refining industry. Leakage from pump and agitator shaft packing glands, piping flanges, and valves generally increases with process equipment size, pressure, and flow rate, as does the travel distance and area of dispersion from the discharge source.

5.7.4 In deciding whether to use an overall plant classification scheme or individual equipment classification, process equipment size, flow rate, and pressure should be taken into consideration. Point-source diagrams can be used in most cases for small or batch chemical plants; for large, high-pressure plants, the API recommendations are more suitable. Table 5.7.4 gives ranges of process equipment size, pressure, and flow rate for equipment and piping that handles combustible material.

5.7.5 The majority of chemical plants fall in the moderate range of size, pressure, and flow rate for equipment and piping that handles combustible materials. However, because all cases are not the same, sound engineering judgment is required.

5.8 Procedure for Classifying Locations.

The procedure described in 5.8.1 through 5.8.4 should be used for each room, section, or area being classified.

5.8.1 Step One — Determining Need for Classification. The area should be classified if a combustible material is processed, handled, or stored there.

Table 5.7.4 Relative Magnitudes of Process Equipment and Piping that Handles Combustible Materials

Process Equipment	Units	Small (Low)	Moderate	Large (High)
Size	gal	<5000	5000–25,000	>25,000
Pressure	psi	<100	100–500	>500
Flow rate	gpm	<100	100–500	>500

5.8.2 Step Two — Gathering Information.

5.8.2.1 Proposed Facility Information. For a proposed facility that exists only in drawings, a preliminary area classification can be done so that suitable electrical equipment and instrumentation can be purchased. Plants are rarely built exactly as the drawings portray them, so the area classification should be modified later based on the actual facility.

5.8.2.2 Existing Facility History. For an existing facility, the individual plant experience is extremely important in classifying areas within the plant. Both operation and maintenance personnel in the actual plant should be asked the following questions:

(1) Have there been instances of leaks?
(2) Do leaks occur frequently?
(3) Do leaks occur during normal or abnormal operation?
(4) Is the equipment in good condition, questionable condition, or in need of repair?
(5) Do maintenance practices result in the formation of ignitible mixtures?
(6) Does routine flushing of process lines, changing of filters, opening of equipment, and so forth result in the formation of ignitible mixtures?

5.8.2.3 Process Flow Diagram. A process flow diagram showing the pressure, temperature, flow rates, composition, and quantities of various materials (i.e., mass flow balance sheets) passing through the process is needed.

5.8.2.4 Plot Plan. A plot plan (or similar drawing) is needed showing all vessels, tanks, trenches, lagoons, sumps, building structures, dikes, partitions, levees, ditches, and similar items that would affect dispersion of any liquid, gas, or vapor. The plot plan should include the prevailing wind direction.

5.8.2.5* Fire Hazard Properties of Combustible Material. The properties needed for determining area classification for many materials are shown in Table 4.4.2.

5.8.2.5.1 A material could be listed in Table 4.4.2 under a chemical name different from the chemical name used at a facility. Table 4.4.3 is provided to cross-reference the CAS number of the material to the chemical name used in Table 4.4.2.

5.8.2.5.2 Where materials being used are not listed in Table 4.4.2 or in other reputable chemical references, the necessary information can be obtained by the following:

(1) Contact the material supplier to determine if the material has been tested or group-classified. If tested, estimate

the group classification using the criteria shown in Annex A.

(2) Have the material tested and estimate the group classification using the criteria shown in Annex A.

(3) Refer to Annex B for a method for determining the group classification for some mixed combustible material streams.

5.8.3 Step Three — Selecting the Appropriate Classification Diagram.

5.8.3.1 The list of combustible materials from the process flow diagram and the material mass balance data should be correlated with the quantities, pressures, flow rates *(see Table 5.7.4)*, and temperatures to determine the following:

(1) Whether the process equipment size is low, moderate, or high
(2) Whether the pressure is low, moderate, or high
(3) Whether the flow rate is low, moderate, or high
(4) Whether the combustible material is lighter than air (vapor density < 1) or heavier than air (vapor density > 1)
(5) Whether the source of leaks is above or below grade
(6) Whether the process is a loading/unloading station, product dryer, filter press, compressor shelter, hydrogen storage, or marine terminal

5.8.3.2 Use Table 5.9 and the information in 5.8.3.1 to select the appropriate classification diagram(s).

5.8.4 Step Four — Determining the Extent of the Classified Location. The extent of the classified area can be determined by using sound engineering judgment to apply the methods discussed in 5.5.2 and the diagrams contained in this chapter.

5.8.4.1 The potential sources of leaks should be located on the plan drawing or at the actual location. These sources can include rotating or reciprocating shafts (e.g., pumps, compressors, and control valves) and atmospheric discharges from pressure relief devices.

5.8.4.2 For each leakage source, an equivalent example should be found from the selected classification diagram to determine the minimum extent of classification around the leakage source. The extent can be modified by considering the following:

(1) Whether an ignitible mixture is likely to occur frequently due to repair, maintenance, or leakage
(2) Where conditions of maintenance and supervision are such that leaks are likely to occur in process equipment, storage vessels, and piping systems containing combustible material
(3) Whether the combustible material could be transmitted by trenches, pipes, conduits, or ducts
(4) Ventilation or prevailing wind in the specific area, and the dispersion rates of the combustible materials

5.8.4.3 Once the minimum extent is determined, utilize distinct landmarks (e.g., curbs, dikes, walls, structural supports, edges of roads) for the actual boundaries of the area classification. These landmarks permit easy identification of the boundaries of the hazardous (classified) locations for electricians, instrument technicians, operators, and other personnel.

Table 5.9 Matrix of Diagrams Versus Material/Property/Application

Figure Number for Class I													
Division	**Zone**	**Special Condition**	**VD > 1**	**VD < 1**	**Cryogenic**	**Indoor**	**Indoor, Poor Ventilation**	**Outdoor**	**Above Grade**	**At Grade**	**Size**	**Pressure**	**Flow**
5.9.1(a)	5.10.1(a)		X					X		X	S/M	S/M	S/M
5.9.1(b)	5.10.1(b)		X					X	X		S/M	S/M	S/M
5.9.1(c)	5.10.1(c)		X			X				X	S/M	S/M	S/M
5.9.1(d)	5.10.1(d)		X			X			X		S/M	S/M	S/M
5.9.1(e)	5.10.1(e)		X			X				X	S/M	S/M	S/M
5.9.1(f)	5.10.1(f)		X				X			X	S/M	S/M	S/M
5.9.1(g)	5.10.1(g)		X					X		X	L	M/L	L
5.9.1(h)	5.10.1(h)		X					X	X		L	M/L	L
5.9.1(i)	5.10.1(i)		X				X		X		M/L	L	M/L

(continues)

Table 5.9 Continued

Figure Number for Class I													
Division	**Zone**	**Special Condition**	**VD > 1**	**VD < 1**	**Cryogenic**	**Indoor**	**Indoor, Poor Ventilation**	**Outdoor**	**Above Grade**	**At Grade**	**Size**	**Pressure**	**Flow**
5.9.1(j)	5.10.1(j)		X			X			X		M/L	L	M/L
5.9.1(k)	5.10.1(k)		X					X	X	X	S/M	S/M	S/M
5.9.1(l)	5.10.1(l)		X					X	X	X	M/L	M/L	M/L
5.9.1(m)	5.10.1(m)		X					X	X	X	S/M	S/M	S/M
5.9.1(n)	5.10.1(n)		X			X			X	X	S/M	S/M	S/M
5.9.2(a)	5.10.2(a)		X		X			X		X	S/M	M/H	S/M
5.9.2(b)	5.10.2(b)		X		X			X	X		S/M	M/H	S/M
5.9.3(a)	5.10.3(a)	Product dryer	FL			X		X	X				
5.9.3(b)	5.10.3(b)	Filter press	FL			X			X				
5.9.4(a)	5.10.4(a)	Storage tank	FL					X		X	M/L	L	M/L
5.9.4(b)	5.10.4(b)	Tank car loading	FL					X	X				
5.9.4(c)	5.10.4(c)	Tank car loading	FL					X	X	X			
5.9.4(d)	5.10.4(d)	Tank truck loading	FL					X	X	X			
5.9.4(e)	5.10.4(e)	Tank car loading/ tank truck loading	FL					X	X	X			
5.9.5	5.10.5	Tank car loading/ tank truck loading	FL		X			X	X				
5.9.6	5.10.6	Drum filling station	FL			X		X	X				
5.9.7	5.10.7	Emergency basin	FL					X	X	X			
5.9.8(a)	5.10.8(a)	Liquid H_2 storage		X	X	X		X	X	X			
5.9.8(b)	5.10.8(b)	Gaseous H_2 storage		X		X		X	X	X			
5.9.9(a)	5.10.9(a)	Compressor shelter		X		X			X	X			
5.9.9(b)	5.10.9(b)	Compressor shelter		X			X		X	X			
5.9.10(a)	5.10.10(a)	Cryogenic storage			X			X	X	X			
5.9.10(b)	5.10.10(b)	Cryogenic storage			X			X	X	X			
5.9.10(c)	5.10.10(c)	Cryogenic storage			X			X	X	X			
5.9.11	5.10.11		LNG					X	X	X			
5.9.12	5.10.12		LNG			X			X	X			
5.9.13	5.10.13		LNG						X				
5.9.14	5.10.14	Marine terminal	FL			X		X	X				

Notes: FL: Flammable Liquid. LNG: Liquefied Natural Gas. X: Diagram Applies.
L: Large. M: Moderate. S: Small. H: High.

5.9 Classification Diagrams for Class I, Divisions.

Most diagrams in Section 5.9 and Section 5.10 include tables of "suggested applicability" and use check marks to show the ranges of process equipment size, pressure, and flow rates. *(See Table 5.7.4.)* Unless otherwise stated, these diagrams assume that the material being handled is a flammable liquid. Table 5.9 provides a summary of where each diagram is intended to apply. Class I, Division diagrams include Figure 5.9.1(a) through Figure 5.9.14.

5.9.1 Indoor and Outdoor Process — Flammable Liquids. *[See Figure 5.9.1(a), Figure 5.9.1(b), Figure 5.9.1(c), Figure 5.9.1(d), Figure 5.9.1(e), Figure 5.9.1(f), Figure 5.9.1(g), Figure 5.9.1(h), Figure 5.9.1(i), Figure 5.9.1(j), Figure 5.9.1(k), Figure 5.9.1(l), Figure 5.9.1(m), and Figure 5.9.1(n).]*

5.9.2 Outdoor Process — Flammable Liquid, Flammable Gas, Compressed Flammable Gas, or Cryogenic Liquid. *[See Figure 5.9.2(a) and Figure 5.9.2(b).]*

5.9.3 Product Dryer and Plate and Frame Filter Press — Solids Wet with Flammable Liquids. *[See Figure 5.9.3(a) and Figure 5.9.3(b).]*

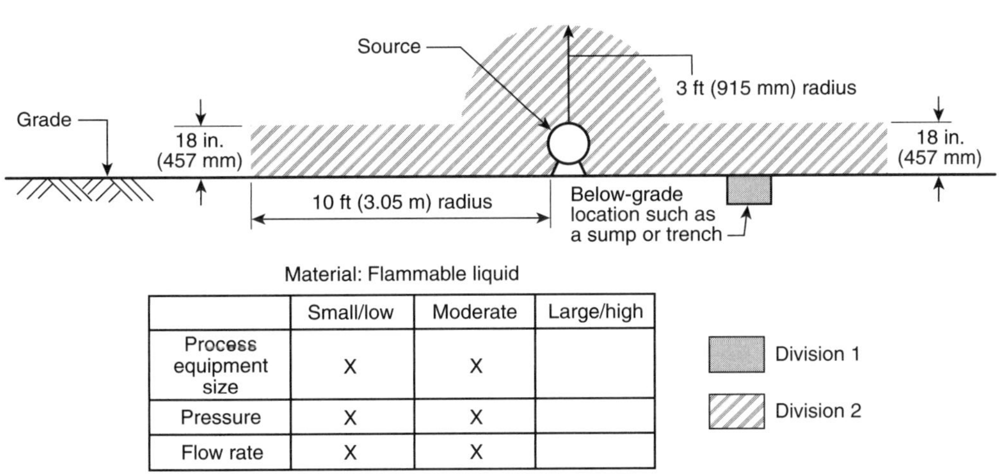

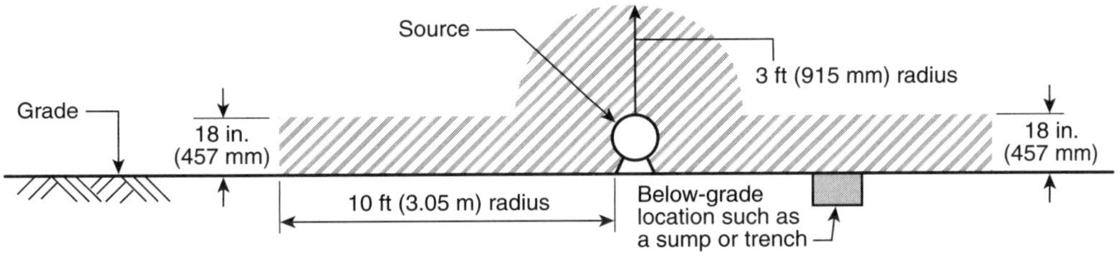

Material: Flammable liquid

	Small/low	Moderate	Large/high
Process equipment size	X	X	
Pressure	X	X	
Flow rate	X	X	

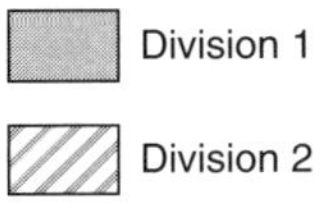

FIGURE 5.9.1(a) Leakage Located Outdoors, at Grade. The material being handled is a flammable liquid.

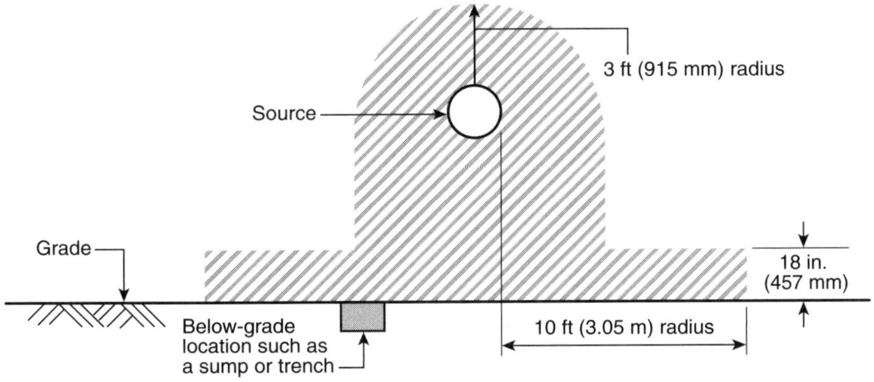

Material: Flammable liquid

	Small/low	Moderate	Large/high
Process equipment size	X	X	
Pressure	X	X	
Flow rate	X	X	

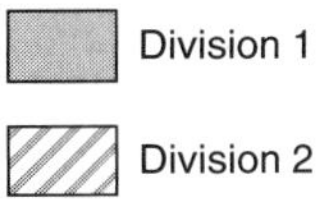

FIGURE 5.9.1(b) Leakage Located Outdoors, above Grade. The material being handled is a flammable liquid.

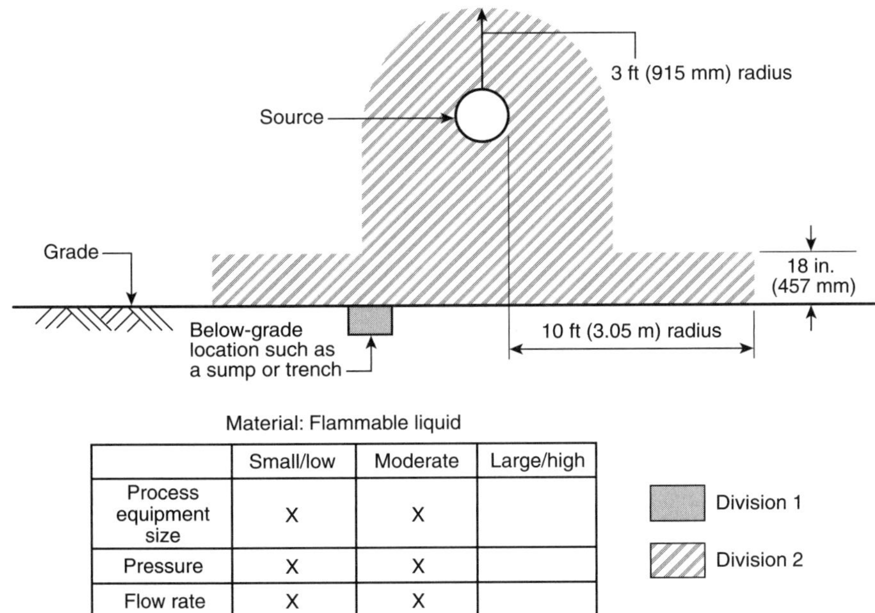

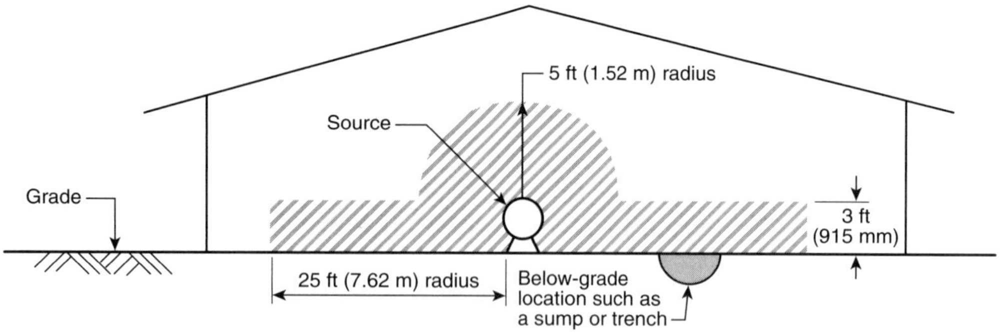

Material: Flammable liquid

	Small/low	Moderate	Large/high
Process equipment size	X	X	
Pressure	X	X	
Flow rate	X	X	

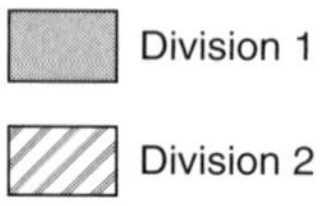

FIGURE 5.9.1(c) Leakage Located Indoors, at Floor Level. Adequate ventilation is provided. The material being handled is a flammable liquid.

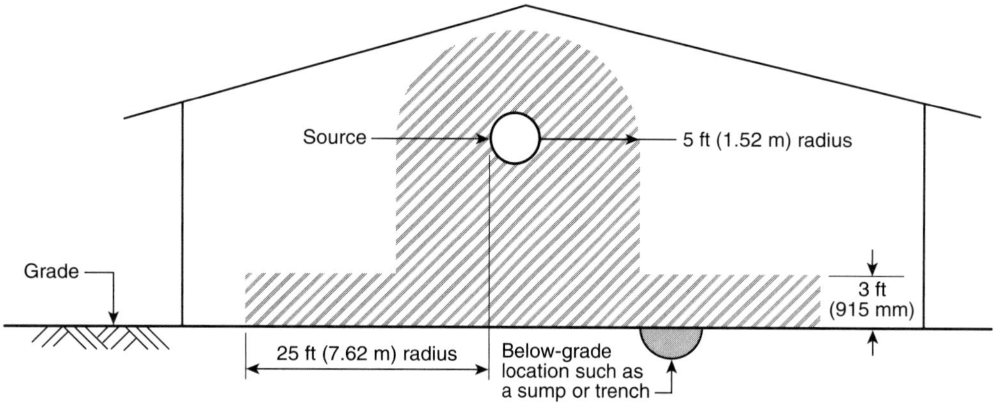

Material: Flammable liquid

	Small/low	Moderate	Large/high
Process equipment size	X	X	
Pressure	X	X	
Flow rate	X	X	

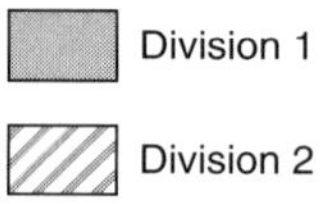

FIGURE 5.9.1(d) Leakage Located Indoors, Above Floor Level. Adequate ventilation is provided. The material being handled is a flammable liquid.

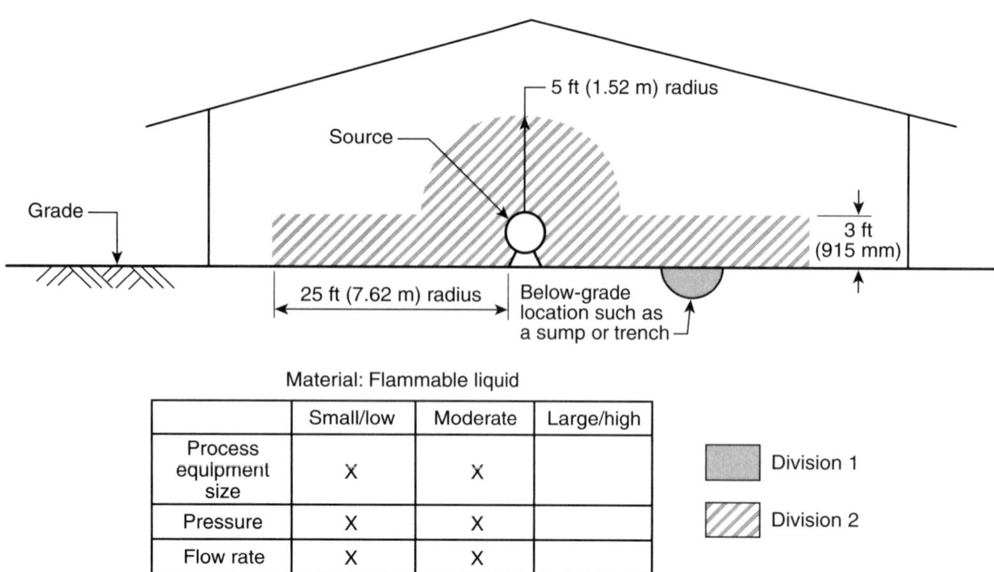

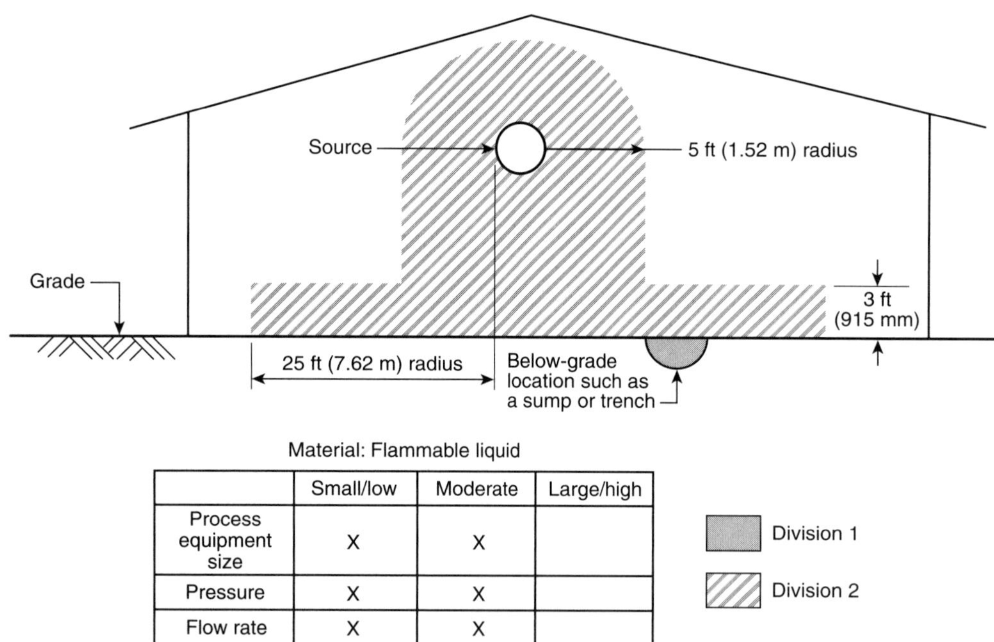

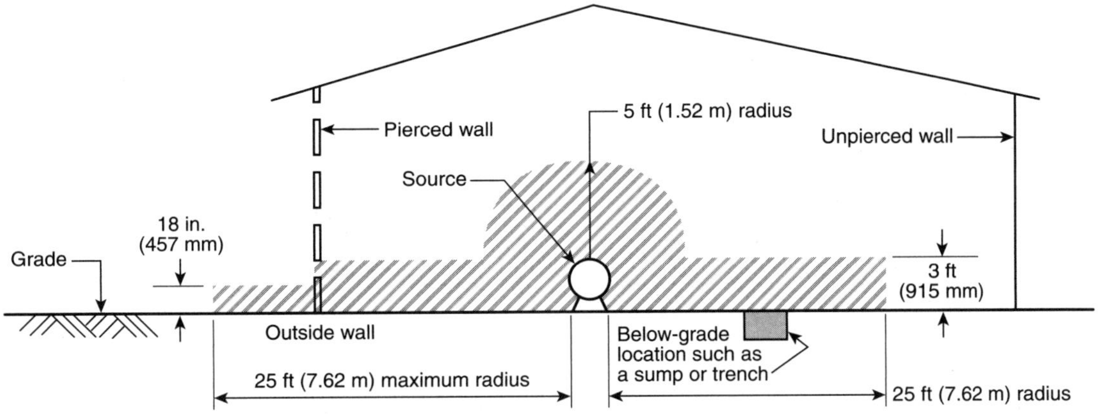

Material: Flammable liquid

	Small/low	Moderate	Large/high
Process equipment size	X	X	
Pressure	X	X	
Flow rate	X	X	

Division 1

Division 2

FIGURE 5.9.1(e) Leakage Located Indoors, at Floor Level, Adjacent to an Opening in an Exterior Wall. Adequate ventilation is provided. The material being handled is a flammable liquid.

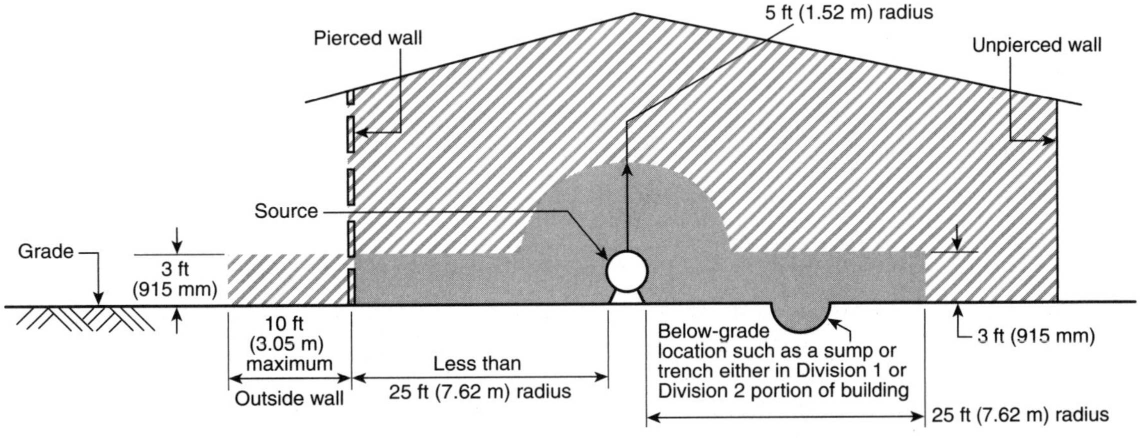

Note: If building is small compared to size of equipment, and leakage can fill the building, the entire building interior is classified Division 1.

Material: Flammable liquid

	Small/low	Moderate	Large/high
Process equipment size	X	X	
Pressure	X	X	
Flow rate	X	X	

Division 1

Division 2

FIGURE 5.9.1(f) Leakage Located Indoors, at Floor Level, Adjacent to an Opening in an Exterior Wall. Ventilation is not adequate. The material being handled is a flammable liquid.

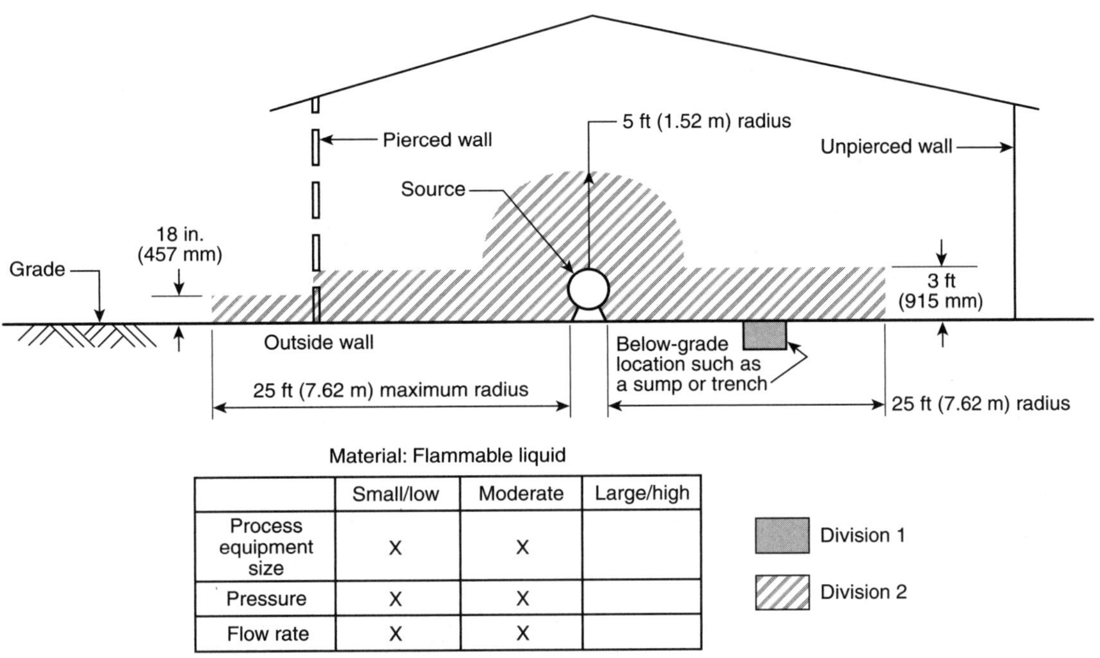

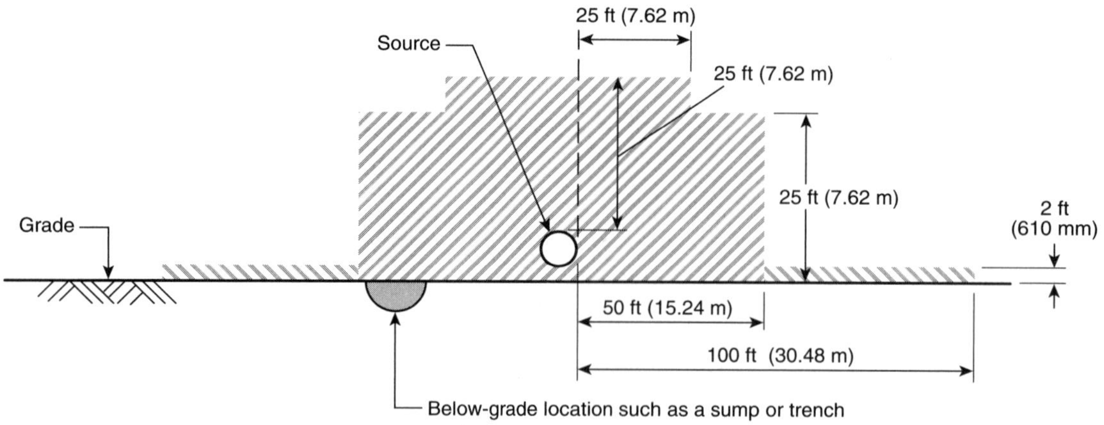

Material: Flammable liquid

	Small/low	Moderate	Large/high
Process equipment size			X
Pressure		X	X
Flow rate			X

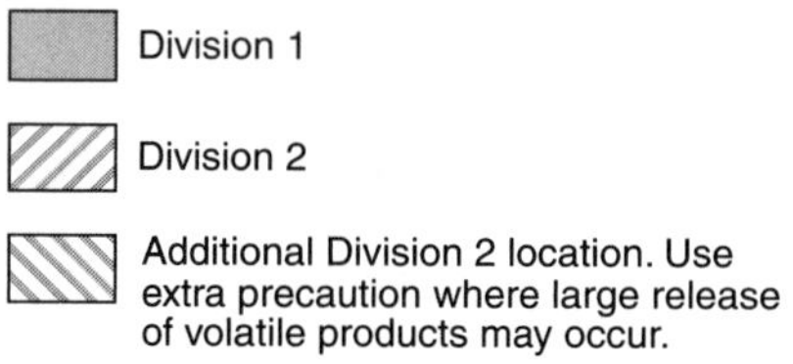

FIGURE 5.9.1(g) Leakage Located Outdoors, at Grade. The material being handled is a flammable liquid.

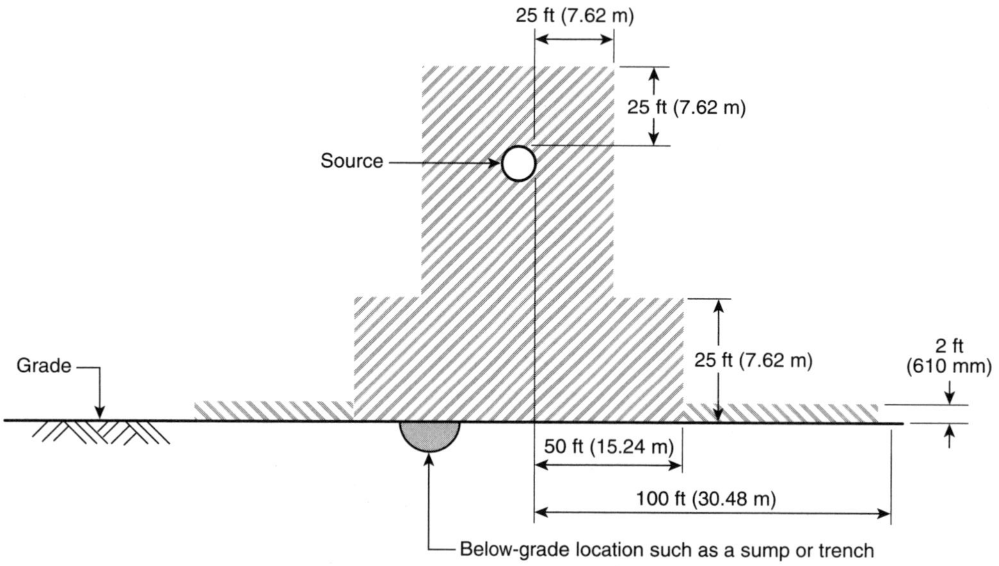

Material: Flammable liquid

	Small/low	Moderate	Large/high
Process equipment size			X
Pressure		X	X
Flow rate			X

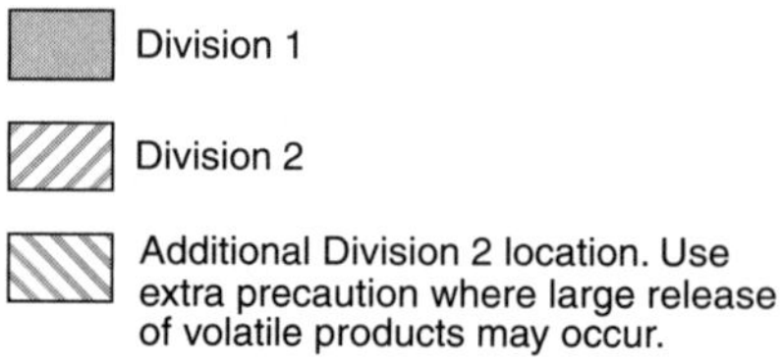

FIGURE 5.9.1(h) Leakage Located Outdoors, Above Grade. The material being handled is a flammable liquid.

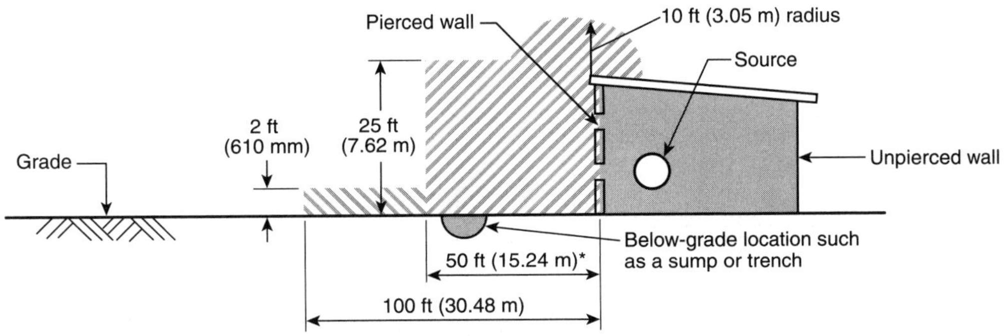

* "Apply" horizontal distances of 50 ft from the source of vapor or 10 ft beyond the perimeter of the building, whichever is greater, except that beyond unpierced vaportight walls the area is unclassified.

Material: Flammable liquid

	Small/low	Moderate	Large/high
Process equipment size		X	X
Pressure			X
Flow rate		X	X

Division 1

Division 2

Additional Division 2 location. Use extra precaution where large release of volatile products may occur.

FIGURE 5.9.1(i) Leakage Located Indoors, Adjacent to an Opening in an Exterior Wall. Ventilation is not adequate. The material being handled is a flammable liquid.

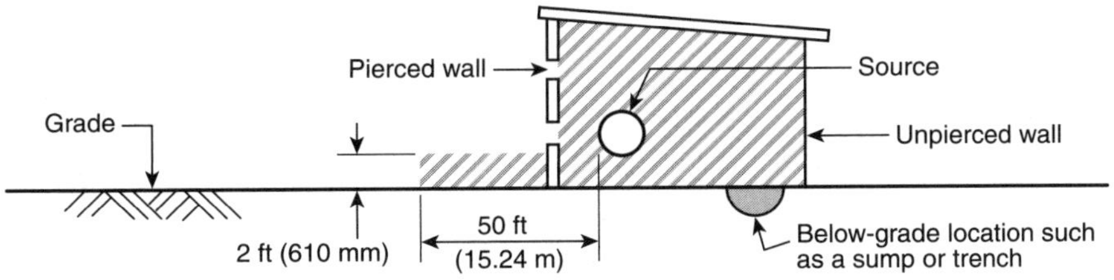

Material: Flammable liquid

	Small/low	Moderate	Large/high
Process equipment size		X	X
Pressure			X
Flow rate		X	X

Division 1

Division 2

FIGURE 5.9.1(j) Leakage Located Indoors, Adjacent to an Opening in an Exterior Wall. Adequate ventilation is provided. The material being handled is a flammable liquid.

5.9.4 Storage Tanks and Tank Vehicles — Flammable Liquids. *[See Figure 5.9.4(a), Figure 5.9.4(b), Figure 5.9.4(c), Figure 5.9.4(d), and Figure 5.9.4(e).]*

5.9.5 Tank Vehicle — Flammable Liquefied Gas, Flammable Compressed Gas, or Flammable Cryogenic Liquid. *(See Figure 5.9.5.)*

5.9.6 Indoor or Outdoor Drum Filling Station — Flammable Liquids. *(See Figure 5.9.6.)*

5.9.7 Emergency Impounding Basins, Emergency Drainage Ditches, or Oil/Water Separators — Flammable Liquids. *(See Figure 5.9.7.)*

5.9.8 Storage of Liquid or Gaseous Hydrogen. *[See Figure 5.9.8(a) and Figure 5.9.8(b).]*

5.9.9 Compressor Shelters — Lighter-than-Air Gas. *[See Figure 5.9.9(a) and Figure 5.9.9(b).]*

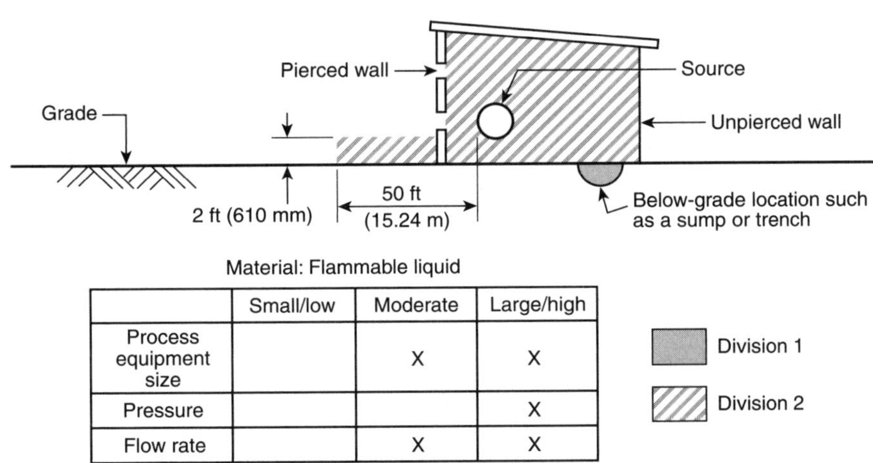

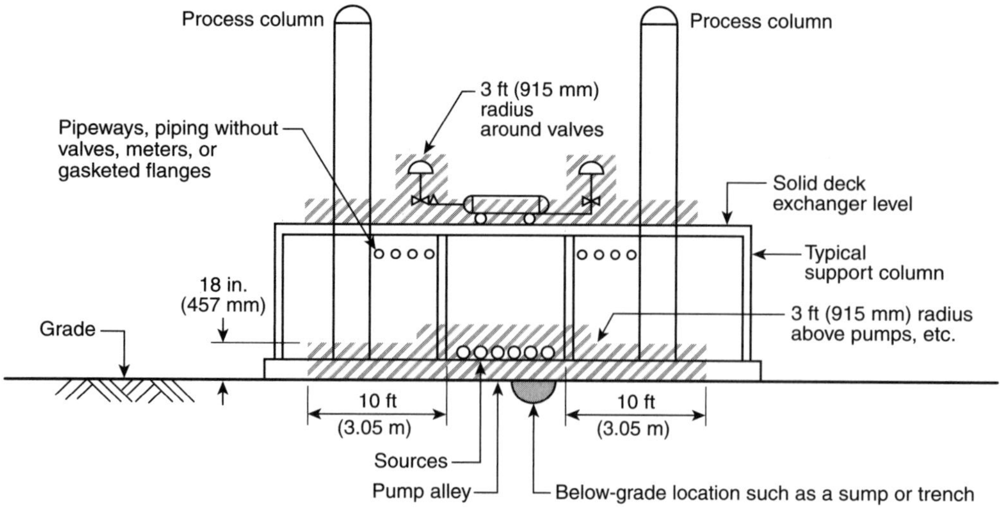

Material: Flammable liquid

	Small/low	Moderate	Large/high
Process equipment size	X	X	
Pressure	X	X	
Flow rate	X	X	

Division 1

Division 2

FIGURE 5.9.1(k) Leakage, Located Both at Grade and Above Grade, in an Outdoor Process Area. The material being handled is a flammable liquid.

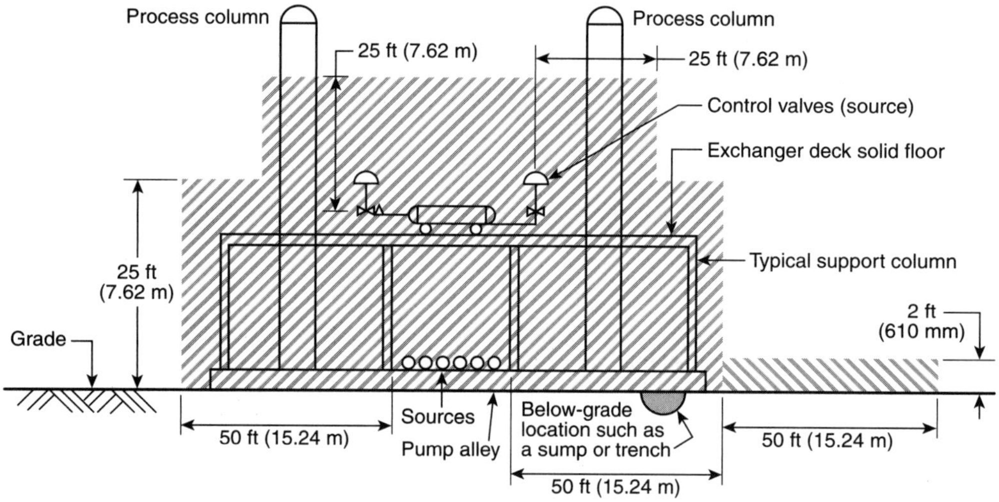

Material: Flammable liquid

	Small/low	Moderate	Large/high
Process equipment size		X	X
Pressure		X	X
Flow rate		X	X

Division 1

Division 2

Additional Division 2 area where release may be large

FIGURE 5.9.1(l) Multiple Sources of Leakage, Located Both at Grade and Above Grade, in an Outdoor Process Area. The material being handled is a flammable liquid.

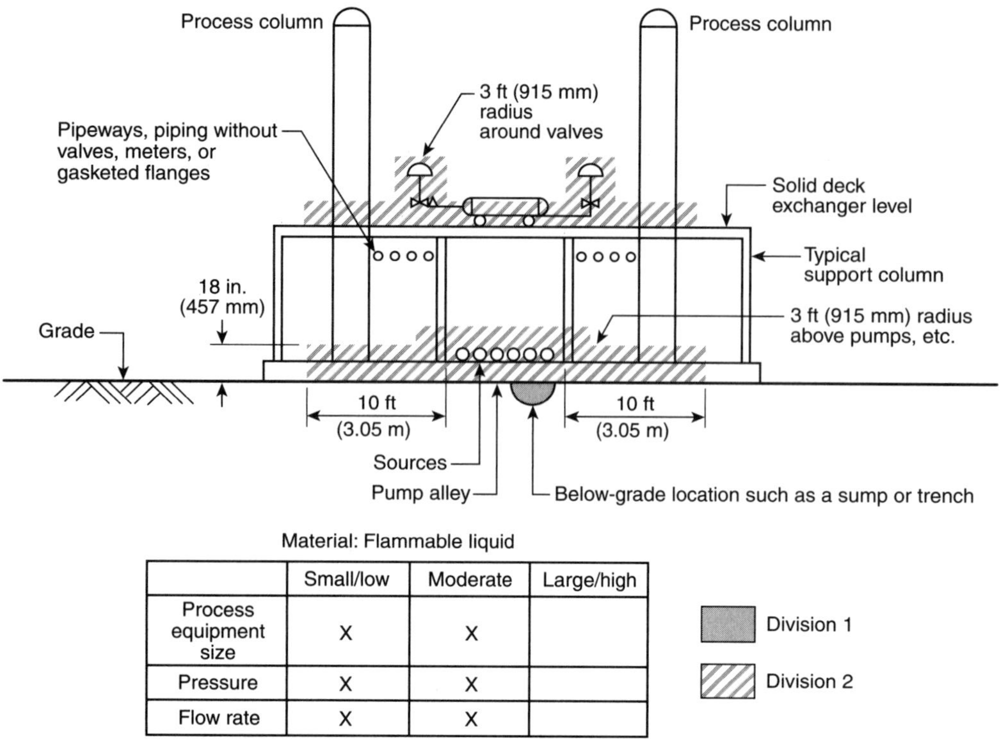

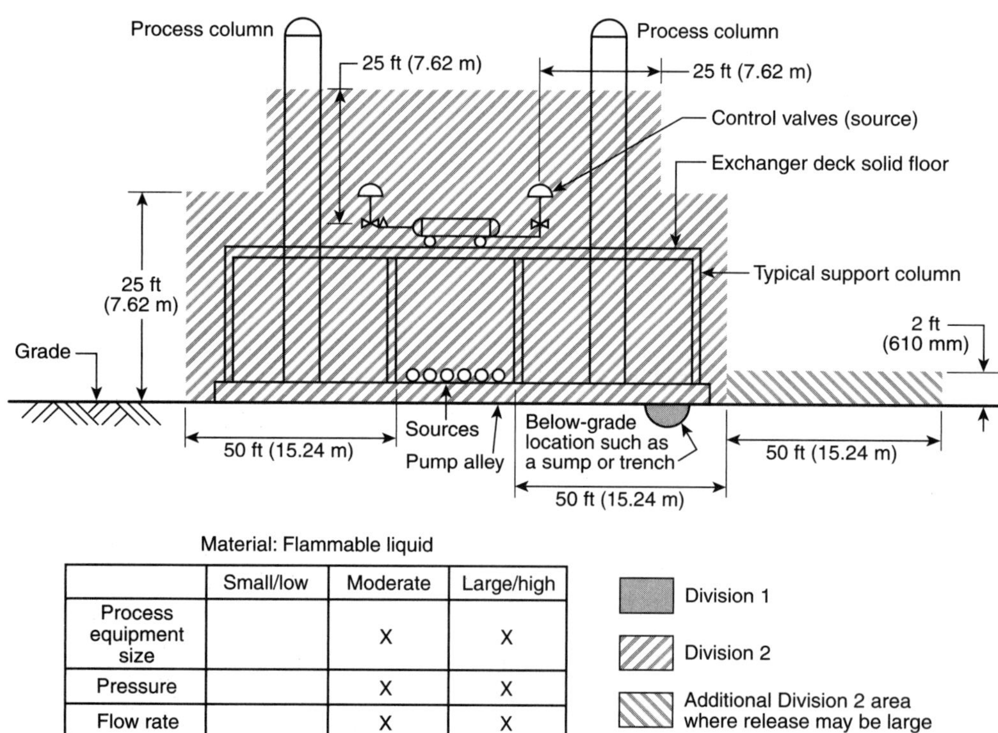

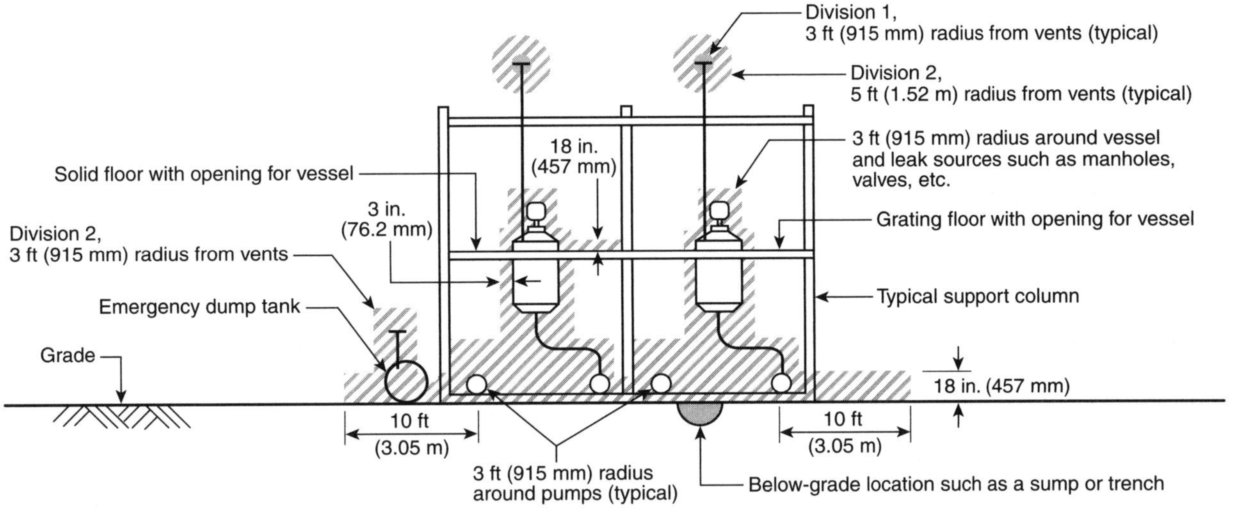

Material: Flammable liquid

	Small/low	Moderate	Large/high
Process equipment size	X	X	
Pressure	X	X	
Flow rate	X	X	

FIGURE 5.9.1(m) Multiple Sources of Leakage, Located Both at and Above Grade, in an Outdoor Process Area. The material being handled is a flammable liquid.

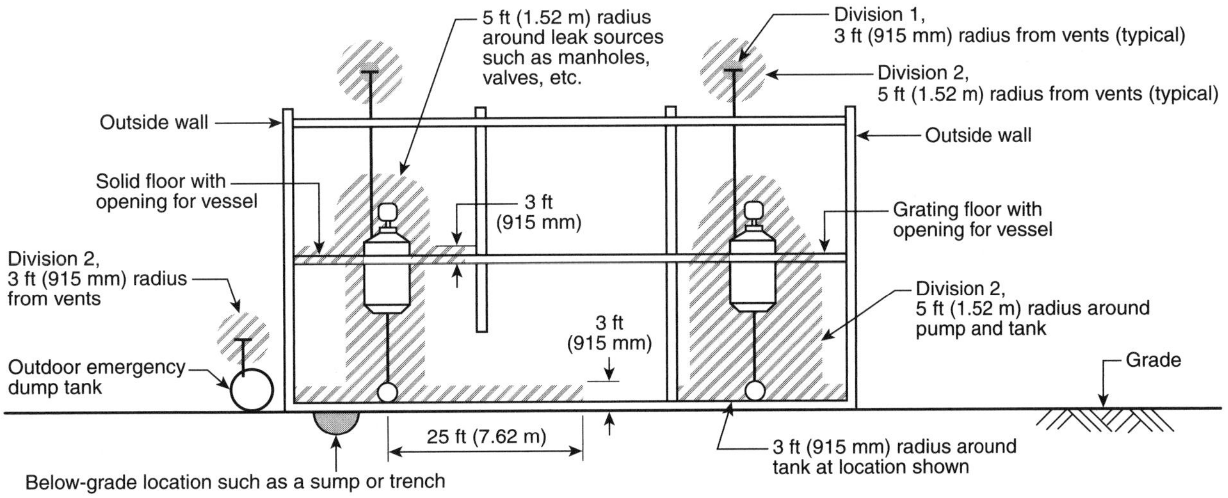

Material: Flammable liquid

	Small/low	Moderate	Large/high
Process equipment size	X	X	
Pressure	X	X	
Flow rate	X	X	

FIGURE 5.9.1(n) Multiple Sources of Leakage, Located Both at and Above Floor Level, in an Adequately Ventilated Building. The material being handled is a flammable liquid.

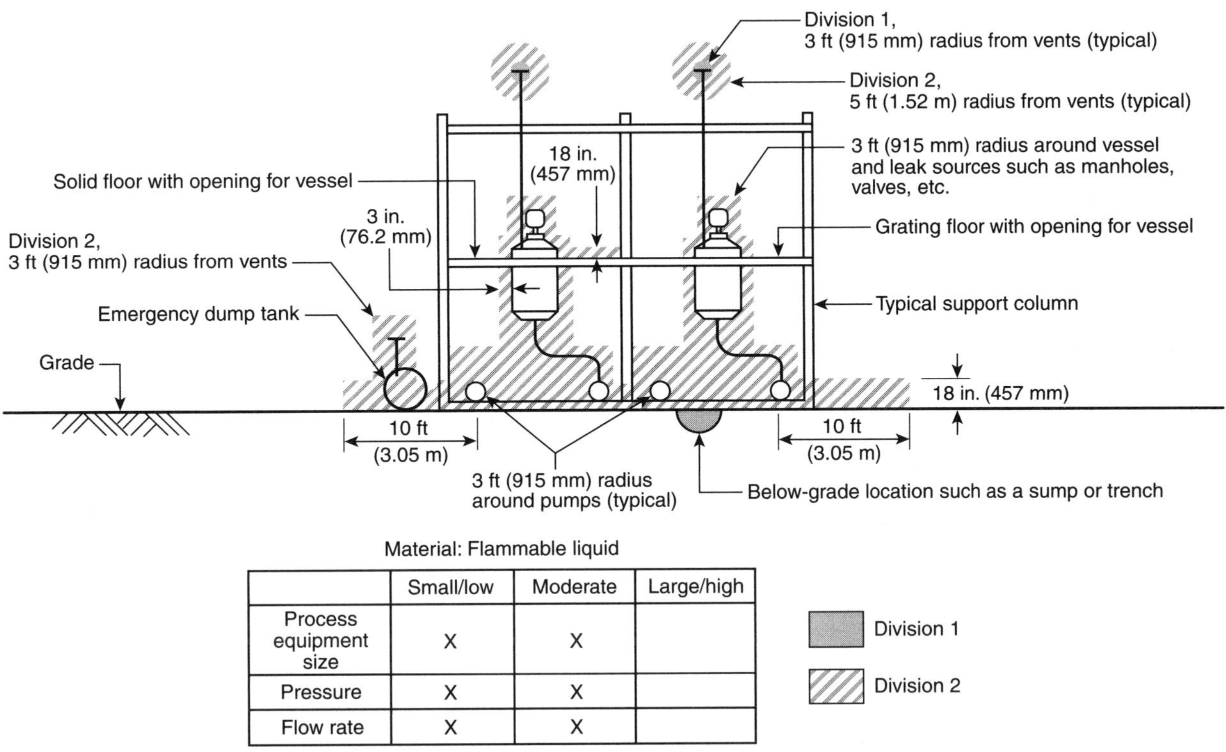

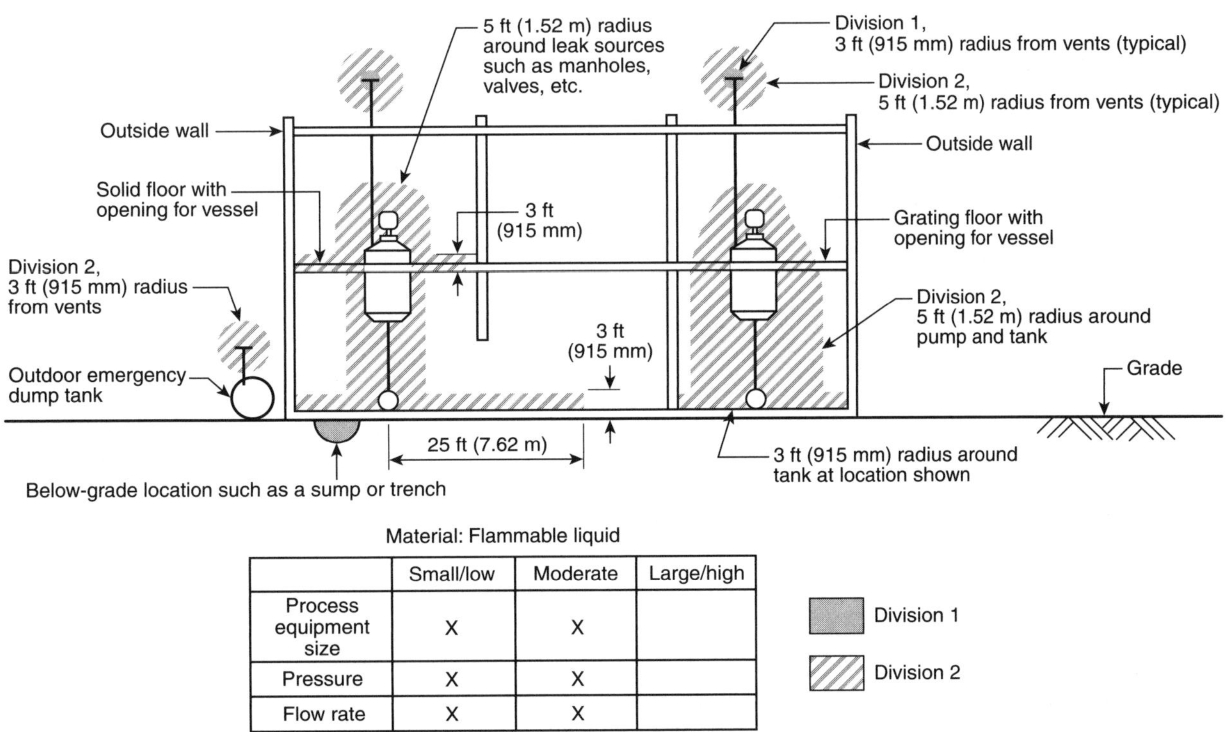

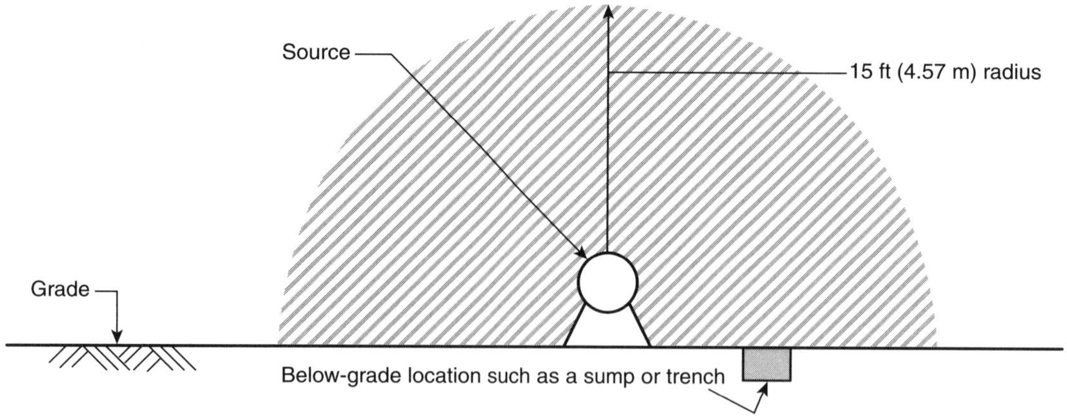

Material: Flammable liquid, liquefied flammable gas, compressed flammable gas, and cryogenic liquid

	Small/low	Moderate	Large/high
Process equipment size	X	X	
Pressure		X	X
Flow rate	X	X	

Division 1

Division 2

FIGURE 5.9.2(a) Leakage Located Outdoors, at Grade. The material being handled could be a flammable liquid, a liquefied or compressed flammable gas, or a flammable cryogenic liquid.

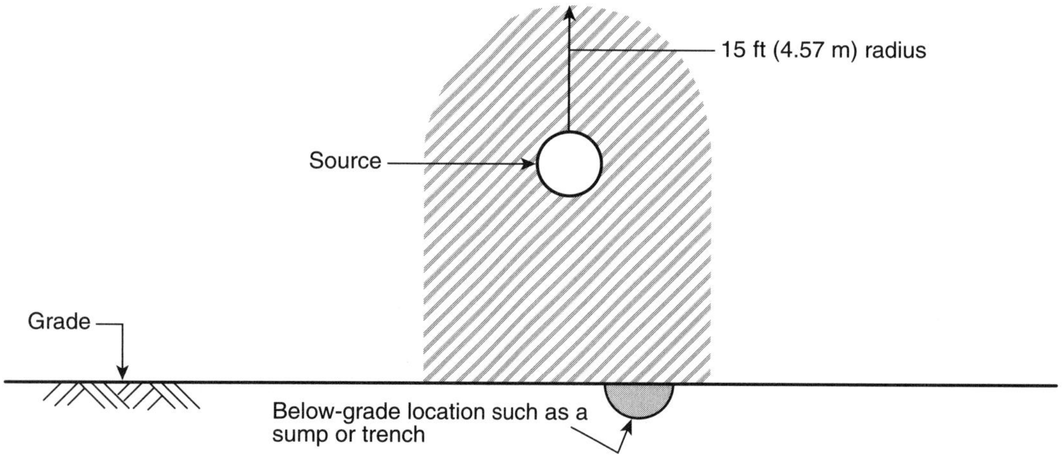

Material: Flammable liquid, liquefied flammable gas, compressed flammable gas, and cryogenic liquid

	Small/low	Moderate	Large/high
Process equipment size	X	X	
Pressure		X	X
Flow rate	X	X	

Division 1

Division 2

FIGURE 5.9.2(b) Leakage Located Outdoors, Above Grade. The material being handled could be a flammable liquid, a liquefied or compressed flammable gas, or a flammable cryogenic liquid.

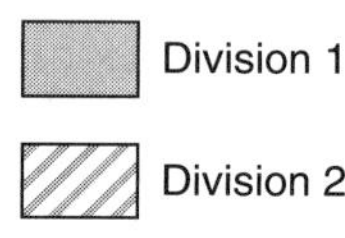

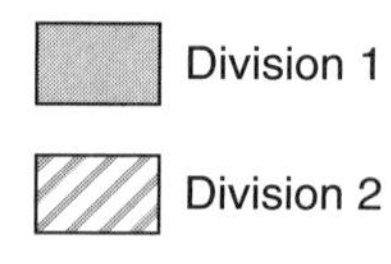

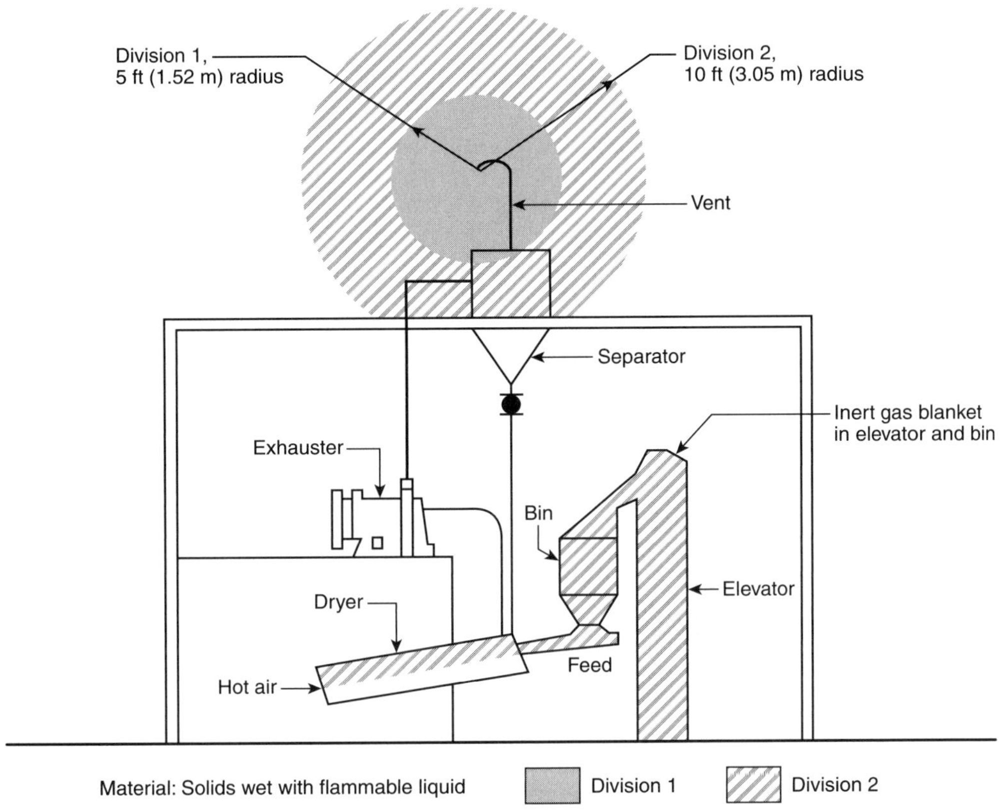

FIGURE 5.9.3(a) Product Dryer Located in an Adequately Ventilated Building. The product dryer system is totally enclosed. The material being handled is a solid wet with a flammable liquid.

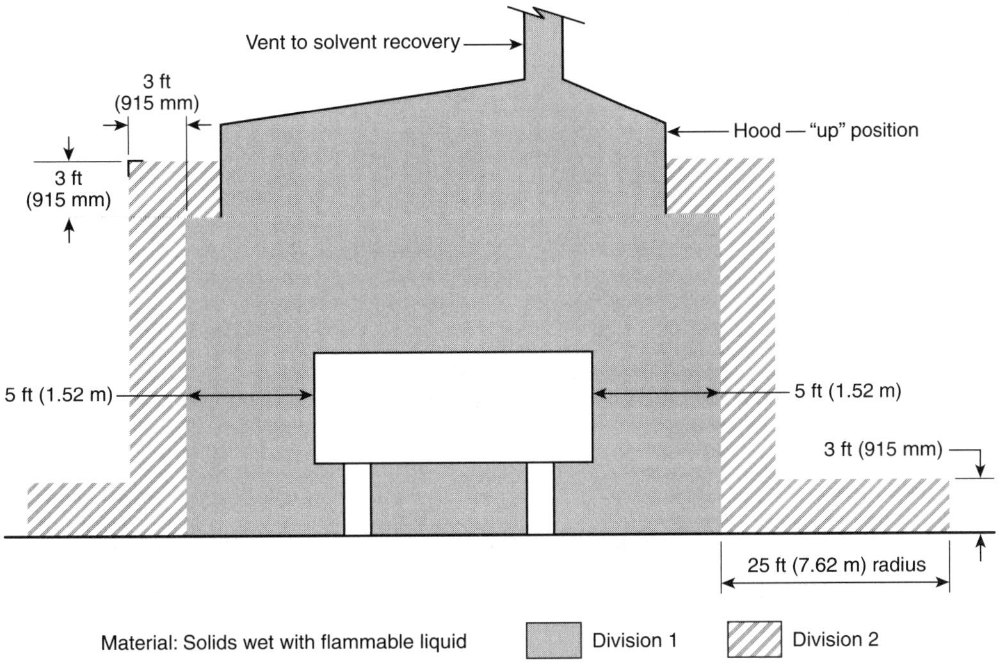

FIGURE 5.9.3(b) Plate and Frame Filter Press. Adequate ventilation is provided. The material being handled is a solid wet with a flammable liquid.

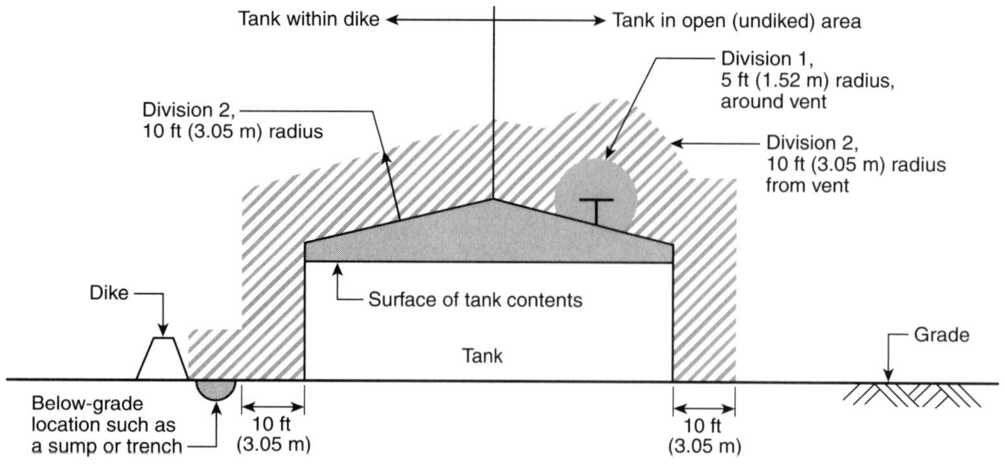

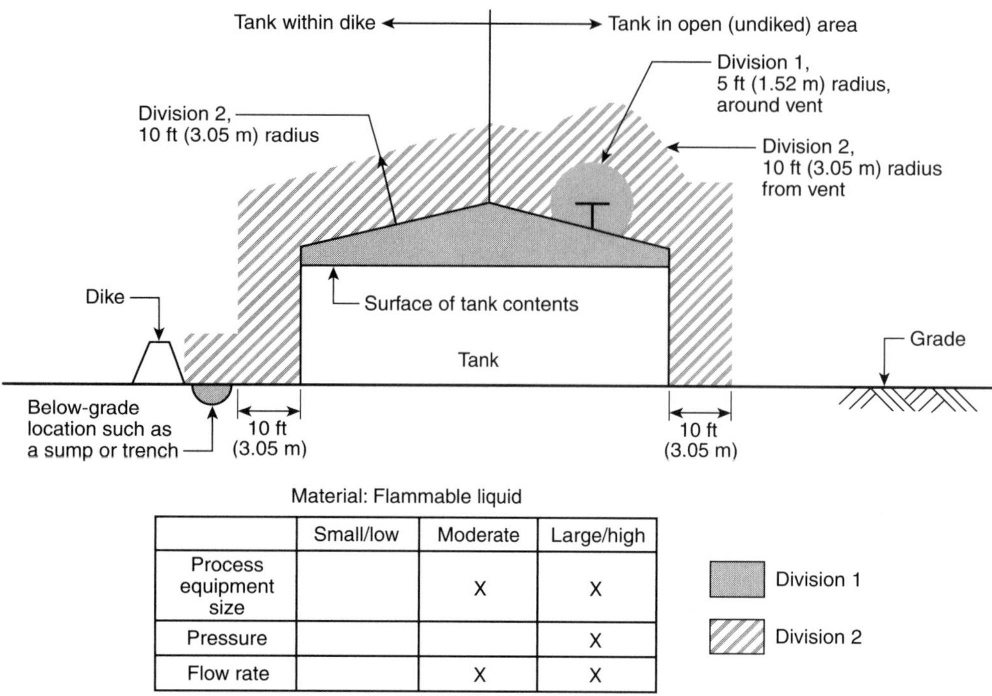

Material: Flammable liquid

	Small/low	Moderate	Large/high
Process equipment size		X	X
Pressure			X
Flow rate		X	X

Division 1
Division 2

FIGURE 5.9.4(a) Product Storage Tank Located Outdoors, at Grade. The material that is being stored is a flammable liquid.

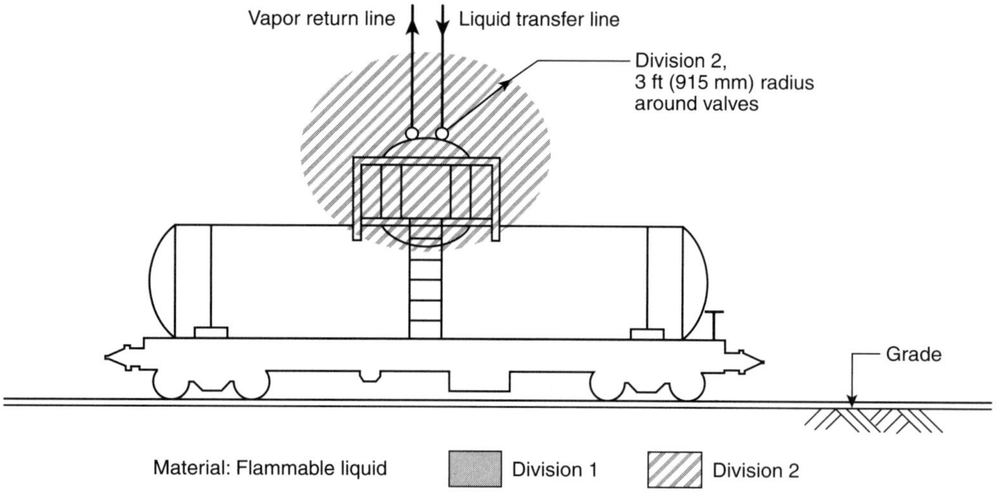

FIGURE 5.9.4(b) Tank Car Loading and Unloading via a Closed Transfer System. Material is transferred only through the dome. The material being transferred is a flammable liquid.

5.9.10 Storage Tanks for Cryogenic Liquids. *[See Figure 5.9.10(a), Figure 5.9.10(b), and Figure 5.9.10(c).]*

5.9.11 Outdoor Handling — Liquefied Natural Gas or Other Cryogenic Flammable Gas. *(See Figure 5.9.11.)*

5.9.12 Indoor Handling — Liquefied Natural Gas or Other Cryogenic Flammable Gas. *(See Figure 5.9.12.)*

5.9.13 Routinely Operating Bleeds — Liquefied Natural Gas or Other Cryogenic Flammable Gas. *(See Figure 5.9.13.)*

5.9.14 Marine Terminal — Flammable Liquids. *(See Figure 5.9.14.)*

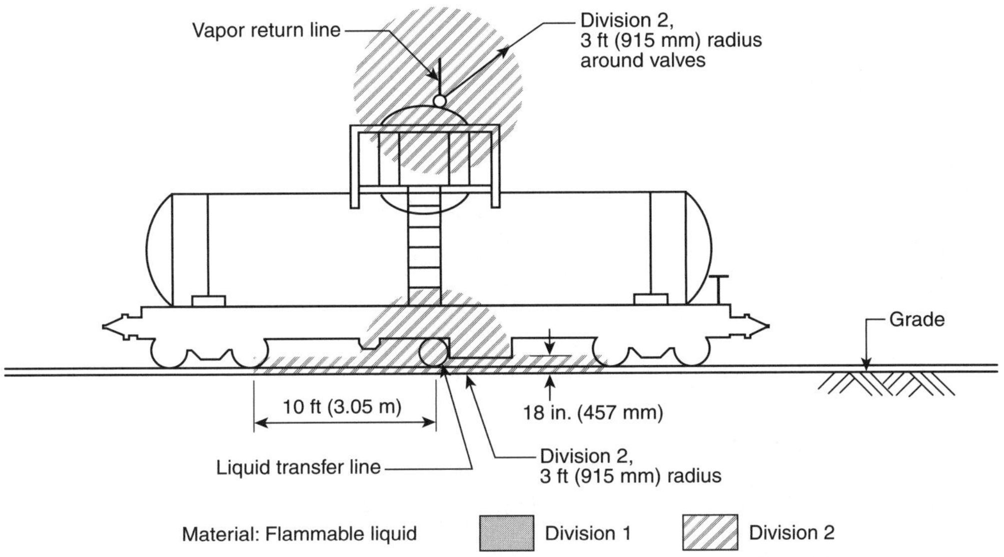

FIGURE 5.9.4(c) Tank Car Loading and Unloading via a Closed Transfer System. Material is transferred through the bottom fittings. The material being transferred is a flammable liquid.

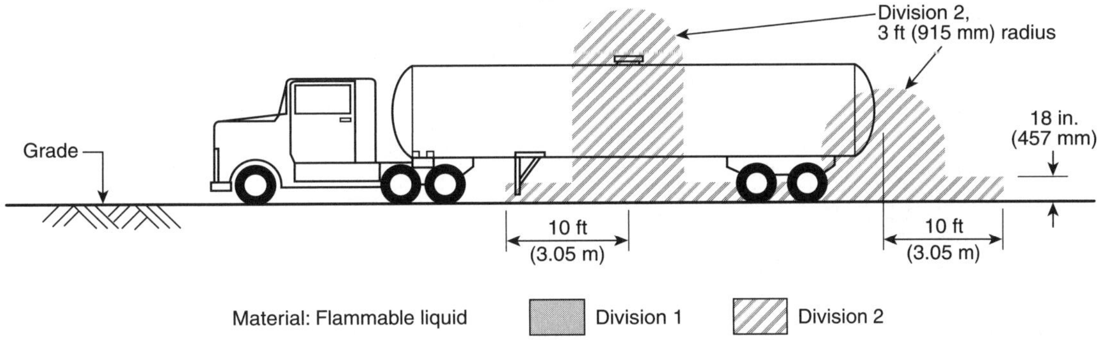

FIGURE 5.9.4(d) Tank Truck Loading and Unloading via a Closed Transfer System. Material is transferred through the bottom fittings. The material being transferred is a flammable liquid.

5.10 Classification Diagrams for Class I, Zones.

Class I, Zone diagrams include Figure 5.10.1(a) through Figure 5.10.1(n). Table 5.9 provides a summary of where each diagram is intended to apply.

5.10.1 Indoor and Outdoor Process — Flammable Liquids. *[See Figure 5.10.1(a), Figure 5.10.1(b), Figure 5.10.1(c), Figure 5.10.1(d), Figure 5.10.1(e), Figure 5.10.1(f), Figure 5.10.1(g), Figure 5.10.1(h), Figure 5.10.1(i), Figure 5.10.1(j), Figure 5.10.1(k), Figure 5.10.1(l), Figure 5.10.1(m), and Figure 5.10.1(n).]*

5.10.2 Outdoor Process — Flammable Liquid, Flammable Gas, Compressed Flammable Gas, or Cryogenic Liquid. *[See Figure 5.10.2(a) and Figure 5.10.2(b).]*

5.10.3 Product Dryer and Plate and Frame Filter Press — Solids Wet with Flammable Liquids. *[See Figure 5.10.3(a) and Figure 5.10.3(b).]*

5.10.4 Storage Tanks and Tank Vehicles — Flammable Liquids. *[See Figure 5.10.4(a), Figure 5.10.4(b), Figure 5.10.4(c), Figure 5.10.4(d), and Figure 5.10.4(e).]*

5.10.5 Tank Vehicle — Flammable Liquefied Gas, Flammable Compressed Gas, or Flammable Cryogenic Liquid. *(See Figure 5.10.5.)*

5.10.6 Indoor or Outdoor Drum Filling Station — Flammable Liquids. *(See Figure 5.10.6.)*

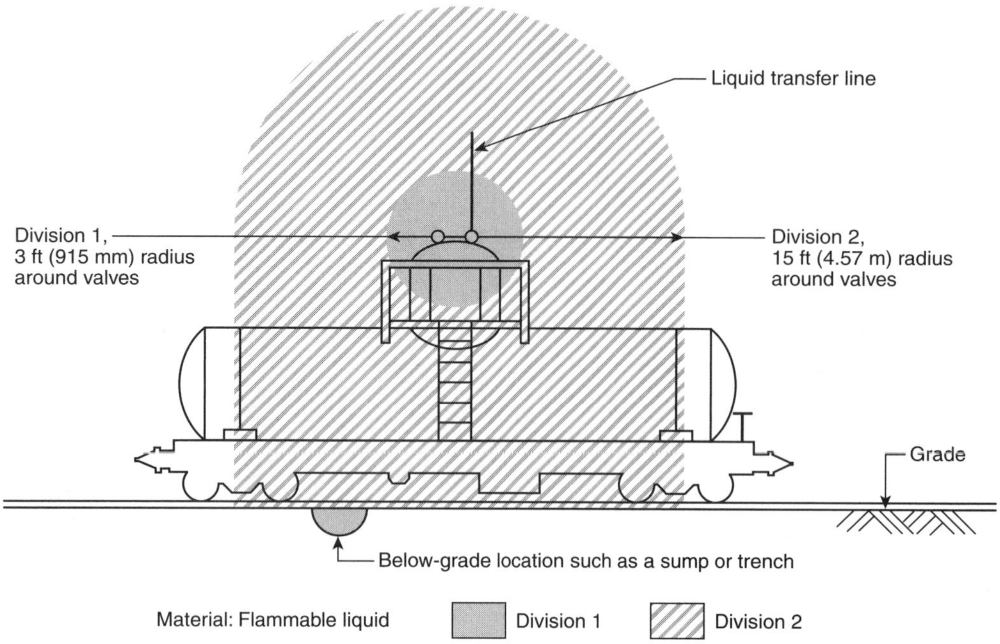

FIGURE 5.9.4(e) Tank Car (or Tank Truck) Loading and Unloading via an Open Transfer System. Material is transferred either through the dome or the bottom fittings. The material being transferred is a flammable liquid.

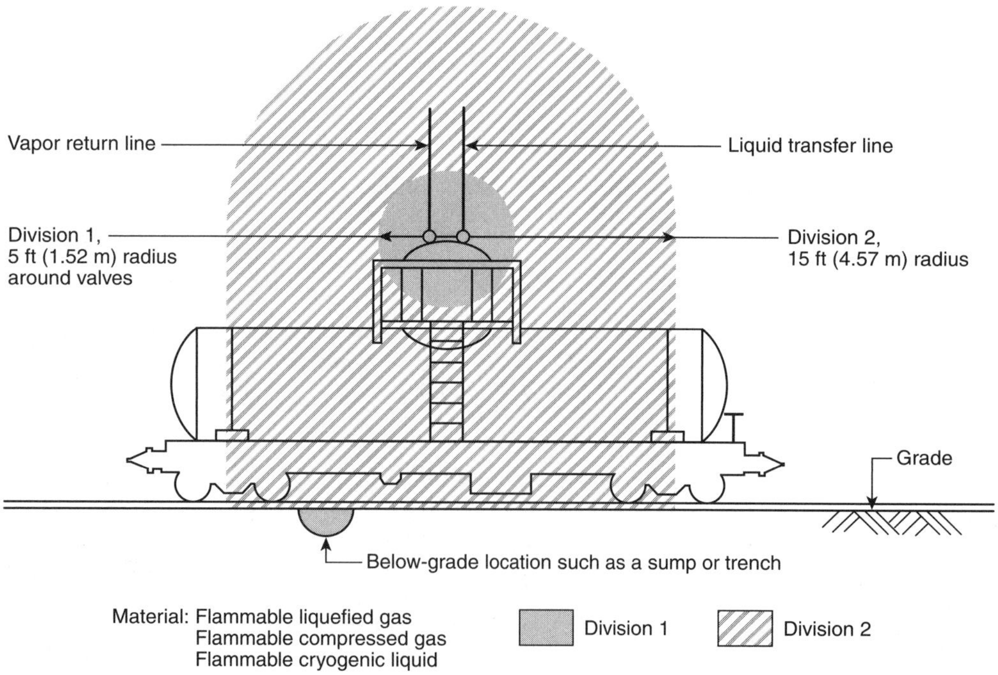

FIGURE 5.9.5 Tank Car (or Tank Truck) Loading and Unloading via a Closed Transfer System. Material is transferred only through the dome. The material being transferred may be a liquefied or compressed flammable gas or a flammable cryogenic liquid.

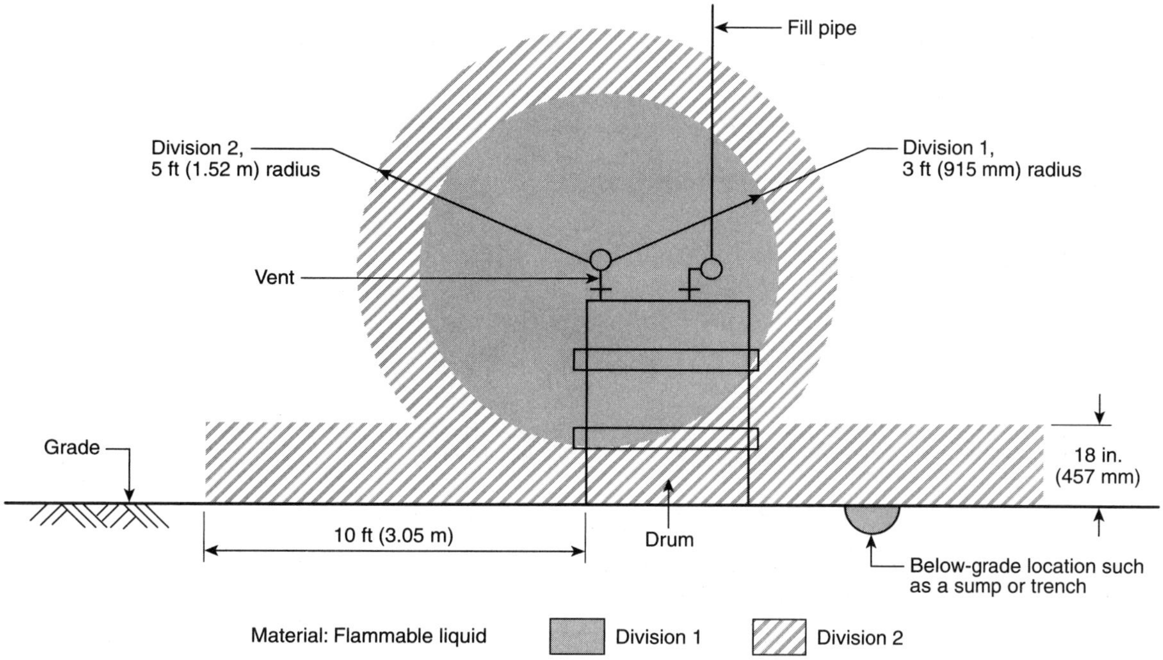

FIGURE 5.9.6 Drum Filling Station Located Either Outdoors or Indoors in an Adequately Ventilated Building. The material being handled is a flammable liquid.

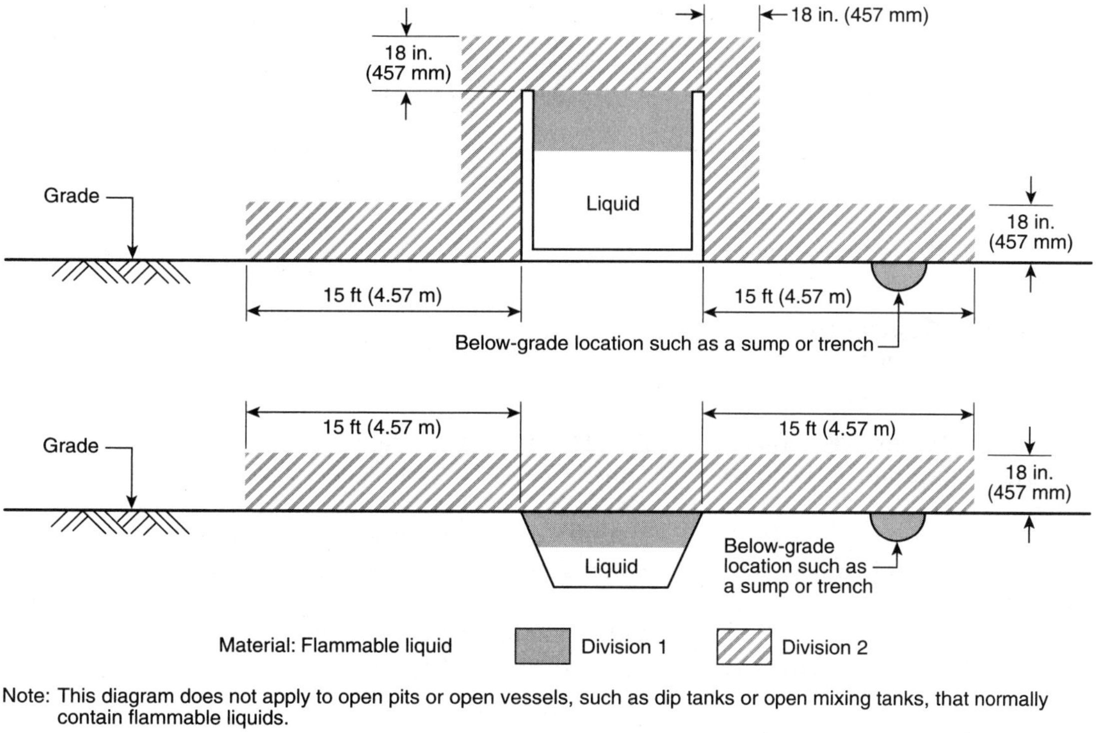

FIGURE 5.9.7 Emergency Impounding Basin or Oil/Water Separator and an Emergency or Temporary Drainage Ditch or Oil/Water Separator. The material being handled is a flammable liquid.

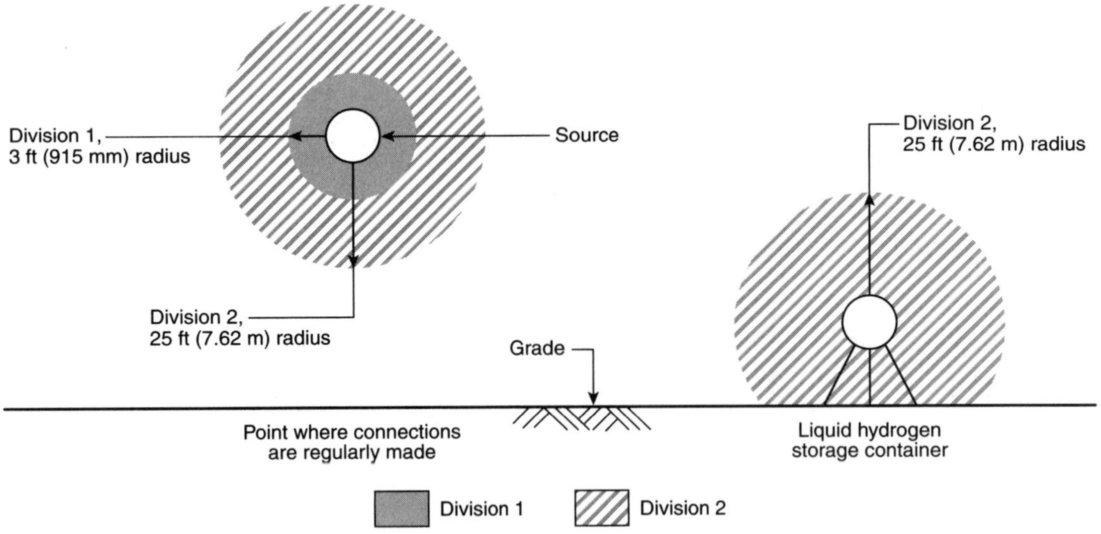

FIGURE 5.9.8(a) Liquid Hydrogen Storage Located Outdoors or Indoors in an Adequately Ventilated Building. This diagram applies to liquid hydrogen only.

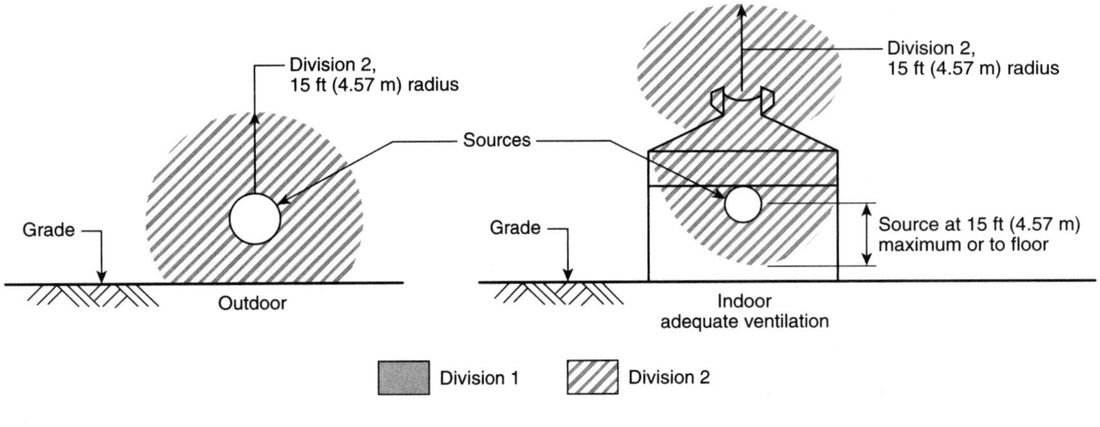

FIGURE 5.9.8(b)
Gaseous Hydrogen Storage Located Outdoors, or Indoors in an Adequately Ventilated Building. This diagram applies to gaseous hydrogen only.

5.10.7 Emergency Impounding Basins, Emergency Drainage Ditches, or Oil/Water Separators — Flammable Liquids. *(See Figure 5.10.7.)*

5.10.8 Storage of Liquid or Gaseous Hydrogen. *[See Figure 5.10.8(a) and Figure 5.10.8(b).]*

5.10.9 Compressor Shelters — Lighter-than-Air Gas. *[See Figure 5.10.9(a) and Figure 5.10.9(b).]*

5.10.10 Storage Tanks for Cryogenic Liquids. *[See Figure 5.10.10(a), Figure 5.10.10(b), and Figure 5.10.10(c).]*

5.10.11 Outdoor Handling — Liquefied Natural Gas or Other Cryogenic Flammable Gas. *(See Figure 5.10.11.)*

5.10.12 Indoor Handling — Liquefied Natural Gas or Other Cryogenic Flammable Gas. *(See Figure 5.10.12.)*

5.10.13 Routinely Operating Bleeds — Liquefied Natural Gas or Other Cryogenic Flammable Gas. *(See Figure 5.10.13.)*

5.10.14 Marine Terminal — Flammable Liquids. *(See Figure 5.10.14.)*

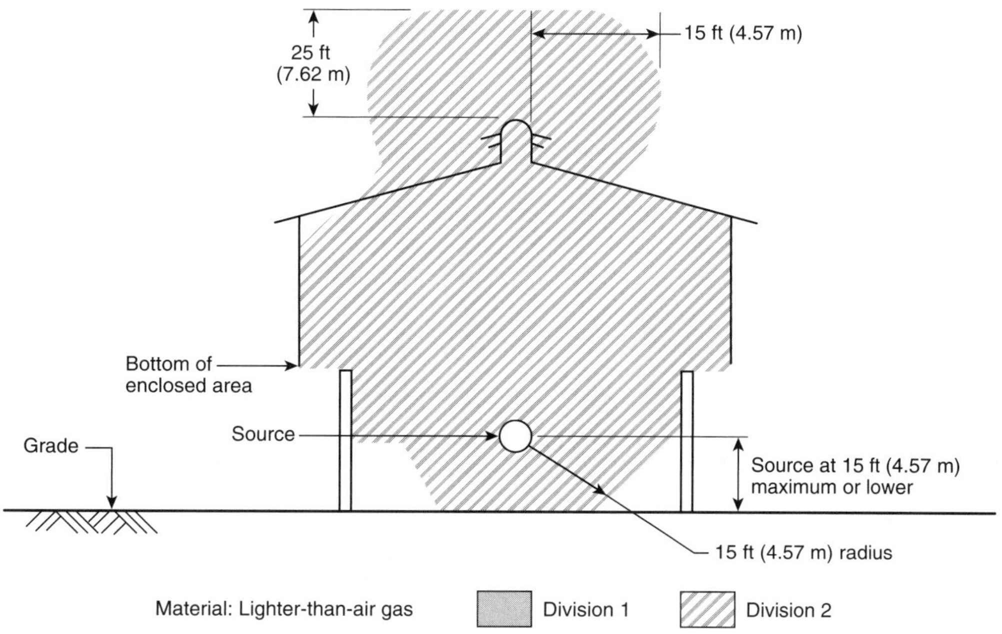

FIGURE 5.9.9(a) Adequately Ventilated Compressor Shelter. The material being handled is a lighter-than-air gas.

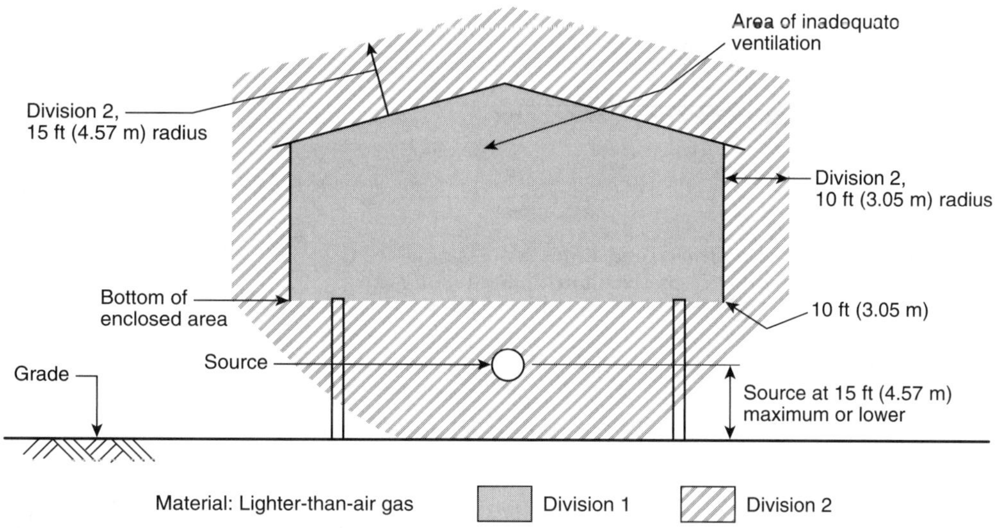

FIGURE 5.9.9(b)
Inadequately Ventilated Compressor Shelter. The material being handled is a lighter-than-air gas.

Annex A Explanatory Material

Annex A is not a part of the recommendations of this NFPA document but is included for informational purposes only. This annex contains explanatory material, numbered to correspond with the applicable text paragraphs.

A.3.3.2 Autoignition Temperature (AIT). See *NFPA Fire Protection Guide to Hazardous Materials.*

A.3.3.5.1 Combustible Material (Class I, Division). (Groups A, B, C, and D). Historically, the topic of hazardous (classified) locations first appeared in the *NEC* in 1920, when a new article entitled "Extra-Hazardous Locations" was accepted. This article addressed rooms or compartments in which highly flammable gases, liquids, mixtures, or other substances were manufactured, used, or stored. In 1931, "classifications" consisting of Class I, Class II, and so on, were defined for hazardous locations. However, it was not

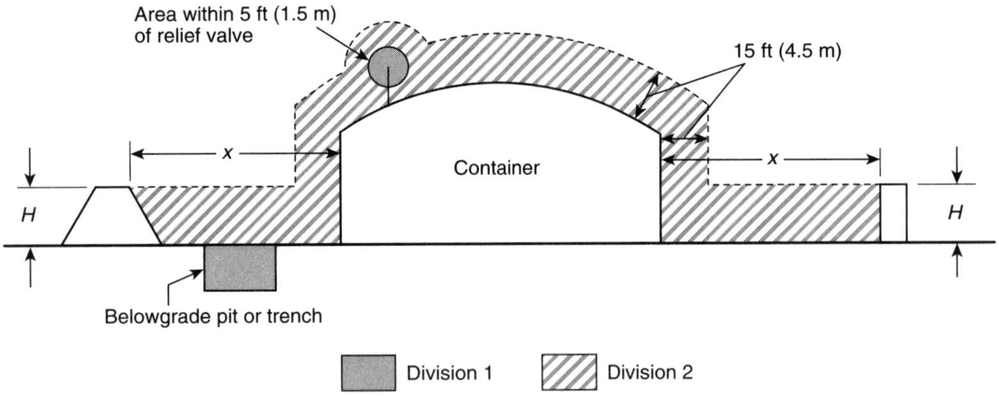

FIGURE 5.9.10(a) Tank for the Storage of Cryogenic and Other Cold Liquefied Flammable Gases. Dike height is less than distance from container to dike (*H* less than *x*). [Source: Figure 7.6.2(b) of NFPA 59A.]

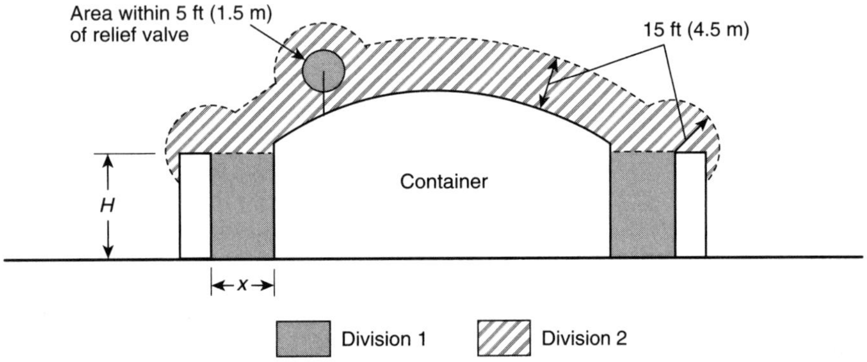

FIGURE 5.9.10(b) Tank for the Storage of Cryogenic and Other Cold Liquefied Flammable Gases. Dike height is less than distance from container to dike (*H* greater than *x*). [Source: Figure 7.6.2(c) of NFPA 59A.]

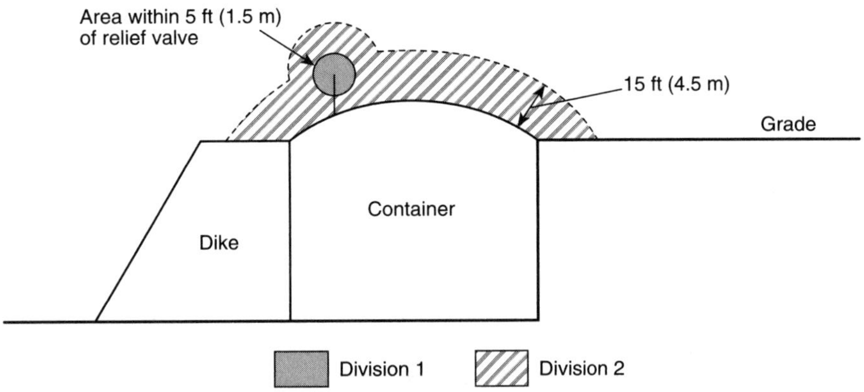

FIGURE 5.9.10(c) Tank for the Storage of Cryogenic and Other Cold Liquefied Flammable Gases. Container with liquid below grade or top of dike. [Source: Figure 7.6.2(d) of NFPA 59A.]

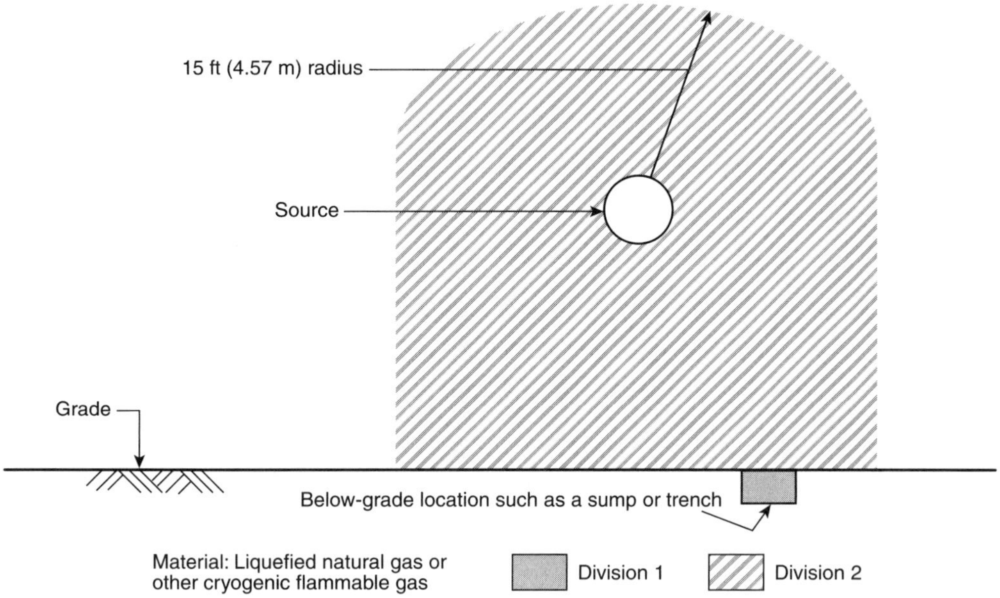

FIGURE 5.9.11 Source of Leakage from Equipment Handling Liquefied Natural Gas or Other Cold Liquefied Flammable Gas, and Located Outdoors, at or Above Grade.

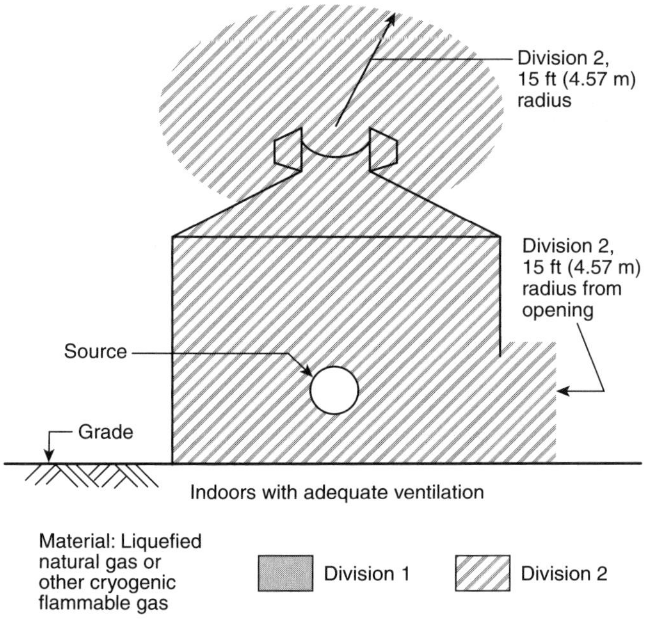

FIGURE 5.9.12 Source of Leakage from Equipment Handling Liquefied Natural Gas or Other Cold Liquefied Flammable Gas, and Located Indoors, in an Adequately Ventilated Building.

until 1935 that "groups" were introduced into the *NEC*. (Note: "Divisions" were introduced into the *NEC* in 1947.) The four gas groups, Groups A, B, C, and D, complemented the design of electrical equipment used in hazardous (classified) locations and were defined based on the level of hazard associated with explosion pressures of specific atmospheres and the likelihood that the effects of that explosion could be transmitted outside the enclosure. Group A was defined as atmospheres containing acetylene. Group B was defined as atmospheres containing hydrogen or gas or vapors of equivalent hazard. Group C was defined as atmospheres containing ethyl ether vapor. Group D was defined as atmospheres containing gasoline, petroleum, naphtha, alcohols, acetone, lacquers, solvent vapors, and natural gas. Despite the fact that the introduction of these groups was done without standardized testing and without the advantage of today's technological advances or equipment, these definitions have changed little since that time. The first major testing, in fact, was only conducted in the late 1950s, when engineers at Underwriters Laboratories Inc. developed a test apparatus that provided a means to determine how various materials behaved with respect to explosion pressures and transmission when the specific combustible material was ignited in the test vessel. This apparatus, called the Westerberg Explosion Test Vessel, provided standardized documentation of a factor called the "Maximum Experimental Safe Gap" (MESG) and permitted other materials to be "classified by test" into one of the four gas groups. The results of these tests are contained in Underwriters Laboratories Inc. (UL) Bulletin Nos. 58 and 58A (reissued in July 1993, as UL Technical Report No. 58). In 1971, the International Electrotechnical Commission (IEC) published IEC 79-1A (reissued in July 2002 as 60079-1-1) defining a different type of apparatus for obtaining MESG results. While the two MESG test apparatuses are different physically in both size and

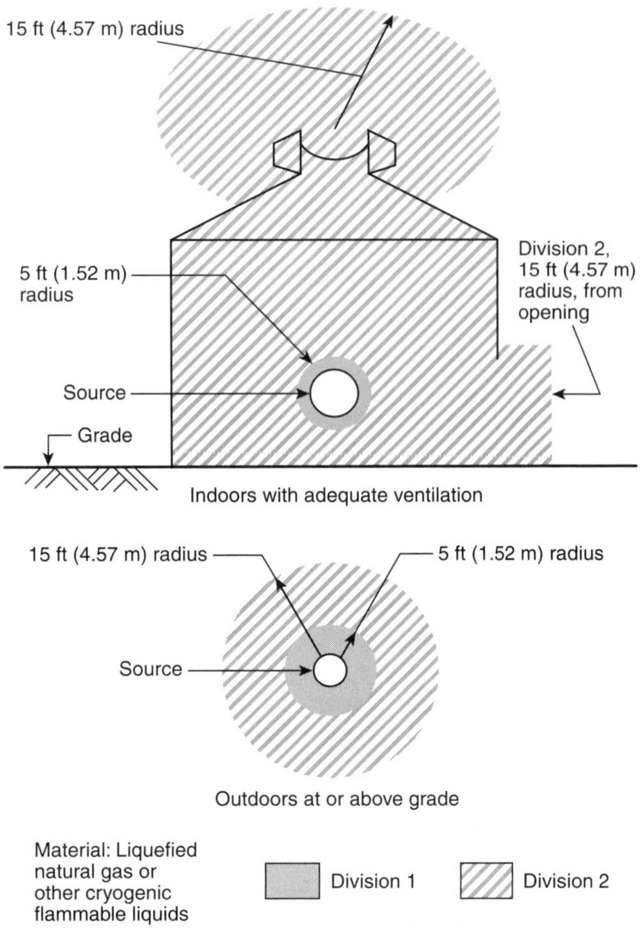

FIGURE 5.9.13 Classified Zones around Liquefied Natural Gas Routinely Operating Bleeds, Drips, Vents, and Drains Both Outdoors, at or above Grade, and Indoors, in an Adequately Ventilated Building. This diagram also applies to other cold liquefied flammable gases. (Source: NFPA 59A.)

shape, the results are statistically comparative, although in some cases differences have been observed. A sample of values is shown in Table A.3.3.5.1.

Papers have been written to attempt to explain the reasons for these differences in the test data. A paper written by H. Phillips, entitled "Differences between Determinations of Maximum Experimental Safe Gaps in Europe and U.S.A.," appeared in a 1981 edition of the *Journal of Hazardous Materials*. The paper cites a condition of spontaneous combustion in one portion of the Westerberg Apparatus, which was reflected in materials (such as diethyl ether) having low ignition temperatures. Additionally, testing on the Westerberg Apparatus has demonstrated that this theory was true, and the MESG value for diethyl ether more than doubled. Further Westerberg Apparatus testing has also shown that hydrogen MESG value is 0.23 mm.

Although acetylene remains segregated in Group A because of the high explosion pressure that results from its very fast flame speed, newer test methodologies have defined other types of protection methods that now consider acetylene and hydrogen to be of equivalent hazard. One such method examines the "minimum igniting current" required to ignite a specific combustible material. This testing produced more variability when the results of specific combustible materials were compared. However, it was found that the minimum igniting currents of one test could be favorably compared with those of other tests if a ratio value based on methane was applied. This testing has resulted in the generation of MIC ratio data.

Other testing has been performed when it was incorrectly assumed that a factor called "Minimum Ignition Energy" (MIE) and "Autoignition Temperature" (AIT) were related and could be used to place materials into groups. The fact that these were independent factors resulted in deletion of AITs as a basis for group determination in the 1971 *NEC*.

MIEs have been found to exhibit theoretical results that do not translate into practical designs that can be applied to actual electrical devices with their associated energy levels.

Because the primary concern is to have electrical devices that can safely operate when used in locations classified by class, group, and division, the definitions for the four gas groups have been defined on the basis of the parameters providing the most significant basis for that design, which are MESG and MIC ratio. Lacking these values, experience-based data indicating equivalency to atmospheres providing similar hazards may be used.

The Class I Division group designation of selected combustible materials is listed in Table 4.4.2.

A.3.3.5.2 Combustible Material (Class I, Zone). The Class I, Zone group designation of selected combustible materials is also listed in Table 4.4.2.

A.3.3.10 Minimum Igniting Current (MIC) Ratio. The test apparatus is defined by IEC 60079-11, *Electrical apparatus for explosive gas atmospheres, Part 11 — Intrinsic safety "i."*

A.4.1.2 Article 500 also defines two other hazardous (classified) locations: Class II and Class III. In a Class II hazardous (classified) location, the combustible material present is a combustible dust. In a Class III hazardous (classified) location, the combustible material present is an ignitible fiber or flying. This recommended practice covers Class I hazardous (classified) locations.

A.4.4.2 Combustible materials shown in Table 4.4.2 have been classified into groups. Some of these combustible materials are indicated in groups that may not seem to agree with defined MESG or MIC ratio values, but have been continued within groups due to historical experiences and specific

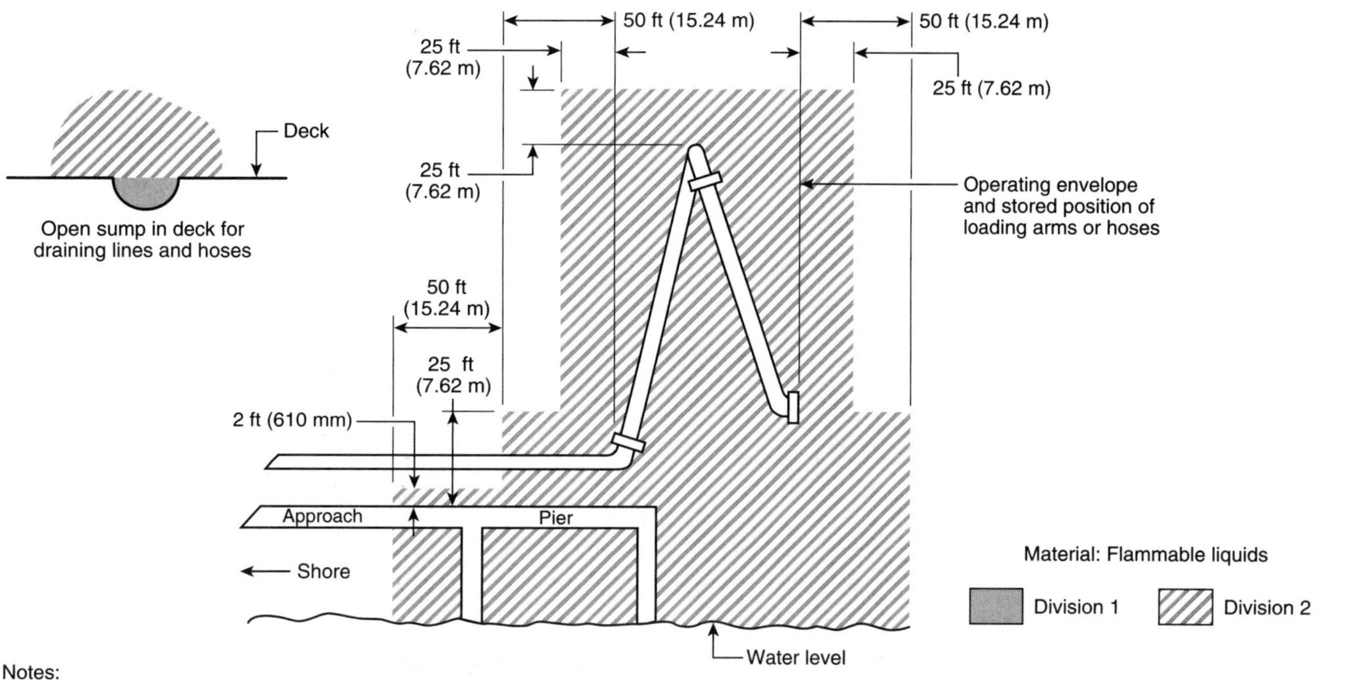

Notes:
1. The "source of vapor" is the operating envelope and stored position of the outboard flange connection of the loading arm (or hose).
2. The berth area adjacent to tanker and barge cargo tanks is to be Division 2 to the following extent:
   (a) 25 ft (7.62 m) horizontally in all directions on the pier side from that portion of the hull containing cargo tanks.
   (b) From the water level to 25 ft (7.62 m) above the cargo tanks at their highest position.
3. Additional locations may have to be classified as required by the presence of other sources of flammable liquids or by Coast Guard or other regulations.

FIGURE 5.9.14 Classified Locations at a Marine Terminal Handling Flammable Liquids; Includes the Area Around the Stored Position of Loading Arms and Hoses.

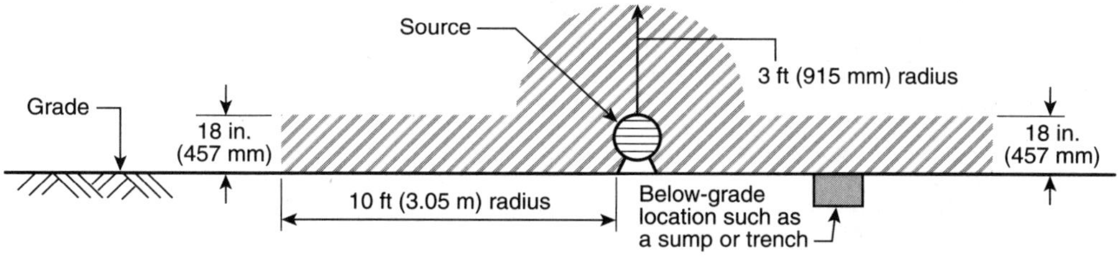

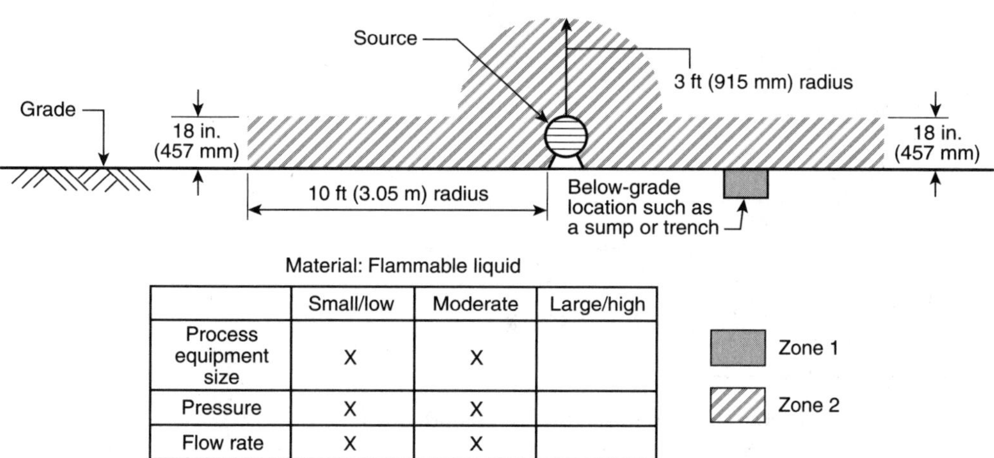

Material: Flammable liquid

	Small/low	Moderate	Large/high
Process equipment size	X	X	
Pressure	X	X	
Flow rate	X	X	

Zone 1

Zone 2

FIGURE 5.10.1(a) Leakage Located Outdoors, at Grade. The material being handled is a flammable liquid.

properties that are not reflected in MESG or MIC ratio testing. For example, one source for group classifications was NMAB 353-1, *Matrix of Combustion-Relevant Properties and Classification of Gases, Vapors, and Selected Solids*, published by the National Academy of Sciences. Those materials whose group classifications are marked with asterisks were previously assigned group classifications based on tests conducted in the Westerberg Apparatus at Underwriters Laboratories Inc. (See UL Bulletin of Research No. 58, *An Investigation of Flammable Gases or Vapors with Respect to Explosion-Proof Electrical Equipment*, and subsequent Bulletin Nos. 58A and 58B, reissued July 23, 1993 as UL

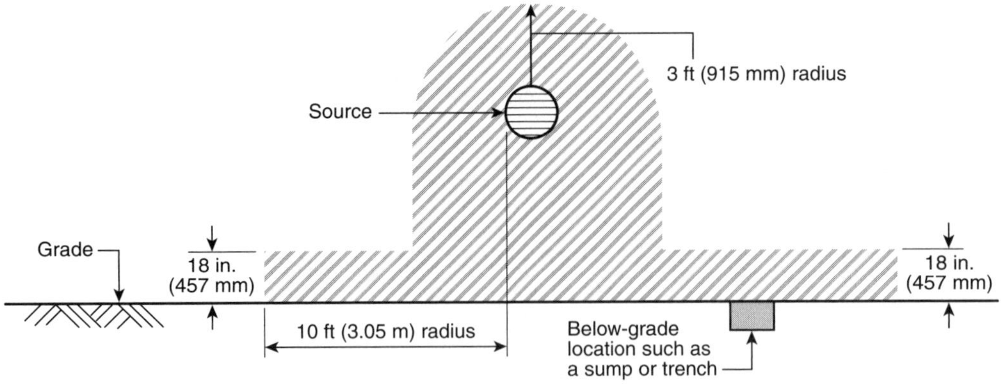

Material: Flammable liquid

	Small/low	Moderate	Large/high
Process equipment size	X	X	
Pressure	X	X	
Flow rate	X	X	

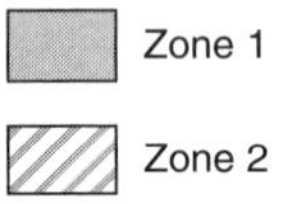

FIGURE 5.10.1(b) Leakage Located Outdoors, Above Grade. The material being handled is a flammable liquid.

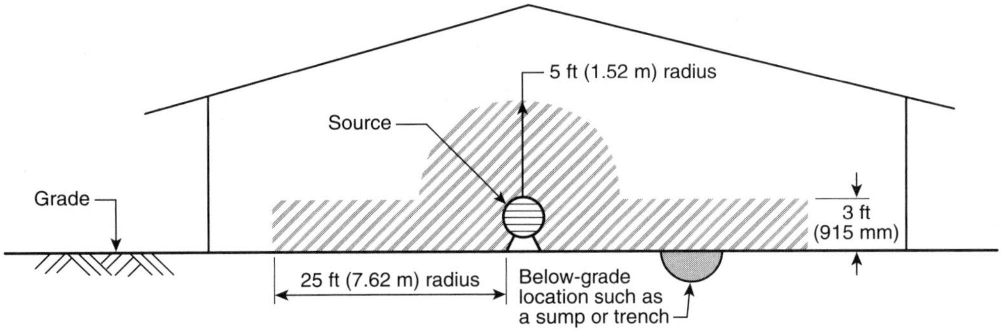

Material: Flammable liquid

	Small/low	Moderate	Large/high
Process equipment size	X	X	
Pressure	X	X	
Flow rate	X	X	

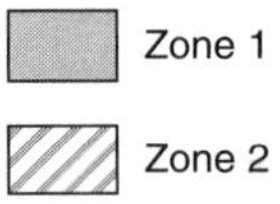

FIGURE 5.10.1(c) Leakage Located Indoors, at Floor Level. Adequate ventilation is provided. The material being handled is a flammable liquid.

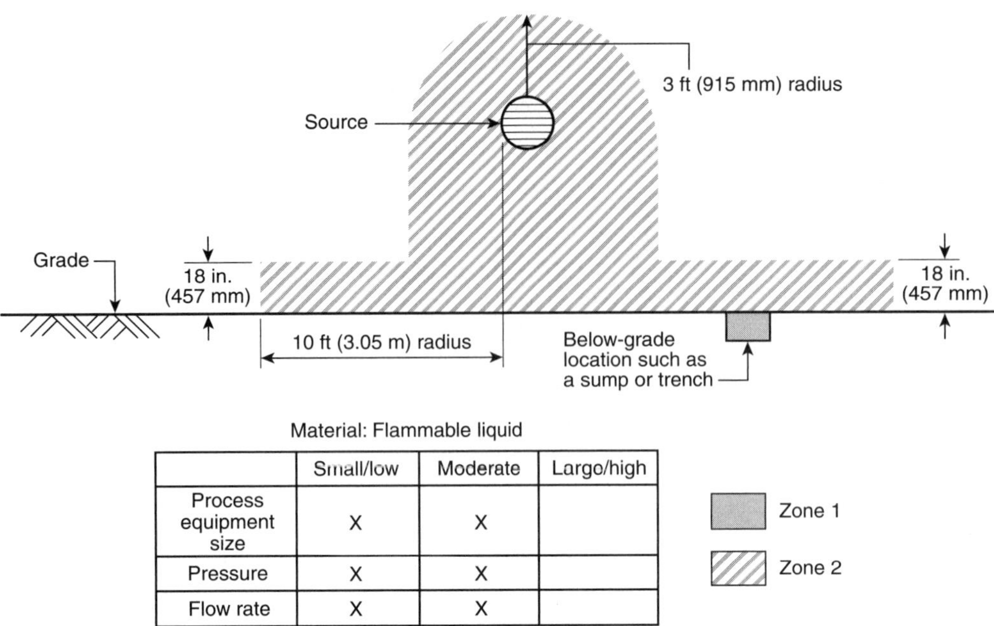

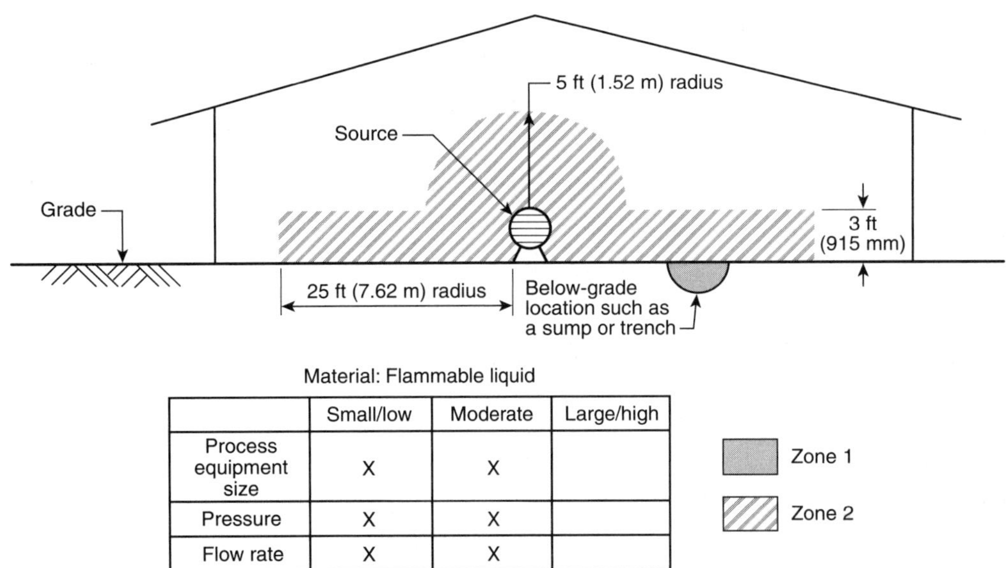

Technical Report No. 58.) All other materials were assigned group classifications based on analogy with tested materials and on chemical structure, or on reputable published data reflecting MESG or MIC ratio values. (See, for example, IEC 60079-20.) Although the classifications of these latter materials represent the best judgment of three groups of experts, it is conceivable that the group classification of any particular untested material may be incorrect. Users of these data should be aware that the data are the result of experimental determination, and as such are influenced by variation in experimental apparatus and procedures and in the accuracy of the instrumentation. Additionally, some of the data have been determined at an above-ambient temperature so that the vapor is within the flammable region. Variation in the temperature for the determination would be expected to influence the result of the determination. Also, the values shown generally represent the lowest reported in reputable, documented studies. In certain instances, therefore, it may

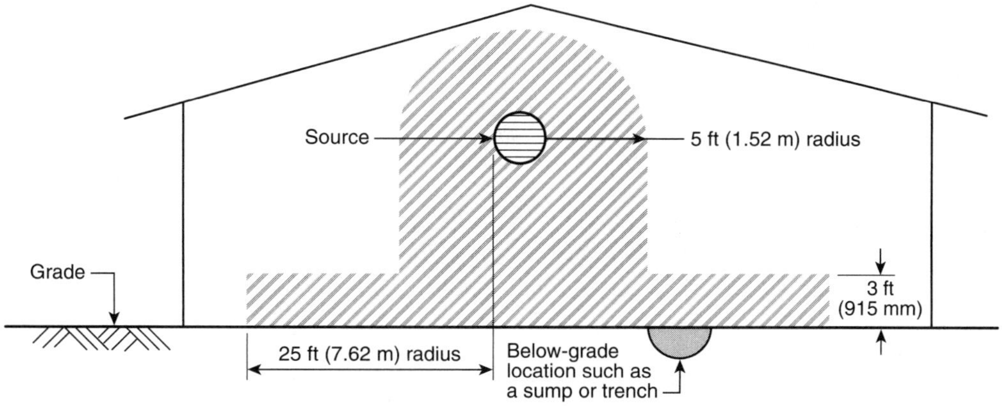

Material: Flammable liquid

	Small/low	Moderate	Large/high
Process equipment size	X	X	
Pressure	X	X	
Flow rate	X	X	

Zone 1
Zone 2

FIGURE 5.10.1(d) Leakage Located Indoors, Above Floor Level. Adequate ventilation is provided. The material being handled is a flammable liquid.

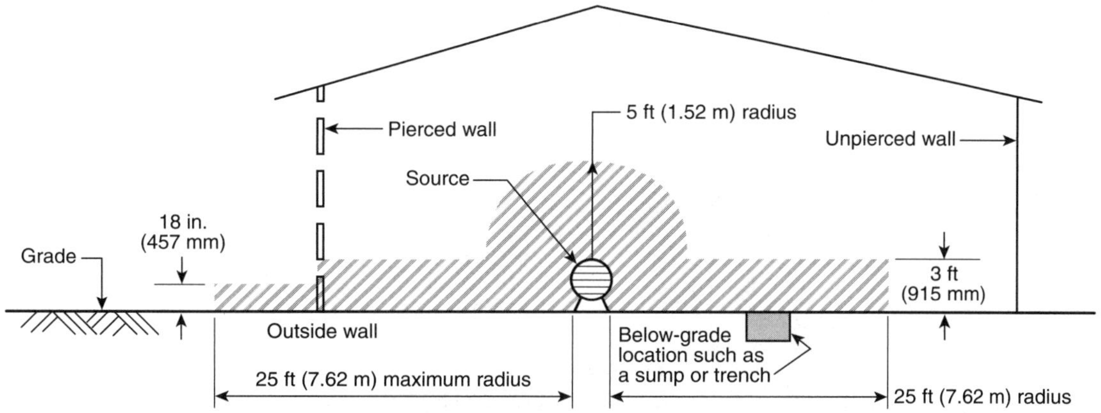

Material: Flammable liquid

	Small/low	Moderate	Large/high
Process equipment size	X	X	
Pressure	X	X	
Flow rate	X	X	

Zone 1
Zone 2

FIGURE 5.10.1(e) Leakage Located Indoors, at Floor Level, Adjacent to an Opening in an Exterior Wall. Adequate ventilation is provided. The material being handled is a flammable liquid.

be advisable to submit an untested material to a qualified testing laboratory for verification of the assigned group classification.

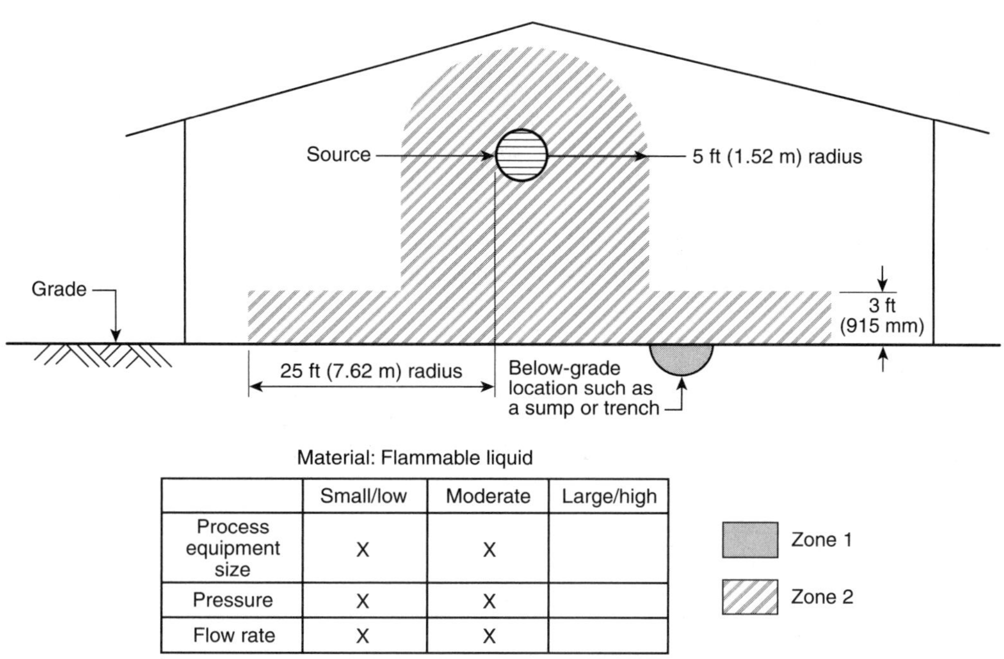

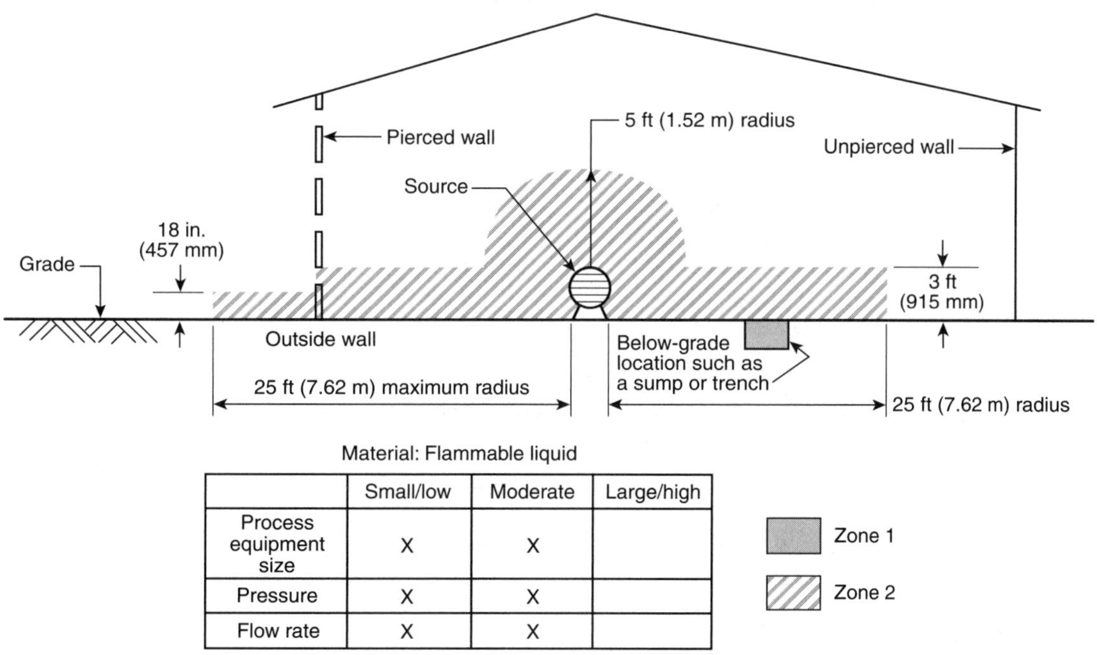

Regarding Note (g) of Table 4.4.2, selecting electrical equipment based on the lowest AIT shown in Table 4.4.2 could be unnecessarily restrictive. As an example, in an area handling a commercial grade of hexane, Table 4.4.2 tabulates five different isomers of hexane solvent (C_6H_{14}), with the AIT ranging from a low of 225°C to a high of 405°C. The AIT of the solvent mixtures should be determined either

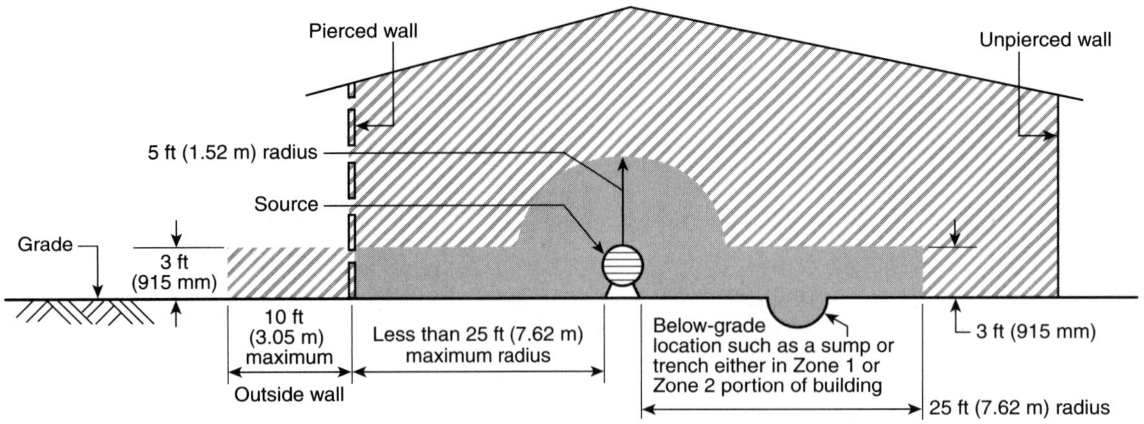

Note: If building is small compared to size of equipment, and leakage can fill the building, the entire building interior is classified Zone 1.

Material: Flammable liquid

	Small/low	Moderate	Large/high
Process equipment size	X	X	
Pressure	X	X	
Flow rate	X	X	

Zone 1

Zone 2

FIGURE 5.10.1(f) Leakage Located Indoors, at Floor Level, Adjacent to an Opening in an Exterior Wall. Ventilation is not adequate. The material being handled is a flammable liquid.

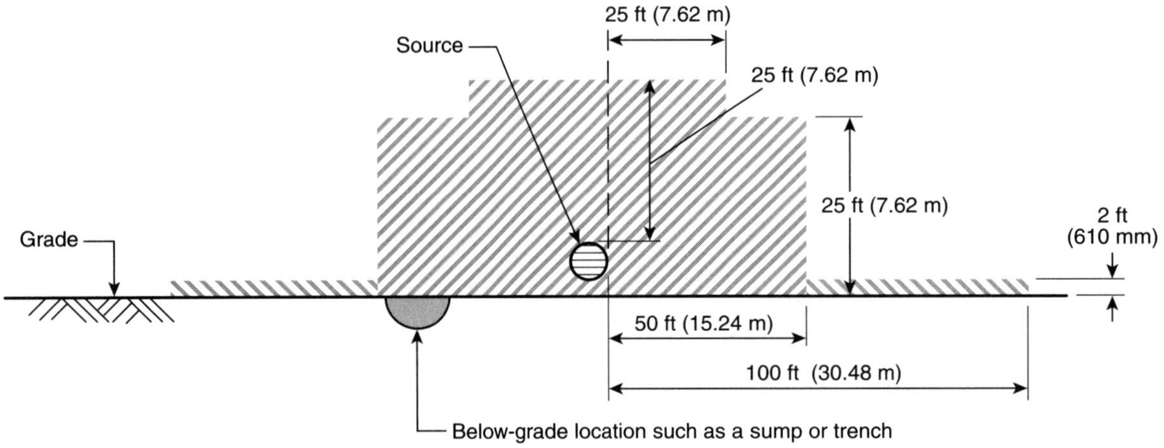

Material: Flammable liquid

	Small/low	Moderate	Large/high
Process equipment size			X
Pressure		X	X
Flow rate			X

Zone 1

Zone 2

Additional Zone 2 location. Use extra precaution where large release of volatile products may occur.

FIGURE 5.10.1(g) Leakage Located Outdoors, at Grade. The material being handled is a flammable liquid.

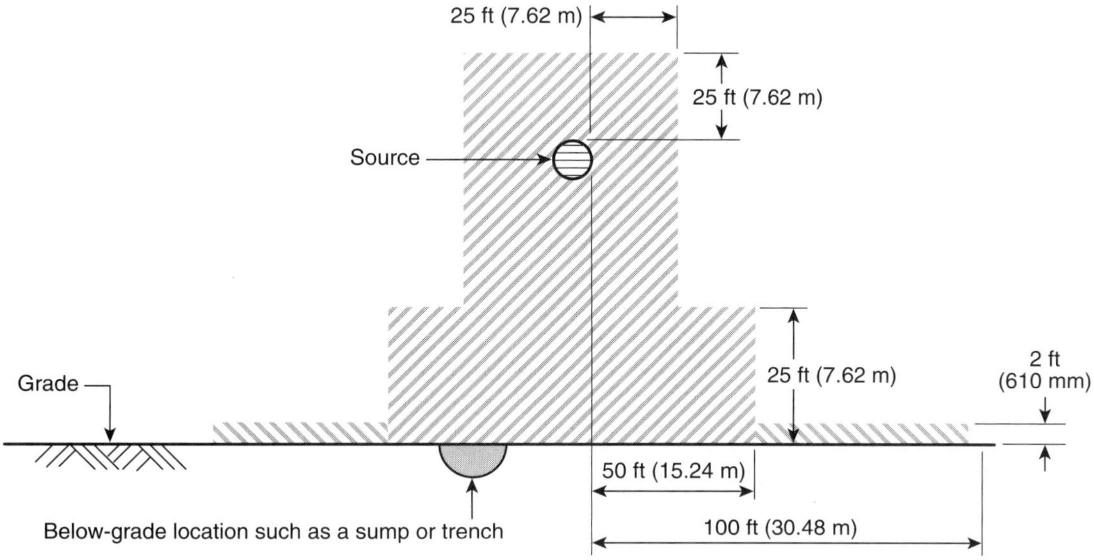

Material: Flammable liquid

	Small/low	Moderate	Large/high
Process equipment size			X
Pressure		X	X
Flow rate			X

Zone 1

Zone 2

Additional Zone 2 location. Use extra precaution where large release of volatile products may occur.

FIGURE 5.10.1(h) Leakage Located Outdoors, Above Grade. The material being handled is a flammable liquid.

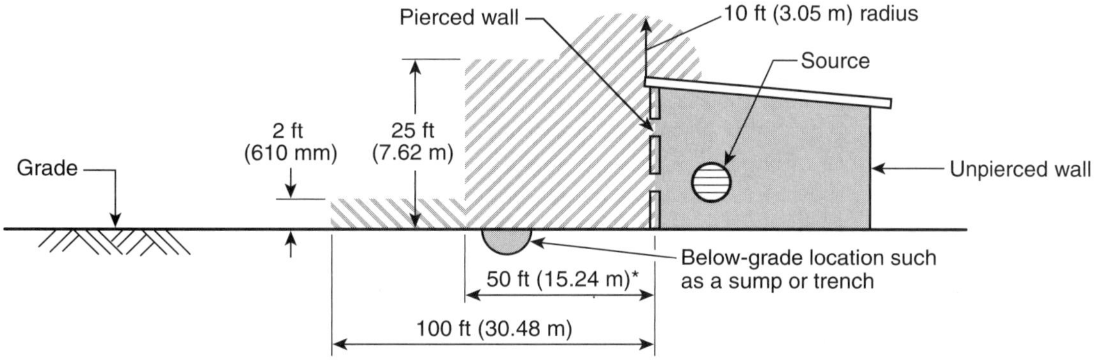

* "Apply" horizontal distances of 50 ft from the source of vapor or 10 ft beyond the perimeter of the building, whichever is greater, except that beyond unpierced vaportight walls the area is unclassified.

Material: Flammable liquid

	Small/low	Moderate	Large/high
Process equipment size		X	X
Pressure			X
Flow rate		X	X

Zone 1

Zone 2

Additional Zone 2 location. Use extra precaution where large release of volatile products may occur.

FIGURE 5.10.1(i) Leakage Located Indoors, Adjacent to an Opening in an Exterior Wall. Ventilation is not adequate. The material being handled is a flammable liquid.

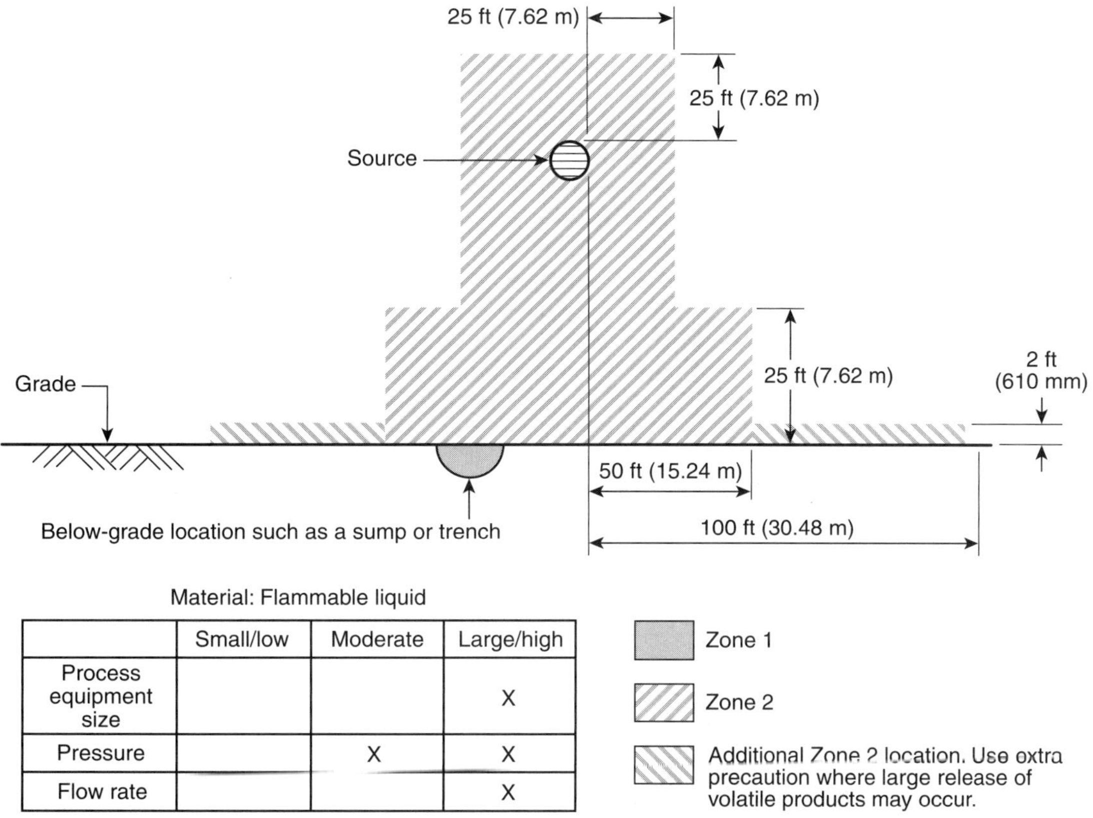

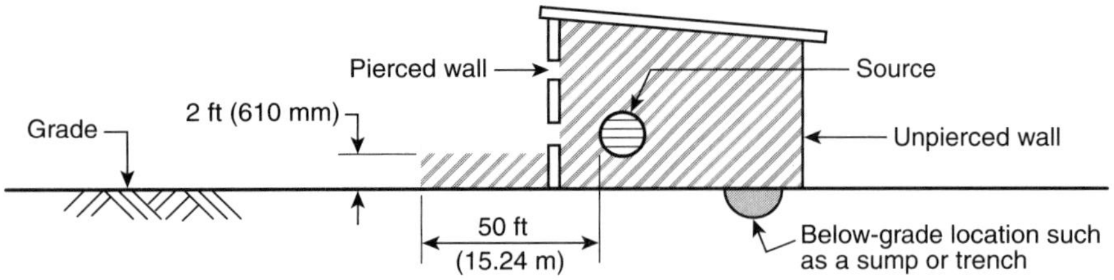

Material: Flammable liquid

	Small/low	Moderate	Large/high
Process equipment size		X	X
Pressure			X
Flow rate		X	X

FIGURE 5.10.1(j) Leakage Located Indoors, Adjacent to an Opening in an Exterior Wall. Adequate ventilation is provided. The material being handled is a flammable liquid.

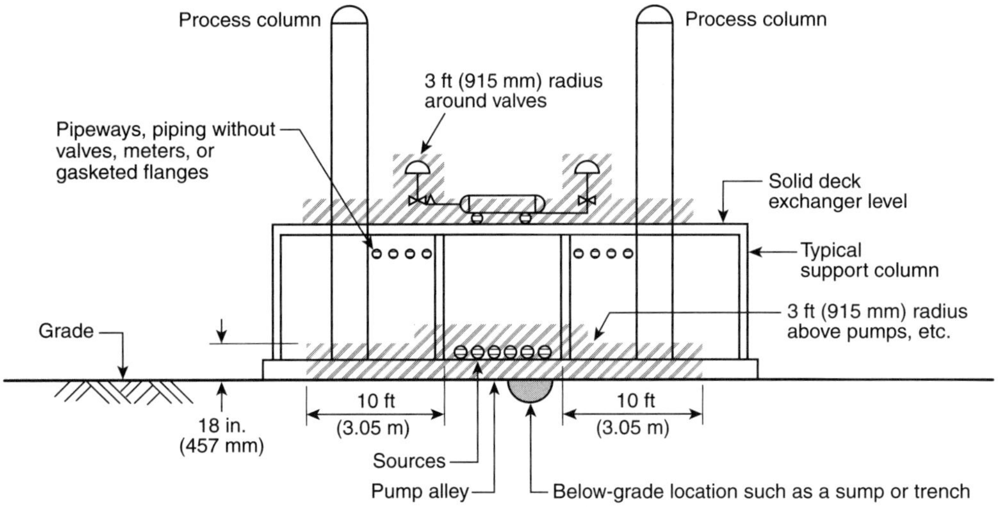

Material: Flammable liquid

	Small/low	Moderate	Large/high
Process equipment size	X	X	
Pressure	X	X	
Flow rate	X	X	

FIGURE 5.10.1(k) Leakage, Located Both at Grade and Above Grade, in an Outdoor Process Area. The material being handled is a flammable liquid.

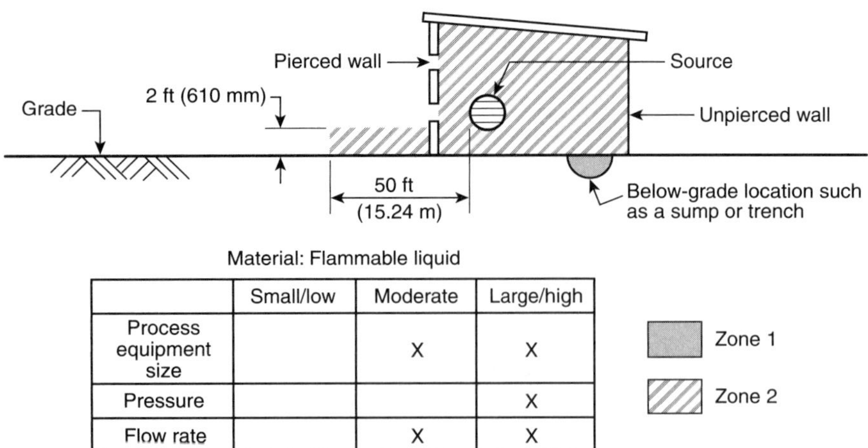

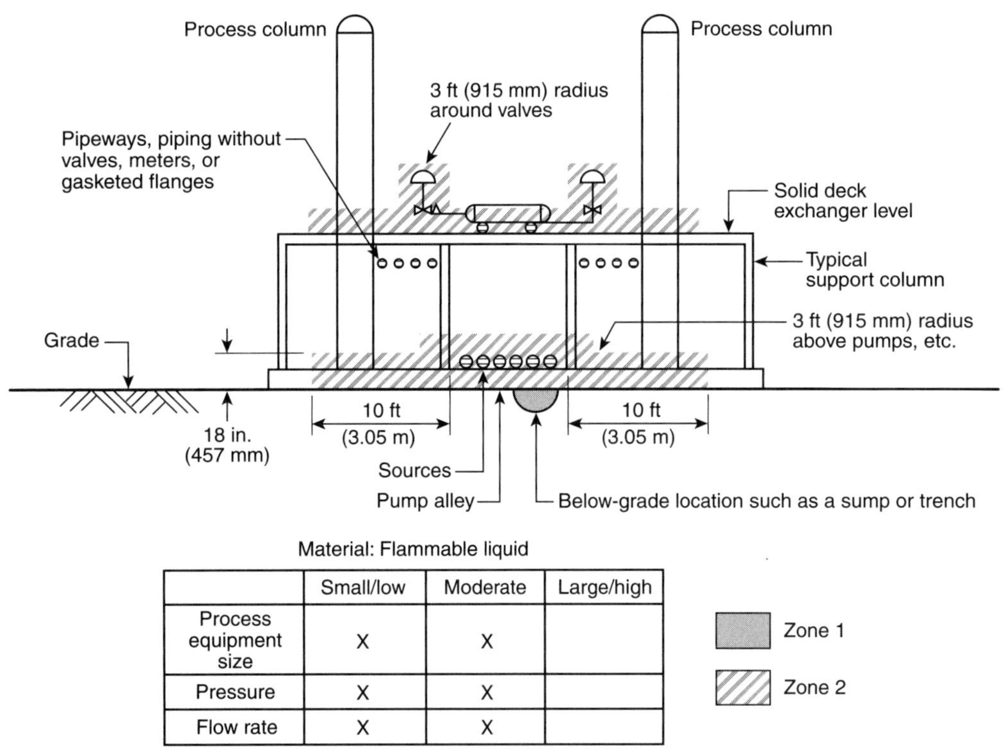

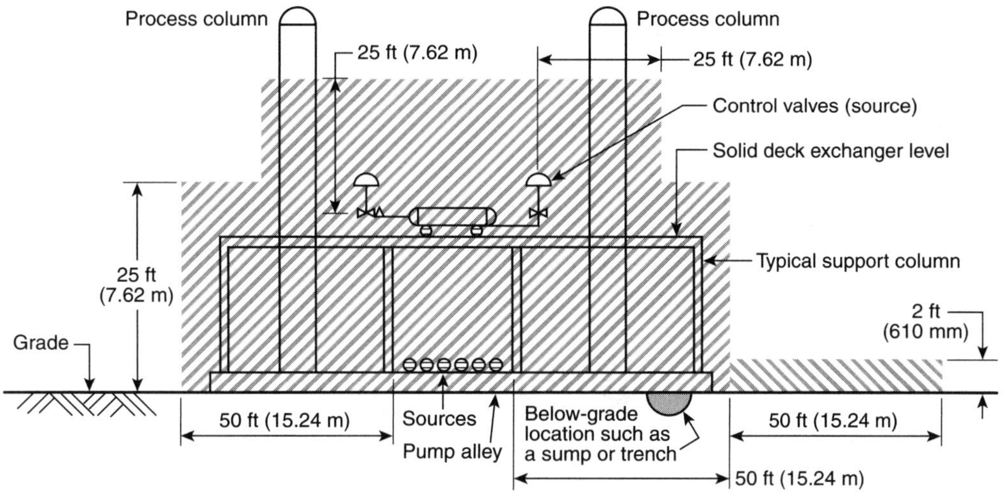

Material: Flammable liquid

	Small/low	Moderate	Large/high
Process equipment size		X	X
Pressure		X	X
Flow rate		X	X

FIGURE 5.10.1(l) Multiple Sources of Leakage, Located Both at Grade and Above Grade, in an Outdoor Process Area. The material being handled is a flammable liquid.

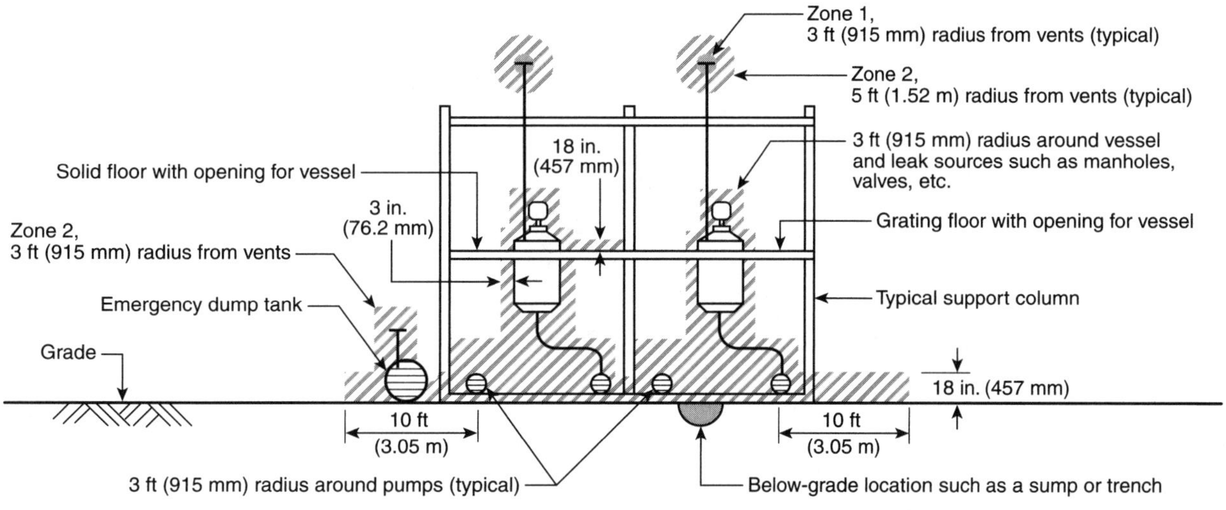

Material: Flammable liquid

	Small/low	Moderate	Large/high
Process equipment size	X	X	
Pressure	X	X	
Flow rate	X	X	

FIGURE 5.10.1(m) Multiple Sources of Leakage, Located Both at and Above Grade, in an Outdoor Process Area. The material being handled is a flammable liquid.

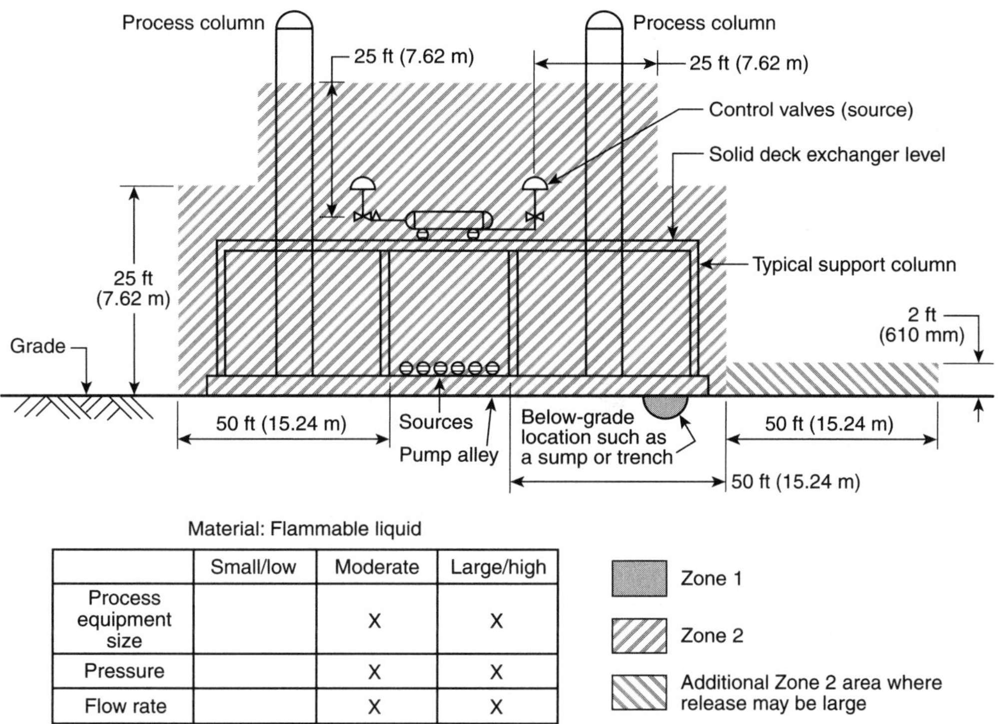

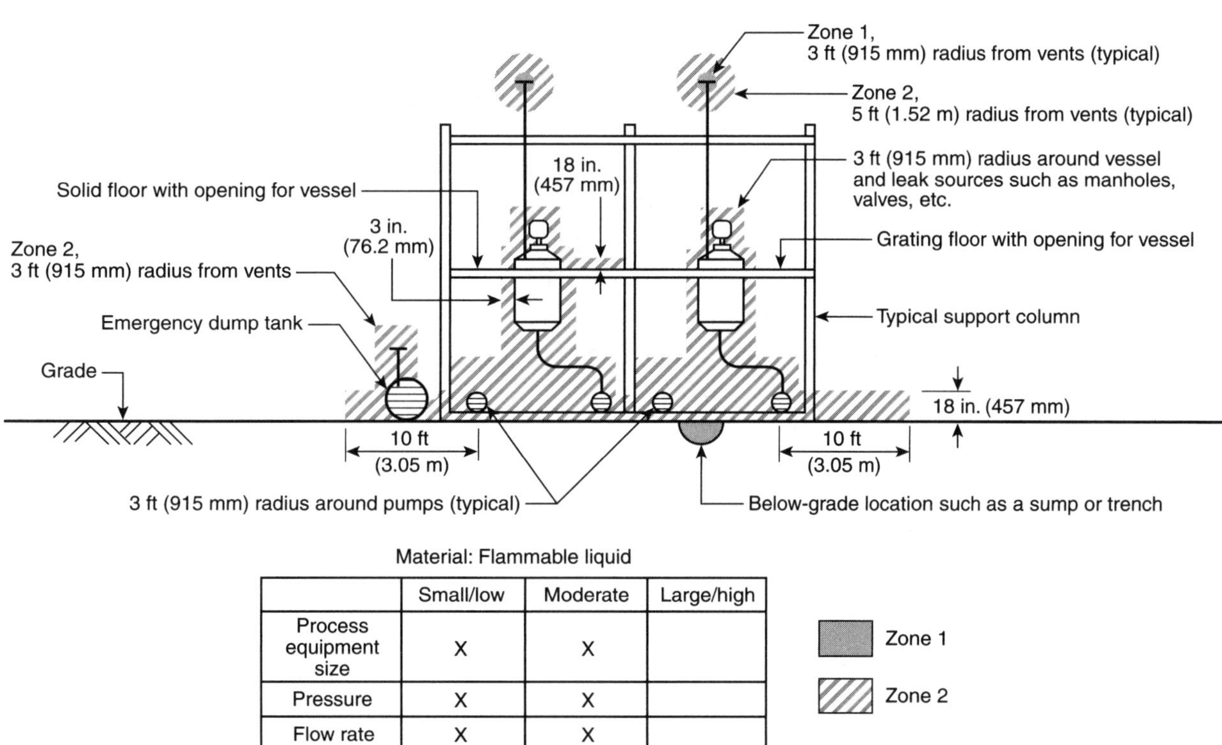

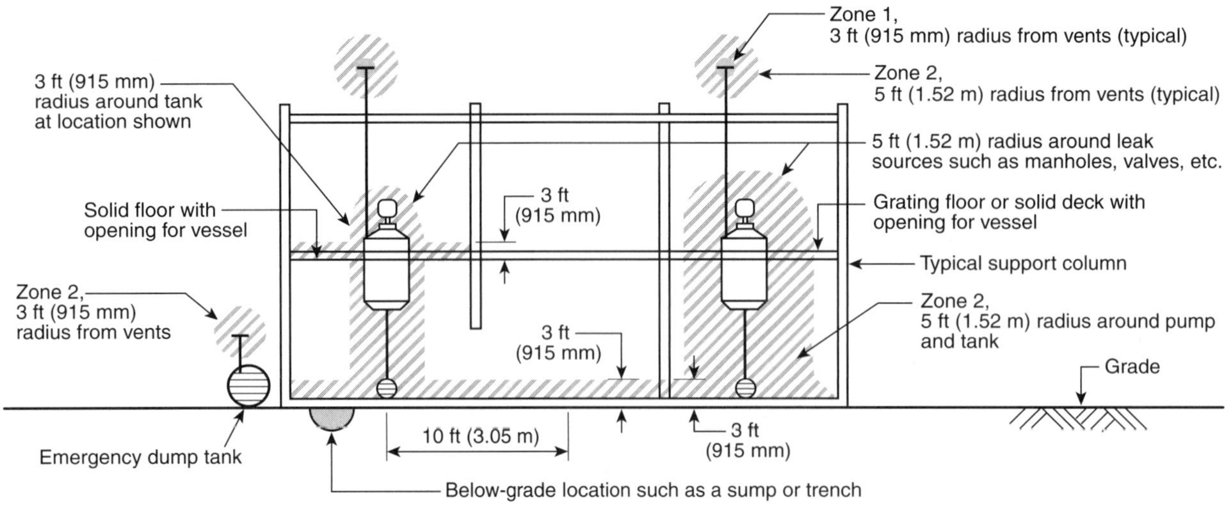

Material: Flammable liquid

	Small/low	Moderate	Large/high
Process equipment size	X	X	
Pressure	X	X	
Flow rate	X	X	

FIGURE 5.10.1(n) Multiple Sources of Leakage, Located Both at and Above Floor Level, in an Adequately Ventilated Building. The material being handled is a flammable liquid.

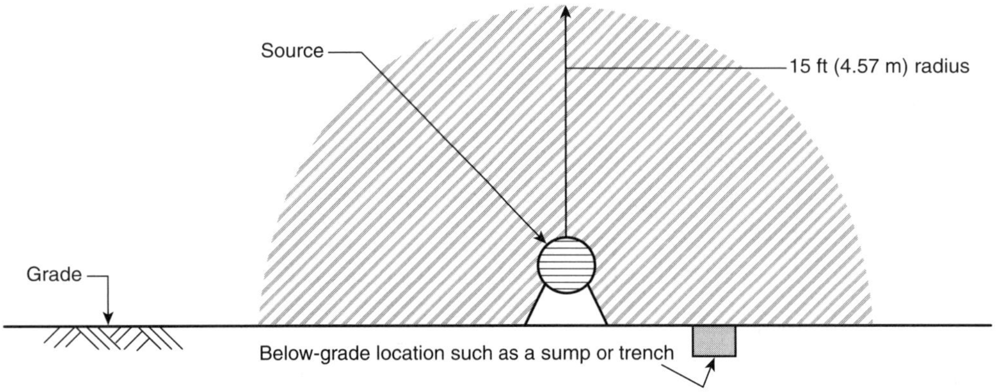

Material: Flammable liquid, liquefied flammable gas, compressed flammable gas, and cryogenic liquid

	Small/low	Moderate	Large/high
Process equipment size	X	X	
Pressure		X	X
Flow rate	X	X	

FIGURE 5.10.2(a) Leakage Located Outdoors, at Grade. The material being handled could be a flammable liquid, a liquefied or compressed flammable gas, or a flammable cryogenic liquid.

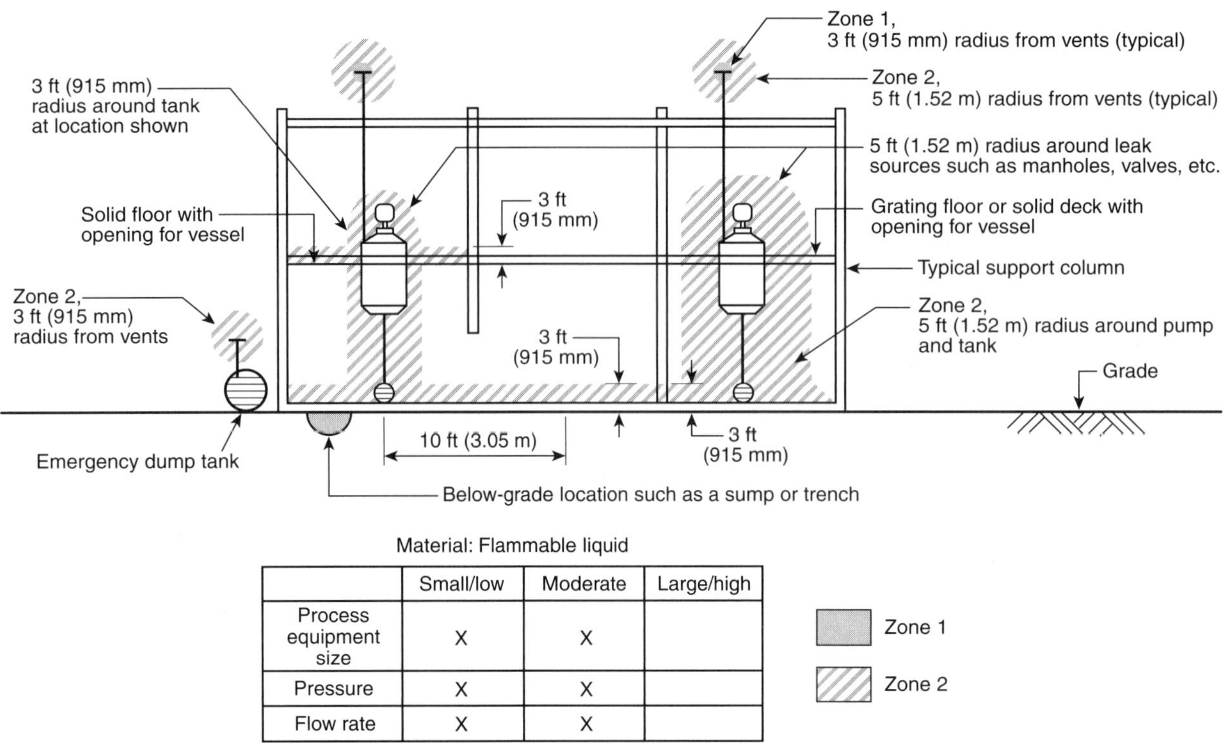

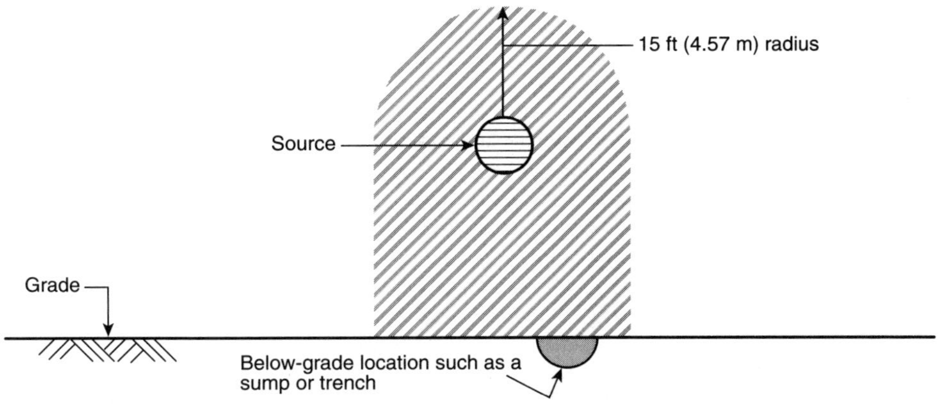

Material: Flammable liquid, liquefied flammable gas, compressed flammable gas, and cryogenic liquid

	Small/low	Moderate	Large/high
Process equipment size	X	X	
Pressure		X	X
Flow rate	X	X	

Zone 1

Zone 2

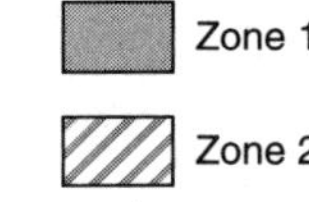

FIGURE 5.10.2(b) Leakage Located Outdoors, Above Grade. The material being handled could be a flammable liquid, a liquefied or compressed flammable gas, or a flammable cryogenic liquid.

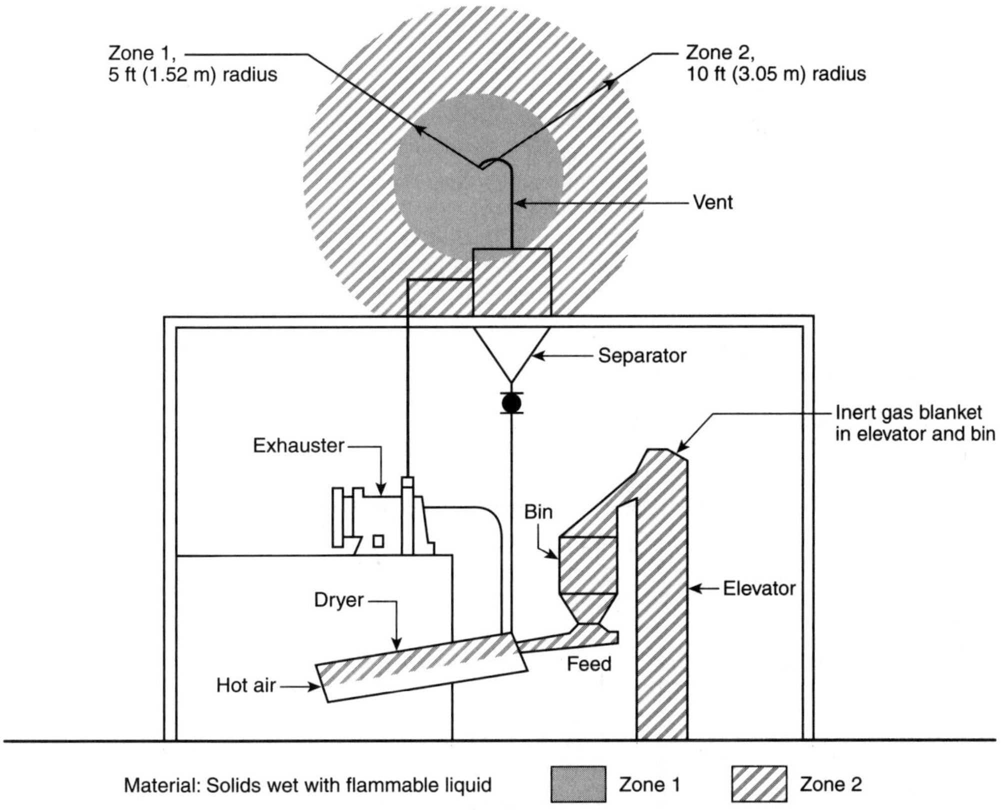

FIGURE 5.10.3(a) Product Dryer Located in an Adequately Ventilated Building. The product dryer system is totally enclosed. The material being handled is a solid wet with a flammable liquid.

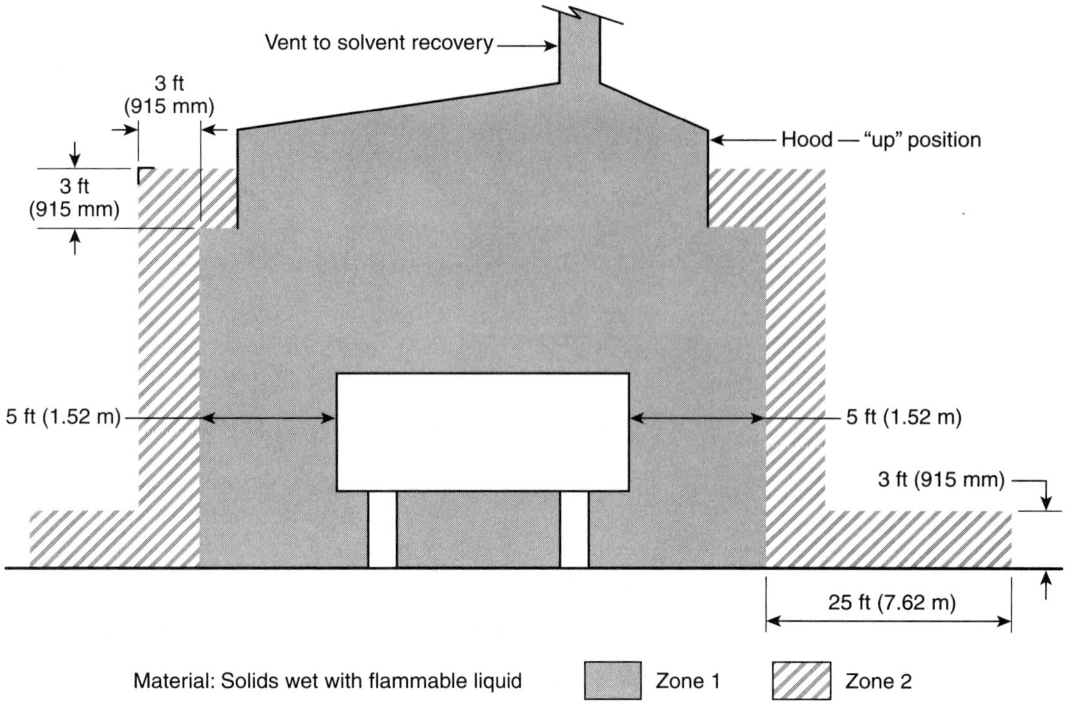

FIGURE 5.10.3(b) Plate and Frame Filter Press. Adequate ventilation is provided. The material being handled is a solid wet with a flammable liquid.

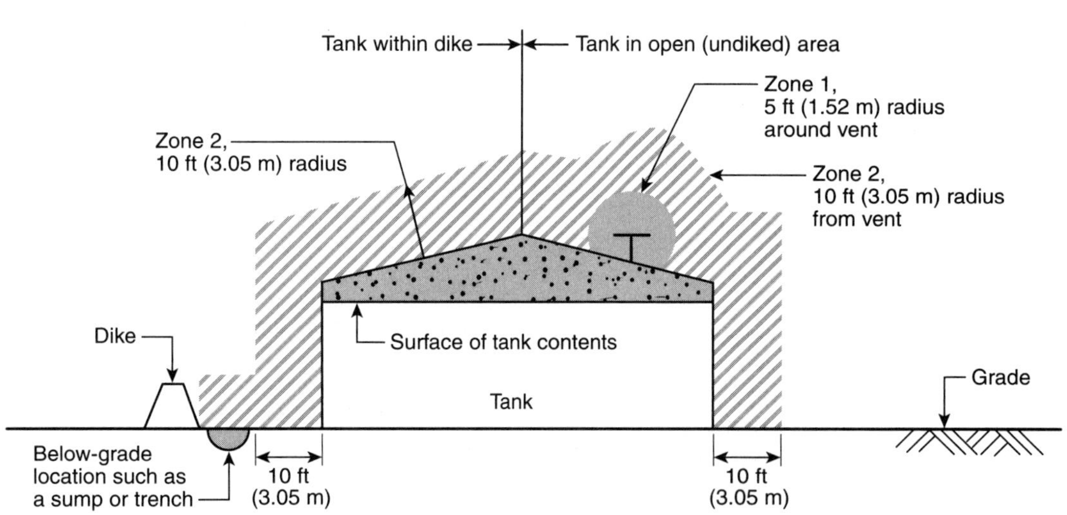

Material: Flammable liquid

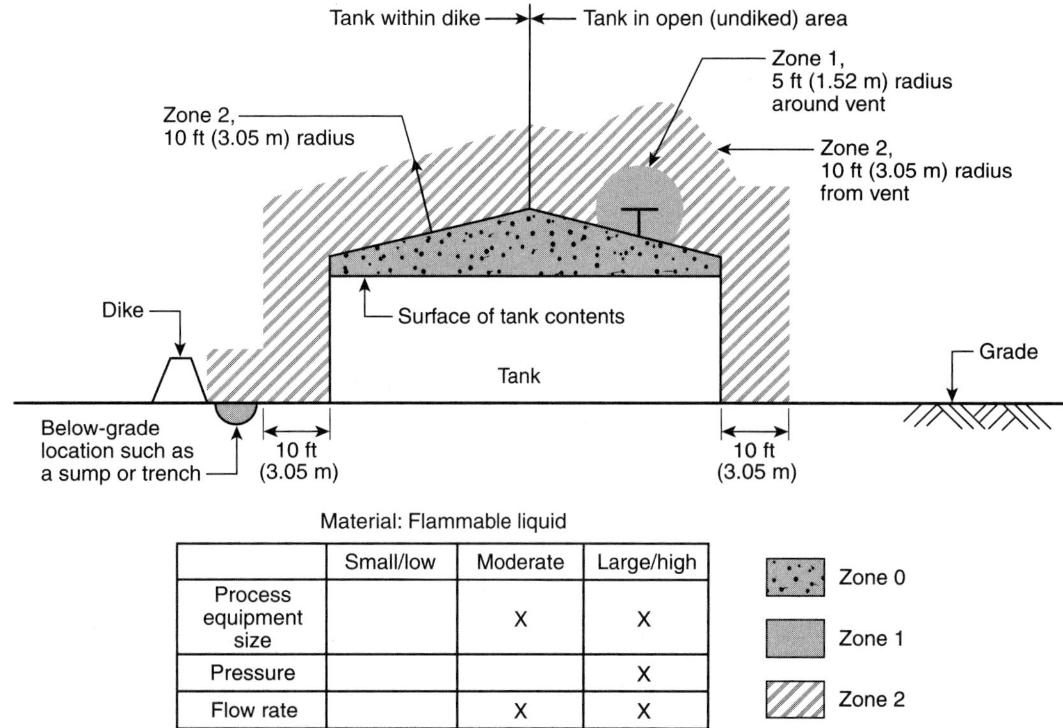

	Small/low	Moderate	Large/high
Process equipment size		X	X
Pressure			X
Flow rate		X	X

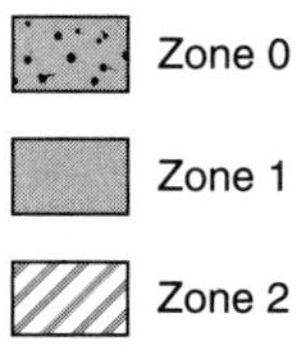

FIGURE 5.10.4(a) Product Storage Tank Located Outdoors, at Grade. The material being stored is a flammable liquid.

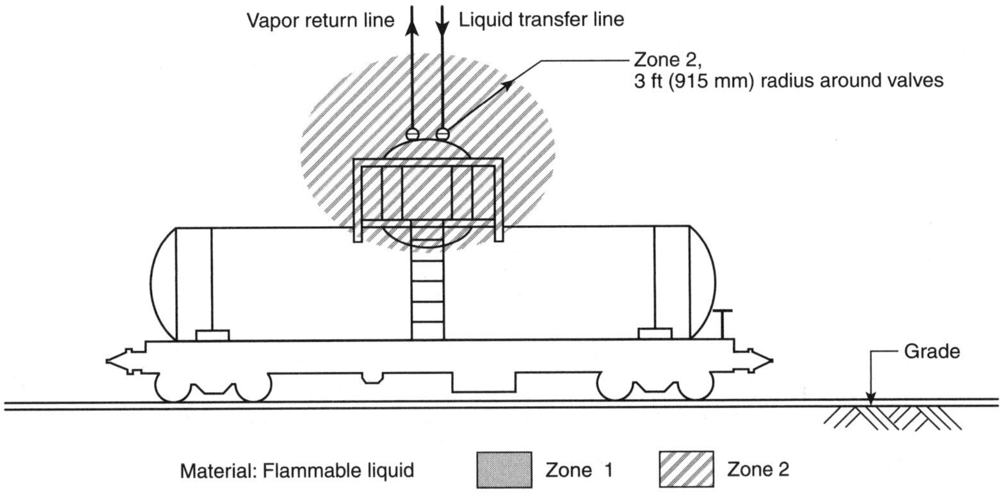

FIGURE 5.10.4(b) Tank Car Loading and Unloading via a Closed Transfer System. Material is transferred only through the dome. The material being transferred is a flammable liquid.

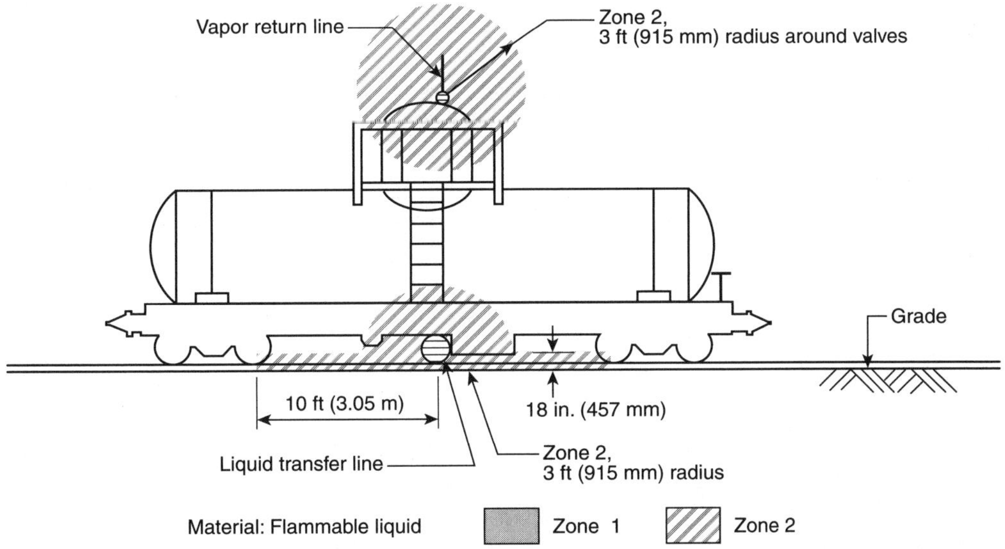

FIGURE 5.10.4(c) Tank Car Loading and Unloading via a Closed Transfer System. Material is transferred through the bottom fittings. The material being transferred is a flammable liquid.

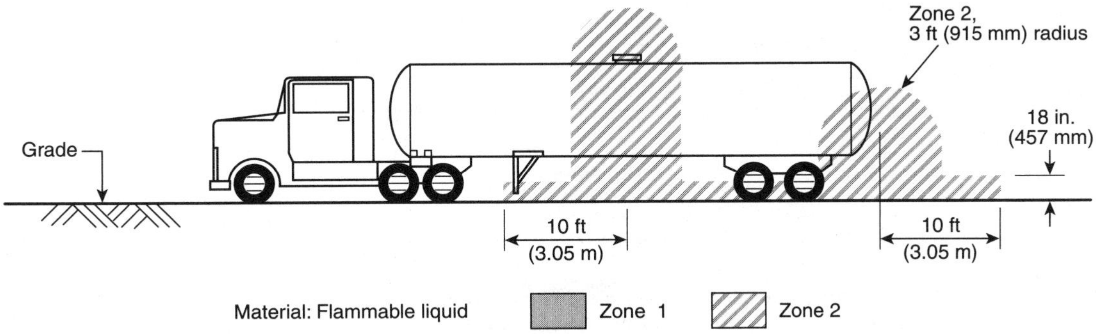

FIGURE 5.10.4(d) Tank Truck Loading and Unloading via a Closed Transfer System. Material is transferred through the bottom fittings. The material being transferred is a flammable liquid.

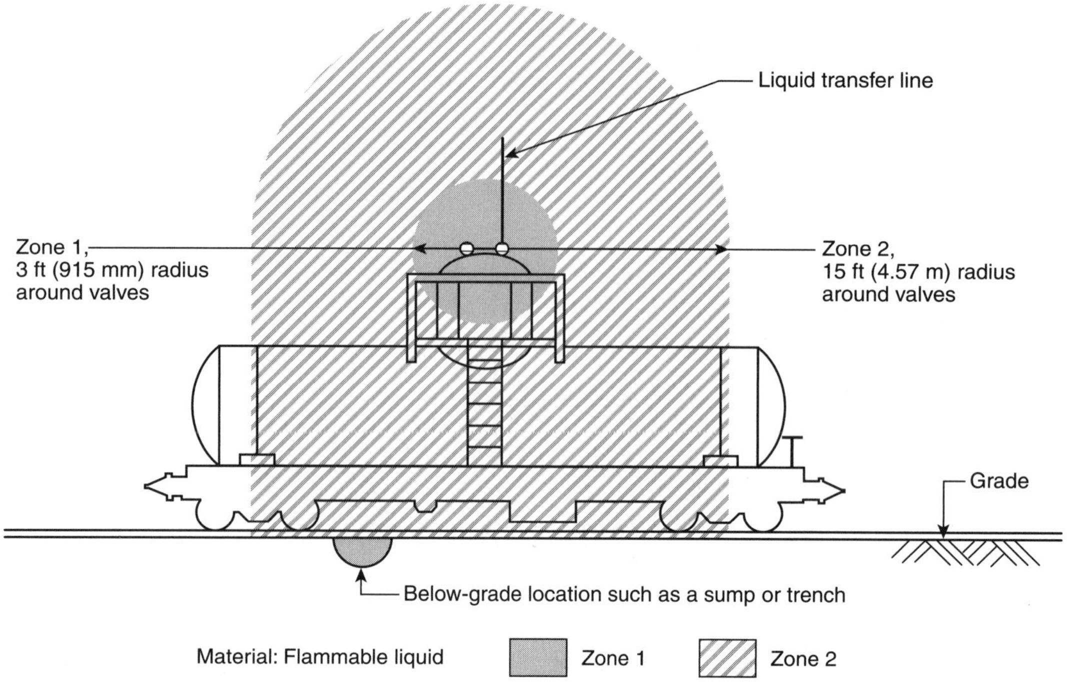

FIGURE 5.10.4(e) Tank Car (or Tank Truck) Loading and Unloading via an Open Transfer System. Material is transferred either through the dome or the bottom fittings. The material being transferred is a flammable liquid.

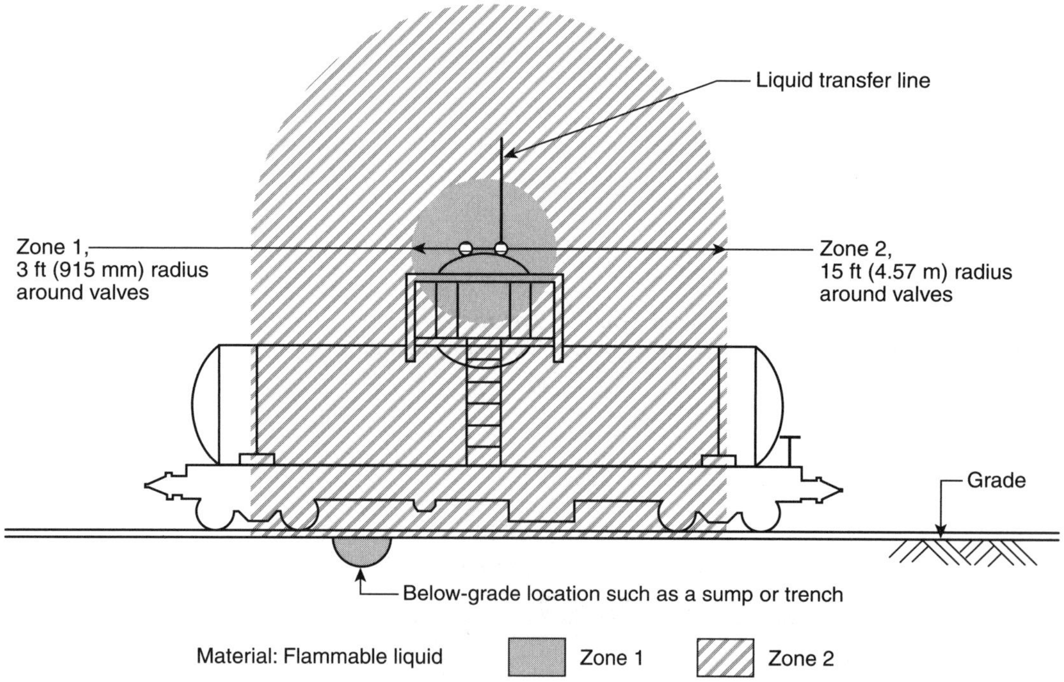

FIGURE 5.10.5 Tank Car (or Tank Truck) Loading and Unloading via a Closed Transfer System. Material is transferred only through the dome. The material being transferred may be a liquefied or compressed flammable gas or a flammable cryogenic liquid.

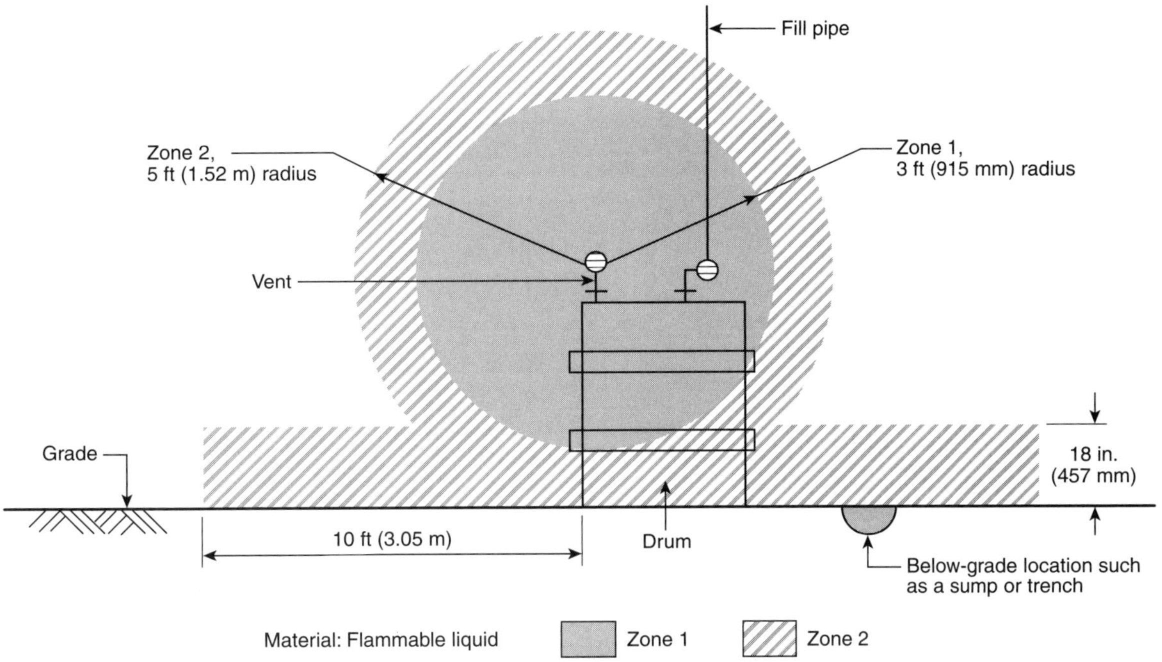

FIGURE 5.10.6 Drum Filling Station Located Either Outdoors or Indoors in an Adequately Ventilated Building. The material being handled is a flammable liquid.

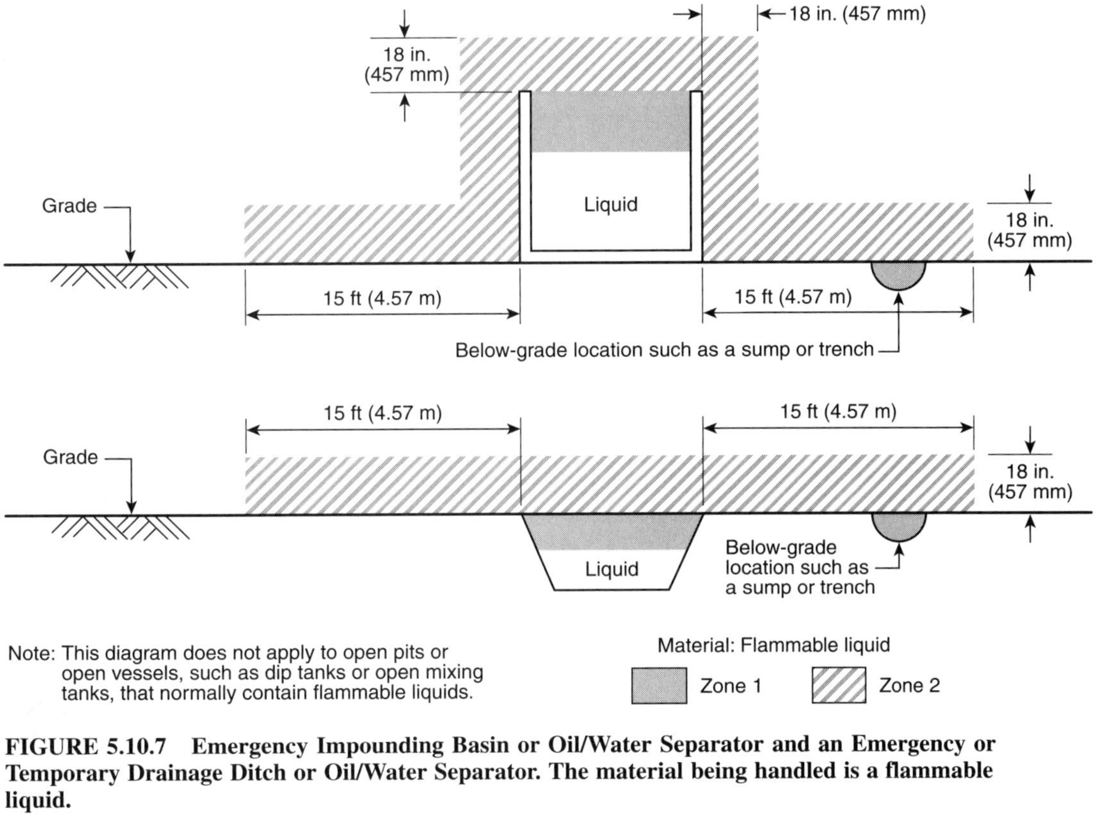

FIGURE 5.10.7 Emergency Impounding Basin or Oil/Water Separator and an Emergency or Temporary Drainage Ditch or Oil/Water Separator. The material being handled is a flammable liquid.

experimentally or from the supplier. It would be expected that the commercial grade of hexane would have an AIT ranging from 265°C to 290°C.

A.5.5.2 The degree to which air movement and material volatility combine to affect the extent of the classified area can be illustrated by two experiences monitored by combus-

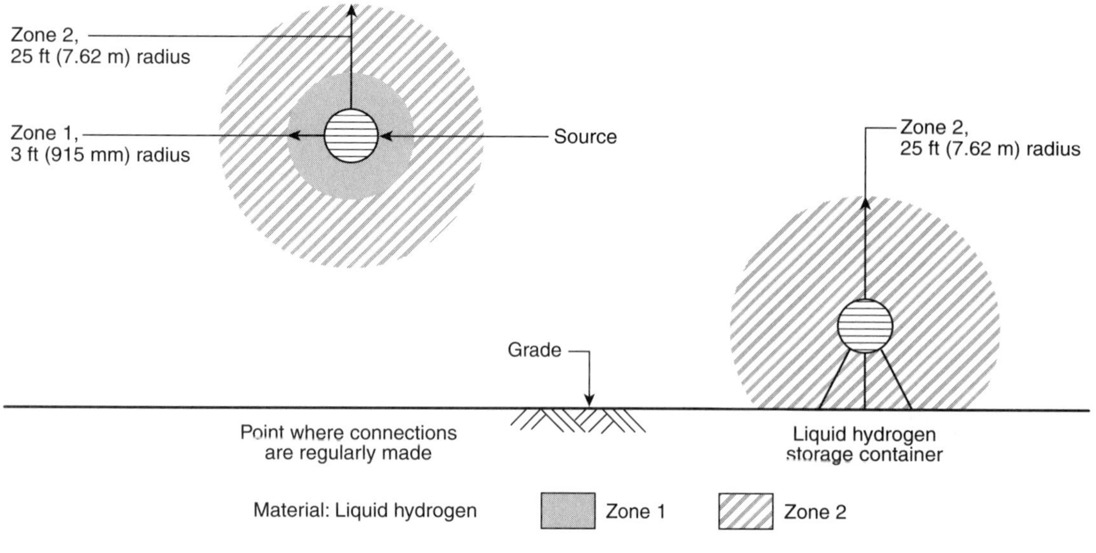

FIGURE 5.10.8(a)
Liquid Hydrogen Storage Located Outdoors or Indoors in an Adequately Ventilated Building. This diagram applies to liquid hydrogen only.

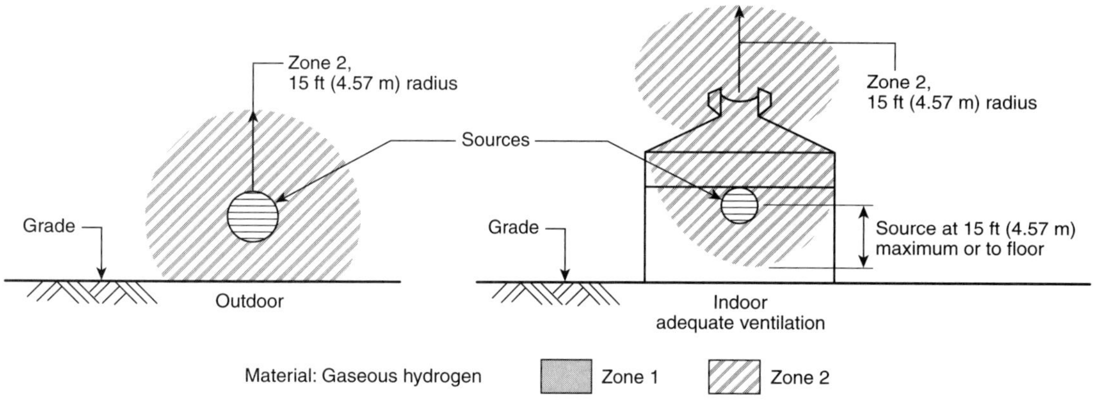

FIGURE 5.10.8(b)
Gaseous Hydrogen Storage Located Outdoors, or Indoors in an Adequately Ventilated Building. This diagram applies to gaseous hydrogen only.

tible gas detectors. Gasoline spilled in a sizable open manifold pit gave no indication of ignitible mixtures beyond 3 ft to 4 ft (0.9 m to 1.2 m) from the pit when the breeze was 8 to 10 mph (13 to 16 km/hr). A slightly smaller pool of a more volatile material, blocked on one side, was monitored during a gentle breeze. At grade, vapors could be detected for approximately 100 ft (30 m) downwind; however, at 18 in. (46 cm) above grade, there was no indication of vapor as close as 30 ft (9 m) from the pool.

These examples show the great variability that may be present in situations of this type, and point out again that careful consideration must be given to a large number of factors when classifying areas.

A.5.8.2.5 When fire hazard properties of a combustible material are not available, the appropriate group may be estimated using the following information:

(1) Minimum igniting current ratio (MIC ratio)
(2) Ratio of upper flammable limit (UFL) to lower flammable limit (LFL)

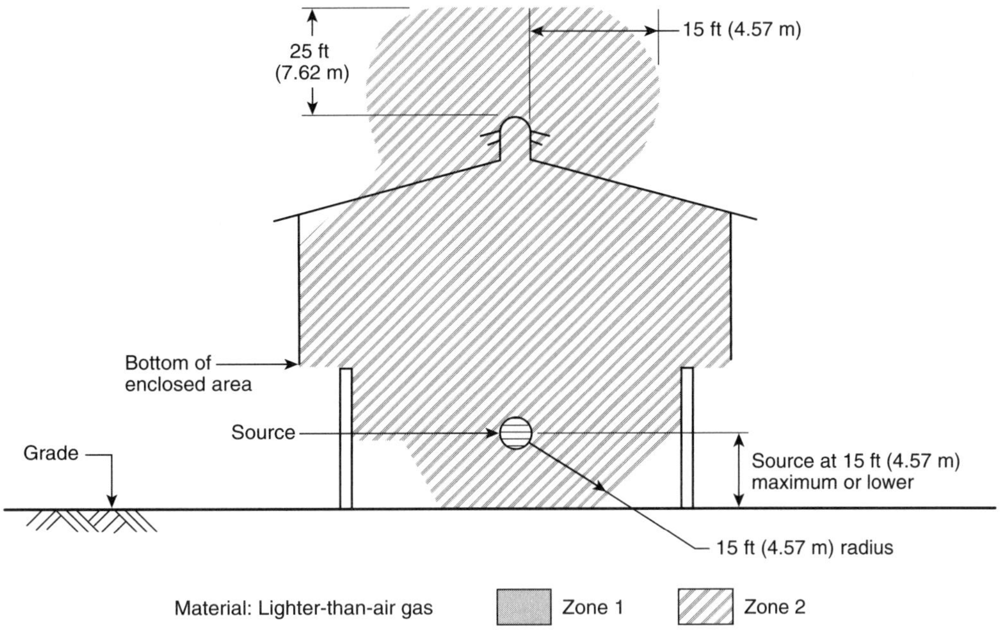

FIGURE 5.10.9(a)
Adequately Ventilated Compressor Shelter. The material being handled is a lighter-than-air gas.

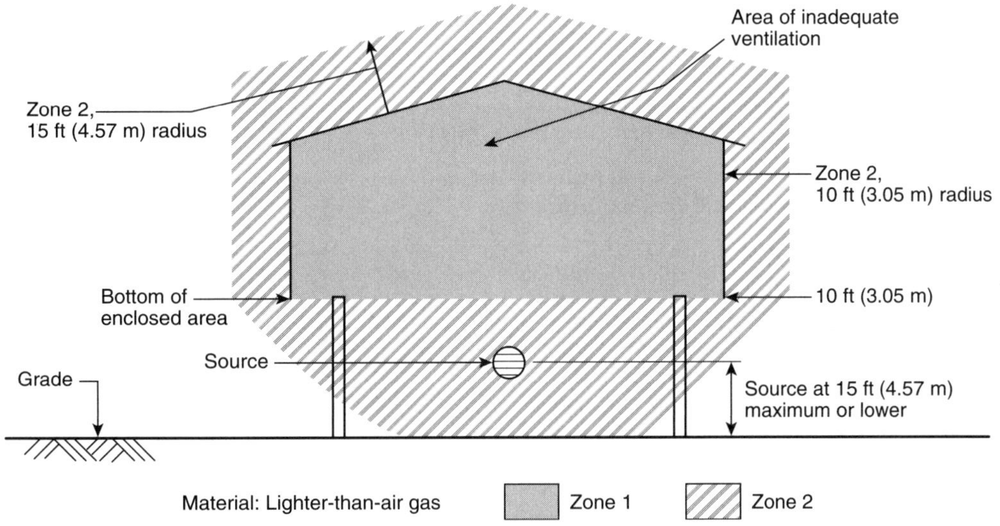

FIGURE 5.10.9(b) Inadequately Ventilated Compressor Shelter. The material being handled is a lighter-than-air gas.

(3) Molar heat of combustion multiplied by the lower flammable limit (LFL)
(4) Ratio of the lower flammable limit (LFL) to the stoichiometric concentration
(5) Maximum experimental safe gap (MESG)
(6) Minimum ignition energy (MIE)
(7) Stoichiometric flame temperature
(8) Knowledge of the chemical structure

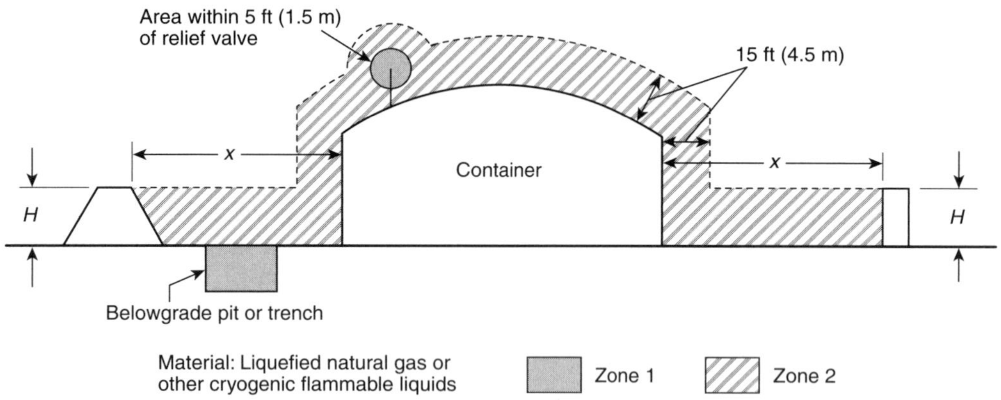

FIGURE 5.10.10(a) Tank for the Storage of Cryogenic and Other Cold Liquefied Flammable Gases. Dike height less than distance from container to dike (*H* less than *x*).

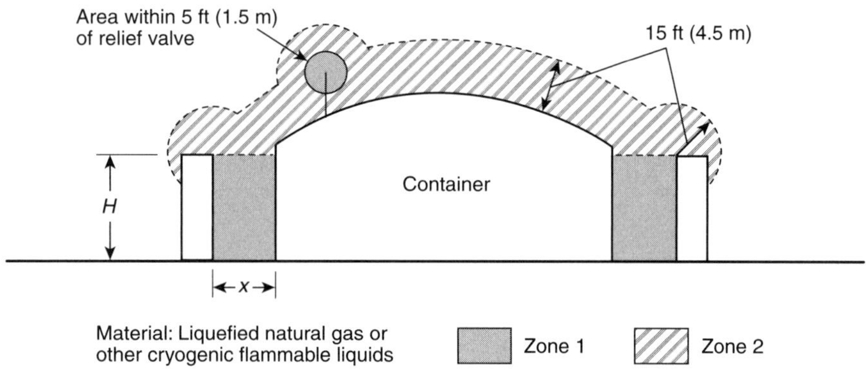

FIGURE 5.10.10(b) Tank for the Storage of Cryogenic and Other Cold Liquefied Flammable Gases. Dike height less than distance from container to dike (*H* greater than *x*).

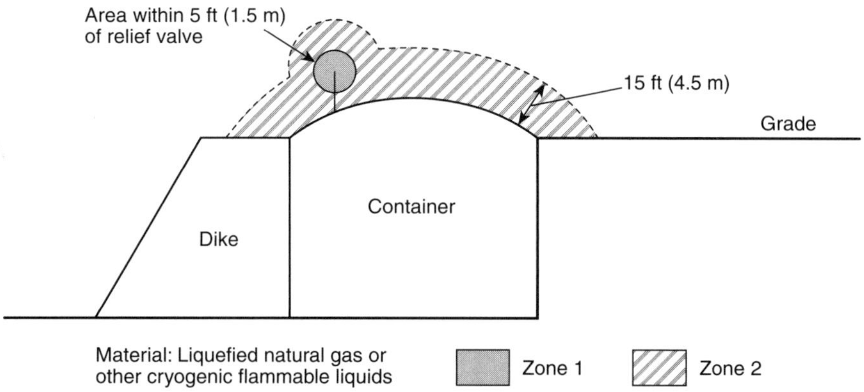

FIGURE 5.10.10(c) Tank for the Storage of Cryogenic and Other Cold Liquefied Flammable Gases. Container with liquid level below grade or top of dike.

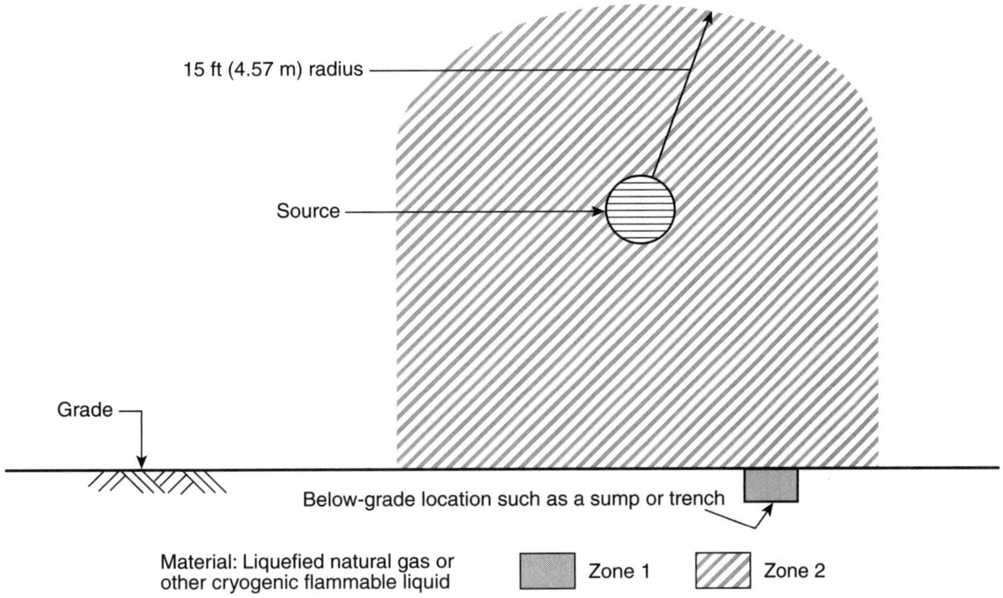

FIGURE 5.10.11 Source of Leakage from Equipment Handling Liquefied Natural Gas or Other Cold Liquefied Flammable Gas, and Located Outdoors, at or Above Grade.

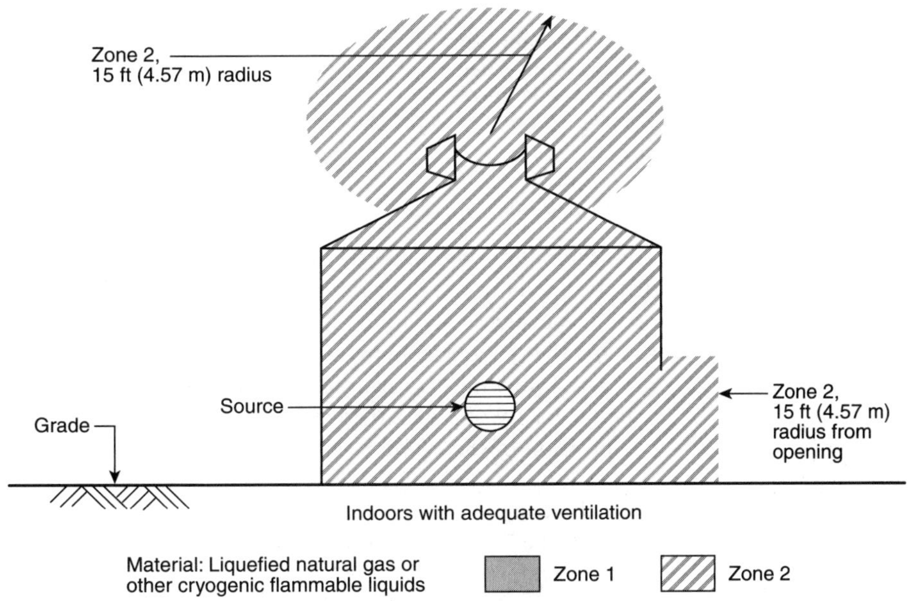

FIGURE 5.10.12 Source of Leakage from Equipment Handling Liquefied Natural Gas or Other Cold Liquefied Flammable Gas and Located Indoors in an Adequately Ventilated Building.

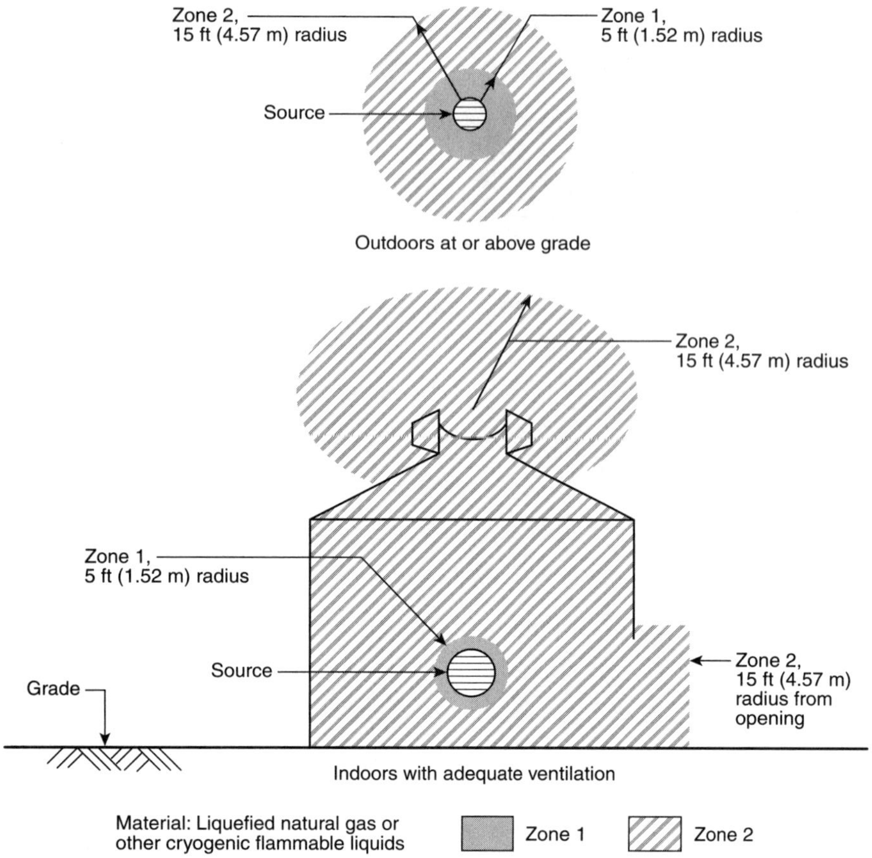

FIGURE 5.10.13 Classified Zones Around Liquefied Natural Gas Routinely Operating Bleeds, Drips, Vents, and Drains Both Outdoors, at or Above Grade, and Indoors, in an Adequately Ventilated Building. This diagram also applies to other cold liquefied flammable gases.

Table A.3.3.5.1 Comparison of MESG Test Results for Selected Materials

Material	Westerberg Apparatus, MESG (mm)	IEC Apparatus, MESG (mm)
Propane	0.92	0.94
Ethylene	0.69	0.65
Butadiene	0.79	0.79
Diethyl ether	0.30	0.87
Hydrogen	0.08	0.29

Annex B Example of a Method for Determining *NEC* Group Classification for Mixtures

This annex is not a part of the recommendations of this NFPA document but is included for informational purposes only.

B.1

This example is provided to show how the information in this document can help determine an *NEC* group classification. The example should not be applied to mixtures and/or streams that have acetylene or its equivalent hazard.

The following six materials are used in a tank having a total capacity of 100,000 kg as a single mixed stream. The composition of the mixture is:

Ethylene	45% by vol.
Propane	12% by vol.
Nitrogen	20% by vol.
Methane	3% by vol.
Isopropyl ether	17.5% by vol.
Diethyl ether	2.5% by vol.

Question: What is the appropriate *NEC* group to use for the mixture?

Table B.1 is a list of the physical properties for the materials contained in the tank. This information is from Table 4.4.2 and other references.

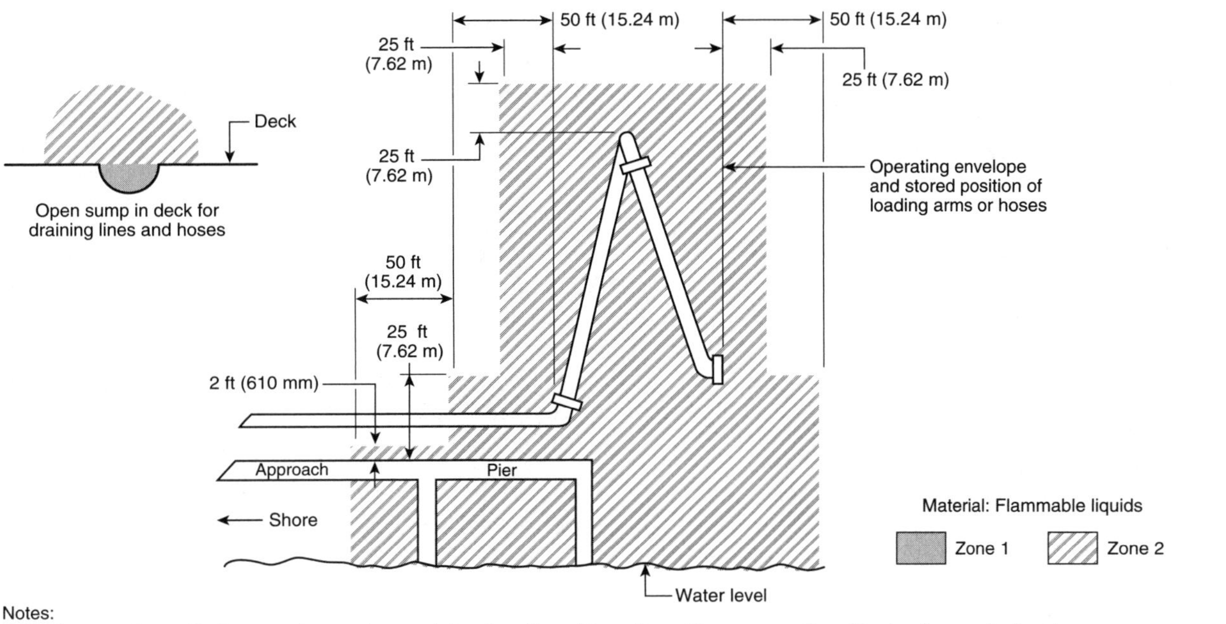

Notes:
1. The "source of vapor" is the operating envelope and stored position of the outboard flange connection of the loading arm (or hose).
2. The berth area adjacent to tanker and barge cargo tanks is to be Zone 2 to the following extent:
   (a) 25 ft (7.62 m) horizontally in all directions on the pier side from that portion of the hull containing cargo tanks
   (b) From the water level to 25 ft (7.62 m) above the cargo tanks at their highest position
3. Additional locations may have to be classified as required by the presence of other sources of flammable liquids or by coast guard or other regulations.

FIGURE 5.10.14 Classified Locations at a Marine Terminal Handling Flammable Liquids; Includes the Area Around the Stored Position of Loading Arms and Hoses.

Table B.1 Physical Properties of Selected Materials

Material	Mol. wt.	%LFL	%UFL	AIT (°C)	Vapor pressure. (mm Hg at 25°C)	BP (°C)	*NEC* Group	MESG (mm)	MIC ratio
Ethylene	28.05	2.7	36	450	52,320	−104	C	0.65	0.53
Propane	44.09	2.1	9.5	450	7150	−42	D	0.97	0.82
Nitrogen	14.0							∞	
Methane	16.04	5.0	15	600	463,800	−162	D	1.12	1.0
Isopropyl ether	102.17	1.4	21	443	148.7	69	D	0.94	
Diethyl ether	74.12	1.9	36	160	38.2	34.5	C	0.83	0.88

One method to estimate the *NEC* group is to determine the MESG of the mixture by applying a form of Le Châtelier relationship shown below:

$$MESG_{mix} = \frac{1}{\sum_{1}\left(\frac{X_i}{MESG_i}\right)}$$

Applying the volume percent and property information above results in the following equation:

$$\frac{1}{\frac{0.45}{0.65} + \frac{0.12}{0.97} + \frac{0.20}{\infty} + \frac{0.03}{1.12} + \frac{0.175}{0.94} + \frac{0.025}{0.83}} = 0.9442$$

Solving the equation above results in an MESG mixture value of 0.9442. From the definitions in 3.3.6 for Groups A, B, C, and D, this calculated MESG value would fall under Group D. Thus this mixture may be considered a Group D material.

Annex C Informational References

C.1 Reference Publications.

The following documents or portions thereof are referenced within this recommended practice for informational purposes only and are thus not part of the recommendations of this document unless also listed in Chapter 2.

C.1.1 NFPA Publications. National Fire Protection Association, 1 Batterymarch Park, Quincy, MA 02169-7471.

NFPA 70, *National Electrical Code®*, 2002 edition.

NFPA Fire Protection Guide to Hazardous Materials, 2001 edition.

C.1.2 Other Publications.

C.1.2.1 IEC Publications. International Electrotechnical Commission, 3 rue de Varembé, P.O. Box 131, CH-1211 Geneva 20, Switzerland.

IEC 60079-1-1, *Electrical apparatus for explosive gas atmospheres, Part 1-1: Flameproof enclosures "d" — Method of test for ascertainment of maximum experimental safe gap*, 2002.

IEC 60079-11, *Electrical apparatus for explosive gas atmospheres, Part 11: Intrinsic safety "i"*, 1999.

IEC/TR3 60079-20, *Electrical apparatus for explosive gas atmospheres, Part 20: Data for flammable gases and vapours, relating to the use of electrical apparatus*, 1996.

C.1.2.2 National Academy of Sciences Publication. National Materials Advisory Board of the National Academy of Sciences, 500 Fifth Street, N.W., Washington, DC 20001.

NMAB 353-1, *Matrix of Combustion-Relevant Properties and Classification of Gases, Vapors and Selected Solids.*

C.1.2.3 UL Publication. Underwriters Laboratories Inc., 333 Pfingsten Road, Northbrook, IL 60062-2096.

Technical Report No. 58 (TR 58), *An Investigation of Flammable Gases or Vapors with Respect to Explosion-Proof Electrical Equipment*, 1993.

C.1.2.4 Phillips, H. "Differences between Determinations of Maximum Experimental Safe Gaps in Europe and U.S.A." *Journal of Hazardous Materials*, 1981.

C.2 Informational References.

The following documents or portions thereof are listed here as informational resources only. They are not a part of the recommendations of this document.

C.2.1 NFPA Publications. National Fire Protection Association, 1 Batterymarch Park, Quincy, MA 02169-7471.

NFPA 30, *Flammable and Combustible Liquids Code*, 2003 edition.

NFPA 36, *Standard for Solvent Extraction Plants*, 2004 edition.

C.2.2 ASHRAE Publication. American Society of Heating, Refrigeration and Air-Conditioning Engineers, Inc., 1791 Tullie Circle, N.E., Atlanta, GA 30329.

ASHRAE 15, *Safety Code for Mechanical Refrigeration*, 1994.

C.2.3 ASTM Publications. American Society for Testing and Materials, 100 Barr Harbor Drive, West Conshohocken, PA 19428-2959.

ASTM D 56-02a, *Standard Method of Test for Flash Point by the Tag Closed Tester*, 2002.

ASTM D 93-02a, *Standard Test Method for Flash Point by Pensky-Martens Closed Cup Tester*, 2002.

ASTM D 3278-96el, *Standard Method of Tests for Flash Point of Liquids by Small Scale Closed-Cup Apparatus*, 1997.

ASTM E 659-78, *Standard Test Method for Autoignition Temperature of Liquid Chemicals*, 2000.

ASTM E 681-98, *Standard Test Method for Concentration Limits of Flammability of Chemicals*, 1998.

C.2.4 Bureau of Mines Publication. U.S. Government Printing Office, Washington, DC 20402.

RI 7009, *Minimum Ignition Energy and Quenching Distance in Gaseous Mixture.*

C.2.5 CGA Publication. Compressed Gas Association, 4221 Walney Rd, 5th Floor, Chantilly, VA 20151-2923.

ANSI/CGA G-2.1, *Safety Requirements for the Storage and Handling of Anhydrous Ammonia*, 1989.

C.2.6 IEC Publication. International Electrotechnical Commission, 3 rue de Varembé, P.O. Box 131, CH-1211 Geneva 20, Switzerland.

IEC 60079-12, *Electrical apparatus for explosive gas atmospheres, Part 12: Classification of mixtures of gases or vapours with air according to their maximum experimental safe gaps and minimum igniting currents*, 1978.

C.2.7 Other Publications. Energy Institute. (Institute of Petroleum.) *Model Code of Safe Practice for the Petroleum Industry, Part 15: Area Classification Code for Installations Handling Flammable Fluids*, 2002.

Hilado, C. J. and S. W. Clark. "Autoignition Temperatures of Organic Chemicals." *Chemical Engineering*, September 4, 1972.

C.3 References for Extracts.

The following documents are listed here to provide reference information, including title and edition, for extracts given throughout this recommended practice as indicated by a reference in brackets [ ] following a section or paragraph. These documents are not a part of the recommendations of this document unless also listed in Chapter 2 for other reasons.

NFPA 30, *Flammable and Combustible Liquids Code*, 2003 edition.

NFPA 70, *National Electrical Code®*, 2002 edition.

NFPA 499, *Recommended Practice for the Classification of Combustible Dusts and of Hazardous (Classified) Locations for Electrical Installations in Chemical Process Areas,* 2004

Editor's Note: *This document is reprinted here in its entirety, including its annexes.*

Chapter 1 Administration

1.1 Scope.

1.1.1 This recommended practice applies to those locations where combustible dusts are produced, processed, or handled, and where dust released into the atmosphere or accumulated on surfaces could be ignited by electrical systems or equipment.

1.1.2 This recommended practice provides information on specific combustible dusts whose relevant combustion properties have been sufficiently identified to allow their classification into the groups established by NFPA 70, *National Electrical Code®* (*NEC®*), for proper selection of electrical equipment in hazardous (classified) locations. The tables of selected combustible materials contained in this document are not intended to be all-inclusive.

1.1.3 This recommended practice also applies to chemical process areas. As used in this document, a chemical process area could be a chemical process plant, or it could be a part of such a plant. A chemical process area could be a part of a manufacturing facility where combustible dusts are produced or used in chemical reactions, or are handled or used in operations such as mixing, coating, extrusion, conveying, drying, and/or grinding.

1.1.4 This recommended practice does not apply to agricultural grain-handling facilities except where powdered grain is used in a chemical reaction or mixture.

1.1.5 This recommended practice does not apply to situations that could involve catastrophic failure of, or catastrophic discharge from, silos, process vessels, pipelines, tanks, hoppers, or conveying or elevating systems.

1.1.6 This recommended practice does not apply to oxygen-enriched atmospheres or pyrophoric materials.

1.2 Purpose.

1.2.1 The purpose of this recommended practice is to provide the user with a basic understanding of the parameters that determine the degree and the extent of the hazardous (classified) location. This recommended practice also provides the user with examples of the applications of these parameters.

1.2.2 Information is provided on specific combustible dusts whose relevant properties determine their classification into groups. This will assist in the selection of special electrical equipment for hazardous (classified) locations where such electrical equipment is required.

1.2.3 This recommended practice is intended as a guide and should be applied with sound engineering judgment. Where all factors are properly evaluated, a consistent area classification scheme can be developed.

1.2.4 This recommended practice is based on the criteria established by Articles 500 and 502 of the *NEC*. Once an area is properly classified, the *NEC* specifies the type of equipment and the wiring methods that shall be permitted to be used.

1.3 Relationship to NFPA Codes and Standards.

This recommended practice is not intended to supersede or conflict with the following NFPA standards:

(1) NFPA 36, *Standard for Solvent Extraction Plants*, 2004 edition.
(2) NFPA 61, *Standard for the Prevention of Fires and Dust Explosions in Agricultural and Food Processing Facilities*, 2002 edition.
(3) NFPA 68, *Guide for Venting of Deflagrations*, 2002 edition.
(4) NFPA 69, *Standard on Explosion Prevention Systems*, 2002 edition.
(5) NFPA 484, *Standard for Combustible Metals, Metal Powders, and Metal Dusts*, 2002 edition.

(6) NFPA 654, *Standard for the Prevention of Fire and Dust Explosions from the Manufacturing, Processing, and Handling of Combustible Particulate Solids*, 2000 edition.
(7) NFPA 655, *Standard for Prevention of Sulfur Fires and Explosions*, 2001 edition.
(8) NFPA 664, *Standard for the Prevention of Fires and Explosions in Wood Processing and Woodworking Facilities*, 2002 edition.

Chapter 2 Referenced Publications

2.1 General.

The documents or portions thereof listed in this chapter are referenced within this recommended practice and should be considered part of the recommendations of this document.

2.2 NFPA Publications.

National Fire Protection Association, 1 Batterymarch Park, Quincy, MA 02169-7471.

NFPA 36, *Standard for Solvent Extraction Plants,* 2004 edition.

NFPA 61, *Standard for the Prevention of Fires and Dust Explosions in Agricultural and Food Processing Facilities,* 2002 edition.

NFPA 68, *Guide for Venting of Deflagrations,* 2002 edition.

NFPA 69, *Standard on Explosion Prevention Systems,* 2002 edition.

NFPA 70, *National Electrical Code®*, 2002 edition.

NFPA 484, *Standard for Combustible Metals, Metal Powders, and Metal Dusts,* 2002 edition.

NFPA 654, *Standard for the Prevention of Fire and Dust Explosions from the Manufacturing, Processing, and Handling of Combustible Particulate Solids,* 2000 edition.

NFPA 655, *Standard for Prevention of Sulfur Fires and Explosions,* 2001 edition.

NFPA 664, *Standard for the Prevention of Fires and Explosions in Wood Processing and Woodworking Facilities,* 2002 edition.

2.3 Other Publication.

2.3.1 ASTM Publication.

American Society for Testing and Materials, 100 Barr Harbor Drive, West Conshohocken, PA 19428-2959.

ASTM D 3175, *Standard Test Method for Volatile Matter in the Analysis Sample of Coal and Coke*, 1989.

Chapter 3 Definitions

3.1 General.

The definitions contained in this chapter apply to the terms used in this recommended practice. Where terms are not included, common usage of the terms applies.

3.2 NFPA Official Definitions.

3.2.1 Recommended Practice. A document that is similar in content and structure to a code or standard but that contains only nonmandatory provisions using the word "should" to indicate recommendations in the body of the text.

3.2.2 Should. Indicates a recommendation or that which is advised but not required.

3.3 General Definitions.

3.3.1* Autoignition Temperature (AIT). The minimum temperature required to initiate or cause self-sustained combustion of a solid, liquid, or gas independently of the heating or heated element.

3.3.2 CAS. Chemical Abstract Service.

3.3.3* Combustible Dust. Any finely divided solid material 420 microns or less in diameter (i.e., material passing through a U.S. No. 40 Standard Sieve) that presents a fire or explosion hazard when dispersed.

3.3.4 Combustible Dust, Class II. Class II combustible dusts are divided into Groups E, F, and G.

3.3.4.1 Group E. Atmospheres containing combustible metal dusts, including aluminum, magnesium, and their commercial alloys, or other combustible dusts whose particle size, abrasiveness, and conductivity present similar hazards in the use of electrical equipment.

3.3.4.2 Group F. Atmospheres containing combustible carbonaceous dusts that have more than 8 percent total entrapped volatiles (see ASTM D 3175, *Standard Test Method for Volatile Matter in the Analysis Sample of Coal and Coke*, for coal and coke dusts) or that have been sensitized by other materials so that they present an explosion hazard. Coal, carbon black, charcoal, and coke dusts are examples of carbonaceous dusts.

3.3.4.3 Group G. Atmospheres containing other combustible dusts, including flour, grain, wood flour, plastic, and chemicals.

3.3.5 Combustible Material. A generic term used to describe either a mixture of dust in air, or a hybrid mixture, that may burn, flame, or explode.

3.3.6* Explosion Severity. A measure of the damage potential of the energy released by a dust explosion.

3.3.7 Hybrid Mixture. A mixture of a dust with one or more flammable gases or vapors.

3.3.8 Ignitible Mixture. A combustible material that is within its flammable range.

3.3.9* Ignition Sensitivity. A measure of the ease by which a cloud of combustible dust could be ignited.

Chapter 4 Combustible Dusts

4.1 *National Electrical Code*® Criteria.

4.1.1 Article 500 of the *NEC* classifies any location in which a combustible material is or can be present in the atmosphere in sufficient concentrations to produce an ignitible mixture.

4.1.2* In a Class II hazardous (classified) location, the combustible material present is a combustible dust.

4.1.3 Class II is further subdivided into either Class II, Division 1 or Class II, Division 2 as follows.

4.1.3.1 Class II, Division 1. A Class II, Division 1 location is a location

(1) In which combustible dust is in the air under normal operating conditions in quantities sufficient to produce explosive or ignitible mixtures, or
(2) Where mechanical failure or abnormal operation of machinery or equipment might cause such explosive or ignitible mixtures to be produced, and might also provide a source of ignition through simultaneous failure of electric equipment, through operation of protection devices, or from other causes, or
(3) In which combustible dusts of an electrically conductive nature may be present in hazardous quantities. [**70:**500.5(C)(1)]

4.1.3.2 Class II, Division 2. A Class II, Division 2 location is a location

(1) Where combustible dust is not normally in the air in quantities sufficient to produce explosive or ignitible mixtures, and dust accumulations are normally insufficient to interfere with the normal operation of electrical equipment or other apparatus, but combustible dust may be in suspension in the air as a result of infrequent malfunctioning of handling or processing equipment and
(2) Where combustible dust accumulations on, in, or in the vicinity of the electrical equipment may be sufficient to interfere with the safe dissipation of heat from electrical equipment or may be ignitible by abnormal operation or failure of electrical equipment. [**70:**500.5(C)(2)]

4.1.4 The intent of Article 500 of the *NEC* is to prevent the use of electrical equipment and systems in hazardous (classified) locations that would ignite an ignitible dust cloud or layer.

4.1.5 Electrical installations within hazardous (classified) locations can use various protection techniques. No single protection technique is best in all respects for all types of equipment used in a chemical plant.

4.1.5.1 Dust-ignitionproof electrical equipment, electrical equipment protected by pressurizing, and intrinsically safe electrical equipment are applicable to both Division 1 and Division 2 locations.

4.1.5.2 Other dusttight equipment enclosures, as specified in Article 502 of the *NEC*, are permitted in Division 2 locations.

4.1.5.3 Electrical equipment protected by pressurizing and intrinsically safe electrical equipment are applicable to both Division 1 and Division 2 locations.

4.1.5.4 Equipment and wiring suitable for Class I, Division 1 locations are not required and might not be acceptable in Class II locations.

4.1.6 Electrical equipment in Division 1 locations is enclosed in a manner that will exclude ignitible amounts of dusts or will not permit arcs, sparks, or heat generated or liberated inside the enclosures to cause ignition of dust accumulations or of atmospheric dust suspensions in the vicinity of the equipment.

4.1.7 Electrical equipment in Division 2 locations is designed so that normal operation of the electrical equipment does not provide a source of ignition.

4.1.7.1 Protection against ignition during electrical breakdown is not provided. However, electrical breakdowns are sufficiently rare that the chances of one occurring simultaneously with accidental release of an ignitible mixture are extremely remote.

4.1.7.2 Arcing and sparking devices are permitted only if suitably enclosed or if the sparks are of insufficient energy to ignite the mixture.

4.1.7.3 Electrical installations in Division 2 locations could be designed with dusttight enclosures or other equipment enclosures as specified in Article 502 of the *NEC*.

4.1.8 Where flammable gases or flammable vapors, and combustible dusts are present, electrical equipment and wiring suitable for simultaneous exposure to both Class I and Class II conditions are required.

4.1.9 Where Group E dusts are present in hazardous quantities, there are only Division 1 locations. The *NEC* does not recognize any Division 2 locations for such dusts.

4.1.10 Factors such as corrosion, weather, maintenance, equipment standardization and interchangeability, and possible process changes or expansion frequently dictate the use of special enclosures or installations for electrical systems. However, such factors are outside the scope of this recommended practice.

4.1.11 For the purpose of this recommended practice, locations not classified as Division 1 or Division 2 are "unclassified" locations.

4.2 Behavior of Combustible Dusts.

4.2.1 Dispersion and Explosion.

4.2.1.1 Dust discharged or leaking from equipment into the atmosphere is acted on by gravity and will settle relatively quickly, depending on the size of particles, the internal pressure propelling the particles out of the equipment, and any air currents in the vicinity.

4.2.1.1.1 The result is a layer of dust that settles on horizontal surfaces below the leak opening in a radial or elliptical manner, depending on the location of the opening on the equipment.

4.2.1.1.2 The depth of the layer will be greatest under and close to the source and will taper off to the outside of the circle or ellipse.

4.2.1.1.3 The greater the height of the dust source above the surface, the greater the area covered.

4.2.1.1.4 The internal pressure in the equipment will likewise increase the area covered.

4.2.1.1.5 The size of the leak opening and the elapsed time of emission also affect the quantity of dust on the surface.

4.2.1.1.6 Some dusts have particles that are extremely fine and light (i.e., have a low specific particle density). Such particles could behave more like vapors than like dusts and could remain in suspension for long periods. These particles could travel far from the emitting source and collect on surfaces above the source.

4.2.1.1.7 Although horizontal surfaces accumulate the largest quantities of dust, vertical surfaces could in some instances also accumulate significant quantities.

4.2.1.2 Although a dust cloud will ignite and explode readily in the presence of an open ignition source, dust layers, if undisturbed and not in direct contact with the ignition source, will not explode. However, if a small amount of dust is dispersed in the air at the ignition source, a small explosion will occur. The pressure wave from this explosion blows the dust layer into the air, and a larger explosion then takes place. It is often this secondary explosion that does the most damage.

4.2.2* Hybrid Mixtures. The presence of the flammable gas or vapor, even at concentrations less than their lower flammable limit (LFL), will not only add to the violence of the dust–air combustion but will drastically reduce the ignition energy. This situation is encountered in certain industrial operations, such as fluidized bed dryers and pneumatic conveying systems for plastic dusts from polymerization processes where volatile solvents are used. In such cases, electrical equipment should be specified that is suitable for simultaneous exposure to both the Class I (flammable gas) and the Class II (combustible dust).

4.3* Conditions Necessary for Ignition of a Combustible Dust.

4.3.1 In a Class II location, one of the following sets of conditions must be satisfied for ignition by the electrical installation.

4.3.1.1 In the first set of conditions, the following is true:

(1) A combustible dust must be present.
(2) The dust must be suspended in the air in the proportions required to produce an ignitible mixture. Further, within the context of this recommended practice, a sufficient quantity of this suspension must be present in the vicinity of the electrical equipment.
(3) There must be a source of thermal or electrical energy sufficient to ignite the suspended mixture. Within the context of this recommended practice, the energy source is understood to originate with the electrical system.

4.3.1.2* In the second set of conditions, the following is true:

(1) A combustible dust must be present.
(2) The dust must be layered thickly enough on the electrical equipment to interfere with the dissipation of heat and allow the layer to reach the ignition temperature of the dust.
(3) The external temperature of the electrical equipment must be high enough to cause the dust to reach its ignition temperature directly or to dry out the dust and cause it to self-heat.

4.3.1.3 In the third set of conditions, the following is true:

(1) A Group E dust must be present.
(2) The dust must be layered or in suspension in hazardous quantities.
(3) Current through the dust must be sufficient to cause ignition. *(See 4.4.2.)*

4.3.2 Once ignition has occurred, either in a cloud suspension or in a layer, an explosion is likely.

4.3.2.1 Often the initial explosion is followed by another much more violent explosion fueled from dust accumulations on structural beams and equipment surfaces that are thrown into suspension by the initial blast.

4.3.2.2 For this reason, good housekeeping is vitally important in all areas where dust is handled, and is assumed throughout this recommended practice.

4.3.3 In classifying a particular location, the presence of a combustible dust is significant in determining the correct division.

4.3.3.1 The classification depends both on the presence of dust clouds and on the presence of hazardous accumulations of dust in layer form.

4.3.3.2 As specified in 4.1.3.1, the presence of a combustible dust cloud under normal conditions of operation, or due to frequent repair or maintenance, should be classified as Division 1.

4.3.3.3 Abnormal operation of machinery and equipment, which could simultaneously produce a dust cloud or suspension and a source of ignition, also should be classified as Division 1.

4.3.3.4 In other words, if a dust cloud is present at any time, it is assumed to be ignitible, and all that is necessary for electrical ignition is failure of the electrical system.

4.3.3.5 If dust clouds or hazardous dust accumulations are present only as a result of infrequent malfunctioning of handling or processing equipment, and ignition can result only from abnormal operation or failure of electrical equipment, the location should be classified Division 2.

4.3.4 The presence of an ignitible dust cloud or an ignitible dust layer is important in determining the boundaries of the hazardous (classified) location.

4.3.5 The quantity of dust, its physical and chemical properties, its dispersion properties, and the location of walls and cutoffs must all be considered.

4.4 Dusts with Additional Hazards.

4.4.1 Conductive Dusts. Group E dusts could cause a short in the electrical equipment when exposed to sufficiently high voltages.

4.4.2 Group E dusts are sensitive to a phenomenon whereby an electric current finds the path of least resistance through a dust layer, heating up the dust particles in its path and thus providing a source of ignition. The resulting electric arc could ignite a dust layer or dust cloud.

4.4.3 Magnesium or Aluminum Dust. Dusts containing magnesium or aluminum are particularly hazardous, and the use of extreme precaution will be necessary to avoid ignition and explosion.

4.4.4 Low-Ignition Temperature Dusts. Zirconium, thorium, and uranium dusts have extremely low ignition temperatures [as low as 20°C (68°F)] and extremely low minimum ignition energies.

4.5 Classification of Combustible Dusts.

4.5.1 Combustible dusts are divided into three groups, depending on the nature of the dust: Group E, Group F, and Group G.

4.5.2* A listing of selected combustible dusts with their group classification and relevant physical properties is provided in Table 4.5.2. The chemicals are listed alphabetically.

4.5.3 Table 4.5.3 provides a cross-reference of selected chemicals sorted by their Chemical Abstract Service (CAS) numbers.

4.5.4 References that deal with the testing of various characteristics of combustible materials are listed in Section B.2.1, Section B.2.2, and Section B.2.4.

Table 4.5.2 Selected Combustible Materials

Chemical Name	CAS No.	*NEC* Group	Code	Layer or Cloud Ignition Temp. (°C)
Acetal, Linear		G	NL	440
Acetoacet-p-phenetidide	122-82-7	G	NL	560
Acetoacetanilide	102-01-2	G	M	440
Acetylamino-t-nitrothiazole		G		450
Acrylamide Polymer		G		240
Acrylonitrile Polymer		G		460
Acrylonitrile-Vinyl Chloride-Vinylidenechloride copolymer (70-20-10)		G		210
Acrylonitrile-Vinyl Pyridine Copolymer		G		240
Adipic Acid	124-04-9	G	M	550
Alfalfa Meal		G		200
Alkyl Ketone Dimer Sizing Compound		G		160
Allyl Alcohol Derivative (CR-39)		G	NL	500
Almond Shell		G		200
Aluminum, A422 Flake	7429-90-5	E		320
Aluminum, Atomized Collector Fines		E	CL	550
Aluminum—cobalt alloy (60-40)		E		570
Aluminum—copper alloy (50-50)		E		830
Aluminum—lithium alloy (15% Li)		E		400
Aluminum—magnesium alloy (Dowmetal)		E	CL	430
Aluminum—nickel alloy (58-42)		E		540
Aluminum—silicon alloy (12% Si)		E	NL	670
Amino-5-nitrothiazole	121-66-4	G		460
Anthranilic Acid	118-92-3	G	M	580
Apricot Pit		G		230
Aryl-nitrosomethylamide		G	NL	490
Asphalt	8052-42-4	F		510
Aspirin [acetol (2)]	50-78-2	G	M	660
Azelaic Acid	109-31-9	G	M	610
Azo-bis-butyronitrile	78-67-1	G		350
Benzethonium Chloride		G	CL	380
Benzoic Acid	65-85-0	G	M	440
Benzotriazole	95-14-7	G	M	440
Beta-naphthalene-axo-dimethylaniline		G		175
Bis(2-hydroxy-5-chlorophenyl) Methane	97-23-4	G	NL	570
Bisphenol-A	80-05-7	G	M	570
Boron, Commercial Amorphous (85% B)	7440-42-8	E		400
Calcium Silicide		E		540
Carbon Black (More Than 8% Total Entrapped Volatiles)		F		
Carboxymethyl Cellulose	9000-11-7	G		290
Carboxypolymethylene		G	NL	520
Cashew Oil, Phenolic, Hard		G		180
Cellulose		G		260
Cellulose Acetate		G		340
Cellulose Acetate Butyrate		G	NL	370
Cellulose Triacetate		G	NL	430
Charcoal (Activated)	64365-11-3	F		180
Charcoal (More Than 8% Total Entrapped Volatiles)		F		
Cherry Pit		G		220
Chlorinated Phenol		G	NL	570

Table 4.5.2 Continued

Chemical Name	CAS No.	*NEC* Group	Code	Layer or Cloud Ignition Temp. (°C)
Chlorinated Polyether Alcohol		G		460
Chloroacetoacetanilide	101-92-8	G	M	640
Chromium (97%) Electrolytic, Milled	7440-47-3	E		400
Cinnamon		G		230
Citrus Peel		G		270
Coal, Kentucky Bituminous		F		180
Coal, Pittsburgh Experimental		F		170
Coal, Wyoming		F		
Cocoa Bean Shell		G		370
Cocoa, Natural, 19% Fat		G		240
Coconut Shell		G		220
Coke (More Than 8% Total Entrapped Volatiles)		F		
Cork		G		210
Corn		G		250
Corn Dextrine		G		370
Corncob Grit		G		240
Cornstarch, Commercial		G		330
Cornstarch, Modified		G		200
Cottonseed Meal		G		200
Coumarone-Indene, Hard		G	NL	520
Crag No. 974	533-74-4	G	CL	310
Cube Root, South America	83-79-4	G		230
Di-alphacumyl Peroxide, 40-60 on CA	80-43-3	G		180
Diallyl Phthalate	131-17-9	G	M	480
Dicyclopentadiene Dioxide		G	NL	420
Dieldrin (20%)	60-57-1	G	NL	550
Dihydroacetic Acid		G	NL	430
Dimethyl Isophthalate	1459-93-4	G	M	580
Dimethyl Terephthalate	120-61-6	G	M	570
Dinitro-o-toluamide	148-01-6	G	NL	500
Dinitrobenzoic Acid		G	NL	460
Diphenyl	92-52-4	G	M	630
Ditertiary-butyl-paracresol	128-37-0	G	NL	420
Dithane m-45	8018-01-7	G		180
Epoxy		G	NL	540
Epoxy-bisphenol A		G	NL	510
Ethyl Cellulose		G	CL	320
Ethyl Hydroxyethyl Cellulose		G	NL	390
Ethylene Oxide Polymer		G	NL	350
Ethylene-maleic Anhydride Copolymer		G	NL	540
Ferbam™	14484-64-1	G		150
Ferromanganese, Medium Carbon	12604-53-4	E		290
Ferrosilicon (88% Si, 9% Fe)	8049-17-0	E		800
Ferrotitanium (19% Ti, 74.1% Fe, 0.06% C)		E	CL	380
Flax Shive		G		230
Fumaric Acid	110-17-8	G	M	520
Garlic, Dehydrated		G	NL	360
Gilsonite	12002-43-6	F		500
Green Base Harmon Dye		G		175
Guar Seed		G	NL	500

(continues)

Table 4.5.2 Continued

Chemical Name	CAS No.	*NEC* Group	Code	Layer or Cloud Ignition Temp. (°C)
Gulasonic Acid, Diacetone		G	NL	420
Gum, Arabic		G		260
Gum, Karaya		G		240
Gum, Manila		G	CL	360
Gum, Tragacanth	9000-65-1	G		260
Hemp Hurd		G		220
Hexamethylene Tetramine	100-97-0	G	S	410
Hydroxyethyl Cellulose		G	NL	410
Iron, 98% H_2 Reduced		E		290
Iron, 99% Carbonyl	13463-40-6	E		310
Isotoic Anhydride		G	NL	700
L-sorbose		G	M	370
Lignin, Hydrolized, Wood-type, Fine		G	NL	450
Lignite, California		F		180
Lycopodium		G		190
Malt Barley		G		250
Manganese	7439-96-5	E		240
Magnesium, Grade B, Milled		E		430
Manganese Vancide		G		120
Mannitol	69-65-8	G	M	460
Methacrylic Acid Polymer		G		290
Methionine (l-methionine)	63-68-3	G		360
Methyl Cellulose		G		340
Methyl Methacrylate Polymer	9011-14-7	G	NL	440
Methyl Methacrylate-ethyl Acrylate		G	NL	440
Methyl Methacrylate-styrene-butadiene		G	NL	480
Milk, Skimmed		G		200
N,N-Dimethylthio-formamide		G		230
Nitropyridone	100703-82-0	G	M	430
Nitrosamine		G	NL	270
Nylon Polymer	63428-84-2	G		430
Para-oxy-benzaldehyde	123-08-0	G	CL	380
Paraphenylene Diamine	106-50-3	G	M	620
Paratertiary Butyl Benzoic Acid	98-73-7	G	M	560
Pea Flour		G		260
Peach Pit Shell		G		210
Peanut Hull		G		210
Peat, Sphagnum	94114-14-4	G		240
Pecan Nut Shell	8002-03-7	G		210
Pectin	5328-37-0	G		200
Pentaerythritol	115-77-5	G	M	400
Petrin Acrylate Monomer	7659-34-9	G	NL	220
Petroleum Coke (More Than 8% Total Entrapped Volatiles)		F		
Petroleum Resin	64742-16-1	G		500
Phenol Formaldehyde	9003-35-4	G	NL	580
Phenol Formaldehyde, Polyalkylene-p	9003-35-4	G		290
Phenol Furfural	26338-61-4	G		310
Phenylbetanaphthylamine	135-88-6	G	NL	680
Phthalic Anydride	85-44-9	G	M	650
Phthalimide	85-41-6	G	M	630

Table 4.5.2 Continued

Chemical Name	CAS No.	*NEC* Group	Code	Layer or Cloud Ignition Temp. (°C)
Pitch, Coal Tar	65996-93-2	F	NL	710
Pitch, Petroleum	68187-58-6	F	NL	630
Polycarbonate		G	NL	710
Polyethylene, High Pressure Process	9002-88-4	G		380
Polyethylene, Low Pressure Process	9002-88-4	G	NL	420
Polyethylene Terephthalate	25038-59-9	G	NL	500
Polyethylene Wax	68441-04-8	G	NL	400
Polypropylene (no antioxidant)	9003-07-0	G	NL	420
Polystyrene Latex	9003-53-6	G		500
Polystyrene Molding Compound	9003-53-6	G	NL	560
Polyurethane Foam, Fire Retardant	9009-54-5	G		390
Polyurethane Foam, No Fire Retardant	9009-54-5	G		440
Polyvinyl Acetate	9003-20-7	G	NL	550
Polyvinyl Acetate/Alcohol	9002-89-5	G		440
Polyvinyl Butyral	63148-65-2	G		390
Polyvinyl Chloride-dioctyl Phthalate		G	NL	320
Potato Starch, Dextrinated	9005-25-8	G	NL	440
Pyrethrum	8003-34-7	G		210
Rayon (Viscose) Flock	61788-77-0	G		250
Red Dye Intermediate		G		175
Rice		G		220
Rice Bran		G	NL	490
Rice Hull		G		220
Rosin, DK	8050-09-7	G	NL	390
Rubber, Crude, Hard	9006-04-6	G	NL	350
Rubber, Synthetic, Hard (33% S)	64706-29-2	G	NL	320
Safflower Meal		G		210
Salicylanilide	87-17-2	G	M	610
Sevin	63-25-2	G		140
Shale, Oil	68308-34-9	F		
Shellac	9000-59-3	G	NL	400
Sodium Resinate	61790-51-0	G		220
Sorbic Acid (Copper Sorbate or Potash)	110-44-1	G		460
Soy Flour	68513-95-1	G		190
Soy Protein	9010-10-0	G		260
Stearic Acid, Aluminum Salt	637-12-7	G		300
Stearic Acid, Zinc Salt	557-05-1	G	M	510
Styrene Modified Polyester-Glass Fiber	100-42-5	G		360
Styrene-acrylonitrile (70-30)	9003-54-7	G	NL	500
Styrene-butadiene Latex (>75% styrene)	903-55-8	G	NL	440
Styrene-maleic Anhydride Copolymer	9011-13-6	G	CL	470
Sucrose	57-50-1	G	CL	350
Sugar, Powdered	57-50-1	G	CL	370
Sulfur	7704-34-9	G		220
Tantalum	7440-25-7	E		300
Terephthalic Acid	100-21-0	G	NL	680
Thorium, 1.2% O_2	7440-29-1	E	CL	280
Tin, 96%, Atomized (2% Pb)	7440-31-5	E		430
Titanium, 99% Ti	7440-32-6	E	CL	330

(continues)

Table 4.5.2 Continued

Chemical Name	CAS No.	*NEC* Group	Code	Layer or Cloud Ignition Temp. (°C)
Titanium Hydride (95% Ti, 3.8% H_2)	7704-98-5	E	CL	480
Trithiobisdimethylthio-formamide		G		230
Tung, Kernels, Oil-free	8001-20-5	G		240
Urea Formaldehyde Molding Compound	9011-05-6	G	NL	460
Urea Formaldehyde-phenol Formaldehyde	25104-55-6	G		240
Vanadium, 86.4%	7440-62-2	E		490
Vinyl Chloride-acrylonitrile Copolymer	9003-00-3	G		470
Vinyl Toluene-acrylonitrile Butadiene	76404-69-8	G	NL	530
Violet 200 Dye		G		175
Vitamin B1, Mononitrate	59-43-8	G	NL	360
Vitamin C	50-81-7	G		280
Walnut Shell, Black		G		220
Wheat		G		220
Wheat Flour	130498-22-5	G		360
Wheat Gluten, Gum	100684-25-1	G	NL	520
Wheat Starch		G	NL	380
Wheat Straw		G		220
Wood Flour		G		260
Woodbark, Ground		G		250
Yeast, Torula	68602-94-8	G		260
Zirconium Hydride	7704-99-6	E		270
Zirconium		E	CL	330

Notes:
1. Normally, the minimum ignition temperature of a layer of a specific dust is lower than the minimum ignition temperature of a cloud of that dust. Since this is not universally true, the lower of the two minimum ignition temperatures is listed. If no symbol appears between the two temperature columns, then the layer ignition temperature is shown. "CL" means the cloud ignition temperature is shown. "NL" means that no layer ignition temperature is available, and the cloud ignition temperature is shown. "M" signifies that the dust layer melts before it ignites; the cloud ignition temperature is shown. "S" signifies that the dust layer sublimes before it ignites; the cloud ignition temperature is shown.
2. Certain metal dusts may have characteristics that require safeguards beyond those required for atmospheres containing the dusts of aluminum, magnesium, and their commercial alloys. For example, zirconium, thorium, and uranium dusts have extremely low ignition temperatures [as low as 20°C (68°F)] and minimum ignition energies lower than any material classified in any of the Class I or Class II groups.

4.6 Ignition of Dust Clouds.

4.6.1 The electrical equipment enclosure prevents the dust cloud from being ignited by arcing and sparking parts, or other ignition sources within the enclosure.

4.6.2 The dust cloud could be ignited by hot surface temperatures.

4.6.3 Some dusts that are not normally combustible could form explosive dust clouds when mixed with a flammable gas. See 4.2.2 for discussion on hybrid mixture.

4.7 Ignition of Dust Layers.

4.7.1 The ignition temperature of a dust layer is a function of the type of dust and its physical and chemical properties and is often less than the cloud ignition temperature. The ignition temperature shown in Table 4.5.2 is the lower of the two.

4.7.2 The ignition temperature of a layer of organic dust on heat-producing equipment can decrease over time if the dust dehydrates or carbonizes. For this reason, the heat-producing equipment is not permitted to exceed the lower of either the ignition temperature or 165°C (329°F).

4.7.3 Some dusts in layers could melt before reaching their layer ignition temperatures. This melted material could then act more like a combustible liquid than a dust.

4.7.4 Other dusts, such as some polymers, degrade to a lower molecular weight material or to the monomer itself.

Table 4.5.3 Cross-Reference of Chemical CAS Number to Chemical Name

CAS No.	Chemical Name	CAS No.	Chemical Name
50-78-2	Aspirin [Acetol (2)]	7440-29-1	Thorium, 1.2% O_2
50-81-7	Vitamin C	7440-31-5	Tin, 96, Atomized (2% Pb)
57-50-1	Sucrose	7440-32-6	Titanium, 99% Ti
57-50-1	Sugar, Powdered	7440-42-8	Boron, Commercial Amorphous (85% B)
59-43-8	Vitamin B1, Mononitrate	7440-47-3	Chromium (97%) Electrolytic, Milled
60-57-1	Dieldrin (20%)	7440-62-2	Vanadium, 86.4%
63-25-2	Sevin	7659-34-9	Petrin Acrylate Monomer
63-68-3	Methionine (l-methionine)	7704-34-9	Sulfur
65-85-0	Benzoic Acid	7704-98-5	Titanium Hydride (95% Ti, 3.8% H_2)
69-65-8	Mannitol	7704-99-6	Zirconium Hydride
78-67-1	Azo-bis-butyronitrile	8001-20-5	Tung, Kernels, Oil-free
80-05-7	Bisphenol-A	8002-03-7	Pecan Nut Shell
80-43-3	Di-alphacumyl Peroxide, 40-60 on CA	8003-34-7	Pyrethrum
83-79-4	Cube Root, South America	8018-01-7	Dithane M-45
85-41-6	Phthalimide	8049-17-0	Ferrosilicon (88% Si, 9% Fe)
85-44-9	Phthalic Anydride	8050-09-7	Rosin, DK
87-17-2	Salicylanilide	8052-42-4	Asphalt
92-52-4	Diphenyl	9000-11-7	Carboxymethyl Cellulose
95-14-7	Benzotriazole	9000-59-3	Shellac
97-23-4	Bis(2-hydroxy-5-chlorophenyl) Methane	9000-65-1	Gum, Tragacanth
98-73-7	Paratertiary Butyl Benzoic Acid	9002-88-4	Polyethylene, High Pressure Process
100-21-0	Terephthalic Acid	9002-88-4	Polyethylene, Low Pressure Process
100-42-5	Styrene Modified Polyester-Glass Fiber	9002-89-5	Polyvinyl Acetate/Alcohol
100-97-0	Hexamethylene Tetramine	9003-00-3	Vinyl Chloride-acrylonitrile Copolymer
101-92-8	Chloroacetoacetanilide	9003-07-0	Polypropylene (No Antioxidant)
102-01-2	Acetoacetanilide	9003-20-7	Polyvinyl Acetate
106-50-3	Paraphenylene Diamine	9003-35-4	Phenol Formaldehyde
109-31-9	Azelaic Acid	9003-35-4	Phenol Formaldehyde, Polyalkylene-p
110-17-8	Fumaric Acid	9003-53-6	Polystyrene Latex
110-44-1	Sorbic Acid (Copper Sorbate or Potash)	9003-53-6	Polystyrene Molding Compound
115-77-5	Pentaerythritol	9003-54-7	Styrene-acrylonitrile (70-30)
118-92-3	Anthranilic Acid	9005-25-8	Potato Starch, Dextrinated
120-61-6	Dimethyl Terephthalate	9006-04-6	Rubber, Crude, Hard
121-66-4	Amino-5-nitrothiazole	9009-54-5	Polyurethane Foam, Fire Retardant
122-82-7	Acetoacet-p-phenetidide	9009-54-5	Polyurethane Foam, No Fire Retardant
123-08-0	Para-oxy-benzaldehyde	9010-10-0	Soy Protein
124-04-9	Adipic Acid	9011-05-6	Urea Formaldehyde Molding Compound
128-37-0	Ditertiary-butyl-paracresol	9011-13-6	Styrene-maleic Anhydride Copolymer
131-17-9	Diallyl Phthalate	9011-14-7	Methyl Methacrylate Polymer
135-88-6	Phenylbetanaphthylamine	12002-43-6	Gilsonite
148-01-6	Dinitro-o-toluamide	12604-53-4	Ferromanganese, Medium Carbon
533-74-4	Crag No. 974	13463-40-6	Iron, 99% Carbonyl
557-05-1	Stearic Acid, Zinc Salt	14484-64-1	Ferbam™
637-12-7	Stearic Acid, Aluminum Salt	25038-59-9	Polyethylene Terephthalate
903-55-8	Styrene-butadiene Latex (>75% Styrene)	25104-55-6	Urea Formaldehyde-phenol Formaldehyde
1459-93-4	Dimethyl Isophthalate	26338-61-4	Phenol Furfural
5328-37-0	Pectin	61788-77-0	Rayon (Viscose) Flock
7429-90-5	Aluminum, A422 Flake	61790-51-0	Sodium Resinate
7439-96-5	Manganese	63148-65-2	Polyvinyl Butyral
7440-25-7	Tantalum	63428-84-2	Nylon Polymer

(continues)

Table 4.5.3 Continued

CAS No.	Chemical Name
64365-11-3	Charcoal (Activated)
64706-29-2	Rubber, Synthetic, Hard (33% S)
64742-16-1	Petroleum Resin
65996-93-2	Pitch, Coal Tar
68187-58-6	Pitch, Petroleum
68308-34-9	Shale, Oil
68441-04-8	Polyethylene Wax
68513-95-1	Soy Flour
68602-94-8	Yeast, Torula
76404-69-8	Vinyl Toluene-acrylonitrile Butadiene
94114-14-4	Peat, Sphagnum
100684-25-1	Wheat Gluten, Gum
100703-82-0	Nitropyridone
130498-22-5	Wheat Flour

This dust could act more like a flammable liquid than a dust.

4.7.5 Materials such as unplasticized polyvinyl chloride, sulfur, and zinc stearate melt, but cause only maintenance problems.

4.7.6 An ignited dust layer introduces an open-flame ignition source that can ignite a dust cloud in the vicinity and also can stir up the dust layer, creating a dust cloud.

Chapter 5 Classification of Class II (Combustible Dust) Locations

5.1 General.

5.1.1 The decision to classify an area as hazardous should be based on the probability that a combustible mixture could be present.

5.1.2 Once it is decided that an area should be classified, the next step should be to determine the degree of hazard: Is the area Division 1 or Division 2?

5.2 Division 1 Classified Locations.

5.2.1 Where a dust cloud is likely to be present under normal conditions, the location should be classified as Division 1.

5.2.2* Where a dust layer greater than ⅛ in. (3.0 mm) thick is present under normal conditions, the location should be classified as Division 1.

5.2.3 "Normal" does not necessarily mean the situation that prevails when everything is working properly.

5.2.3.1 For instance, if a bucket elevator requires frequent maintenance and repair, this repair should be viewed as normal.

5.2.3.2 If quantities of ignitible dust are released as a result of the maintenance, the location is Division 1.

5.2.3.3 However, if that elevator is replaced and now repairs are not usually required between turnarounds, the need for repairs is considered abnormal.

5.2.3.4 The classification of the location, therefore, is related to equipment maintenance, both procedures and frequencies.

5.2.3.5 Similarly, if the problem is the buildup of dust layers without the presence of visible dust suspensions, good and frequent cleaning procedures or the lack thereof will influence the classification of the location.

5.3 Division 2 Classified Locations.

5.3.1 The criterion for a Division 2 location is whether the location is likely to have ignitible dust suspensions or hazardous dust accumulations only under abnormal conditions. The term "abnormal" is used here in a limited sense and does not include a major catastrophe.

5.3.2 As an example, consider the replaced bucket elevator of 5.2.3.2, which releases ignitible dust only under abnormal conditions. In this case there is no Division 1 location because the elevator is normally tight. To release dust, the elevator would have to leak, and that would not be normal.

5.3.3 Chemical process equipment does not fail often. Furthermore, the electrical installation requirement of the *NEC* for Division 2 locations is such that an ignition-capable spark or hot surface will occur only in the event of abnormal operation or failure of electrical equipment. Otherwise, sparks and hot surfaces are not present or are contained in enclosures. On a realistic basis, the possibility of process equipment and electrical equipment failing simultaneously is remote.

5.3.4 The Division 2 classification is applicable to conditions not involving equipment failure. For example, consider a location classified as Division 1 because of normal presence of ignitible dust suspension. Obviously, one side of the Division 1 boundary cannot be normally hazardous and the opposite side never hazardous. Similarly, consider a location classified as Division 1 because of the normal presence of hazardous dust accumulations. One side of the division boundary cannot be normally hazardous, with thick layers of dust, and the other side unclassified, with no dust, unless there is an intervening wall.

5.3.5 When there is no wall, a surrounding transition Division 2 location separates a Division 1 location from an unclassified location.

5.3.6 Walls are much more important in separating Division 1 locations from Division 2 and unclassified locations in Class II locations than in Class I locations.

5.3.6.1 Only unpierced solid walls make satisfactory barriers in Class I locations, whereas closed doors, lightweight partitions, or even partial partitions could make satisfactory walls between Class II, Division 1 locations and unclassified locations.

5.3.6.2 Area classification does not extend beyond the wall, provided it is effective in preventing the passage of dust in suspension or layer form.

5.4 Unclassified Locations.

5.4.1 Experience has shown that the release of ignitible dust suspensions from some operations and apparatus is so infrequent that area classification is not necessary. For example, where combustible dusts are processed, stored, or handled, it is usually not necessary to classify the following locations:

(1) Where materials are stored in sealed containers (e.g., bags, drums, or fiber packs on pallets or racks)

(2) Where materials are transported in well-maintained closed piping systems

(3) Where palletized materials with minimal dust are handled or used

(4) Where closed tanks are used for storage and handling

(5) Where dust removal systems prevent the following:
  (a) Visual dust clouds
  (b) Layer accumulations that make surface colors indiscernible *(see A.5.2.2)*

(6) Where excellent housekeeping prevents the following:
  (a) Visual dust clouds
  (b) Layer accumulations that make surface colors indiscernible *(see A.5.2.2)*

5.4.2 Dust removal systems that are provided to allow an unclassified location should have adequate safeguards and warnings against failure.

5.4.3 Open flames and hot surfaces associated with the operation of certain equipment, such as boilers and fired heaters, provide inherent thermal ignition sources.

5.4.3.1 Area classification is not appropriate in the immediate vicinity of inherent thermal ignition sources.

5.4.3.2 Dust-containing operations should be cut off by blank walls or located away from inherent thermal ignition sources.

5.4.3.3 Where pulverized coal or ground-up solid waste is used to fire a boiler or incinerator, it is prudent to avoid installing electrical equipment that could become primary ignition sources for leaks in the fuel feed lines.

5.5 Extent of Hazardous (Classified) Locations.

5.5.1 Careful consideration of the following factors is necessary in determining the extent of the locations:

(1) Combustible material involved
(2) Bulk density of the material
(3) Particle sizes of the material
(4) Particle density
(5) Process or storage pressure
(6) Size of the leak opening
(7) Quantity of the release
(8) Dust removal system
(9) Housekeeping
(10) Presence of any hybrid mixture

5.5.2 The dispersal of dusts and the influence of the factors in 5.5.1 on this dispersal are discussed generally in Sections 4.2 through 4.7. The importance of dust removal and housekeeping are discussed in other paragraphs of this chapter.

5.5.3 In addition, walls, partitions, enclosures, or other barriers and strong air currents will also affect the distance that dust particles will travel and the extent of the Division 1 and Division 2 locations.

5.5.4 Where there are walls that limit the travel of the dust particles, area classifications do not extend beyond the walls. Providing walls and partitions is a primary means of limiting the extent of hazardous (classified) locations.

5.5.5 Where effective walls are not provided, the extent of the Division 1 and Division 2 locations can be estimated as follows:

(1) By visual observation of the existing location using the guidelines of A.5.2.2

(2) By experience with similar dusts and similar operations, and by taking into consideration differences in equipment, enclosures, dust-removal systems, and housekeeping rules and methods
(3) By using the classification diagrams in this chapter

5.5.6 Tight equipment, ventilated hoods and pickup points, good maintenance, and good housekeeping practices should limit Division 1 locations to those inside of process enclosures and equipment and close to openings necessary for transfer of material, as from conveyors to grinders to storage bins to bags. Similarly, the same factors will also limit the Division 2 location surrounding the Division 1 location.

5.5.7 The size of a building and its walls will influence the classification of the enclosed volume. In the case of a small room, it can be appropriate to classify the entire volume as Division 1 or Division 2.

5.5.8 When classifying large buildings, careful evaluation of prior experience with the same or similar installations should be made. Where experience indicates that a particular design concept is sound, continue to follow it. Sound engineering judgment and good housekeeping should be used to minimize the extent of hazardous (classified) locations.

5.5.8.1 Wherever possible with large buildings, walls should be used to cut off dusty operations to minimize the hazardous (classified) location. Where walls are not possible, use the concentric volume approach of a Division 1 location surrounded by a larger Division 2 location, as shown in the diagrams.

5.5.8.2 Where it is necessary to have a number of dusty operations located in a building, there could be a multiplicity of Division 1 locations, with intervening Division 2 and unclassified locations.

5.5.9 The quantity of dust released and its distance of travel is of extreme importance in determining the extent of a hazardous (classified) location. This determination requires sound engineering judgment. However, one cannot lose sight of the purpose of this judgment; the location is classified solely for the installation of electrical equipment.

5.6 Discussion of Diagrams and Recommendations.

5.6.1 The series of diagrams in Section 5.8 illustrate how typical dusty areas should be classified and the recommended extent of classification.

5.6.2 The diagrams should be used as aids in developing electrical classification maps of operating units, storage areas, and process buildings. Most of the maps will be plan views. However, elevations could be necessary to provide the three-dimensional picture of an actual operation.

5.6.3 An operating unit could have many interconnected sources of combustible material, such as storage tanks, bins and silos, piping and ductwork, hammer mills, ball mills, grinders, pulverizers, milling machines, conveyors, bucket elevators, and bagging or other packaging machines. These in turn present sources of leaks, such as flanged and screwed connections, fittings, openings, valves, and metering and weighing devices. Thus, actual diagrams of the equipment could be required so that the necessary engineering judgment to establish the boundaries of Division 1 and Division 2 locations can be applied.

5.6.4 These diagrams apply to operating equipment processing dusts when the specific particle density is greater than 40 lb/ft^3 (640.72 kg/m^3). When dusts with a specific particle density less than 40 lb/ft^3 (640.72 kg/m^3) are being handled, there is a pronounced tendency for the fine dust to drift on air currents normally present in industrial plants for distances considerably farther than those shown on these diagrams. In those cases it will be necessary to extend the hazardous (classified) location using sound engineering judgment and experience.

5.6.5 Good engineering practices, good housekeeping practices, and effective dust removal systems are necessary to limit the extent of the classified areas and to minimize the chances of primary explosions and secondary explosions, which are often more violent.

5.7 Procedure for Classifying Areas.

Paragraphs 5.7.1 through 5.7.4 detail the procedure that should be used for each room, section, or area being classified.

5.7.1 Step One — Need for Classification. The area should be classified if a combustible material is processed, handled, or stored there.

5.7.2 Step Two — Gathering Information.

5.7.2.1 Proposed Facility Information. For a proposed facility that exists on drawings only, a preliminary area classification can be done so that suitable electrical equipment and instrumentation can be purchased. Plants are rarely built exactly as the drawings portray, and the area classification should be modified later based on the actual facility.

5.7.2.2 Existing Facility History. For an existing facility, the individual plant experience is extremely important in classifying areas within the plant. Both operation and maintenance personnel in the actual plant should be asked the following questions:

(1) Is a dust likely to be in suspension in air continuously, periodically, or intermittently under normal conditions in quantities sufficient to produce an ignitible mixture?
(2) Are there dust layers or accumulations on surfaces deeper than ⅛ in. (3.0 mm)?
(3) Are there dust layers or accumulations on surfaces that make the colors of floor or equipment surfaces indiscernible?
(4) What is the dust accumulation after 24 hours?
(5) Is the equipment in good condition, in questionable condition, or in need of repair? Are equipment enclosures in good repair, and do they prevent the entrance of dust?
(6) Do maintenance practices result in the formation of ignitible mixtures?
(7) What equipment is used for dust collection?

5.7.2.3 Material Density. Determine if the specific particle density of the dust is at least 40 lb/ft^3 (640.72 kg/m^3).

5.7.2.4 Plot Plan. A plot plan (or similar drawing) is needed that shows all vessels, tanks, building structures, partitions, and similar items that would affect dispersion or promote accumulation of the dust.

5.7.2.5 Fire Hazard Properties of Combustible Material. The *NEC* group and the layer or cloud ignition temperature are shown in Table 4.5.2 for many materials.

5.7.2.5.1 A material could be listed in Table 4.5.2 under a chemical name different than the chemical name used at the facility. Table 4.5.3 is provided to cross-reference the CAS number of the material to the chemical name used in Table 4.5.2.

5.7.2.5.2 Where materials being used are not listed in Table 4.5.2 or in other reputable chemical references, the information needed to classify the area can be obtained by one of the following methods:

(1) Contact the material supplier to determine if the material has been group classified.
(2) Have the material tested to determine if the ignition sensitivity is less than 0.2 and the explosion severity is less than 0.5. Area classification is not considered necessary for dusts that meet both criteria.

5.7.3 Step Three — Selecting the Appropriate Classification Diagram. Select the appropriate diagrams based on the following:

(1) Whether the process equipment is open or enclosed
(2) Whether the dust is Group E, F, or G
(3) Whether the area is for storage

5.7.4 Step Four — Determining the Extent of the Hazardous (Classified) Location. The extent of the hazardous (classified) location can be determined using sound engineering judgment to apply the methods discussed in Section 5.5 and the diagrams contained in this chapter.

5.7.4.1 Locate the potential sources of leaks on the plan drawing or at the actual location. These sources of leaks could include rotating or reciprocating shafts, doors and covers on process equipment, and so forth.

5.7.4.2 For each leakage source, find an equivalent example on the selected classification diagram to determine the minimum extent of classification around the leakage source. The extent can be modified by considering the following:

(1) Whether an ignitible mixture is likely to occur frequently due to repair, maintenance, or leakage
(2) Where conditions of maintenance and supervision are such that leaks are likely to occur in process equipment, storage vessels, and piping systems containing combustible material
(3) Ventilation or prevailing wind in the specific area and the dispersion rates of the combustible materials

5.7.4.3 Once the minimum extent is determined, for practical reasons utilize distinct landmarks (e.g., curbs, dikes, walls, structural supports, edges of roads) for the actual boundaries of the area classification. These landmarks permit easy identification of the boundaries of the hazardous (classified) locations for electricians, instrument technicians, operators, and other personnel.

5.8 Classification Diagrams.

Classification diagrams are shown in Figure 5.8(a) through Figure 5.8(i). These diagrams assume that the specific particle density is greater than 40 lb/ft^3 (640.72 kg/m^3).

Annex A Explanatory Material

Annex A is not a part of the recommendations of this NFPA document but is included for informational purposes only. This annex contains explanatory material, numbered to correspond with the applicable text paragraphs.

A.3.3.1 Autoignition Temperature (AIT). See NFPA *Fire Protection Guide to Hazardous Materials*, 2001 edition.

A.3.3.3 Combustible Dust. Prior to the 1981 edition of the *National Electrical Code* (*NEC*) (1978 and prior editions), all Group E (metal dusts such as aluminum, magnesium,

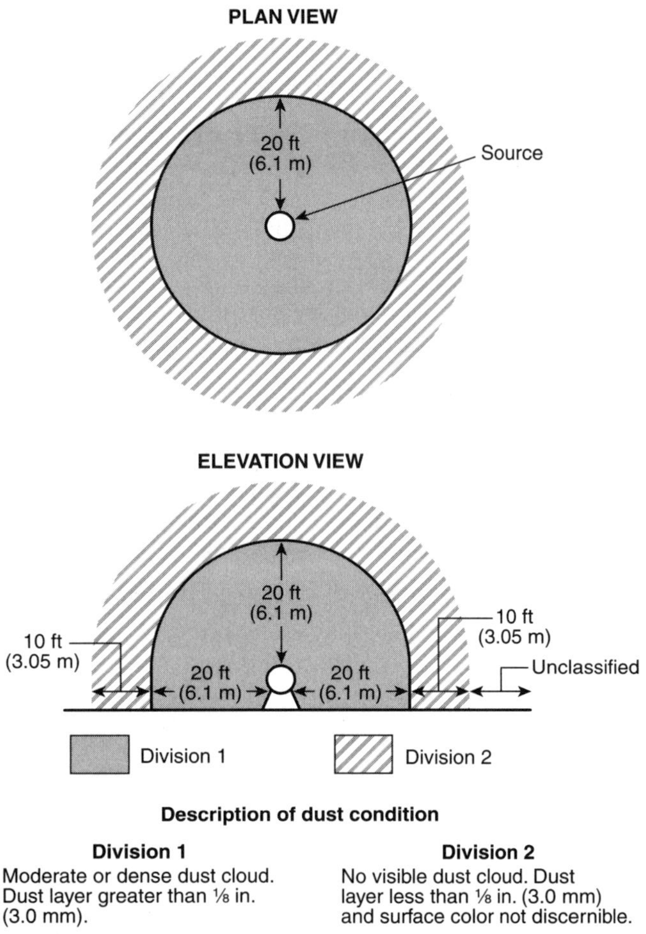

FIGURE 5.8(a) Group F or Group G Dust — Indoor, Unrestricted Area; Open or Semi-Enclosed Operating Equipment.

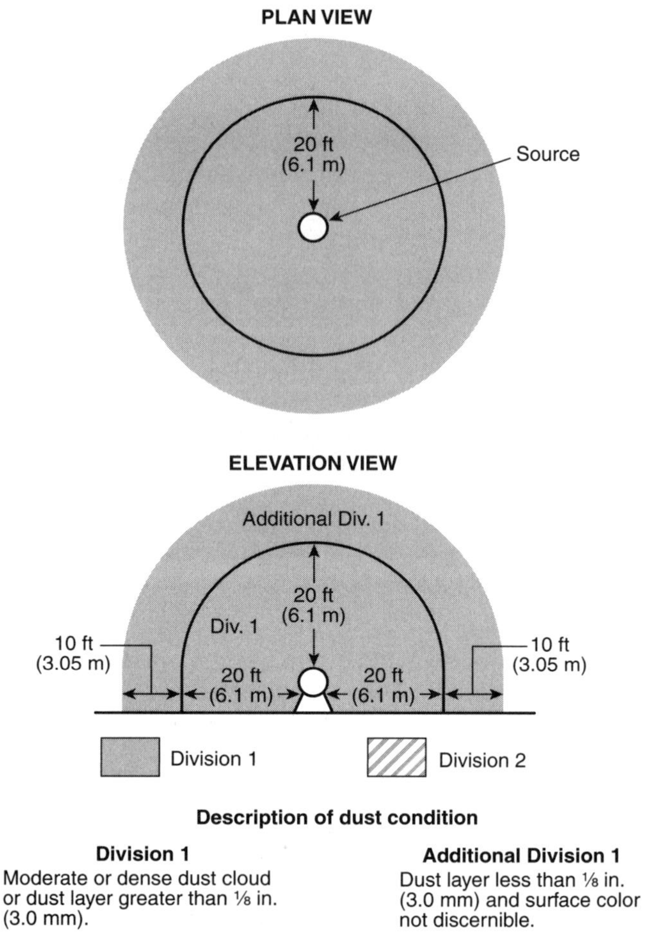

FIGURE 5.8(b) Group E Dust — Indoor, Unrestricted Area; Open or Semi-Enclosed Operating Equipment.

and their commercial alloys) and Group F (carbonaceous dusts such as carbon black, charcoal, or coke dusts having more than 8 percent total volatile materials) were considered to be electrically conductive. As a result, areas containing Group E or Group F dusts were all classified Division 1, as required by the definition of a Class II, Division 1 location. It was only possible to have a Division 2 location for Group G dusts.

The 1984 edition of the *NEC* eliminated Group F altogether. Carbonaceous dusts with resistivity of less than 10^5 ohm/cm were considered conductive and were classified as Group E. Carbonaceous dusts with resistivity of 10^5 ohm/cm or greater were considered nonconductive and were classified as Group G. This reclassification allowed the use of Group G, Division 2 electrical equipment for many carbonaceous materials.

The 1987 edition of the *NEC* reinstated Group F because the close tolerances in Group E motors necessary for metal dusts are unnecessary for conductive carbonaceous dusts, and the low temperature specifications in Group G equipment necessary for grain, flour, and some chemical dusts are unnecessary for nonconductive carbonaceous dusts. This imposed an unwarranted expense on users.

This change allowed the use of Group F, Division 2 electrical equipment for carbonaceous dust with a resistivity greater than 10^5 ohm/cm.

The problem with this work was that the resistivity value, a number that related to the dust's ability to conduct an electric current, was not a constant and varied considerably based on dust particle size and extent of oxidation, the moisture content, voltage applied, temperature, and test apparatus and technique. No standardized test method for the resistivity value considering long-term environmental effects has been developed. Finally, the resistivity value is not directly related to the explosion hazard.

A.3.3.6 Explosion Severity. See A.3.3.9, Ignition Sensitivity.

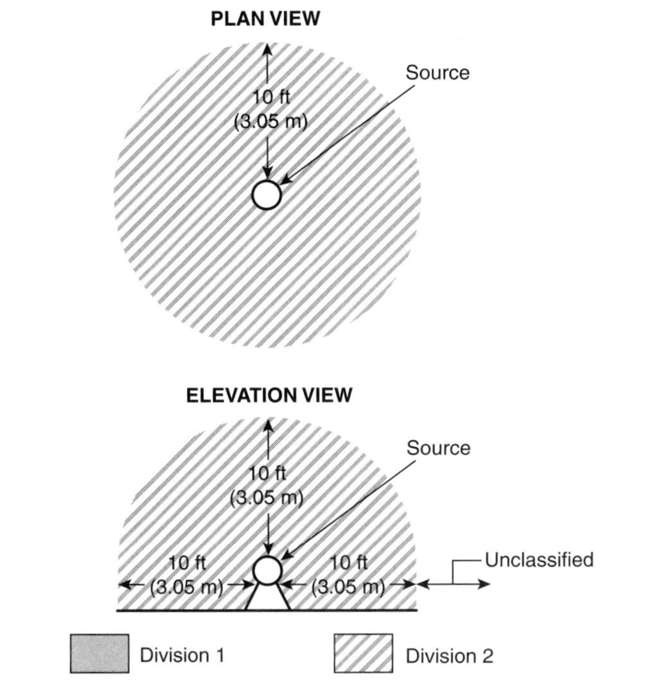

Description of dust condition

Division 1	Division 2
None	No visible dust cloud. Dust layer less than ⅛ in. (3.0 mm) and surface color not discernible.

FIGURE 5.8(c) Group F or Group G Dust — Indoor, Unrestricted Area; Operating Equipment Enclosed; Area Classified as a Class II, Division 2 Location.

A.3.3.9 Ignition Sensitivity. The U.S. Bureau of Mines has defined ignition sensitivity and explosion severity as follows:

$$\text{Explosion severity} = \frac{(P_{max} \times P)_2}{(P_{max} \times P)_1}$$

$$\text{Ignition sensitivity} = \frac{(T_c \times E \times M_c)_1}{(T_c \times E \times M_c)_2}$$

Where:

P_{max} = maximum explosive pressure
P = maximum rate of pressure rise
T_c = minimum ignition temperature
E = minimum ignition energy
M_c = minimum explosive concentration

Subscript 1 refers to the appropriate values for Pittsburgh seam coal, the standard dust used by the U.S. Bureau of Mines.

Subscript 2 refers to the values for the specific dust in question.

Note that units must be consistent in both numerators and denominators.

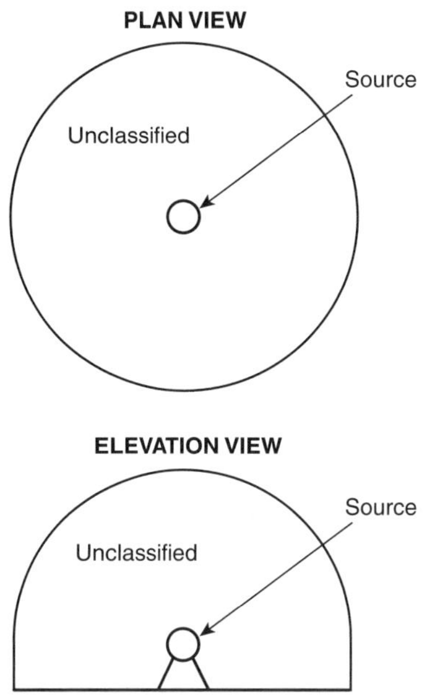

Description of dust condition

Division 1	Division 2	Unclassified
None	None	Surface color discernible.

FIGURE 5.8(d) Group F or Group G Dust — Indoor, Unrestricted Area; Operating Equipment Enclosed; Area Is an Unclassified Location.

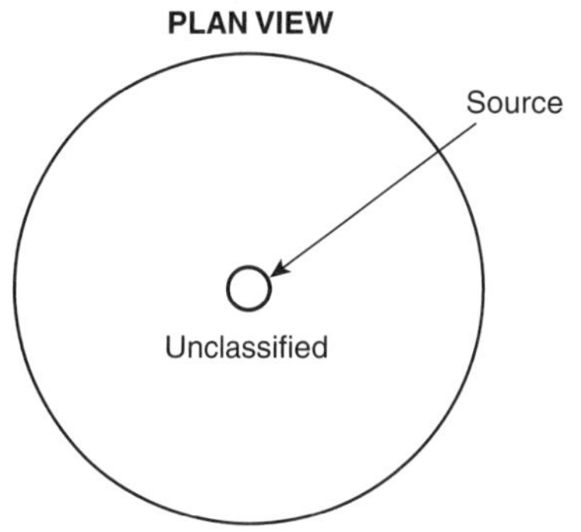

Description of dust condition

Division 1	Division 2	Unclassified
None	None	Dust layer not apparent. Surface color discernible.

FIGURE 5.8(e) Groups E, F, or G Dusts — Storage Area Bags, Drums, or Closed Hoppers.

A.4.1.2 Article 500 also defines two other hazardous (classified) locations: Class I and Class III. In a Class I hazardous (classified) location, the combustible material present is a flammable gas or vapor. In a Class III hazardous (classified) location, the combustible material present is an ignitible fiber or flying. This recommended practice covers Class II hazardous (classified) locations.

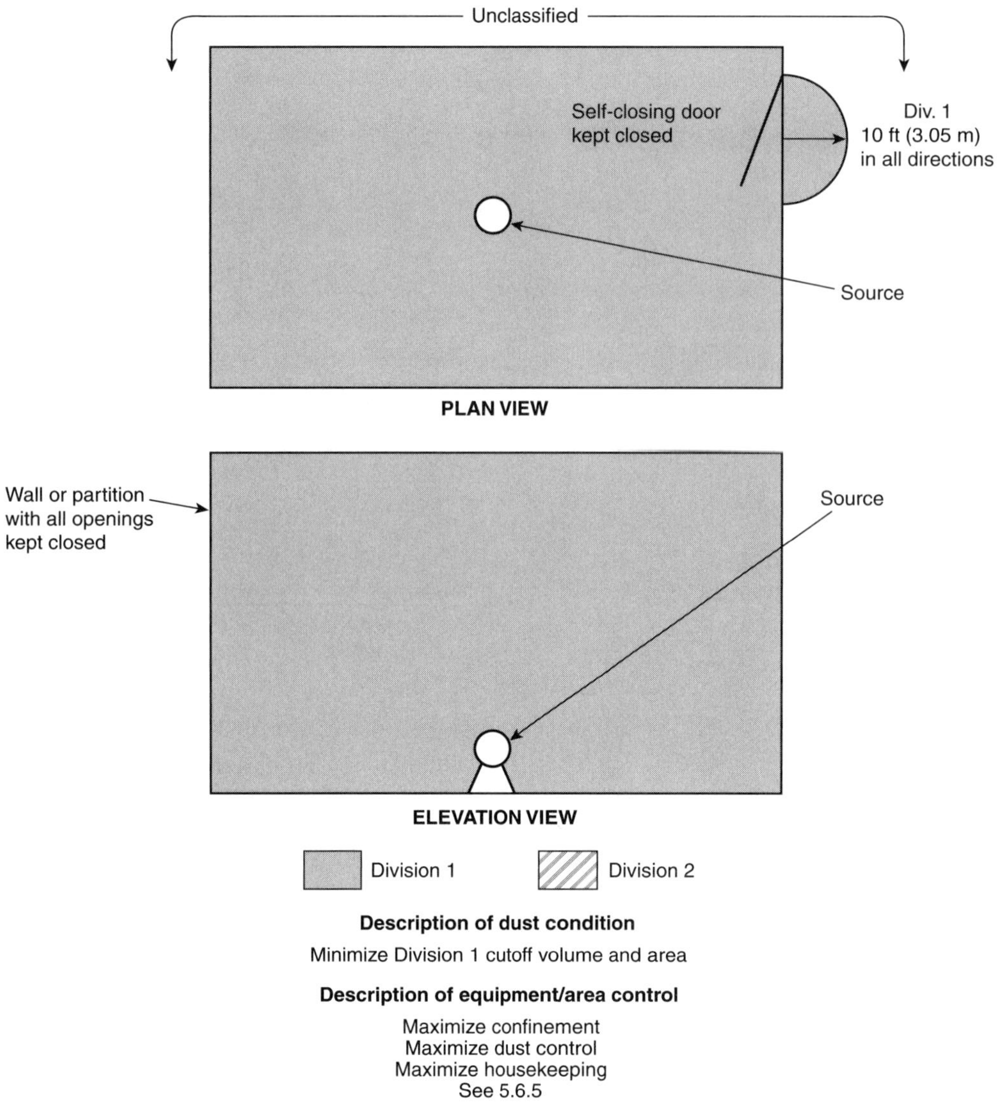

FIGURE 5.8(f) Group E Dust — Indoor, Walled-Off Area; Operating Equipment Enclosed.

A.4.2.2 The presence of flammable gas in a combustible dust cloud drastically reduces the ignition energy. The flammable gas need not be present in amounts sufficient to reach the lower flammable limit (considering the gas phase alone) to exhibit this phenomenon.

A.4.3 Open flames, and welding and cutting operations have far more energy and heat than most electrical fault sparks and arcs and are quite capable of igniting dusts. Hot surfaces such as those in some heaters, or those caused by continuous friction, can also have sufficient heat to ignite dusts. Such sources of ignition therefore should be carefully controlled.

A.4.3.1.2 When subjected to heat, dusts of thermosetting plastics, such as phenol formaldehyde resins, tend to polymerize (''set up'') and become hard. Continued heat buildup in the polymerized material ultimately leads to carbonization (degradation) of the material and a significantly lower ignition temperature. Although this phenomenon is well known, there is no standardized test to define the precise parameters. Nonplastic materials such as sugar, cornstarch, and dextrine also carbonize and ignite at lower-than-expected temperatures.

A.4.5.2 The materials, and their group classifications, listed in Table 4.5.2, were taken from NMAB 353-3, *Classification of Combustible Dusts in Accordance with the National Electrical Code*, published by the National Academy of Sciences. Dusts having ignition sensitivities equal to or greater than 0.2, or explosion severities equal to or greater than 0.5, are

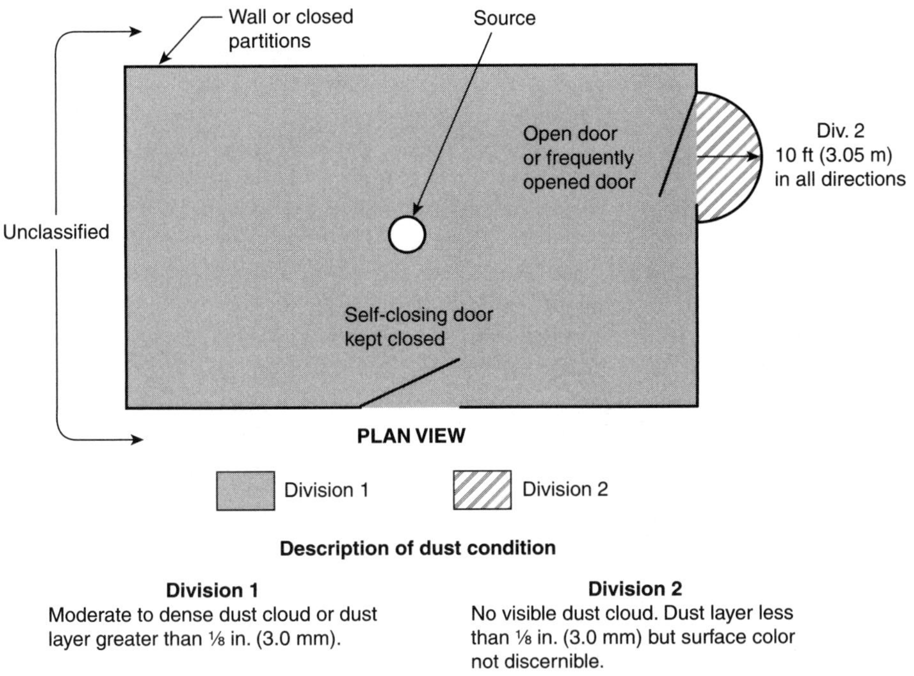

FIGURE 5.8(g) Group F or G — Indoor, Walled-Off Area; Operating Equipment Open or Semi-Enclosed.

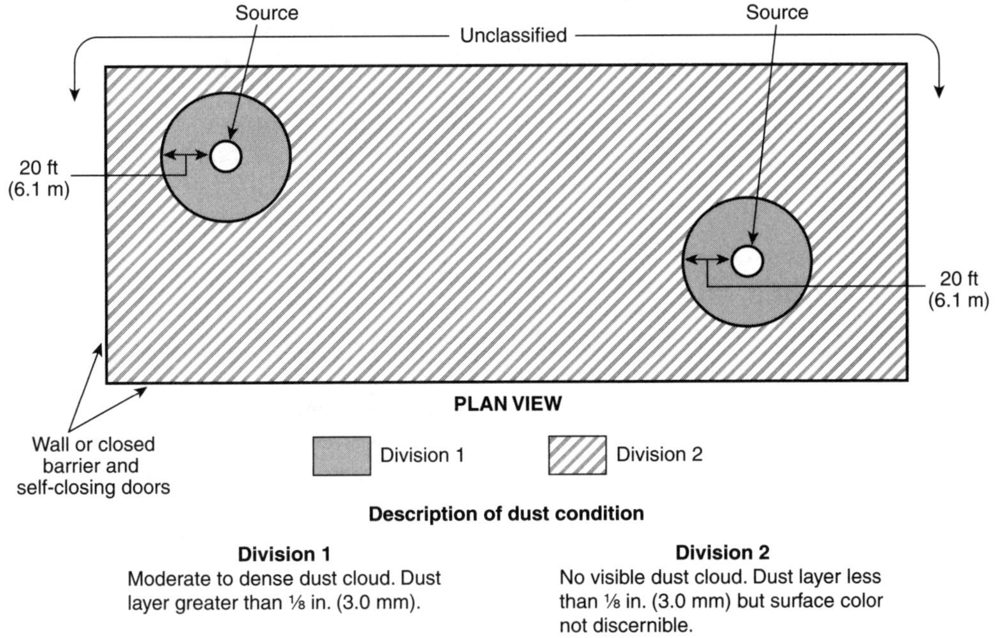

FIGURE 5.8(h) Group F or Group G — Indoor, Walled-Off Area; Multiple Pieces of Operating Equipment.

listed. Dusts with explosibility parameters that fall below these limits are generally not considered to be significant explosion hazards and, therefore, are not included in this table. Selection of electrical equipment for dusts that sublime or melt below the operating temperature of the equipment requires additional consideration of the properties of the

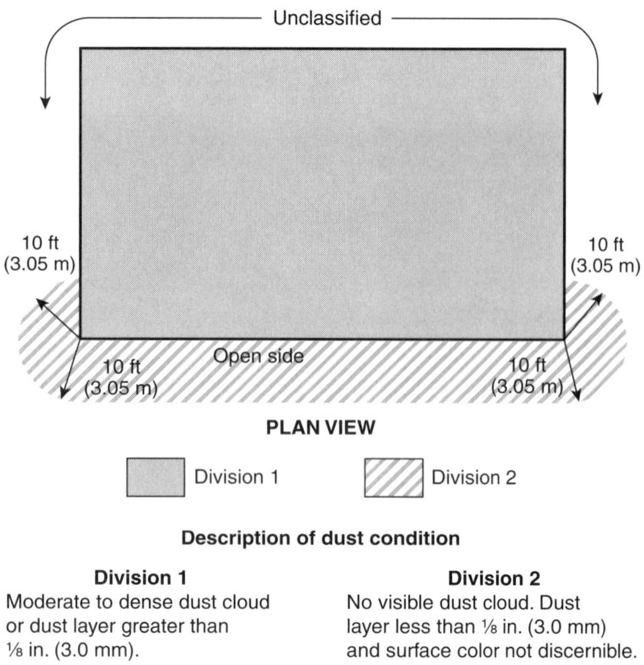

FIGURE 5.8(i) Group F or Group G — Indoor, Unrestricted Area; Ventilated Bagging Head.

specific dust. Electrical equipment evaluated and found acceptable for use in the presence of dusts might not be acceptable when exposed to molten material.

A.5.2.2 Generally speaking, the *NEC* indicates that (1) if there are explosive dust clouds under normal operating conditions, or (2) if such explosive dust clouds can be produced at the same time that a source of ignition is produced, then the area is a Division 1 location. The dust described in (2) can be provided directly by some malfunction of machinery or equipment or can be provided by accumulations of dust that are thrown into the air. Presumably, if all the dust on all surfaces in a room is sufficient to produce a dust concentration above the minimum explosive concentration, then that quantity of dust should define a Division 1 location.

From a practical point of view, a room with a concentration of dust that is above the minimum explosive concentration [criterion (1)] results in an atmosphere so dense that visibility beyond 3–5 ft (0.9–1.5 m) is impossible. Such a condition is unacceptable under today's standards for chemical plant workplaces. If such a situation were encountered, accumulations on horizontal surfaces would build up very rapidly.

On the other hand, working back from dust layers on horizontal surfaces in a room to a minimum explosive concentration in the room, based on laboratory dust explosion tests, shows a very thin layer of dust on the order of ⅛ in. (3.0 mm) to be hazardous. This is an equally impractical answer, because one of the most difficult experimental problems in dust explosion test work is to obtain a reasonably uniform cloud for ignition. As a result, the test apparatus is designed specifically to obtain uniform dust distribution. For dust lying on horizontal surfaces in a room or factory to attain such an efficient uniform distribution during an upset condition obviously is impossible.

A typical calculation considers cornstarch with a powder bulk density of approximately 25 lb/ft^3 (400 kg/m^3). The minimum explosive concentration is 0.04 oz/ft^3 (40 g/m^3). In a room 10 ft (3.05 m) high × 10 ft (3.05 m) wide × 10 ft (3.05 m) long, the depth of dust that would accumulate on the floor if the room were completely filled with a cornstarch cloud at the minimum concentration can be calculated as follows:

$$\left(\frac{0.04\text{ oz}}{\text{ft}^3}\right) \times 1000\text{ ft}^3 \times \left(\frac{1\text{ lb}}{16\text{ oz}}\right) \times \left(\frac{1\text{ ft}^3}{25\text{ lb}}\right) = 0.1\text{ ft}^3\text{ dust on floor}$$

Evenly distributed over 100 ft^2, the depth of dust would be as follows:

$$\frac{0.1\text{ ft}^3}{100\text{ ft}^2} = 0.001\text{ ft} = 0.012\text{ in. }(\tfrac{1}{84}\text{ in. thick})$$

Theoretically, throwing this amount of dust from the floor and ledges into the room volume would create a hazardous condition. Accomplishing such a feat, even experimentally, would be virtually impossible.

The optimum concentration is that in which the maximum rate of pressure rise is obtained under test conditions. Because the optimum concentration is far higher than the minimum explosive concentration, the layer thicknesses necessary to produce an optimum concentration range from 0.075 to 0.5 in. (1.9 to 12.7 mm). There is then much more dust available to be thrown into uniform suspension without postulating a 100-percent efficiency of dispersal and distribution. In addition, there are a number of factors such as particle size and shape, moisture content, uniformity of distribution, and so on that negatively affect the susceptibility of a dust to ignition. Thus, dusts encountered in industrial plants tend to be less susceptible to ignition than those used in the laboratory to obtain explosion concentration data. The classifications of areas in accordance with Table A.5.2.2(a)

Table A.5.2.2(a) Division Determination Guidelines Based on Dust Layer Thickness

Thickness of Dust Layer	Classification
Greater than ⅛ in. (3.0 mm)	Division 1
Less than ⅛ in. (3.0 mm), but surface color not discernible	Division 2
Surface color discernible under the dust layer	Unclassified

are recommended, based on a buildup of the dust level in a 24-hour period on the major portions of the horizontal surfaces.

Based on these thicknesses of dust, good housekeeping can determine the difference between a classification of Division 1 and a classification of Division 2, and a classification of Division 2 and unclassified. It should be emphasized, however, that housekeeping is a supplement to dust source elimination and ventilation. It is not a primary method of dust control.

Table A.5.2.2(b) shows the theoretical thickness of dust on the floor of a 10 ft (3.05 m) × 10 ft (3.05 m) × 10 ft (3.05 m) room necessary to satisfy the concentration requirements for a uniform dust cloud of minimum explosive concentration and for a uniform dust cloud of optimum concentration for four dusts.

Annex B Informational References

B.1 Referenced Publications.

The following documents or portions thereof are referenced within this recommended practice for informational purposes only and are thus not part of the recommendations of this document unless also listed in Chapter 2.

B.1.1 NFPA Publications. National Fire Protection Association, 1 Batterymarch Park, Quincy, MA 02169-7471.

NFPA 70, *National Electrical Code®*, 2002 edition.

NFPA *Fire Protection Guide to Hazardous Materials*, 2001 edition.

B.1.2 Other Publication.

B.1.2.1 National Academy of Sciences Publications. National Materials Advisory Board, National Academy of Sciences, 500 Fifth Street, N.W., Washington, DC 20001.

NMAB 353-3, *Classification of Combustible Dusts in Accordance with the National Electrical Code.*

B.2 Informational References.

The following documents or portions thereof are listed here as informational resources only. They are not a part of the recommendations of this document.

B.2.1 ASTM Publications. American Society for Testing and Materials, 100 Barr Harbor Drive, West Conshohocken, PA 19428-2959.

ASTM E 789, *Standard Test Method for Dust Explosions in a 1.2-Litre Closed Cylindrical Vessel*, 1995.

ASTM D 3175, *Standard Test Method for Volatile Matter in the Analysis Sample of Coal and Coke*, 1989.

B.2.2 Bureau of Mines Publications. U.S. Government Printing Office, Washington, DC 20402.

RI 5624, *Laboratory Equipment and Test Procedures for Evaluating Explosibility of Dusts.*

RI 5753, *Explosibility of Agricultural Dusts.*

RI 5971, *Explosibility of Dusts Used in the Plastics Industry.*

RI 6516, *Explosibility of Metal Powders.*

RI 6597, *Explosibility of Carbonaceous Dusts.*

RI 7009, *Minimum Ignition Energy and Quenching Distance in Gaseous Mixture.*

RI 7132, *Dust Explosibility of Chemicals, Drugs, Dyes, and Pesticides.*

RI 7208, *Explosibility of Miscellaneous Dusts.*

B.2.3 National Academy of Sciences Publications. National Materials Advisory Board, National Academy of Sciences, 500 Fifth Street, N.W., Washington, DC 20001.

NMAB 353-1, *Matrix of Combustion-Relevant Properties and Classifications of Gases, Vapors, and Selected Solids.*

Table A.5.2.2(b) Dust Thickness

Material	Minimum Conc. (oz/ft^3)	Depth of Dust (in.)	Optimum Conc. (oz/ft^3)	Depth of Dust (in.)	Bulk Density (lb/ft^3)
Cornstarch	0.04	0.012	0.5	0.15	25–50
Cork	0.035	0.022	0.2	0.125	12–15
Sugar	0.045	0.0068	0.5	0.075	50–55
Wood Flour	0.035	0.016	1.0	0.47	16–36
Polyethylene (Low Density)	0.020	0.0072	0.5	0.180	21–35

NMAB 353-2, *Test Equipment for Use in Determining Classifications of Combustible Dusts.*

NMAB 353-3, *Classification of Combustible Dusts in Accordance with the National Electrical Code.*

B.2.4 Other Publication. Miron, Y. and Lazzara, C. P. "Hot Surface Ignition Temperatures of Dust Layers." *Fire and Materials* 12: 1988; 115-126.

B.3 References for Extracts.

The following documents are listed here to provide reference information, including title and edition, for extracts given throughout this recommended practice as indicated by a reference in brackets [ ] following a section or paragraph. These documents are not a part of the recommendations of this document unless also listed in Chapter 2 for other reasons.

NFPA 70, *National Electrical Code®*, 2002 edition.

NFPA 505, *Fire Safety Standard for Powered Industrial Trucks Including Type Designations, Areas of Use, Conversions, Maintenance, and Operation,* 2002

1.1 Scope.

1.1.1 This standard shall apply to fork trucks, tractors, platform lift trucks, motorized hand trucks, and other specialized industrial trucks powered by electric motors or internal combustion engines.

1.1.2 This standard shall not apply to compressed air-operated or nonflammable compressed gas-operated industrial trucks, farm vehicles, or automotive vehicles for highway use.

Chapter 4 Hazardous (Classified) Locations

4.1 General.

4.1.1* The authority having jurisdiction shall determine the hazard classification for any specific location as defined in NFPA 70, *National Electrical Code.*

A.4.1.1 Recent developments in the classification of hazardous areas have led to the class/zone criteria that are identified in the 1996 edition of NFPA 70, *National Electrical Code.* The Technical Committee on Industrial Trucks is endeavoring to incorporate the developments into the next revision of NFPA 505. In the interim, NFPA 497, *Recommended Practice for the Classification of Flammable Liquids, Gases, or Vapors and of Hazardous (Classified) Locations for Electrical Installations in Chemical Process Areas,* and NFPA 499, *Recommended Practice for the Classification of Combustible Dusts and of Hazardous (Classified) Locations for Electrical Installations in Chemical Process Areas,* should be referenced to determine the classification established by NFPA 70.

4.1.2 The location shall be classified prior to considering the use of industrial trucks therein, and the type of industrial truck required shall be as specified in Section 4.2 for the given location.

4.1.3 Different areas of any single plant or building shall be permitted to be classified differently.

4.2 Specific Areas of Use.

4.2.10 Hazardous Areas Not Otherwise Classified. The authority having jurisdiction shall determine which types of approved power-operated industrial trucks shall be used following an engineering survey of the property and an evaluation of the fire and explosion hazards.

NFPA 560, *Standard for the Storage, Handling, and Use of Ethylene Oxide for Sterilization and Fumigation,* 2002

1.1 Scope.

This standard shall apply to the storage and handling of ethylene oxide in portable containers for its use in sterilization and fumigation. It also shall apply to flammable mixtures of ethylene oxide with other chemicals.

1.1.1 This standard shall not apply to the following:

(1) Nonflammable mixtures of ethylene oxide with other chemicals
(2) Ethylene oxide manufacturing facilities, and container filling, refilling, or transfilling facilities
(3)* The off-site transportation of portable containers of ethylene oxide

A.1.1.1(3) For regulations on the transportation of gases, see 49 CFR 100–179 (*Transportation*), and *Transportation of Dangerous Goods* Regulations of Transport Canada.

(4) Facilities using ethylene oxide as a chemical feedstock
(5) Ethylene oxide in chambers 0.283 m^3 (10 ft^3) or less in volume, or for containers holding 200 g (7.05 oz) of ethylene oxide or less

Chapter 9 Electrical Installation

9.1 Area Electrical Classification.

9.1.1 The sterilizer room, gas dispensing room, ethylene oxide container storage area, aeration rooms, and emission control area shall be classified as a Class I, Division 2, Group B area in accordance with NFPA 70, *National Electrical Code®*.

Exception: Aeration rooms shall be permitted to be unclassified electrically where it is demonstrated that flammable concentrations of ethylene oxide cannot occur during all normal and nonroutine operating conditions, including power failure.

9.1.2 The interior of the sterilization vessel shall be classified Class I, Division 1, Group B.

NFPA 654, *Standard for the Prevention of Fire and Dust Explosions from the Manufacturing, Processing, and Handling of Combustible Particulate Solids,* 2002

1.1 Scope.

1.1.1* This standard shall apply to all phases of the manufacture, processing, blending, pneumatic conveying, repackaging, and handling of combustible particulate solids or hybrid mixtures, regardless of concentration or particle size, where the materials present a fire or explosion hazard.

A.1.1.1 Examples of industries that handle combustible particulate solids, either as a process material or as a fugitive or nuisance dust, include but are not limited to the following:

(1) Agricultural, chemical, and food commodities, fibers, and textile materials
(2) Forest and furniture products industries
(3) Metals processing
(4) Paper products
(5) Pharmaceuticals
(6) Resource recovery operations (tires, municipal solid waste, metal, paper, or plastic recycling operations)
(7) Wood, metal, or plastic fabricators

1.1.2 This standard shall apply to systems that convey combustible particulate solids that are produced as a result of a principal or incidental activity, regardless of concentration or particle size, where the materials present a fire or explosion hazard.

1.1.3 This standard shall not apply to materials covered by the following documents, unless specifically referenced by the applicable document:

(1) NFPA 30B, *Code for the Manufacture and Storage of Aerosol Products*
(2) NFPA 61, *Standard for the Prevention of Fires and Dust Explosions in Agricultural and Food Products Facilities*
(3) NFPA 120, *Standard for Coal Preparation Plants*
(4) NFPA 432, *Code for the Storage of Organic Peroxide Formulations*
(5) NFPA 480, *Standard for the Storage, Handling, and Processing of Magnesium Solids and Powders*
(6) NFPA 481, *Standard for the Production, Processing, Handling, and Storage of Titanium*
(7) NFPA 482, *Standard for the Production, Processing, Handling, and Storage of Zirconium*
(8) NFPA 485, *Standard for the Storage, Handling, Processing, and Use of Lithium Metal*
(9) NFPA 495, *Explosive Materials Code*
(10) NFPA 651, *Standard for the Machining and Finishing of Aluminum and the Production and Handling of Aluminum Powders*
(11) NFPA 655, *Standard for Prevention of Sulfur Fires and Explosions*
(12) NFPA 664, *Standard for the Prevention of Fires and Explosions in Wood Processing and Woodworking Facilities*
(13) NFPA 1124, *Code for the Manufacture, Transportation, and Storage of Fireworks and Pyrotechnic Articles*
(14) NFPA 1125, *Code for the Manufacture of Model Rocket and High Power Rocket Motors*
(15) NFPA 8503, *Standard for Pulverized Fuel Systems*

1.1.4 In the event of a conflict between this standard and a specific occupancy standard, the specific occupancy standard requirements shall apply.

Chapter 2 Facility and Systems Design

2.7.2* In local areas of a plant where a hazardous quantity of dust accumulates or is suspended in air, the area shall be classified and all electrical equipment and installations in those local areas shall comply with Article 502 or Article 503 of NFPA 70, *National Electrical Code*, as applicable.

A.2.7.2 Refer to NFPA 499, *Recommended Practice for the Classification of Combustible Dusts and of Hazardous (Classified) Locations for Electrical Installations in Chemical Process Areas.*

Chapter 5 Ignition Sources

5.2 Electrical Equipment.

All electrical equipment and installations shall comply with the requirements of Section 2.7.

NFPA 655, *Standard for Prevention of Sulfur Fires and Explosions,* 2001

1.1 Scope.

1.1.1* This standard shall apply to the crushing, grinding, or pulverizing of sulfur and to the handling of sulfur.

A.1.1.1 Sulfur differs from most other combustible dusts found in industry in that it has relatively low melting and ignition points. Depending on purity, sulfur melts at or slightly below 119°C (246°F). The ignition temperature of a dust cloud is 190°C (374°F); the ignition temperature of a dust layer is 220°C (428°F). Dilution of sulfur with inert solids is not effective in raising the ignition temperature.

Sulfur is handled and processed in the liquid and vapor states in some cases. The liquid is highly combustible, and the vapor is explosive when mixed with air in the proper proportions.

The finely divided sulfur produced during crushing and pulverizing is the most hazardous from an explosion standpoint. Also, mixtures containing finely divided elemental sulfur can be just as hazardous if the sulfur is present in sufficient quantity. Some explosion and fire hazards also accompany the handling and processing of sulfur in bulk in coarse sizes due to the fine dust present.

1.1.2 This standard shall not apply to the mining or transportation of sulfur.

Chapter 4 Handling Finely Divided Sulfur in Bulk

4.3* Electrical Wiring and Equipment.

All electrical wiring and equipment shall comply with NFPA 70, *National Electrical Code®*. In areas where a dust explosion hazard exists, electrical wiring and equipment shall comply with Article 502 of NFPA 70.

A.4.3 Although sulfur is not now included in atmospheres classified as Class II, Group G, it has been the experience of the sulfur industry that such equipment can be suitable. However, consideration should be given to the melting point of sulfur 112°C to 119°C (233°F to 246°F) in the selection of heat-producing electrical equipment.

Chapter 7 Handling of Liquid Sulfur and Sulfur Vapor at Temperatures above 154°C (309°F)

7.2.5* All electrical wiring and equipment installed in areas handling liquid sulfur shall meet the requirements of Article 501 of NFPA 70, *National Electrical Code*.

A.7.2.5 Due to the potential for release of dissolved hydrogen sulfide, molten sulfur handling systems require a Class I, Group C, classification for confined areas.

NFPA 664, *Standard for the Prevention of Fires and Explosions in Wood Processing and Woodworking Facilities,* 2002

1.1 Scope.

This standard shall establish the minimum requirements for fire and explosion prevention and protection of industrial, commercial, or institutional facilities that process wood or manufacture wood products, using wood or other cellulosic fiber as a substitute for or additive to wood fiber, and that process wood, creating wood chips, particles, or dust.

1.1.1 Woodworking and wood processing facilities shall include, but are not limited to, wood flour plants, industrial woodworking plants, furniture plants, plywood plants, composite board plants, lumber mills, and production-type woodworking shops and carpentry shops that are incidental to facilities that would not otherwise fall within the purview of this standard.

1.1.2* This standard shall apply to woodworking operations that occupy areas of more than 465 m^2 (5000 ft^2) or where dust-producing equipment requires an aggregate dust collection flow rate of more than 2549 m^3/hr (1500 ft^3/min).

A.1.1.2 Specific criteria in this standard are advisable for facilities that fall outside this document's scope. A hazard and risk analysis should be performed to identify areas where specific criteria are appropriate.

Chapter 3 Definitions

3.3.6 Deflagration Hazard. A situation where deflagrable wood dust is normally in suspension or can be placed in suspension at concentrations at or above the minimum explosible concentration (MEC).

Chapter 7 Prevention of Ignition and Control of Ignition Sources

7.3 Electrical Systems.

7.3.1 All electrical systems and system components shall be installed in accordance with NFPA 70, *National Electrical Code®*.

7.3.2* Portions of the facility where dust accumulations occur or where suspensions of wood dust in air could occur shall be equipped with electrical systems and equipment per Article 502 or 503 of NFPA 70, *National Electrical Code*.

A.7.3.2 Refer to NFPA 497, *Recommended Practice for the Classification of Flammable Liquids, Gases, or Vapors and of Hazardous (Classified) Locations for Electrical Installations in Chemical Process Areas*.

7.18* Portable Electric Equipment and Appliances.

Portable electric equipment and appliances used in hazardous areas shall be listed for the area in which they are to be used.

A.7.18 Electric appliances, including but not limited to coffee pots and portable space heaters, have been found to cause fires in industrial occupancies. The use of these appliances should be controlled by management to limit the probability of ignition.

Chapter 8 Processes, Operations, and Special Systems

8.4* Particulate Size Reduction Equipment.

A.8.4 This equipment typically consists of high-speed rotating machinery that cuts, shears, breaks, or pulverizes wood fractions into smaller pieces. Equipment in this category includes, but is not limited to, hogs, chippers, stranders, flakers, disk refiners, hammermills, and pulverizers.

The size and power of the equipment is sufficient to create high heat and showers of sparks if any foreign material enters the equipment. Tramp metal in the material being processed is a common problem, as are rocks or other foreign material.

Fires and explosions can occur in the equipment and propagate rapidly into downstream equipment. The degree of hazard is primarily a function of the moisture content of the material being processed and the size distribution of the particulate produced by the size reduction equipment.

8.4.2 Deemed to Satisfy Prescriptive Requirements.

8.4.2.4* Size Reduction Equipment with a Deflagration Hazard.

A.8.4.2.4 This section includes additional requirements for size reduction equipment deemed to have a deflagration hazard.

8.4.2.4.3 Rooms containing the size reduction equipment shall be considered a Class II, Division 2 hazardous area as defined in Article 500 of NFPA 70, *National Electrical Code*.

Chapter 11 Housekeeping

11.2 Cleanup Methods.

11.2.1 Removal of Dust.

11.2.1.1* Surfaces shall be cleaned in a manner that minimizes the generation of dust clouds. Blowing down with steam or compressed air or even vigorous sweeping shall be permitted only if the following requirements are met:

A.11.2.1.1 Sweeping and/or vacuuming are the preferred methods to be utilized. Blowing down with steam or compressed air, or even vigorous sweeping, produces dust clouds. Facilities should not be operating during blowdown. Blowdown should be done in individual sections of the building, starting near the center and working out, in order to prevent filling the entire building with dust-laden air. Blowdown should be frequent enough that large amounts of dust are not blown into suspension.

11.2.1.2* Powered sweepers, vacuum cleaning equipment, and other powered cleaning apparatus used in dusty areas shall be approved for Class II, Division 1, Group G locations as defined in Article 502 of NFPA 70, *National Electrical Code*.

A.11.2.1.2 Unapproved vacuum-cleaning equipment can be used if the powered suction source is located in a remote, nondusty area.

NFPA 820, *Standard for Fire Protection in Wastewater Treatment and Collection Facilities,* 2003

Editor's Note: *For the reader's convenience, all pertinent figures from Annex A sections of this document have been gathered into one section at the end of the document, under the heading* ***Annex A Figures.***

1.1 Scope.

1.1.1* General. This standard shall establish minimum requirements for protection against fire and explosion hazards in wastewater treatment plants and associated collection systems, including the hazard classification of specific areas and processes.

A.1.1.1 Other NFPA standards should be consulted for additional requirements relating to wastewater treatment and collection facilities.

1.1.2 This standard shall apply to the following:

(1) Collection sewers
(2) Trunk sewers
(3) Intercepting sewers
(4) Combined sewers
(5) Storm sewers
(6) Pumping stations
(7) Wastewater treatment plants
(8) Sludge-handling facilities
(9) Chemical-handling facilities
(10) Treatment facilities
(11) Ancillary structures *(see 3.3.67.1)*

1.1.3 This standard shall not apply to the following:

(1) Collection, treatment, or disposal of industrial wastes or manufactured by-products that are treated on-site and not discharged to a publicly or privately operated municipal facility
(2) On-site treatment systems *(see 3.3.68.1)*
(3) Pressure sewer systems *(see 3.3.61.8, definition of Pressure Sewer)*
(4) Building drain systems and appurtenances *(see 3.3.8, definition of Building Drain)*
(5) Industrial sewer systems and appurtenances *(see 3.3.61.5, definition of Industrial Sewer)*
(6) Personnel safety from toxic and hazardous materials or products of combustion
(7) Separate nonprocess-related structures *(see 3.3.67.2)*

Chapter 4 Collection Systems

4.1* General.

A.4.1 Additional information on sources of hazards, sources of ignition, and mitigation measures associated with the collection and transmission of municipal wastewater is contained in Annex D.

4.1.1 This chapter shall establish minimum criteria for protection against fire and explosion hazards in the collection and transportation of municipal wastewater.

4.1.2 This chapter shall not apply to on-site systems, force mains, or those sewers that principally convey industrial wastes.

4.2* Design and Construction.

The design and construction of collection system facilities shall conform to Table 4.2, which summarizes the various components associated with wastewater collection and transport systems.

A.4.2 See Figure A.4.2(a) through Figure A.4.2(g), which provide examples for Table 4.2.

Chapter 5 Liquid Stream Treatment Processes

5.1* General.

A.5.1 Additional information on sources of hazards, sources of ignition, and mitigation measures associated with liquid stream treatment processes is contained in Annex D.

5.1.1 This chapter shall establish minimum criteria for protection against fire and explosion hazards associated with liquid stream treatment processes.

5.1.2 This chapter shall not apply to treatment systems serving individual structures or treatment systems that principally treat industrial wastes.

5.2* Design and Construction.

The design and construction of liquid stream treatment processes shall conform to Table 5.2.

A.5.2 See Figure A.5.2, which provides an example for Table 5.2.

Chapter 6 Solids Treatment Processes

6.1* General.

A.6.1 Additional information on sources of hazards, sources of ignition, and mitigation measures associated with solids treatment processes is contained in Annex D.

6.1.1 This chapter shall establish minimum criteria for protection against fire and explosion hazards associated with solids treatment processes.

6.1.2 This chapter shall not apply to the treatment of solids from industrial waste treatment processes.

6.2* Design and Construction.

The design and construction of solids treatment processes shall conform to Table 6.2(a) and Table 6.2(b).

Table 4.2 Collection Systems

Row	Line	Location and Function	Fire and Explosion Hazard	Ventilation	Extent of Classified Area	*NEC*-Area Electrical Classification (All Class I, Group D)	Material of Construction for Buildings or Structures	Fire Protection Measures
1		MATERIALS USED IN REHABILITATION, RECONSTRUCTION, OR SLIP-LINING OF SEWERS	NA	NA	NA	NA	In accordance with 8.3.1	NA
2		INDUSTRIAL SEWER Sewer transporting industrial wastewater only (no sanitary wastewater)	Not included within the scope of this standard					
3		STORM SEWER Sewer transporting storm water only (no sanitary wastewater)	Possible ignition of flammable gases and floating flammable liquids	NNV	Inside of sewer	Division 2	In accordance with 8.3.1	NR
4		STORM WATER PUMPING STATION WET WELLS Liquid side of pumping station serving only a storm sewer system	Possible ignition of flammable gases and floating flammable liquids	NNV	Entire room or space	Division 2	NC, LC, or LFS	CGD if enclosed
5	a	STORM WATER PUMPING STATION DRY WELLS Dry side of a pumping station serving only a storm sewer system and physically separated from wet well	Buildup of vapors from flammable or combustible liquids	D	Entire dry well	Division 2, or unclassified, if space provided with pressuri-zation in accordance with NFPA 496	NC, LC, or LFS	FE
	b			C		Unclassified		
6		PRESSURE SEWER (Force main) Sewer under pressure (flooded discharge pipe from pump or tank)	Not included within the scope of this standard					
7		BUILDING SEWER (Lateral sewer or drain) Sewer serving a house or single building (plumbing)	Not included within the scope of this standard					
8		INDIVIDUAL RESIDENTIAL SEWER Sewer serving one but not more than five dwellings	NA	NNV	Within enclosed space	Unclassified	NR	NR
9		INDIVIDUAL RESIDENTIAL PUMPING UNITS Pumping units serving one but not more than five dwellings (e.g., grinder pumps, septic tank effluent pumps, ejector pumps)	NA	NNV	Within enclosed space	Unclassified	NR	NR

(continues)

Table 4.2 Continued

Row	Line	Location and Function	Fire and Explosion Hazard	Ventilation	Extent of Classified Area	*NEC*-Area Electrical Classification (All Class I, Group D)	Material of Construction for Buildings or Structures	Fire Protection Measures
10	a	RESIDENTIAL SEWER Sewer transporting primarily residential wastewater	Possible ignition of flammable gases and floating flammable liquids	NNV	Within enclosed space	Division 2	In accordance with 8.3.1	NR
	b			B		Unclassified		
11	a	RESIDENTIAL WASTEWATER PUMPING STATION WET WELL Pumping station transporting primarily residential wastewater	Possible ignition of flammable gases and floating flammable liquids	A	Entire room or space	Division 2	NC, LC, or LFS	CGD
	b			B		Unclassified		
12	a	RESIDENTIAL WASTEWATER PUMPING STATION DRY WELL Dry side of a pumping station transporting primarily residential wastewater	Buildup of vapors from flammable or combustible liquids	D	Entire room or space	Division 2	NC, LC, or LFS	FE
	b			C		Unclassified		
13		OUTFALL SEWER Final discharge pipe from a treatment plant, transporting treated wastewater	NA	NNV	NA	Unclassified	NR	NR
14	a	SANITARY SEWER Sewer transporting domestic, commercial, and industrial wastewater	Possible ignition of flammable gases and floating flammable liquids	NNV	Inside of sewer	Division 1	In accordance with 8.3.1	NR
	b			B		Division 2		
15	a	COMBINED SEWER Sewer transporting domestic, commercial, and industrial wastewater and storm water	Possible ignition of flammable gases and floating flammable liquids	NNV	Inside of sewer	Division 1	In accordance with 8.3.1	NR
	b			B		Division 2		
16	a	WASTEWATER PUMPING STATION WET WELLS Liquid side of a pumping station serving a sanitary sewer or combined system	Possible ignition of flammable gases and floating flammable liquids	A	Entire room or space	Division 1	NC, LC, or LFS	CGD
	b			B		Division 2		
17	a	BELOWGRADE OR PARTIALLY BELOWGRADE WASTEWATER PUMPING STATION DRY WELL Pump room physically separated from wet well; pumping of wastewater from a sanitary or combined sewer system through closed pumps and pipes	Buildup of vapors from flammable or combustible liquids	C	Entire space or room	Unclassified	NC, LC, or LFS	FE
	b			D		Division 2, or unclassified, if space provided with pressurization in accordance with NFPA 496		

Table 4.2 Continued

Row	Line	Location and Function	Fire and Explosion Hazard	Ventilation	Extent of Classified Area	*NEC*-Area Electrical Classification (All Class I, Group D)	Material of Construction for Buildings or Structures	Fire Protection Measures
18		ABOVEGRADE WASTEWATER PUMPING STATION Pump room physically separated with no personnel access to wet well; pumping of wastewater from a sanitary or combined sewer system through closed pumps and pipes	NA	NR	NA	Unclassified	NC, LC, or LFS	FE
19	a	ABOVEGRADE WASTEWATER PUMPING STATION Pump room not physically separated from wet well; pumping of wastewater from a sanitary or combined sewer system through closed pumps and pipes	Possible ignition of flammable gases and floating flammable liquids	A	Entire space or room	Division 1	NC	FE
	b			B		Division 2	NC, LC, or LFS	
20	a	ODOR-CONTROL SYSTEM AREAS Areas physically separated from wet well that house systems handling wet well gases	Leakage and ignition of sewage gases	D	Entire area if enclosed	Division 2	NC, LC, or LFS	CGD and FDS
	b			C, or outdoors	Areas within 0.9 m (3 ft) of leakage sources such as fans, dampers, flexible connections, flanges, pressurized unwelded ductwork, and odor-control vessels	Division 2		
	c				Areas beyond 0.9 m (3 ft)	Unclassified		
21	a	MAINTENANCE HOLES Access to sewer for personnel entry	Possible ignition of flammable gases and floating flammable liquids	NNV	Inside	Division 1	In accordance with 8.3.1	NR
	b			B		Division 2		
22	a	JUNCTION CHAMBERS Structure where sewers intersect	Buildup of vapors from flammable or combustible liquids	NNV	Inside	Division 1	In accordance with 8.3.1	NR
	b			B	Open and above grade or inside and ventilated	Division 2		
23		INVERTED SIPHONS Depressed section of gravity sewer	Possible ignition of flammable gases and floating flammable liquids	NNV	Interior of inlet and outlet structures	Division 1	NC	NR

(continues)

Table 4.2 Continued

Row	Line	Location and Function	Fire and Explosion Hazard	Ventilation	Extent of Classified Area	*NEC*-Area Electrical Classification (All Class I, Group D)	Material of Construction for Buildings or Structures	Fire Protection Measures
24		CATCH BASINS (Curb inlet) Inlet where street water enters a storm or combined sewer	Buildup of vapors from flammable or combustible liquids	NNV	Enclosed space	Division 1	In accordance with 8.3.1	NR
25	a	RESIDENTIAL DIVERSION STRUCTURES Enclosed structures where residential wastewater can be diverted	Buildup of vapors from flammable or combustible liquids	NNV	Enclosed space	Division 2	In accordance with Chapter 8	NR
	b			B		Unclassified		
26	a	RESIDENTIAL BELOWGRADE VALVE VAULT With an exposed residential wastewater surface	Possible ignition of gases and floating flammable liquids	NNV	Enclosed space	Division 2	In accordance with 8.3.1	NR
	b			B		Unclassified		
27	a	RESIDENTIAL CONTROL STRUCTURES Enclosed structures where residential wastewater flow is regulated	Buildup of vapors from flammable or combustible liquids	A	Enclosed space	Division 2	In accordance with Chapter 8	NR
	b			B		Unclassified		
28	a	RESIDENTIAL BELOWGRADE METERING VAULT With an exposed residential wastewater surface	Possible ignition of flammable gases and floating flammable liquids	NNV	Enclosed space	Division 2	In accordance with 8.3.1	NR
	b			B		Unclassified		
29	a	DIVERSION STRUCTURES Enclosed structures where wastewater can be diverted	Buildup of vapors from flammable or combustible liquids	NNV	Enclosed space	Division 1	In accordance with Chapter 8	NR
	b			B		Division 2		
30		ABOVEGRADE VALVE VAULT Physically separated from the wet well; valves in vault in closed piping system	NA	NR	NA	Unclassified	NC, LC, or LFS	NR
31	a	BELOWGRADE VALVE VAULT Physically separated from the wet well and with closed piping system	Buildup of vapors from flammable or combustible liquids	NNV	Enclosed space	Division 2	NC, LC, or LFS	NR
	b			C		Unclassified		
32	a	BELOWGRADE VALVE VAULT With an exposed wastewater surface	Possible ignition of gases and floating flammable liquids	NNV	Enclosed space	Division 1	NC	NR
	b			B		Division 2	NC, LC, or LFS	
33	a	CONTROL STRUCTURES Enclosed structures where wastewater or storm water flow is regulated	Buildup of vapors from flammable or combustible liquids	A	Enclosed space	Division 1	In accordance with Chapter 8	NR
	b			B		Division 2		

Table 4.2 Continued

Row	Line	Location and Function	Fire and Explosion Hazard	Ventilation	Extent of Classified Area	*NEC*-Area Electrical Classification (All Class I, Group D)	Material of Construction for Buildings or Structures	Fire Protection Measures
34	a	WASTEWATER HOLDING BASINS Enclosed structures temporarily holding untreated or partially treated wastewater	Possible ignition of flammable gases and floating flammable liquids	A	Enclosed space	Division 1	NC	NR
	b			B		Division 2	NC, LC, or LFS	
35		WASTEWATER HOLDING BASINS, LINED OR UNLINED Open structures holding storm water, combined wastewater, untreated or partially treated wastewater	NR	NR	NR	NR	NR	NR
36	a	BELOWGRADE METERING VAULT Physically separated from the wet well and with closed piping system	Buildup of vapors from flammable or combustible liquids	NNV	Enclosed space	Division 2	NC, LC, or LFS	NR
	b			C		Unclassified		
37	a	BELOWGRADE METERING VAULT With an exposed wastewater surface	Possible ignition of flammable gases and floating flammable liquids	NNV	Enclosed space	Division 1	NC	NR
	b			B		Division 2	NC, LC, or LFS	
38		COARSE AND FINE SCREEN FACILITIES *(See Coarse and Fine Screen Facilities in Table 5.2.)*						

Notes:
(1) The NR designation in the ventilation column indicates that no ventilation requirements are established for the space and, therefore, Table 9.1.1.4 also has no requirements.
(2) Row and Line columns are used to refer to specific figures in A.4.2 and specific requirements for each location and function.
(3) The following codes are used in this table:
A — No ventilation or ventilated at less than 12 air changes per hour
B — Continuously ventilated at 12 changes per hour or in accordance with Chapter 9
C — Continuously ventilated at 6 air changes per hour or in accordance with Chapter 9
CGD — Combustible gas detection system
D — No ventilation or ventilated at less than 6 air changes per hour
FDS — Fire detection system
FE — Portable fire extinguisher
LC — Limited-combustible material
LFS — Low flame spread material
NA — Not applicable
NC — Noncombustible material
NEC — In accordance with NFPA 70
NNV — Not normally ventilated
NR — No requirement

Table 5.2 Liquid Stream Treatment Processes

Row	Line	Location and Function	Fire and Explosion Hazard	Ventilation	Extent of Classified Area[1]	*NEC*-Area Electrical Classification (All Class I, Group D)	Material of Construction for Buildings or Structures	Fire Protection Measures
1	a	COARSE AND FINE SCREEN FACILITIES Removal of screenings from raw wastewater	Possible ignition of flammable gases and floating flammable liquids	A	Enclosed — entire space	Division 1	NC	FE, H, and CGD if enclosed
	b			B	Enclosed — entire space	Division 2	NC, LC, or LFS	
	c			Not enclosed, open to atmosphere	Within a 3-m (10-ft) envelope around equipment and open channel[2,3]	Division 2	NC, LC, or LFS	
2		PUMPING STATIONS *(See Collection Systems, Table 4.2.)*						
3	a	FLOW EQUALIZATION TANKS Storage of raw or partially treated wastewater	Possible ignition of flammable gases and floating flammable liquids	A	Enclosed — entire space	Division 1	NC	FE, H, and CGD if enclosed
	b			B	Enclosed — entire space	Division 2	NC, LC, or LFS	
	c			Not enclosed, open to atmosphere	Within a 3-m (10-ft) envelope around equipment and open channel[2,3]	Division 2	NC, LC, or LFS	
4	a	GRIT REMOVAL TANKS Separation of grit from raw wastewater	Possible ignition of flammable gases and floating flammable liquids	A	Enclosed — entire space	Division 1	NC	FE, H, and CGD if enclosed
	b			B	Enclosed — entire space	Division 2	NC, LC, or LFS	
	c			Not enclosed, open to atmosphere	Within a 3-m (10-ft) envelope around equipment and open channel[2,3]	Division 2	NC, LC, or LFS	
5	a	PRE-AERATION TANKS Conditioning of wastewater prior to further treatment	Possible ignition of flammable gases and floating flammable liquids	A	Enclosed — entire space	Division 1	NC	H and CGD if enclosed
	b			B	Enclosed — entire space	Division 2	NC, LC, or LFS	
	c			Not enclosed, open to atmosphere	Within a 3-m (10-ft) envelope around equipment and open channel[2,3]	Division 2	NC, LC, or LFS	
6	a	PRIMARY SEDIMENTATION TANKS Separation of floating or settleable solids from raw wastewater	Possible ignition of flammable gases and floating flammable liquids	A	Enclosed — entire space	Division 1	NC	H and CGD if enclosed
	b			B	Enclosed — entire space	Division 2	NC, LC, or LFS	
	c			Not enclosed, open to atmosphere	Interior of the tank from the minimum operating water surface to the top of the tank wall; envelope 0.46 m (18 in.) above the top of the tank and	Division 2	NC, LC, or LFS	

Table 5.2 Continued

Row	Line	Location and Function	Fire and Explosion Hazard	Ventilation	Extent of Classified Area[1]	*NEC*-Area Electrical Classification (All Class I, Group D)	Material of Construction for Buildings or Structures	Fire Protection Measures
					extending 0.46 m (18 in.) beyond the exterior wall; envelope 0.46 m (18 in.) above grade extending 3 m (10 ft) horizontally from the exterior tank walls			
7		AERATION BASIN, POND, LAGOON, OXIDATION DITCH, AEROBIC SUSPENDED GROWTH SYSTEMS, SEQUENCING BATCH REACTORS Aerobic treatment of wastewater open to the atmosphere		NA		Unclassified (If process is not preceded by primary sedimentation, see Primary Sedimentation Tanks in Table 5.2 for classification.)	NR	H
8	a	ENCLOSED AERATION BASIN OR AEROBIC SUSPENDED GROWTH SYSTEMS Aerobic treatment of wastewater not preceded by primary treatment	Possible ignition of flammable gases or floating flammable liquids	A	Entire enclosed space not routinely entered by personnel	Division 1	NC	NR
	b			B		Division 2	NC, LC, or LFS	
9		ENCLOSED AERATION BASIN OR AEROBIC SUSPENDED GROWTH SYSTEMS Aerobic treatment of wastewater preceded by primary treatment	NA	NR	Entire enclosed space	Unclassified	NC, LC, or LFS	NR
10		TRICKLING FILTER, BIO-TOWER, AEROBIC FIXED-FILM SYSTEMS Aerobic biological treatment of wastewater	Not normally a significant hazard; however, these processes might contain materials that are combustible under certain conditions	NA		Unclassified (If unit process is not preceded by primary sedimentation, see Primary Sedimentation Tanks in Table 5.2 for classification.)	NR	H

(continues)

Table 5.2 Continued

Row	Line	Location and Function	Fire and Explosion Hazard	Ventilation	Extent of Classified Area[1]	*NEC*-Area Electrical Classification (All Class I, Group D)	Material of Construction for Buildings or Structures	Fire Protection Measures
11	a	ANAEROBIC TOWERS, ANAEROBIC FIXED-FILM SYSTEM Anaerobic biological treatment if sealed from atmosphere	Normally produces combustible gas as treatment process by-product	NA	Tank interior	Division 1	NC	FE and H
	b				3-m (10-ft) envelope around tank	Division 2	NC, LC, or LFS	
12	a	GAS-HANDLING SYSTEMS FOR LIQUID TREATMENT PROCESSES	Combustible gas, often under pressure	A	Enclosed — entire space	Division 1	NC	FE and H
	b			B		Division 2	NC, LC, or LFS	
	c			Not enclosed, open to atmosphere	Within a 3-m (10-ft) envelope around equipment[2]			
13		OXYGEN AERATION TANKS Tanks for aerobic treatment of wastewater using high-purity oxygen rather than air	Ignition of flammable gases and floating flammable liquids in an oxygen-enriched environment	NA	Enclosed space	Division 2 (If unit process is not preceded by primary sedimentation, see Primary Sedimentation Tanks in Table 5.2 for classification.)	Any equipment or material within the reactor space shall be safe for exposure to volatile hydrocarbons in an oxygen-enriched atmosphere	Special provision for LEL monitoring and automatic isolation of equipment and oxygen supply
14		INTERMEDIATE, SECONDARY, OR TERTIARY SEDIMENTATION TANKS Separate floating and settleable solids from wastewater at various treatment stages		NA	NA	Unclassified (If unit process is not preceded by primary sedimentation, see Primary Sedimentation Tanks in Table 5.2 for classification.)	NR	H
15		FLASH MIXER OR FLOCCULATION TANKS Tanks for mixing various treatment chemicals with wastewater		NA	NA	Unclassified (If unit process is not preceded by primary sedimentation, see Primary Sedimentation Tanks in Table 5.2 for classification.)	NR	H
16		NITRIFICATION AND DENITRIFICATION TANKS Tertiary treatment of wastewater to reduce or remove nitrogen		NA	NA	Unclassified (If unit process is not preceded by primary sedimentation, see Primary Sedimentation Tanks in Table 5.2 for classification.)	NR	H

Table 5.2 Continued

Row	Line	Location and Function	Fire and Explosion Hazard	Ventilation	Extent of Classified Area[1]	*NEC*-Area Electrical Classification (All Class I, Group D)	Material of Construction for Buildings or Structures	Fire Protection Measures
17		BREAKPOINT CHLORINATION TANKS AND CHLORINE CONTACT TANKS Application of chlorine in aqueous solution to wastewater		NA	NA	Unclassified	NR (These unit processes use corrosive chemicals that require the use of specific materials of construction. Special consideration shall be given to these materials of construction.)	H
18		AMMONIA STRIPPING TOWERS	*(See Trickling Filter in Table 5.2.)*	NA	NA	Unclassified	NR (These unit processes use corrosive chemicals. Special consideration shall be given to these materials of construction.)	H
19		INTERMEDIATE OR FINAL PUMPING STATIONS Pump(s) at intermediate stage or end of the treatment process		NA	NA	Unclassified	NR	H
20		GRAVITY AND PRESSURE FILTERS Filtering of treated wastewater through sand or other media		NA	NA	Unclassified	NR	H
21		CARBON COLUMN OR TANKS Vessels containing carbon for tertiary treatment of wastewater	Significant hazard from combustible carbon material	NA	NA	Unclassified	NR	H
22		ON-SITE OZONE GENERATION SYSTEM AND OZONE CONTACT TANKS Ozone generation and purification for disinfection of wastewater	Similar to oxygen generation with addition of being highly corrosive *(see Table D.1.1)*	NA	NA	Not covered in this standard	NR	NR

(continues)

Table 5.2 Continued

Row	Line	Location and Function	Fire and Explosion Hazard	Ventilation	Extent of Classified Area[1]	*NEC*-Area Electrical Classification (All Class I, Group D)	Material of Construction for Buildings or Structures	Fire Protection Measures
23		BACKWASH WATER AND WASTE BACKWASH WATER HOLDING TANKS Tanks for temporary storage of backwash water	NA	NA	NA	Unclassified	NR	H
24		ULTRAVIOLET DISINFECTION UNIT Disinfection of wastewater by ultraviolet radiation		NA	NA	Unclassified	NR	H
25		EFFLUENT STRUCTURES Various structures conveying treated wastewater away from treatment processes		NA	NA	Unclassified	NR	H
26	a	ODOR-CONTROL SYSTEM AREAS Areas physically separated from processes that house systems handling flammable gases	Leakage and ignition of flammable gases	D	Entire area if enclosed	Division 2	NC, LC, or LFS	CGD, FDS, and FE
	b			C	Areas within 0.9 m (3 ft) of leakage sources such as fans, dampers, flexible connections, flanges, pressurized unwelded ductwork, and odor-control vessels	Division 2		
	c				Areas beyond 0.9 m (3 ft)	Unclassified		

Notes:
(1) The NR designation in the ventilation column indicates that no ventilation requirements are established for the space and, therefore, Table 9.1.1.4 also has no requirements.
(2) Row and Line columns are used to refer to the figure in A.5.2 and specific requirements for each location and function.
(3) The following codes are used in this table:
A — No ventilation or ventilated at less than 12 air changes per hour
B — Continuously ventilated at 12 changes per hour or in accordance with Chapter 9
C — Continuously ventilated at 6 air changes per hour or in accordance with Chapter 9
CGD — Combustible gas detection system
D — No ventilation or ventilated at less than 6 air changes per hour
FE — Portable fire extinguisher
H — Hydrant protection in accordance with 7.2.4
LC — Limited-combustible material
LFS — Low flame spread material
NA — Not applicable
NC — Noncombustible material
NEC — In accordance with NFPA 70
NR — No requirement
[1]Open channels and open structures upstream from the unit processes are classified the same as the downstream processes they supply.
[2]The area beyond the envelope is unclassified.
[3]Where liquid turbulence is not induced by aeration or other factors, the following criteria apply: (1) Interior of the tank from the minimum operating water surface to the top of the tank wall; (2) Envelope 0.46 m (18 in.) above the top of the tank and extending 0.46 m (18 in.) beyond the exterior wall; (3) Envelope 0.46 m (18 in.) above grade extending 3 m (10 ft) horizontally from the exterior tank walls.

Table 6.2(a) Solids Treatment Processes

Row	Line	Location and Function	Fire and Explosion Hazard	Ventilation	Extent of Classified Area	*NEC*-Area Electrical Classification (All Class I, Group D)	Material of Construction for Buildings or Structures	Fire Protection Measures
1		COARSE AND FINE SCREENINGS-HANDLING BUILDINGS Storage, conveying, or dewatering of screenings (no exposed flow of wastewater through building or area)	NA	NR	NA	Unclassified	NC, LC, or LFS	H, FE, and FAS
2		GRIT-HANDLING BUILDING Storage, conveying, and dewatering of heavy small screenings and grit (no exposed flow of wastewater through building or area)	NA	NR	NA	Unclassified	NC, LC, or LFS	H, FE, and FAS
3	a	SCUM-HANDLING BUILDING OR AREA Holding, dewatering, or storage	Possible grease or flammable liquids carryover	A	Enclosed space	Division 2	NC, LC, or LFS	H, FE, and CGD if enclosed
	b			B	NA	Unclassified		
	c			Not enclosed, open to atmosphere				
4	a	SCUM PITS	Buildup of vapors from flammable or combustible liquids	A	Enclosed — entire space	Division 1	NC	H, FE, and CGD if enclosed
	b			B				
	c			Not enclosed, open to atmosphere	Within a 3-m (10-ft) envelope around equipment and open channel[1]	Division 2	NC, LC, or LFS	
5	a	SCUM-PUMPING AREAS Pumping of scum, wet side of pumping station	Carryover of floating flammable liquids	A	Enclosed — entire space	Division 1	NC	H, FE, and CGD if enclosed
	b			B				
	c			Not enclosed, open to atmosphere	Within a 3-m (10-ft) envelope around equipment and open channel[1]	Division 2	NC, LC, or LFS	
6	a	SCUM-PUMPING AREAS Pumping of scum, dry side of pumping station	Not significant	D	Enclosed space	Division 2	NC, LC, or LFS	FE
	b			C		Unclassified		
	c			Not enclosed, open to atmosphere	NA			
7		SCUM INCINERATORS[2] Elimination of scum through burning	Firebox explosion from possible carryover of flammable scum	NR	Incinerator area if separated from scum storage	Unclassified	NC, LC, or LFS	FSS (if indoors), H, and FE

(continues)

Table 6.2(a) Continued

Row	Line	Location and Function	Fire and Explosion Hazard	Ventilation	Extent of Classified Area	*NEC*-Area Electrical Classification (All Class I, Group D)	Material of Construction for Buildings or Structures	Fire Protection Measures
8	a	SLUDGE THICKENER (CLARIFIER) Sludge concentration and removal, gravity, or dissolved air flotation	Possible generation of methane from sludge; carryover of floating flammable liquids	A	Enclosed — entire space	Division 1	NC	H, FE, and CGD if enclosed
	b			B		Division 2	NC, LC, or LFS	
	c			Not enclosed, open to atmosphere	Envelope 0.46 m (18 in.) above water surface and 3 m (10 ft) horizontally from wetted walls[1]			
9	a	SLUDGE PUMPING STATION DRY WELLS Dry side of a sludge pumping station	Buildup of methane gas or flammable vapors	D	Entire dry well when physically separated from a wet well or separate structures	Division 2	NC, LC, or LFS	H and FE
	b			C	Entire dry well when physically separated from a wet well or separate structures	Unclassified		
10	a	SLUDGE STORAGE WET WELLS, PITS, AND HOLDING TANKS Retaining of sludge	Possible generation of methane gas in explosive concentrations; carryover of floating flammable liquids	A	Enclosed — entire space	Division 1	NC	CGD, H, and FE if tank enclosed in structure
	b			B		Division 2	NC, LC, or LFS	NR
	c			Not enclosed, open to atmosphere	Envelope 0.46 m (18 in.) above water surface and 3 m (10 ft) horizontally from wetted walls[1]			
11	a	SLUDGE-BLENDING TANKS AND HOLDING WELLS Retaining of sludge with some agitation	Possible generation of methane gas in explosive concentrations; carryover of floating flammable liquids	A	Enclosed — entire space	Division 1	NC	H, FE, and CGD if tank enclosed in structure
	b			B		Division 2	NC, LC, or LFS	
	c			Not enclosed, open to atmosphere	Envelope 0.46 m (18 in.) above water surface and 3 m (10 ft) horizontally from wetted walls[1]			NR
12		DEWATERING BUILDINGS CONTAINING CENTRIFUGES, GRAVITY BELT THICKENERS, BELT AND VACUUM FILTERS, AND FILTER PRESSES Removal of water from sludge and the conveyance of sludge or sludge cake	NA	NR	NA	Unclassified	NC, LC, or LFS	FE, FDS, and FAS

Table 6.2(a) Continued

Row	Line	Location and Function	Fire and Explosion Hazard	Ventilation	Extent of Classified Area	*NEC*-Area Electrical Classification (All Class I, Group D)	Material of Construction for Buildings or Structures	Fire Protection Measures
13		INCINERATORS[2] AND INCINERATOR BUILDINGS Conveying and burning of sludge cake	Firebox explosion	NR	NA	Unclassified	NC, LC, or LFS	FSS (if indoors), H, and FE
14		HEAT TREATMENT UNITS, LOW- OR HIGH-PRESSURE OXIDATION UNITS Closed oxidation of sludge	None, other than in high-pressure systems	NR	NA	Unclassified	NC, LC, or LFS	H and FE
15	a	ANAEROBIC DIGESTERS, BOTH FIXED ROOF AND FLOATING COVER Generation of sludge gas from digesting sludge	Leakage of gas from cover, piping, emergency relief valves, and appurtenances	Not enclosed, open to atmosphere	Tank interior; areas above and around digester cover; envelope 3 m (10 ft) above the highest point of cover, when cover is at its maximum elevation, and 1.5 m (5 ft) from any wall	Division 1	NC	H and FE
	b				Envelope 4.6 m (15 ft) above Division 1 area over cover and 1.5 m (5 ft) beyond Divison 1 area around tank walls	Division 2		
	c			A	For digester tanks enclosed in a building: tank interior; entire area inside building	Division 1		CGD if enclosed
	d			B	For digester tanks enclosed in a building: tank interior; areas above and around digester cover; envelope 3 m (10 ft) above highest point of cover, when cover is at its maximum elevation, and 1.5 m (5 ft) from any wall of digester tank	Division 1	NC	CGD if enclosed
	e				Remaining space in enclosed area	Division 2	NC, LC, or LFS	

(continues)

Table 6.2(a) Continued

Row	Line	Location and Function	Fire and Explosion Hazard	Ventilation	Extent of Classified Area	*NEC*-Area Electrical Classification (All Class I, Group D)	Material of Construction for Buildings or Structures	Fire Protection Measures
16	a	ANAEROBIC DIGESTER CONTROL BUILDING Storage, handling, or burning of sludge gas	Leaking and ignition of sludge gas	A	Entire building	Division 1	NC	CGD, H, and FE
	b			B	Enclosed areas that contain gas-handling equipment	Division 2	NC, LC, or LFS	
	c			C	Physically separated from gas-handling equipment	Unclassified		
17	a	DIGESTER GAS-PROCESSING ROOMS Gas compression, handling, and processing	Sludge gas ignition	A	Entire room	Division 1	NC	CGD, H, and FE
	b			B		Division 2	NC, LC, or LFS	
	c			B	Within 1.5 m (5 ft) of equipment	Division 1		
18		ANAEROBIC DIGESTER GAS STORAGE Storage of sludge gas	Gas storage piping and handling	NNV	Within a 3-m (10-ft) envelope of tanks, valves, and appurtenances	Division 1	NC, LC, or LFS	H, FE, and CGD
19		CHLORINE OXIDATION UNITS Chlorine reaction with sludge	Chlorine is a very strong oxidizing agent	NR	NA	Unclassified	NR (These unit processes use corrosive chemicals that require the use of specific materials of construction. Special consideration shall be given to such materials of construction.)	H and FE
20	a	UNDERGROUND (PIPING) TUNNELS CONTAINING NATURAL GAS PIPING OR SLUDGE GAS PIPING Transmission of gas, sludge, water, air, and steam via piping; also might contain power cable and conduit	Ignition of natural gas or sludge gases	D	Within 3 m (10 ft) of valves and appurtenances	Division 1	NC, LC, or LFS	CGD, FDS, and FE
	b				Entire tunnel	Division 2		
	c			C	Areas within 3 m (10 ft) of valves, meters, gas check valves, condensate traps, and other piping appurtenances			
	d				Areas beyond 3 m (10 ft)	Unclassified		

Table 6.2(a) Continued

Row	Line	Location and Function	Fire and Explosion Hazard	Ventilation	Extent of Classified Area	*NEC*-Area Electrical Classification (All Class I, Group D)	Material of Construction for Buildings or Structures	Fire Protection Measures
21		UNDERGROUND (PIPING) TUNNELS NOT CONTAINING NATURAL GAS PIPING OR SLUDGE GAS PIPING Transmission of sludge, water, air, and steam piping; also might contain power cable and conduit	NA	NR	NA	Unclassified	NC, LC, or LFS	FDS and FE
22	a	COMPOSTING PILES Aerobic sludge reduction	Liberation of ammonia and toxic gas (composting materials can self-ignite)	D	Enclosed area	Division 2	NC, LC, or LFS	H and FDS
	b			C		Unclassified		
23	a	IN-VESSEL COMPOSTING Aerobic sludge reduction	Liberation of ammonia and toxic gas (composting materials can self-ignite)	As required by process	If enclosed, interior of reactor vessel plus a 3-m (10-ft) envelope around reactor vessel	Division 2	NC	H and FDS
	b				Areas beyond 3 m (10 ft)	Unclassified		
24	a	ODOR-CONTROL SYSTEM AREAS Areas physically separated from processes that house systems handling flammable gases	Leakage and ignition of flammable gases	D	Entire area if enclosed	Division 2	NC, LC, or LFS	CGD, FDS, and FE
	b			C	Areas within 1.5 m (3 ft) of leakage sources such as fans, dampers, flexible connections, flanges, pressurized unwelded ductwork, and odor-control vessels	Division 2		
	c				Areas beyond 1.5 m (3 ft)	Unclassified		

(continues)

Table 6.2(a) Continued

Row	Line	Location and Function	Fire and Explosion Hazard	Ventilation	Extent of Classified Area	*NEC*-Area Electrical Classification (All Class I, Group D)	Material of Construction for Buildings or Structures	Fire Protection Measures
25		PUMPING OF DRAINAGE FROM DIGESTED SLUDGE-DEWATERING PROCESSES Pumping of centrate, filtrate, leachate, drying beds, and so forth	NA	NR	NA	Unclassified	NC, LC, or LFS	H

Notes:
(1) The NR designation in the ventilation column indicates that no ventilation requirements are established for the space and, therefore, Table 9.1.1.4 also has no requirements.
(2) Row and Line columns are used to refer to the figure in A.6.2 and for specific requirements for each location and function.
(3) The following codes are used in this table:
A — No ventilation or ventilated at less than 12 air changes per hour
B — Continuously ventilated at 12 air changes per hour or in accordance with Chapter 9
C — Continuously ventilated at 6 air changes per hour or in accordance with Chapter 9
CGD — Combustible gas detection system
D — No ventilation or ventilated at less than 6 air changes per hour
FAS — Fire alarm system
FDS — Fire detection system
FE — Portable fire extinguisher
FSS — Fire suppression system (e.g., automatic sprinkler, water spray, foam, gaseous, or dry chemical)
H — Hydrant protection in accordance with 7.2.4
LC — Limited-combustible material
LFS — Low flame spread material
NA — Not applicable
NC — Noncombustible material
NEC — In accordance with NFPA 70
NVV — Not normally ventilated
NR — No requirement
[1]The area beyond the envelope is unclassified.
[2]See NFPA 54, NFPA 82, and NFPA 85.

A.6.2 See Figure A.6.2(a) through Figure A.6.2(g), which provide examples for Table 6.2(a).

Chapter 7 Fire and Explosion Prevention and Protection

7.4 Combustible Gas Detection.

7.4.5 Combustible gas detection equipment located in hazardous (classified) locations, as defined in accordance with NFPA 70 shall be listed for use in such atmospheres.

7.4.5.1 Detectors located in hazardous (classified) locations shall be set to alarm at 10 percent of the lower explosive limit (LEL) in accordance with the manufacturers' calibration instructions and shall be connected to alarm signaling systems.

7.4.5.2 Where permitted by the authority having jurisdiction, the alarm limits shall be permitted to be set at higher than 10 percent of the explosive limit where experience indicates ambient levels would produce spurious alarms.

Chapter 8 Materials of Construction

8.2 Materials Selection.

8.2.4 Materials of construction used for unit processes located in areas with an NFPA 70 classification of Class I, Division 1 or Division 2, and Class II shall be selected based on an overall evaluation, including the following:

(1) Fire risk of the material attributes
(2) Economic impact of replacing the unit process
(3) Potential environmental dangers caused by having the

Table 6.2(b) Solids Treatment Processes — Sludge Drying

	Location and Function	Fire and Explosion Hazard	Ventilation	Extent of Classified Area	*NEC*-Area Electrical Classification (All Class II, Group G)	Material of Construction for Buildings or Structures	Fire Protection Measures
1	SLUDGE-DRYING PROCESSES[1]	Potential for ignition of dust	NR	Entire room[2]	Division 1, or if acceptable to the authority having jurisdiction with classifications in NFPA 499	NC (Construction in accordance with NFPA 68 and NFPA 69)	H, FAS, and FSS *(See NFPA 61 and NFPA 69.)*
2	DRIED SLUDGE STORAGE AREAS, IF ENCLOSED	Potential for ignition of dust	NR	Entire room[2]	Division 1, or if acceptable to the authority having jurisdiction with classifications in NFPA 499	NC (Construction in accordance with NFPA 68 and NFPA 69)	H, FAS *(See NFPA 61 and NFPA 69.)*

Notes:
(1) The NR designation in the ventilation column indicates that no ventilation requirements are established for the space and, therefore, Table 9.1.1.4 also has no requirements.
(2) The following codes are used in this table:
FAS — Fire alarm system
FSS — Fire suppression system (e.g., automatic sprinkler, water spray, foam, gaseous, or dry chemical)
H — Hydrant protection in accordance with 7.2.4
NEC — In accordance with NFPA 70
NC — Noncombustible material
NR — No requirement
[1] See NFPA 54, NFPA 82, and NFPA 85.
[2] The area beyond the envelope is unclassified.

unit process out of service for an extended period of time due to fire or explosion

Chapter 9 Ventilation

9.1.2 Hazardous Classifications. Hazardous classifications as established in Table 4.2, Table 5.2, Table 6.2(a), and Table 6.2(b) shall be permitted to be reduced to a lower classification, including unclassified, with positive pressurization in accordance with both of the following:

(1) Article 500 of NFPA 70
(2) NFPA 496

9.2 Installation.

9.2.3 Ventilation systems serving hazardous areas classified under the provisions of Article 500 of NFPA 70 shall incorporate fans fabricated in accordance with Air Moving and Control Association (AMCA) Type A or Type B spark-resistant construction.

Annex A Figures

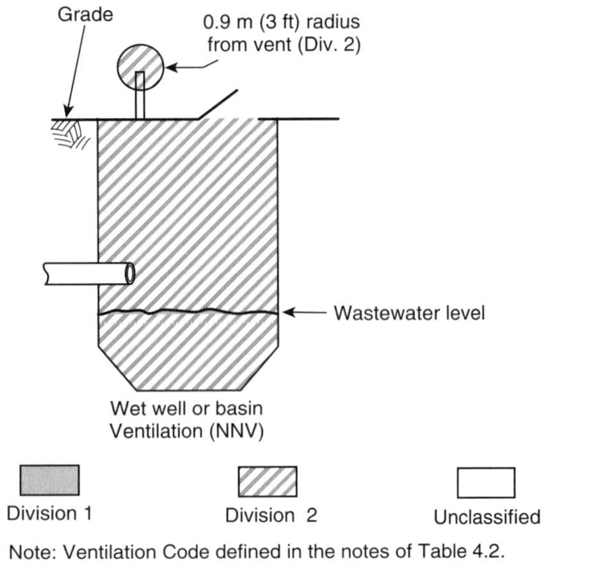

FIGURE A.4.2(a) Wet Well or Basin Serving a Storm Sewer; Illustration of Table 4.2, Row 4.

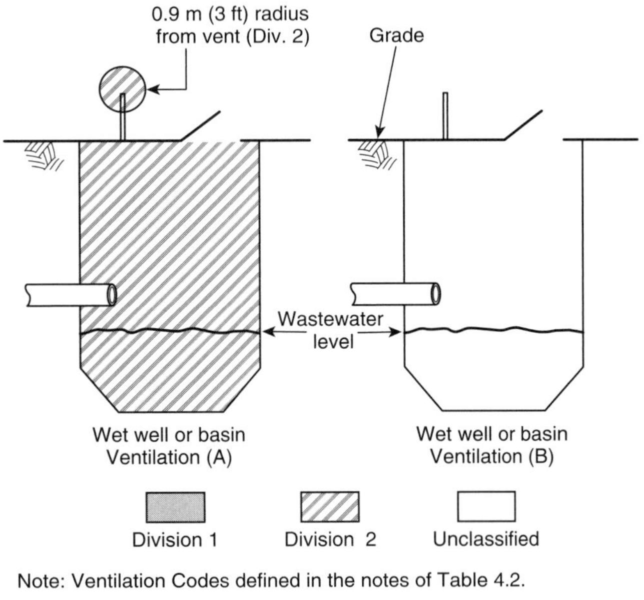

FIGURE A.4.2(b) Wet Well or Basin Serving a Residential Sewer; Illustration of Table 4.2, Row 11.

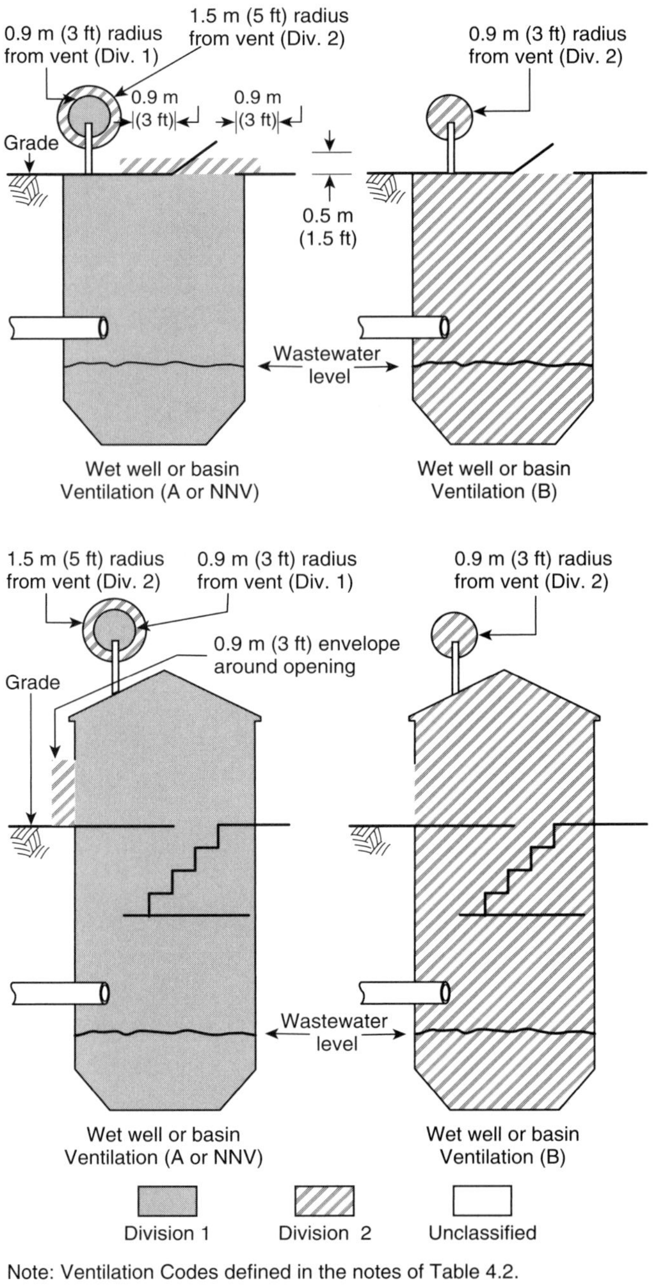

FIGURE A.4.2(c) Wet Well or Basin Serving Separate or Combined Sanitary Sewer; Illustration of Table 4.2, Rows 16 and 34.

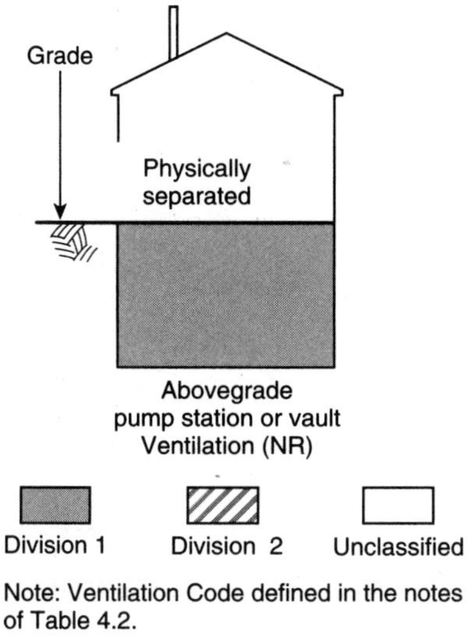

FIGURE A.4.2(d) Abovegrade Equipment Housing or Vault Physically Separated from Wet Well or Basin; Illustration of Table 4.2, Rows 18 and 30.

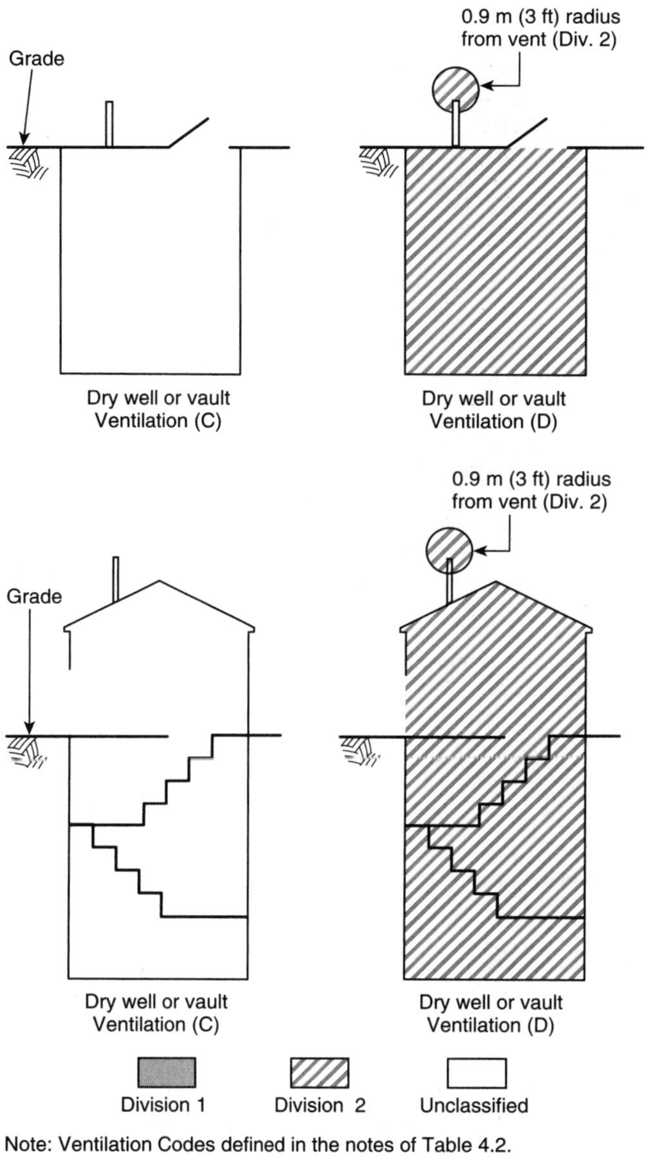

FIGURE A.4.2(e) Belowgrade or Partially Belowgrade Equipment Housing or Vault Physically Separated from Wet Well or Basin; Illustration of Table 4.2, Rows 5, 12, 17, 31, and 36.

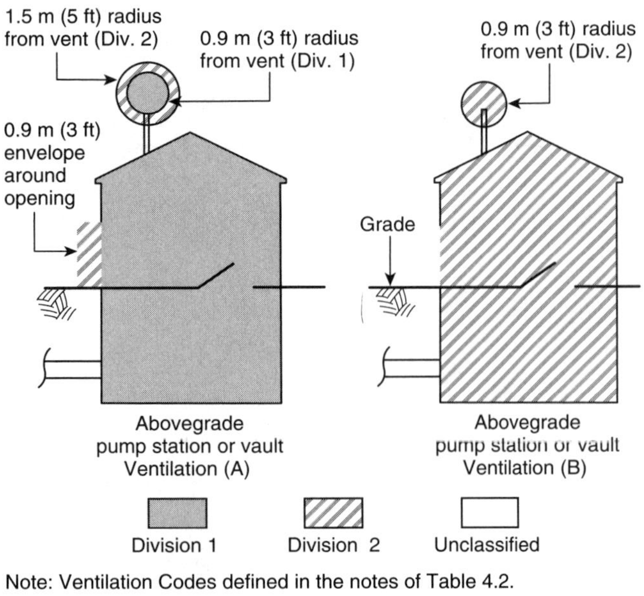

FIGURE A.4.2(f) Abovegrade Equipment Housing or Vault not Physically Separated from Wet Well or Basin; Illustration of Table 4.2, Row 19.

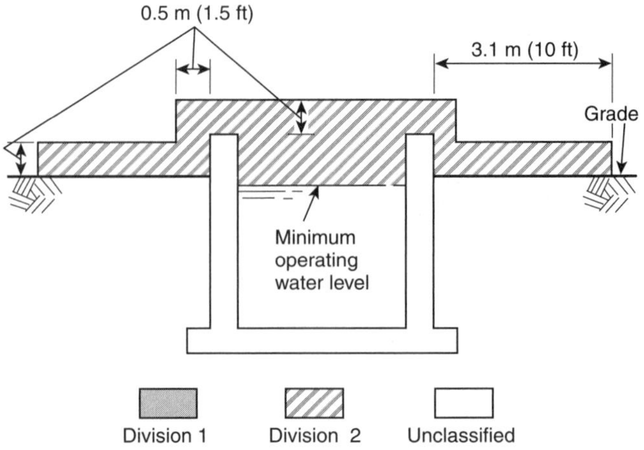

FIGURE A.5.2 Primary Sedimentation Tank; Illustration of Table 5.2, Row 6.

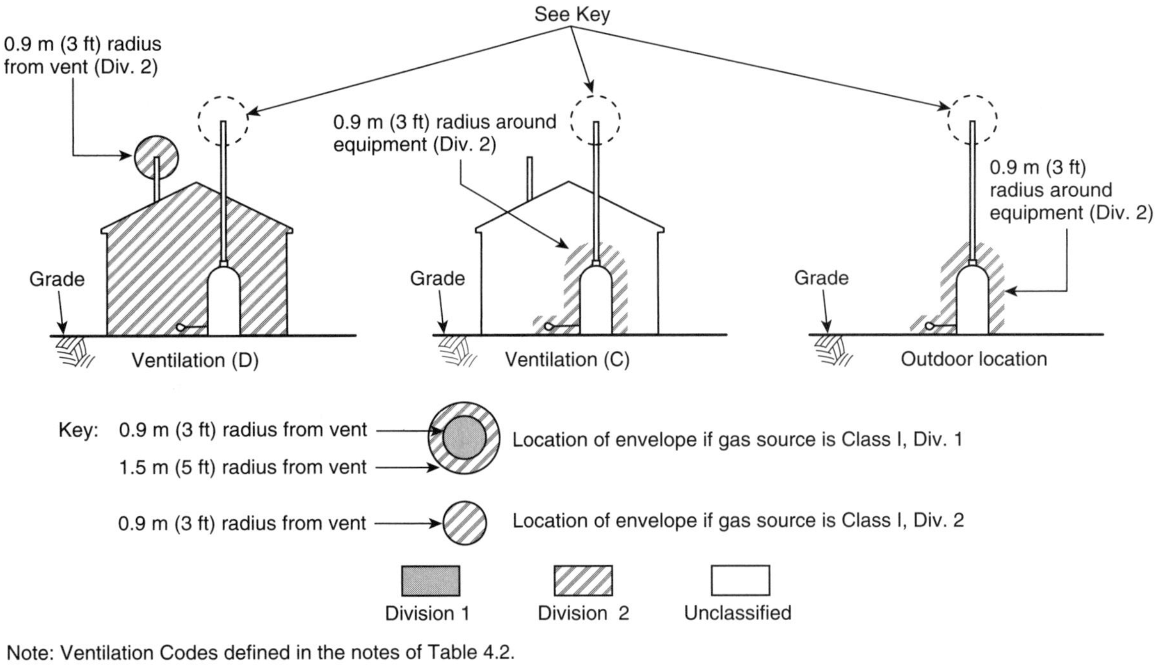

FIGURE A.4.2(g) Odor-Control System Location Physically Separated from Wet Well; Illustration of Table 4.2, Row 20.

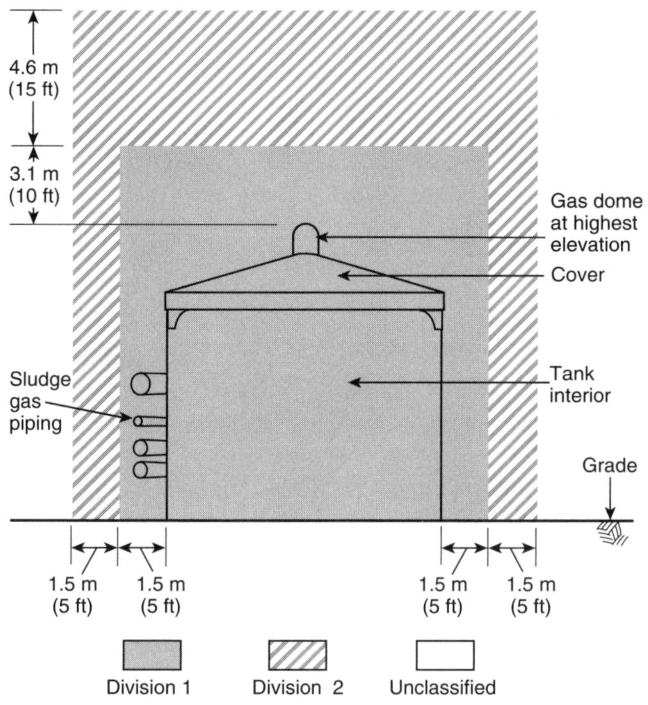

FIGURE A.6.2(a) Anaerobic Digester with Fixed or Floating Cover Abovegrade not Enclosed in a Building; Illustration of Table 6.2(a), Rows 15a and 15b.

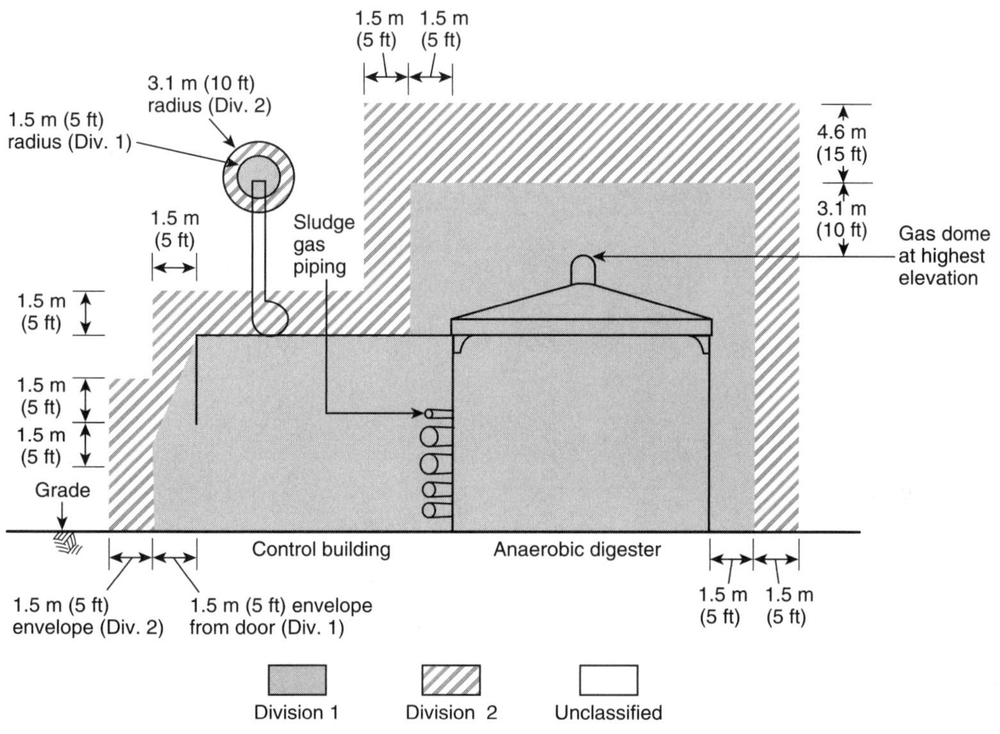

FIGURE A.6.2(b) Anaerobic Digester Control Building Containing Sludge Gas Piping and Using Ventilation Method (A); Illustration of Table 6.2(a), Row 16a.

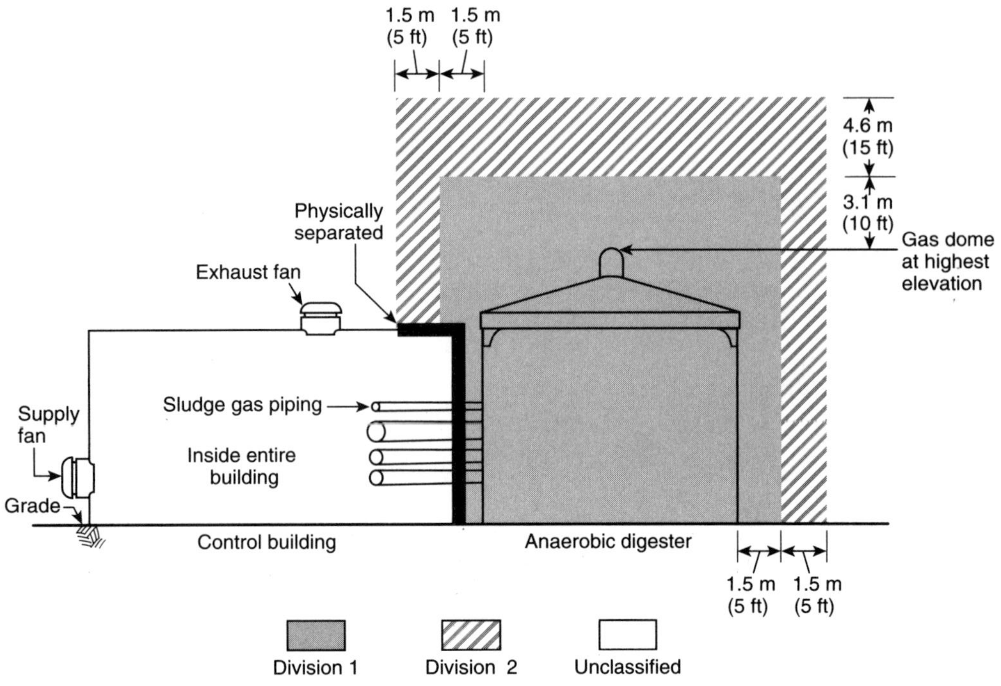

FIGURE A.6.2(c) Anaerobic Digester Control Building Containing Sludge Gas Piping and Using Ventilation Method (C); Illustration of Table 6.2(a), Row 16c.

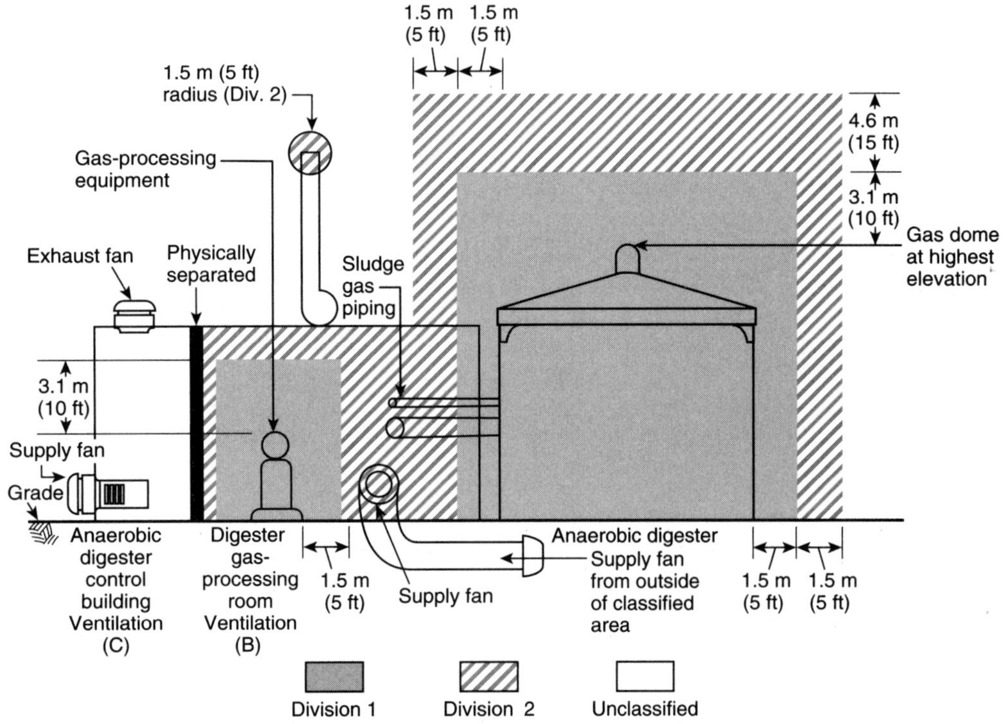

Note: Ventilation Codes defined in the notes of Table 6.2(a).

FIGURE A.6.2(d) Anaerobic Digester Control Building Containing Sludge Gas–Processing Equipment Physically Separated and Using Ventilation Method (B) for the Processing Room and Ventilation Method (C) for the Control Building; Illustration of Table 6.2(a), Rows 16c and 17b.

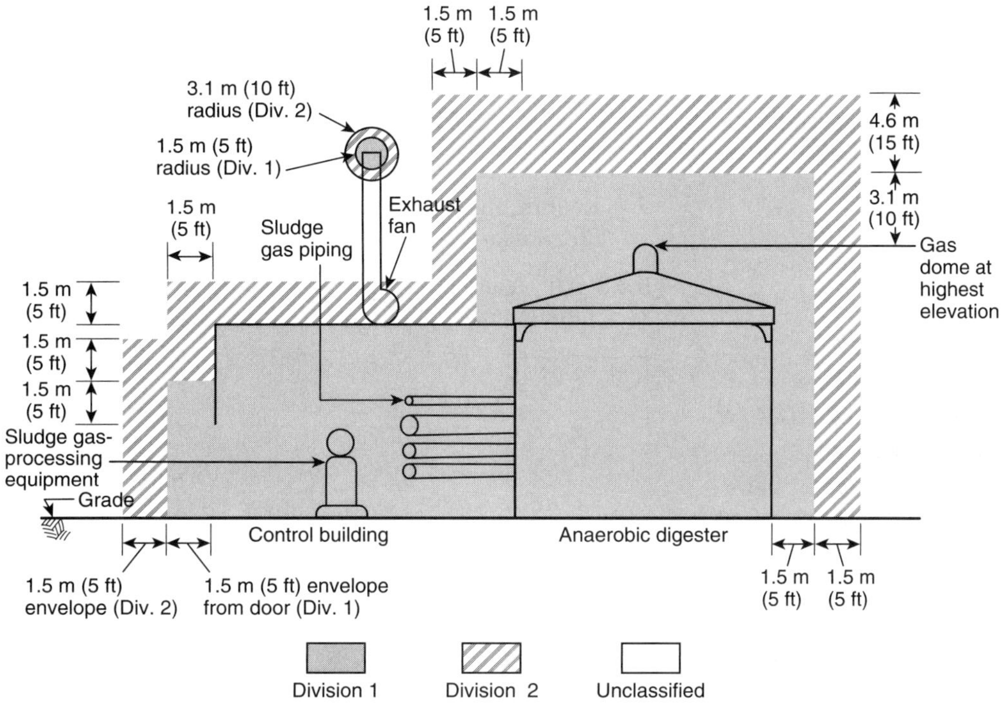

FIGURE A.6.2(e) Anaerobic Digester Control Building Containing Sludge Gas–Processing Equipment not Physically Separated and Using Ventilation Method (A); Illustration of Table 6.2(a), Row 16.

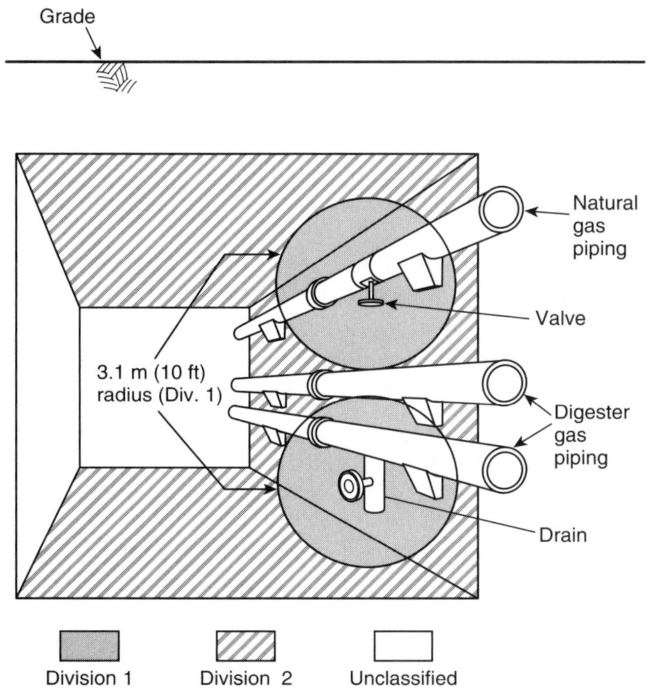

FIGURE A.6.2(f) Underground Tunnel Containing Natural Gas or Sludge Gas Piping and Using Ventilation Method (D); Illustration of Table 6.2(a), Rows 20a and 20b.

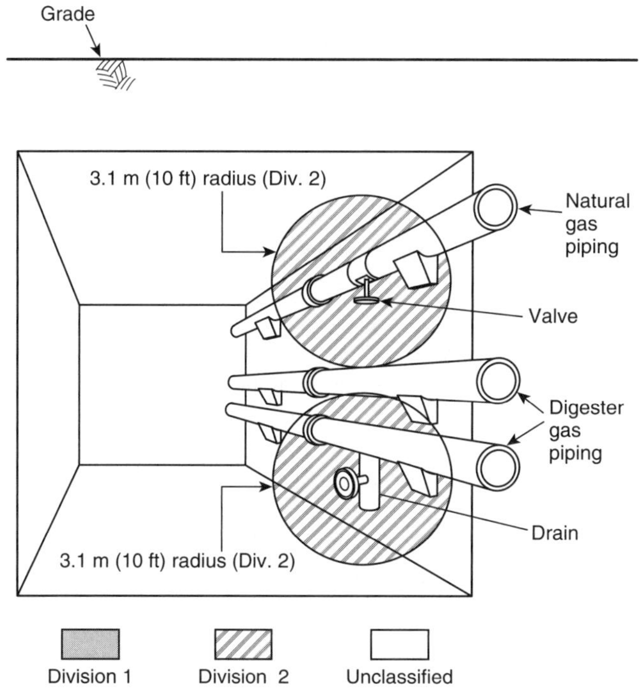

FIGURE A.6.2(g) Underground Tunnel Containing Natural Gas or Sludge Gas Piping and Using Ventilation Method (C); Illustration of Table 6.2(a), Rows 20c and 20d.

NFPA 1124, *Code for the Manufacture, Transportation, Storage, and Retail Sales of Fireworks and Pyrotechnic Articles,* 2003

1.1 Scope.

This code regulates the construction, use, and maintenance of buildings and facilities for the following:

(1) The manufacture and storage of fireworks at fireworks manufacturing facilities
(2) The storage of display fireworks, pyrotechnic articles, salute powder, pyrotechnic and explosive compositions, and black powder at other than display sites
(3) The retail sales and related storage of consumer fireworks in consumer fireworks retail sales facilities and stores
(4) The transportation of fireworks, pyrotechnic articles, and components thereof containing pyrotechnic or explosive materials on public highways

Chapter 4 Manufacturing Operations

4.5 Construction.

4.5.3 Heating, Lighting, and Electrical Equipment.

4.5.3.3 Unit heaters located in buildings that contain exposed explosive or pyrotechnic composition shall be equipped with motors and electrical devices for use in hazardous locations in accordance with Article 502 of NFPA 70, *National Electrical Code®*.

4.5.3.4 All wiring, switches, and electrical fixtures in process buildings shall meet the requirements for hazardous locations in accordance with Article 502 of NFPA 70, *National Electrical Code®*.

4.5.3.4.1 Portable lighting equipment shall not be used, unless both of the following criteria are met:

(1) Listed portable lighting equipment shall be permitted to be used during repair operations.
(2) The repair operations area shall be cleared of all pyrotechnic or explosive material, and all dust or residue shall be removed before portable lighting equipment is used.

4.5.3.4.2 All presses and other such mechanical devices used in the vicinity of exposed explosive or pyrotechnic composition shall be electrically bonded and grounded.

Chapter 5 Storage of Display Fireworks, Pyrotechnic Articles, Salute Powder, Pyrotechnic and Explosive Compositions, and Black Powder

5.4 Requirements for Shipping Buildings for Display Fireworks.

5.4.5 All electrical equipment and fixtures in a shipping building shall meet the requirements for hazardous locations in accordance with NFPA 70, *National Electrical Code®*.

NFPA 1125, *Code for the Manufacture of Model Rocket and High Power Rocket Motors,* 2001

1.1 Scope.

1.1.1* This code shall apply to the manufacture of model and high power rocket motors designed, sold, and used for the purpose of propelling recoverable aero models.

A.1.1.1 For further information on propelling recoverable aero models, see NFPA 1122, *Code for Model Rocketry,* and NFPA 1127, *Code for High Power Rocketry.*

1.1.2 This code shall apply to the design, construction, limitation of propellant mass and power, and reliability of model and high power rocket motors and model rocket and high power motor-reloading kits and their components.

1.1.3 This code shall not apply to the sale and use of the following:

(1) Model rocket motors (covered by NFPA 1122, *Code for Model Rocketry*)
(2) High power rocket motors (covered by NFPA 1127, *Code for High Power Rocketry*)

1.1.4* This code shall not apply to the manufacture, transportation, and storage of fireworks.

A.1.1.4 For further information on fireworks, see NFPA 1124, *Code for the Manufacture, Transportation, and Storage of Fireworks and Pyrotechnic Articles.*

1.1.5 This code shall not apply to the manufacture, transportation, and storage of rocket motors by the United States military or other agencies or political subdivisions of the United States.

1.1.6 This code shall not apply to the assembly of reloadable model or high power rocket motors by the user.

Chapter 4 Manufacturing Operations for Rocket Motors

4.9 Heat, Light, and Electrical Equipment.

4.9.2.1 Unit heaters located in rooms where propellant, delay, or ejection composition or flammable materials are being processed or stored shall be equipped with motors and electrical devices in accordance with Article 500, NFPA 70, *National Electrical Code®*.

Chapter 5 Limited Quantity Manufacturing Operations for Composite Propellant Rocket Motors

5.8 Heat, Light, and Electrical Equipment.

5.8.2.1 Unit heaters located in rooms where propellant, delay, or ejection composition or flammable materials are being processed or stored shall be equipped with motors and electrical devices in accordance with Article 500, NFPA 70, *National Electrical Code*.

Exception: Buildings or rooms that are used only for static motor tests.

INDEX

Aboveground tanks
Solvent storage **36:**5.5.1.2, **36:**A.5.5.1.2
Acetylene
Static electricity and **77:**7.15
Air ionizers 77:4.2.6, **77:**6.1.2, **77:**6.5.1
Definition **77:**D.1.1
Aircraft fuel
Systems
Aircraft engine test facilities **423:**5.3.1, **423:**5.3.2
Aircraft storage and servicing area
Landing gear pits, ducts, and tunnels **409:**5.9, **409:**A.5.9.1
Alarms
Definition **496:**3.3.1, **496:**A.3.3.1
Aluminum
Dust, combustible **499:**4.4.3
Paste **484:**4.2
Powder **484:**4.2,**484:**4.1
Aluminum powder plants 484:4.1
Analyzer rooms or buildings
Definition **496:**3.3.2
Pressurized **496:**1.3.4
Antistatic
Definition **77:**3.1.1
Antistatic additives 77:7.6.5
Definition **77:**D.1.2
Apparatus (nonfire)
Associated **70:**504.50(A)
Definition **70:**504.2
Intrinsically safe **70:**504.50(A)
Definition **70:**504.2
Arcs, electrical 70:511.7(B)(1), **70:**513.7(C), **70:**514.7, **70:**515.7(B), **70:**516.7(B), **70:**517.61(B)(2)
Autoignition temperature
Chemicals **497:**A.4.4.2
Definition **497:**3.3.2, **499:**3.3.1, **499:**A.3.3.1

Batteries (power)
Aircraft **70:**513.10
Aircraft hangars **70:**513.10
Equipment **70:**503.160, **70:**511.10(A), **70:**513.10(B)
Garages **70:**511.10
Battery chargers 70:503.160, **70:**511.10(A)
Aircraft hangars **70:**513.10(B)
Belt conveyors
Metal mineral processing facilities **122:**A.1.1
Static electricity generated by **77:**9.4
Bonding, electrical
Definition **77:**3.1.2
Hazardous (classified) locations **70E:**440.4
Intrinsically safe systems **70:**504.60
Static electricity, to control **77:**5.1.2, **77:**6.4.1, **77:**7.4, **77:**7.6, **77:**8.8.3
Bonding jumpers
Hazardous (classified) locations **70:**501.30(A), **70:**502.30(A), **70:**503.30(A)
Boxes, electrical
Outlet, device, pull and junction
Hazardous (classified) locations **70E:**440.3(B)(1), **70E:**440.3(D)
Branch circuits
Multiwire **70:**501.40, **70:**502.40, **70:**505.21
Breakdown strength
Definition **77:**3.1.3
Breakdown voltage
Definition **77:**3.1.4
Brush discharges 77:4.3.2.1, **77:**8.4.2.3, **77:**8.5.2, **77:**8.8.1, **77:**8.12, **77:**9.1.2.2
Definition **77:**D.1.3
Bucket elevators
Chemical process areas **499:**5.2.3.1, **499:**5.3.2, **499:**5.6.3
Buildings *see also* Construction
Explosive materials, used for preparation of **495:**5.2
LP-Gas
Heating **58:**10.2.3
Bulk oxygen systems 55:1.1.2(4)
Bulk plants
Definition **30A:**3.3.2
Bulk terminals
Definition **30A:**3.3.2
Bulking brush discharges 77:4.3.8, **77:**8.4.2.2 to 8.4.2.3, **77:**8.5.4, **77:**8.12, **77:**9.1.2.4
Definition **77:**D.1.4

Cables, electric
Sealing **70:**501.15(D), **70:**501.15(E)
Calcium carbide
Storage **51:**1.1.1(3)(b)
Capacitance 77:4.2.1, **77:**4.2.4, **77:**4.3.3, **77:**5.4.3, **77:**5.9, **77:**8.5.1
Definition **77:**3.1.5, **77:**A.3.1.5

Carbon dioxide
Static electricity and **77:**7.15
Cargo tank vehicles
LP-Gas **58:**6.20.2.6, **59:**4.5.2.6, **59:**A.4.5.2.2
Cathode ray tube (CRT) video display terminal 77:9.6
Charge decay time 77:4.2.4
Definition **77:**D.1.6
Charge density 77:6.5, **77:**8.3, **77:**8.12
Definition **77:**D.1.7
Surface-charge density **77:**8.3.2
Definition **77:**D.1.52
Charge relaxation (dissipation) 77:6.1.2(1), **77:**7.3.2, **77:**A.7.3.2
Definition **77:**D.1.8
In intermediate bulk containers **77:**9.1.1, **77:**9.1.5.2, **77:**9.1.6.2, **77:**9.1.7.1
Time **77:**4.2.4, **77:**8.4.1
Time constant
Definition **77:**D.1.45
Chemical Abstract Service (CAS)
Definition **497:**3.3.3, **499:**3.3.2
Numbers **499:**4.5.3
Chemical plants
Electrical installations in **497:**5.7.3
Chemicals
Wastewater treatment, use in **820:**1.1.2(9)
Circuit breakers 70:514.11(A)
Hazardous (classified) locations **70:**501.115, **70:**502.115, **70:**503.115
Circuits
Intrinsically safe **70:**504.30
Definition **70:**504.2
Nonincendive
Definition **70:**506.2
Closed containers
Definition **34:**3.3.2
Clothing
Static electricity and **77:**6.6.4
Clothing manufacturing plants, electrical systems 70:500.5(D), **70:**506
Coal
Piles **120:**9.5.3, **120:**A.9.1
Silos **120:**A.9.1
Coating processes
Static electricity and **77:**6.4.3.6, **77:**7.5, **77:**9.2.3, **77:**9.2.4.2
Coatings
Curtain
Definition **34:**3.3.3.1
Flow
Definition **34:**3.3.3.2
Roll
Definition **34:**3.3.3.3
Collector sewers 820:1.1.2(1)
Combined sewers 820:1.1.2(4)
Combustible 70:506
Definition **77:**3.1.6
Combustible gas
Detectors and detection systems
Definition **70:**505.2
Wastewater treatment and collection facilities **820:**7.4
Combustible liquids
Classification **30:**1.7.3, **497:**4.2.7, **497:**5.4.4, **497:**5.7.2
Definition **497:**3.3.4
Definition **30:**1.7.3.1
Storage
Large storage areas
Definition **122:**3.3.4.1
Combustible materials
Classification, chemical process areas **497:**5.6, **497:**5.8.2.5, **497:**5.8.3, **497:**5.9, **497:**5.10
Definition **497:**3.3.5.1, **497:**A.3.3.5.1, **499:**3.3.5
Combustible powders
Definition **33:**3.3.3
Composition
Safety precautions **1125:**4.9.2.1
Compressed air
Dust removal use **664:**11.2.1.1
Compressed natural gas (CNG)
Properties **52:**A.1.1
Compressed natural gas (CNG) systems
Vehicular fuel systems
Motor fuel dispensing facilities **30A:**1.1.2, **30A:**A.1.1.2
Condensate (condensation)
Definition **36:**3.3.1
Conductive
Definition **77:**3.1.8
Conductive hose 77:7.4.3, **77:**A.7.4.3
Definition **77:**D.1.14
Conductivity 77:4.1.4, **77:**4.1.9, **77:**4.2.2
Definition **77:**D.1.15
Increasing to control static electricity **77:**6.4.3.2
Of liquids **77:**7.3.3, **77:**A.7.3.3
Conductor fill
Hazardous (classified) locations, sealing fittings **70:**501.15(C)(6)
Conductors, electrical
Charging of **77:**4.1.9
Copper **70:**505.18(A
Corrosive conditions **70:**505.18(B
Definition **77:**3.1.9
Insulated **77:**4.3.6
Hazardous (classified) locations **70:**501.20, **70:**505.18(B)
Intrinsically safe **70:**504.30, **70:**504.50
Sparks between **77:**4.3.3, **77:**A.4.3.3.4

Construction
Magazines
Explosive materials **495:**8.5
Wastewater treatment and collection facilities **820:**4.2, **820:**5.2, **820:**6.2, **820:**A.4.2, **820:**A.6.2
Containers
Cleaning
Static electricity control **77:**7.13.7
Metal, static electricity control **77:**7.13.3
Portable
Static electricity control **77:**7.13
Powder
Static electricity control **77:**8.4.2.2
Static electricity control
During filling/emptying **77:**7.13
Grounded containers **77:**8.4.2.2
Linings **77:**7.13.4
Nonconductive containers **77:**8.4.2.2, **77:**8.11.2
Control rooms
Pressurized **496:**1.3.2
Controlled areas
Solvent extraction plants **36:**7.7.3
Controllers
Motor
Hazardous (classified) locations **70:**501.115, **70:**502.115, **70:**503.115
Conveyor ducts
Static electricity control **77:**8.6
Conveyors
Dipping and coating processes **34:**11.4, **34:**11.6, **34:**A.11.4.2
Corona discharges 77:4.1.10, **77:**4.3.2, **77:**6.5.2.2, **77:**6.5.3.1 **77:**7.15, **77:**8.4.2.3 **77:**8.5.2
Definition **77:**D.1.17
Coulomb
Definition **77:**D.1.18
Cranes
Hazardous (classified) locations **70:**503.155
Cryogenic liquids/fluids 497:4.2.5, **497:**5.6.5
Current
Definition **77:**D.1.19
Curtain coating
Definition **34:**3.3.3.1
Cylinders *see also* Containers
Storage
Oxygen-fuel gas **51:**1.1.1(3)(a)

Deenergizing (electricity)
Hazardous (classified) location electrical equipment **70B:**22.2.3
Density
Bulk **77:**8.12
Definition **77:**D.1.20
Detearing
Definition **34:**3.3.4
Detection systems
Mines **122:**A.1.1
Dielectric breakdown
Definition **77:**D.1.21
Dielectric constant 77:8.4.1
Definition **77:**D.1.22
Dielectric strength
Definition **77:**D.1.23
Diesel equipment
Underground mines, in **122:**1.1.2
Diesel fuel tanks
Fixed storage in underground mines **122:**A.1.1
Dip tanks
Definition **34:**3.3.5
Direct fired vaporizers 59:4.7.2
Dissipative
Definition **77:**D.1.24
Drainage
Equipment **70:**501.15(F)
Drycleaning plants 497:5.3.2.2(4)
Drycleaning rooms 32:7.3.2
Ducts and duct systems
Aircraft hangars **409:**5.9
Dust
Accumulations
Combustible particulate solids, handling of **654:**2.7.2, **654:**A.2.7.2
Wood processing and woodworking facilities **664:**7.3.2, **664:**A.7.3.2
Clouds **499:**4.1.3.1(1), **499:**4.1.3.2(1), **499:**4.2.1.2, **499:**5.2.1 **499:**5.3.4, **499:**5.3.6.2 **499:**5.4.1, **499:**5.7.2.2(1)
Ignition of **499:**4.1.4, **499:**4.3.1.1(2), **499:**4.3.1.3(2), **499:**4.3.2, **499:**4.6
Static electricity and **77:**8.2, **77:**9.1.3, **77:**9.1.7.3, **77:**A.8.2.4
Combustible
Definition **77:**3.1.7, **499:**3.3.3, **499:**A.3.3.3
Hazardous (classified) locations, in **70E:**440.1, **70E:**440.3(B)(6)
Static electricity and **77:**4.3, **77:**9.1.3, **77:**9.1.7.3
Conductive **499:**4.4.1
Dispersion **499:**4.2.1
Explosions **499:**4.2.1, **499:**4.3.2, **499:**4.6.3
Hazardous (classified) locations, in **70E:**440.1, **70E:**440.3(B)(6)
Hybrid **499:**4.2, **499:**4.6.3
Definition **499:**3.3.7
Ignition properties of **496:**6.2.5, **496:**A.6.2.5, **499:**4.1.4 **499:**4.4.4, **499:**4.6, **499:**4.7, **499:**A.4.3, **499:**A.4.5.2

Layers **499:**4.1.3.2, **499:**4.2.1.1, **499:**4.2.1.2, **499:**5.2.3.5, **499:**5.3.4, **499:**5.3.6.2, **499:**5.4.1(5), (6), **499:**5.7.2.2(2)
Ignition of **499:**4.1.4, **499:**4.3.1.2(2), **499:**4.3.1.3(2), **499:**4.3.2, **499:**4.3.3.1, **499:**4.3.3.5, **499:**4.3.4, **499:**4.7
Thickness **499:**5.2.2
Static electricity and **77:**4.3.3.6, **77:**9.1.3, **77:**9.1.7.3
Wood processing and woodworking facilities **664:**11.2
Dust collection systems
Chemical process areas **499:**5.4.1(5), **499:**5.4.2, **499:**5.5.1(8), **499:**5.5.2, **499:**5.5.5(2), **499:**5.5.6, **499:**5.6.5
Dust-ignitionproof 70:500.7(B), **70:**506.8(A)
Definition **70:**500.2, **70:**506.2
Equipment **70B:**22.1, **70B:**22.2.10.1, **499:**4.1.5.1
Dusttight 70:500.7(C), **70:**506.8(D)
Definition **70:**500.2, **70:**506.2
Enclosures **499:**4.1.6, **499:**4.1.7.3
Equipment **70B:**22.1, **70B:**22.2.10.1, **70E:**440.3(D)
Dye plants, electrical installations 70:500.5(B)(1), **497:**5.3.2.2(4)

Electric meters
Hazardous (classified) locations **70:**502.150
Electrical equipment
Aerosol manufacture and storage **30B:**5.5
Aircraft fuel servicing **407:**4.3.7, **407:**A.4.3.7.4
Combustible particulate solids, systems handling **654:**5.2, **654:**A.2.7.2
Compressed natural gas vehicular fuel systems **52:**6.4.3.8, **52:**6.12, **52:**A.6.12
Definition **70:**505.2
Fireworks facilities **1124:**4.5.3, **1124:**5.4.5
Flammable and combustible liquids storage and handling areas **30:**7.9.5
Wharves **30:**7.7.16
LP-Gas systems **58:**A.6.20.2.2, **58:**A.6.20.2.3, **59:**4.5.2, **59:**4.5.2.2, **59:**4.5.2.4, **59:**4.7, **59:**A.4.5.2.2
Model rocket motor plants **1125:**4.9, **1125:**5.8
Motor fuel dispensing facilities **30A:**12.6.2, **30A:**8.5, **30A:**A.8.5.2
Safety, personnel and equipment **70B:**22.2.2, **70B:**22.2.8
Solvent extraction plants **36:**4.3.1, **36:**5.5.1, **36:**6.3.1, **36:**A.5.5.1.2, **36:**A.6.3.1, **36:**A.7.7.1
Spray application **33:**11.3.1, **33:**12.4, **33:**17.5
Testing **70B:**22.2.1
Electrical installations
Agricultural and food products facilities **61:**4.1.3, **61:**11.2.3
Aircraft engine test facilities **423:**5.3
Coal mines and coal preparation plants **120:**7.1.5.6 **120:**7.5.1.1, **120:**9.1.1(8), **120:**A.7.5.1.1
Dipping and coating processes **34:**11.6, **34:**11.8, **34:**A.11.8
Drycleaning plants **32:**7.3.2
Electrostatic detearing apparatus **34:**11.6, **34:**11.8, **34:**A.11.8
Hypobaric facilities **99B:**4.7, **99B:**A.4
Laboratories **45:**A.5.6.2
Liquefied natural gas (LNG) fueling facilities **57:**A.5.12.4
Liquefied natural gas (LNG) process systems **59A:**7.6
Metals, combustible, handling and processing operations **484:**4.1.6, **484:**4.2.4.2, **484:**6.1.6.1, **484:**6.3.4, **484:**7.1.3, **484:**8.6.2.4, **484:**9.6.2.4
Semiconductor fabrication facilities **318:**6.5, **318:**A.6.5
Wastewater treatment and collection facilities **820:**7.4.5.1, **820:**9.1.2, **820:**9.2.3
Wood processing and woodworking facilities **664:**A.7.3.2, **664:**A.7.18
Electrical neutralizers 77:6.5.3
Electrical preventive maintenance (EPM) programs 70B:22.2.1
Electrical systems *see also* Electrical installations
Oxygen-enriched atmospheres, use in **53:**A.7.7.2
Electricity
Basic explanation of **77:**4.1.2
Electrometers 77:5.6.3
Definition **77:**D.1.26
Electrostatic equipment 70:516.4(E), **70:**516.10
Fluidized beds
Definition **33:**3.3.5.1
Spray equipment
Automated **33:**15.10
Hand spraying **33:**15.11
Electrostatic field 77:5.4.2 **77:**8.3.2
Definition **77:**D.1.2
Electrostatic field meters 77:5.4.2, **77:**5.5.1, **77:**5.6.2
Definition **77:**D.1.28
Enclosures
Coal mines and coal preparation plants **120:**6.2.1.3, **120:**A.6.2.1.4.1, **120:**A.6.2.1.6
Dusttight **499:**4.1 6, **499:**4.1.7.3
Hazardous (classified) locations **70B:**22.2.5, **70B:**22.2.10.1, **70B:**22.2.10.3
Intrinsically safe conductors in **70:**504.30(A)(2)
Volume
Definition **496:**3.3.3
Engine test cells (aircraft)
Construction and systems **423:**5.3.1
Exhaust systems
Dipping and coating processes **34:**11.6(1)
Explosion hazards
Definition **664:**3.3.6
Explosion severity
Definition **499:**3.3.6, **499:**A.3.3.9

Explosionproof apparatus 70:500.2, **70:**500.7(A), **70:**502.5
Explosive compositions 1124:4.5.3.3, **1124:**4.5.3.4.2
Explosive materials/explosives
Static electricity control **77:**9.5
Exposed
Hazardous (classified) locations **70:**501.25, **70:**502.25, **70:**503.25, **70:**505.19
Extinguishers, portable fire
Metal mineral processing plants **122:**A.1.1
Mines **122:**A.1.1

Fabrication areas, semiconductor fabrication facilities 318:6.5.1
Faraday cage
Definition **77:**D.1.29
Fibers
Combustible **70E:**440.1
Hazardous (classified) locations **70:**506
Static electricity and **77:**4.3.3.6, **77:**6.6.4.1
Field mills 77:5.4.2
Field suppression
Definition **77:**D.1.30
Fire protection
Spray application operations **33:**A.1.1
Fire suppression systems
Mines
Metal and nonmetal **122:**A.1.1
Fittings
Electrical systems **70E:**440.3(B)(1), **70E:**440.3(D), **70E:**440.4(B)
Flammable gas 55:1.1.2(8), **55:**1.1.2(13), **55:**1.1.2(14), **55:**A.1.1.2(8)
Hazardous (classified) locations, in **70E:**440.1, **70E:**440.3(B)(6)
Pressurized analyzer rooms **496:**1.3.4
Pressurized enclosures **496:**1.3.3
Static electricity and **77:**4.1.7, **77:**4.3.3.7, **77:**5.10, **77:**7.5.2.5, **77:**7.15, **77:**B.1
Flammable limits 77:7.2.2, **77:**A.7.2.2
Flammable liquids
Aerosol manufacture and storage **30B:**4.3.1.1, **30B:**5.5.2
Classification **30:**1.7.3
Coal mines
Storage and use **120:**7.5, **120:**A.7.5.1.1
Definition **30:**1.7.3.2, **497:**3.3.6
Hazardous (classified) locations, in **70E:**440.1
Mines
Coal **120:**7.5, **120:**A.7.5.1.1
Pressurized analyzer rooms **496:**1.3.4
Storage **122:**1.1.2
Areas **120:**7.5.1.1, **120:**A.7.5.1.1
Laboratories **45:**12.2.2.2, **45:**A.12.2.2.2
Large storage areas **122:**9.4, **122:**A.9.4.2.1
Underground
Storage in mines **120:**7.5, **120:**A.7.5.1.1, **122:**1.1.2, **122:**A.1.1.7
Flammable vapor-air mixtures
Static electricity and **77:**7.2.3
Flammable vapors
Pressurized analyzer rooms **496:**1.3.4
Pressurized enclosures **496:**1.3.3
Static electricity and **77:**5.10, **77:**B.1
Flash point
Definition **497:**3.3.7
Liquids **122:**A.1.1.7
Floating roof tanks
Static electricity and **77:**7.5.3
Floors
Conductive **77:**6.6.2
Definition **77:**D.1.13
Flow coating
Definition **34:**3.3.3.2
Fluidized beds
Definition **33:**3.3.5, **33:**3.3.5.1
Flyings *see also* Fibers
Hazardous (classified) locations **70:**506
Foam-water sprinkler and spray systems 13:13.29.1.5, **13:**A.13.29.1.5
Mines, underground **122:**A.1.1.7
Footwear
Conductive **77:**6.6.2
Fuel dispensing systems and devices
Overhead type device
Definition **30A:**3.3.6, **30A:**A.3.3.6

Garages
Commercial
Grounding **70:**511.16
Repair
CNG vehicle storage/repair, electrical installations for **30A:**8.2.1, **30A:**A.8.2.1
Gas *see also* specific gases, e.g., Hydrogen
Heavier-than-air **497:**4.2.2, **497:**4.2.4, **497:**5.5.2, **497:**5.6.5, **497:**5.8.3.1(4), **497:**A.5.5.2
Lighter-than-air **497:**4.2.1, **497:**5.5.2, **497:**5.6.5, **497:**5.8.3.1(4), **497:**A.5.5.2
Protective **496:**4.2.2, **496:**4.4, **496:**4.5.1(3), **496:**4.6, **496:**4.10.3, **496:**7.2, **496:**8.3.1(2), **496:**8.3.3, **496:**8.3.4, **496:**8.3.6, **496:**9.2.8, **496:**A.4.2.2, **496:**A.4.2.3, **496:**A.4.4.1, **496:**A.4.4.4, **496:**A.4.6, **496:**A.7.2.2, **496:**A.7.2.3, **496:**A.9.2.8
Definition **496:**3.3.12
Supply **496:**4.2.1.1, **496:**4.2.1.2, **496:**4.4, **496:**4.8.2, **496:**4.11.5, **496:**A.4.4.1, **496:**A.4.4.4
Definition **496:**3.3.13

Ground-fault circuit interrupters (GFCIs)
Aircraft hangars **70:**513.12
Garages
Commercial **70:**511.12
Grounding
Bulk storage plants **70:**515.16
Hazardous (classified) locations **70:**501.30, **70:**502.30, **70:**503.30, **70:**505.25, **70E:**440.4(G)
Intrinsically safe systems **70:**504.50
Motor fuel dispensing facilities **70:**514.16
Static electricity, to control **77:**5.1.2, **77:**6.1.2, **77:**6.4.1, **77:**6.4.1 7, **77:**6.6.3
Definition **77:**3.1.10
Flammable and combustible liquids and their vapors **77:**7.4.1, **77:**7.4.3, **77:**7.5.2.2, **77:**7.6, **77:**A.7.4.1, **77:**A.7.4.3
Intermediate bulk containers **77:**9.1.1, **77:**9.1.5.1 to 9.1.5.2, **77:**9.1.7.4.1 to 9.1.7.4.2, **77:**9.1.8.1 to 9.1.8.2
Powders and dusts in industrial operations **77:**8.6.2, **77:**8.7, **77:**8.8.3, **77:**8.9, **77:**8.11.1, **77:**A.8.7
Web processes **77:**9.2.5.2

Hangars, aircraft
Electrical installations
Definition **70:**513.2
Hazardous (classified) locations
Aircraft hangars **70:**513.3
Anesthetizing locations **70:**517.60(A), **70:**517.61
Bulk storage plants **70:**515.3
Class I **70:**500, **70:**500, **70:**504.60, **70:**505, **70E:**440.3(B), **70E:**440.4
Division 1
Definition **496:**3.4.1
Purged and pressurized enclosures **496:**4.2.1.2, **496:**4.10.1, **496:**A.4.10.1
Division 2
Definition **496:**3.4.2
Purged and pressurized enclosures **496:**4.9.2
Purged and pressurized enclosures **496:**1.3.1, **496:**1.3.2
Zone 0 **496:**1.1.2(1)
Definition **496:**3.4.5
Zone 1
Definition **496:**3.4.6
Purged and pressurized enclosures **496:**4.10.1
Zone 2
Definition **496:**3.4.7
Purged and pressurized enclosures **496:**4.9.2
Class II **70:**500, **70:**501.120, **70:**506
Division 1
Definition **496:**3.4.3
Purged and pressurized enclosures **496:**4.2.1.2, **496:**4.10.1, **496:**A.4.10.1
Division 2
Definition **496:**3.4.4
Purged and pressurized enclosures **496:**4.9.2
Purged and pressurized enclosures **496:**1.3.1, **496:**1.3.2
Class III **496:**1.1, **70:**506
Combustible particulate solids, facilities **654:**2.7.2, **654:**A.2.7.2
Definition **70E:**440.2
Dusts **70E:**440.3(B)(6)
Dust-ignitionproof protection technique **70:**500.7(B), **70:**506.8(A)
Dusttight protection technique **70:**500.7(C), **70:**506.8(D)
Zone 20, 21, and 22 locations **70:**506
Garages, commercial **70:**511.3, **70:**511.4
Gases **70E:**440.3(B)(6)
Intrinsic safety **70:**500.7(E), **70:**505.2, **70:**506.8(C)
LP-Gas utility plants **59:**4.5.2.2, **59:**A.4.5.2.2
Marinas and boatyards **303:**3.9
Spray application, dipping and coating processes **70:**516.3
Types of protection **70:**505.2
Wastewater treatment and collection facilities **820:**7.4.5.1, **820:**9.1.2, **820:**9.2.3
Heaters *see also* Heating equipment
LP-Gas **58:**10.2.3
Heating equipment *see also* Heaters
Fireworks facilities **1124:**4.5.3
Model rocket motor plants **1125:**4.9 **1125:**5.8
Hermetically sealed 70:500.7(J)
Definition **70:**500.2
Hoists
Hazardous (classified) locations **70:**503.155
Hoods (structural)
Fume, in laboratories **45:**5.6.2, **45:**A.5.6.2
Interior **45:**5.6.2, **45:**A.5.6.2
Hose
Semiconductive **77:**7.4.3, **77:**A.7.4.3
Hot work
Mines
Metal and nonmetal **122:**A.1.1
Humidification
Static electricity, control of **77:**9.2.5.5
Hybrid mixtures 77:4.3.4, **77:**8.10, **77:**9.1.6.3, **77:**9.1.8.5
Hydrant carts, aircraft fuel servicing 407:4.3.7, **407:**A.4.3.7.4
Hydrocarbons
Hazardous (classified) locations, in **497:**5.2.2.1, **497:**5.4.5, **497:**5.5.2
Static electricity and **77:**4.3.3.5, **77:**7.3.3, **77:**A.7.3.3
Hydrogen
Hazardous (classified) locations, in **497:**5.5.2, **497:**5.6.5, **497:**A.3.3.5.1
Static electricity and **77:**7.15

Ignitable mixtures *see also* Flammable vapor-air mixtures
Classification of **497:**4.1.1, **497:**4.1.3.5, **497:**4.2.1, **497:**4.2.5 **497:**4.3(2), **499:**4.1.1 **499:**4.1.3.1, **499:**4.1.3.2
Control by inerting, ventilation, or relocation **77:**6.1.2(1)
Definition **77:**3.1.11, **497:**3.3.8, **499:**3.3.8
Ignition energy 77:4.3.3, **77:**A.4.3.3.4
Definition **77:**D.1.31
Measuring **77:**5.10
Minimum **77:**4.3.3.4, **77:**8.2.4, **77:**8.5, **77:**A.8.2.4
Definition **77:**D.1.41, **497:**3.3.11
Vapor-air mixtures **77:**7.2.3
Ignition sensitivity
Definition **499:**3.3.9, **499:**A.3.3.9
Ignition sources
LP-Gas systems
Utility gas plants, at **59:**A.4.5.2.2
Spray application operations **33:**A.1.1
Static electricity **77:**4.2.2, **77:**4.3.1, **77:**4.3.3, **77:**4.3.6.2, **77:**4.3.7
Ignition temperature
Definition **496:**3.3.5, **496:**A.3.3.5
Dusts **496:**6.2.5, **496:**A.6.2.5
Ignition-capable equipment 496:4.4.6, **496:**9.3.8(2), **496:**A.9.3.8(2)
Definition **496:**3.3.4
Incendive
Definition **77:**D.1.32
Incendive spray equipment 33:11.4
Indicators *see also* Alarms
Purged or pressurized enclosures **496:**4.8.1.1, **496:**4.8.4, **496:**A.4.8.1.1
Definition **496:**3.3.6
Induction
Definition **77:**D.1.33
Induction bars 77:6.5.2.1
Definition **77:**D.1.34
Inductive (static comb) neutralizers 77:6.5.2
Inert gas
Definition **77:**3.1.12
Pressurized enclosures/rooms for electrical equipment, use in **496:**8.3.1(2), **496:**8.3.4, **496:**9.2.8, **496:**A.9.2.8
Inerting
Static electricity control **77:**6.2.1, **77:**7.15, **77:**9.1.6.3, **77:**A.7.2.4
Sulfur, for **655:**A.1.1.1
Inks, printing 77:9.2.3
Instruments
Hazardous (classified) location **70:**501.105, **70:**502.150
Insulators 77:4.1.4, **77:**4.2.3, **77:**4.3.6, **77:**5.1.2
Clothing as **77:**6.6
Humidification and **77:**6.4.2.1, **77:**6.4.2.3
Interceptor sewers 820:1.1.2(3)
Intermediate bulk containers (IBCs)
Static charge and **77:**9.1.1, **77:**9.1.5.2, **77:**9.1.6.2, **77:**9.1.7, **77:**9.1.8.3
Intrinsically safe equipment 120:A.6.2.1.6
Definition **70E:**440.2
Ionization
Air ionizers **77:**4.2.6, **77:**6.1.2, **77:**6.5.1
Definition **77:**D.1.1
Definition **77:**D.1.35
Web processes **77:**9.2.5.3
Isolating switches 70:501.115(B)(2)

Joule
Definition **77:**D.1.36

Laboratory apparatus 45:A.5.6.2
Laboratory work areas 45:12.2.2.3
Electrical classification **45:**5.6.2, **45:**A.5.6.2
Lamps
Portable **70:**511.4(B)(2), **70:**513.10(E)(1), **70:**515.7(C), **70:**516.4(D)
Lighting
Fireworks facilities **1124:**4.5.3
Hypobaric facilities **99B:**4.3
Magazines **495:**8.5.6
Model rocket motor plants **1125:**4.9 **1125:**5.8
Lighting fixtures
Hazardous (classified) locations **70E:**440.3(B)(1), **70E:**440.3(D)
Lightning protection
Hazardous (classified) locations **70:**501.35, **70:**502.35
Limited finishing workstation 33:14.2, **33:**14.3
Definition **33:**3.3.15.1
Limiting oxidant concentration
Definition **77:**D.1.37
Lint *see also* Dust; Fibers
Hazardous (classified) locations **70:**506
Liquefied natural gas (LNG)
Fueling facilities
Motor fuel dispensing facilities **30A:**1.1.2, **30A:**A.1.1.2
Hazardous (classified) locations, in **497:**5.6.5
Vehicular fuel systems
Motor fuel dispensing facilities **30A:**1.1.2, **30A:**A.1.1.2
Live parts
Hazardous (classified) locations **70:**501.25, **70:**502.25, **70:**503.25, **70:**505.19
Lower flammable limit
Definition **77:**D.1.38
LP-Gas
Hazardous (classified) locations, in **497:**5.7.2
LP-Gas containers
Relief devices **59:**A.4.5.2.2
LP-Gas equipment
Installation **58:**A.6.20.2.2, **58:**A.6.20.2.3

LP-Gas systems
Vehicles
Motor fuel dispensing facilities **30A:**1.1.2, **30A:**A.1.1.2
Lubrication of electrical equipment 70B:22.2.8.2
Magazines
Construction **495:**8.5

Magnesium dust 499:4.4.3
Maximum experimental safe gap (MESG)
Definition **497:**3.3.9
Megohmmeter
Definition **77:**D.1.39
Metals
Containers, static electricity control **77:**7.13.3
Methane
Coal mines and coal preparation plants **120:**A.6.2.1.4.1
Minimum explosible concentration (MEC) 77:8.2.3
Definition **77:**D.1.40
Minimum igniting current ratio (MIC)
Definition **497:**3.3.10, **497:**A.3.3.10
Mixers (tank) 77:7.5.2.4
Mixing buildings 495:5.2
Mixing operations (explosives) 495:5.2
Mobile equipment
Mines
Metal and nonmetal **122:**1.1.4(1), **122:**A.1.1
Motor fuel dispensing facilities
Definition **1:**3.3.138.24.2, **30A:**3.3.11
Inside buildings **30A:**8.2.1, **30A:**8.3
Motors 70E:440.3(B)(5)
Hazardous (classified) locations **70:**501.125, **70:**502.125, **70:**505.22

Natural gas
Properties of **52:**A.1.1
Nonconductive
Definition **77:**3.1.13
Nonconductors 77:5.4, **77:**6.4.3.3, **77:**7.3.3, **77:**A.7.3.3
Definition **77:**3.1.14
Nonincendive
Circuits **70:**500.7(F), **70:**506.8(E)
Component **70:**500.7(H)
Definition **70:**500.2
Equipment **70:**500.7(G), **70:**506.8(F)
Definition **70:**500.2

Ohm
Definition **77:**D.1.42
Ohms per square
Definition **77:**D.1.43
Overspray 33:12.5.3, **33:**A.1.1
Oxygen-enriched atmospheres (OEAs) 99B:1.2.4(2)

Piping systems
Flammable and combustible liquids
Static electricity and **77:**7.3.3.1, **77:**7.4, **77:**8.6, **77:**9.1.5.1, **77:**A.7.4.1
Motor fuel dispensing facilities
Grounding/bonding **30A:**8.5.2, **30A:**A.8.5.2
Pits
Landing gear **409:**5.9
Plants
Cleaning and dyeing **70:**500.5(B)(1)
Clothing manufacturing, electrical installations in **70:**500.5(D), **70:**506
Solvent extraction **122:**1.1.7, **122:**A.1.1.7
Plastics
Containers **77:**4.3.6.2, **77:**7.13.4
Dust **499:**4.2.2, **499:**A.4.3.1.2
Films **77:**9.2.2.2
Sheets and wraps **77:**7.16
Static electricity and
Conductive additives **77:**6.4.3.4
Conductive polymers as coatings **77:**6.4.3.7
Containers **77:**4.3.6.2, **77:**7.13.4
Films **77:**9.2.2.2
Pipe/hose **77:**7.3, **77:**A.7.4.2
Sheets and wraps **77:**7.16
Solids in plastic bags **77:**8.11.1
Pneumatic conveying systems
Plastic dusts **499:**4.2.2
Positive pressure air systems (control rooms) 496:7.4
Powders
Static electricity and **77:**4.3.3.6, **77:**4.3.8, **77:**5.4.3, **77:**5.7.1, **77:**5.10, **77:**9.1
Power equipment
Pressurized enclosures, in **496:**4.7, **496:**A.4.7
Definition **496:**3.3.7
Preparation workstations
Definition **33:**3.3.15.2
Pressure relief devices
LP-Gas **59:**A.4.5.2.2
Pressurization
Definition **496:**3.3.8, **70:**506.2
Double **496:**4.4.6
Type X **496:**4.10, **496:**5.5, **496:**6.5, **496:**7.4.5, **496:**A.4.10, **496:**A.5.5.2, **496:**A.6.5
Definition **496:**3.3.8.1
Type Y **496:**4.9, **496:**5.4, **496:**7.4.6, **496:**A.4.9, **496:**A.5.4
Definition **496:**3.3.8.2
Type Z **496:**4.8 **496:**5.4 **496:**7.4.6, **496:**A.4.8, **496:**A.5.4
Definition **496:**3.3.8.3
Pressurizing systems 496:4.3, **496:**A.4.3.1
Definition **496:**3.3.9, **496:**A.3.3.9
Printing
Static electricity and **77:**5.4.3, **77:**9.2

Propagating brush discharges 77:4.3.7, **77:**8.4.2.3, **77:**8.5.3, **77:**8.8.1, **77:**9.1.2.3, **77:**9.1.5.3
Definition **77:**D.1.44
Protected equipment
Definition **496:**3.3.11
Pumps
Solvent transfer equipment **36:**5.5.1.1
Purging
Definition **496:**3.3.14, **70:**506.2
Pyrotechnic compositions 1124:4.5.3.3, **1124:**4.5.3.4.1, **1124:**4.5.3.4.2

Raceways
Bonding **70:**501.30(A), **70:**502.30(A), **70:**503.30(A), **70:**505.25(A)
Radioactive materials
Static electricity neutralizers **77:**6.5.4
Railways
Equipment
LP-Gas tank cars **59:**A.4.5.2.2
Tank cars **77:**7.8, **77:**8.12
Receptacles
Hazardous (classified) locations **70:**501.145, **70:**502.145, **70:**503.145
Refrigerating equipment and systems
Chemical process areas **497:**5.3.2.2(7)
Relays
Hazardous (classified) locations **70:**501.105, **70:**502.150
Resistance 77:4.1.4, **77:**4.2.4
Definition **77:**D.1.46
Resistivity
Surface **77:**5.7
Definition **77:**D.1.47
Volume **77:**5.7
Definition **77:**D.1.48
Rocket motors
Reloadable **1125:**1.1.6
Roll coating
Definition **34:**3.3.3.3

Seals
Conduit systems **70:**501.15, **70:**502.15, **70:**504.70, **70:**506.16
Hermetic **70:**500 7(J)
Definition **70:**500.2
Intrinsically safe systems **70:**504.70
Self-propelled equipment
Metal and nonmetal mines **122:**1.1.4(1), **122:**A.1.1
Semiconductive
Definition **77:**3.1.15
Hose **77:**7.4.3, **77:**A.7.4.3
Liquids **77:**7.3.3, **77:**A.7.3.3
Path **77:**4.2.2
Semiconductors 77:4.1.4
Semitrailers
Aircraft fuel servicing **407:**4.3.7.7
Shipping buildings (fireworks) 1124:5.4
Silos
Coal mines and coal preparation plants **120:**A.9.1
Solvents
Extraction plants **122:**1.1.7, **122:**A.1.1.7
Specific particle density 496:6.2.4, **496:**A.6.2.4
Definition **496:**3.3.15, **496:**A.3.3.15
Spontaneous ignition
Coal mines and coal preparation plants **120:**A.9.1
Spray application
Static electricity, control of **77:**9.3
Spray area
Electrical equipment and **497:**5.3.1.1(2), **497:**5.3.2.2(1)
Unenclosed **33:**1.1.4, **33:**A.1.1.4
Spray booths 33:11.3.7
Definition **33:**3.3.12, **33:**A.3.3.12
Spray rooms
Definition **33:**3.3.13, **33:**A.3.3.13
Static charge 77:4.1.2 to 4.1.3, **77:**4.1.5
Accumulation
On intermediate bulk containers **77:**9.1.2, **77:**9.1.5.2, **77:**9.1.6.2, **77:**9.1.7.1, **77:**9.1.8.3
Measurement of **77:**5.4.4
Nonconductive pipe and lined pipe **77:**7.4.2, **77:**A.7.4.1
In process vessels **77:**7.10.1
Decay monitors **77:**5.6.2
Definition **77:**D.1.5
Generation
Control of **77:**6.3
In intermediate bulk containers **77:**9.1.1, **77:**9.1.7.1
In liquids **77:**7.3.1, **77:**7.3.3, **77:**A.7.3.1, **77:**A.7.3.3
Locators **77:**5.4.2
Static charging
Current
Definition **77:**D.1.9
Field
Definition **77:**D.1.10
Induction **77:**4.1.9
Definition **77:**D.1.11
Triboelectric **77:**4.1.8
Definition **77:**D.1.12
Static electric discharge 77:A.4.3.3.4
Definition **77:**3.1.16
In powder operations **77:**4.3.8, **77:**8.5
Spark discharge **77:**8.5.1, **77:**9.1.2.1
Static electricity
Definition **77:**3.1.17
Flammable and combustible liquids **77:**5.10, **77:**6.4.3.3, **77:**B.2
Conductivity additives **77:**6.4.3.3
Containers **77:**7.13

Powders manually added to flammable liquids **77:**8.11, **77:**9.1.6.3
Process vessels **77:**7.10 **77:**8.11.1
Flammable gas **77:**4.1.7, **77:**4.3.3.7, **77:**5.10, **77:**7.5.2.5, **77:**7.15, **77:**B.1
Solvent extraction plants **36:**6.3.2, **36:**A.6.3.2
Static neutralizers 77:4.3.6.2, **77:**6.1.2, **77:**6.5, **77:**9.2.5.3
Alternating current **77:**6.5.3.2
Definition **77:**D.1.51
Electrically powered **77:**6.5.3
Definition **77:**D.1.50
Static shock, results of 77:6.8
Static sparks 77:4.3.3, **77:**5.9, **77:**A.4.3.3.4
Definition **77:**D.1.49
Ignition by **77:**4.3.3.1
Storage
Calcium carbide **51A:**2.6.2
Coal, at mines and preparation plants **120:**A.9.1
Storm sewers 820:1.1.2(5)
Surface-charge density 77:8.3.2
Definition **77:**D.1.52
Surface streamer 77:4.3.8
Definition **77:**D.1.53

Tank, full trailers 407:4.3.7.7
Tank mixers 77:7.5.2.4
Tank vehicles
Loading and unloading of
LP-Gas **59:**A.4.5.2.2
Static electricity and **77:**7.6, **77:**7.8, **77:**7.12, **77:**8.12
Tanks
Cleaning and safeguarding **77:**7.12
Storage
Solvent extraction plants **36:**5.5.2, **36:**A.5.5.1.2
Temperature
Marking, of pressurized enclosures **496:**4.5, **496:**4.9.3, **496:**4.11.2(4), **496:**A.4.5
Test baths 30B:5.5.3
Trailers (vehicle)
Aircraft fuel servicing **407:**4.3.7.7
Tribocharging 77:4.1.8
Definition **77:**D.1.54, **77:**D.1.55
Trunk sewers 820:1.1.2(2)
Tunnels
Aircraft hangars **409:**5.9

Upper flammable limit
Definition **77:**D.1.56

Vacuum cleaners
Static electricity and **77:**7.14, **77:**A.7.14
Wood processing and woodworking facilities **664:**11.2.1.2
Vacuum pumps
Aerosol manufacture and storage **30B:**5.5.1
Vacuum trucks 77:7.7
Vapor area 34:6.3.1, **34:**A.6.3.1
Definition **34:**3.3.9, **34:**A.3.3.9
Vapor source 34:6.4.1
Definition **34:**3.3.10
Vaporizers
LP-Gas **59:**4.5.2.5, **59:**4.7.2
Ventilated equipment (pressurized enclosures) 496:4.6, **496:**4.9.3, **496:**4.10.3, **496:**6.6, **496:**A.4.6
Ventilation
Adequate
Definition **120:**3.3.1, **497:**3.3.1
Analyzer rooms **496:**9.3.7, **496:**A.9.3.7
Dipping and coating processes using flammable or combustible liquids **34:**6.6, **34:**11.6(1)
Flammable and combustible liquids facilities **30:**8.2.4, **30:**A.8.2.4
Positive pressure (control rooms) **496:**7.3, **496:**A.7.3.2
Spray application operations **33:**11.3.6(1)
Powder coating **33:**15.12.5
Video display terminals, cathode ray tube 77:9.6
Volatile liquids
Flammable
Definition **70:**Art 100-I

Web and sheet processes 77:5.4.3, **77:**9.2
Wiring
Agricultural and food products facilities **61:**4.1.3, **61:**11.2.3
Intrinsically safe systems **70:**504.20
Liquefied natural gas (LNG) fueling facilities **57:**A.5.12.4
Workstations
Spray application operations
Limited finishing workstation **33:**14.3
Definition **33:**3.3.15.1
Preparation workstations
Definition **33:**3.3.15.2

ABOUT THE EDITOR

Peter Schram has spent over 50 years participating in electrical safety and codes- and standards-making activities. He worked in the Engineering Department at Underwriters Laboratories' for 30 years, with ten of them spent managing the group responsible for the evaluation of electrical equipment for use in hazardous locations. During this time, he was the UL representative on a number of NFPA technical committees, including the National Electrical Code Technical Committee's Code-Making Panel 14 and the Technical Committee on Electrical Equipment in Chemical Atmospheres. He was also a member of the United States Advisory Committee to the International Electrotechnical Committee (IEC) committee that developed requirements for equipment in flammable atmospheres, and the National Research Council's Committee on the Evaluation of Industrial Hazards.

Following his career at UL, Schram was NFPA's Chief Electrical Engineer, responsible for the preparation of the *National Electrical Code*® and other electrical standards. He was Secretary to the National Electrical Code Technical Correlating Committee, editor of the *National Electrical Code*® Handbook, and NFPA's *Electrical Equipment in Hazardous Locations.* After retirement from NFPA, he served as a special expert on the *NEC* Code-Making Panel 14 and as an industry consultant on the *NEC* and on electrical equipment in hazardous locations in general. A graduate of the University of Wisconsin (Madison), Schram is a Registered Professional Engineer in Illinois and Massachusetts. He currently resides in Delray Beach, Florida.